Lecture Notes in Production Engineering

Lecture Notes in Production Engineering (LNPE) is a new book series that reports the latest research and developments in Production Engineering, comprising:

- Biomanufacturing
- Control and Management of Processes
- Cutting and Forming
- Design
- Life Cycle Engineering
- Machines and Systems
- Optimization
- Precision Engineering and Metrology
- Surfaces

LNPE publishes authored conference proceedings, contributed volumes and authored monographs that present cutting-edge research information as well as new perspectives on classical fields, while maintaining Springer's high standards of excellence. Also considered for publication are lecture notes and other related material of exceptionally high quality and interest. The subject matter should be original and timely, reporting the latest research and developments in all areas of production engineering.

The target audience of LNPE consists of advanced level students, researchers, as well as industry professionals working at the forefront of their fields. Much like Springer's other Lecture Notes series, LNPE will be distributed through Springer's print and electronic publishing channels.

More information about this series at http://www.springer.com/series/10642

D. Biermann · F. Hollmann
Editors

Thermal Effects in Complex Machining Processes

Final Report of the DFG Priority Programme 1480

Springer

Editors
D. Biermann
Institute of Machining Technology
TU Dortmund University
Dortmund, Nordrhein-Westfalen
Germany

F. Hollmann
Deutsche Forschungsgemeinschaft
Bonn
Germany

Additional material to this book can be downloaded from http://extras.springer.com.

ISSN 2194-0525 ISSN 2194-0533 (electronic)
Lecture Notes in Production Engineering
ISBN 978-3-319-57119-5 ISBN 978-3-319-57120-1 (eBook)
DOI 10.1007/978-3-319-57120-1

Library of Congress Control Number: 2017938536

Printed on acid-free paper

This Springer imprint is published by Springer Nature
The registered company is Springer International Publishing AG
The registered company address is: Gewerbestrasse 11, 6330 Cham, Switzerland

Contents

Introduction

D. Biermann and F. Hollmann

This contributed volume contains the summarised final results of the Priority Programme (PP) 1480 "Modelling, Simulation and Compensation of Thermal Effects for Complex Machining Processes", which was funded by the German Research Foundation (germ.: Deutsche Forschungsgemeinschaft—DFG) in the period from 2010 to 2016. Within the framework of this PP collaborative teams of machining experts on the one hand and researchers from the fields of mechanics, thermodynamics, mathematics as well as materials science on the other hand worked together on an interdisciplinary level to solve the workpiece-related problems resulting from heat effects in complex machining operations (see Fig. 1). Various processes, such as turning, milling, drilling, deep hole drilling, hobbing as well as grinding were investigated by twelve research projects, which developed particular simulation-based prediction and compensation methods for the minimisation of thermally induced impacts on the component machined.

D. Biermann (✉)
Coordinator of the Priority Programme 1480, Institute of Machining Technology,
Baroper Street 303, 44227 Dortmund, Germany
e-mail: biermann@isf.de

F. Hollmann
Programme Director Engineering Sciences, German
Research Foundation, Kennedyallee 40, 53175 Bonn, Germany
e-mail: ferdinand.hollmann@dfg.de

D. Biermann and F. Hollmann (eds.), *Thermal Effects in Complex Machining Processes*, Lecture Notes in Production Engineering,
DOI 10.1007/978-3-319-57120-1_1

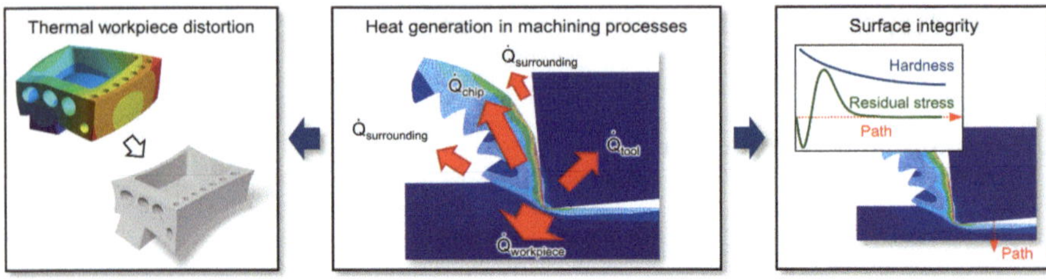

Fig. 1 Negative influences on the workpiece caused by thermal effects during machining

Three working groups, focusing on measurement technology, modelling and simulation techniques as well as on material and friction modelling, were initiated at the beginning of the PP 1480 to promote the scientific exchange among the projects by realising a wide range of activities such as meetings, workshops and round robin tests. Three consecutively stages ensured the structured attainment of the programmes objectives (Fig. 2).

The first stage "feasibility" contains fundamental investigations aiming at the determination of the process heat source as well as the initial modelling and simulation of its impact on the workpiece. Within the second project stage called "flexibility" the models were enhanced and complex part shapes as well as, if necessary, multiple heat sources were considered. The last third stage entitled "compensation" utilised the simulation approaches developed in the first two phases for the prediction of thermal distortion and subsurface damage caused by the corresponding machining operation. The following chapters present a summary of the working group activities (Part 1) and the final results of the individual research projects (Part 2).

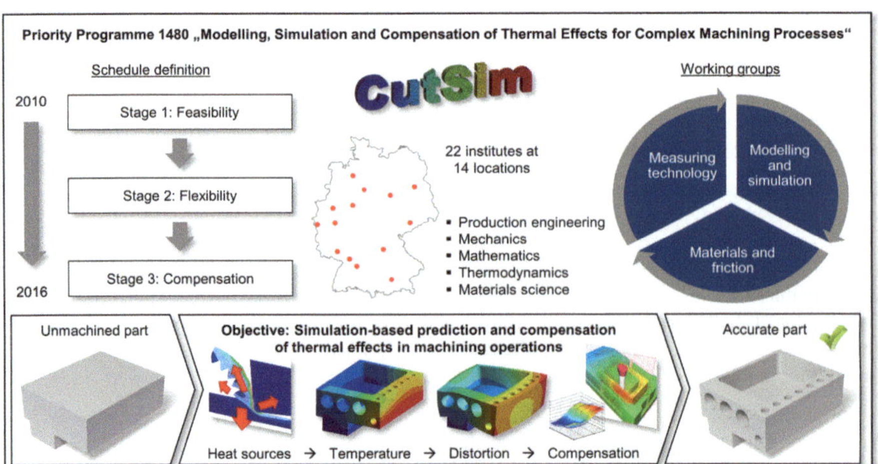

Fig. 2 Structure and objective of the Priority Programme 1480

Part I
Collaboration within the Working Groups

Temperature Measurements and Heat Partitioning in Machining Processes

J. Sölter, R. Frohmüller and H. Wirbser

Abstract Temperature measurements were of special interest in the Priority Programme (SPP) 1480 of the Deutsche Forschungsgemeinschaft (DFG). Therefore the working group on temperature metrology was established. Calibration setups for (1) contacting thermometers developed at the ITT Karlsruhe and (2) radiation thermometers developed at the IFQ Magdeburg were utilized in the projects of the SPP 1480 to provide comparable high quality temperature measurements. Additionally, heat partition data was collected from the involved projects and aggregated by applying the dimensionless thermal number known from orthogonal cutting. Although theoretical calculations qualitatively agree with results from the experiments further research is needed to predict the heat partition to the workpiece for industry relevant processes.

1 Introduction

The thermal impact of machining processes is often critical regarding the achievable geometric workpiece accuracy, especially if coolants are not applied. One reason is the accumulation of heat leading to inhomogeneous temperature distributions in the workpiece. As a result the effective material removal in the process deviates from the nominal material removal given by the machining parameters. Other factors such as the machine tool accuracy or the generation of a locally

J. Sölter (✉)
Department of Manufacturing Technologies, Foundation Institute of Materials Science, University of Bremen, Badgasteiner Str. 3, 28359, Bremen, Germany
e-mail: soelter@iwt.uni-bremen.de

R. Frohmüller
Institut für Fertigungstechnik und Qualitätssicherung, Otto-von-Guericke-Universität Magdeburg, Universitätsplatz 2, 39106 Magdeburg, Germany

H. Wirbser
Institut für Technische Thermodynamik, KIT—Karlsruher Institut für Technologie, Kaiserstraße 12, 76131 Karlsruhe, Germany

D. Biermann and F. Hollmann (eds.), *Thermal Effects in Complex Machining Processes*, Lecture Notes in Production Engineering,
DOI 10.1007/978-3-319-57120-1_2

Table 1 Machining Processes analyzed in the SPP 1480 and applied temperature measurement techniques

technique / process	hobbing	drilling	turning	milling	grinding
thermocouple		IFW, wbk, ISF, IIW	wbk, FBK, [illegible], WZL	ISF, IFW, iwb	ISF
resistance thermometer	ifq				
pyrometer	ifq	IFW, wbk	wbk, [illegible], WZL	iwb	
infrared camera	ifq	IFW, wbk, ISF, IIW	IWF, wbk, [illegible], WZL	IFW, iwb	

plastically deformed surface layer leading to residual stress in the whole workpiece also contribute to size and shape deviations of the workpiece.

In order to be able to quantify the thermal impact of the machining processes analyzed within the framework of the Priority Programme (SPP) 1480 of the Deutsche Forschungsgemeinschaft (DFG), temperature measurements were at the core of the experimental work of the involved projects (Table 1). The project groups supported each other in periodic meetings of the working group on temperature metrology in which mainly the temperature measurement setups and the results were presented and discussed.

2 Objectives

Temperature measurements within the single projects of the Priority Programme mainly aimed at determining the heat partition to the workpiece for different process conditions as input for the process simulations and/or on providing validation data for process simulations. As a result, the main objective of the working group on temperature metrology was to support the projects in their activities and achieve comparable high quality temperature measurements. Moreover, the working group sought to aggregate the heat partition data for different machining processes determined at the laboratories involved in the Priority Programme in a unified way.

Based on the state of the art in temperature measurements, standardized calibration setups and procedures for the utilized measurement techniques were developed and applied. The measured temperatures were then used to determine the heat partition to the workpiece for different processes and process conditions.

3 Temperature Measurement Techniques

Thermally driven mechanisms play a major role in machining processes. Elevated temperatures promote tool wear and affect the generation of material modifications such as residual stress, phase transformations and crack formation as well as the geometric accuracy of the workpiece due to thermal strains. Consequently, temperature measurements in machining have a long history. The first work cited in literature dates back to 1798 [1]. Count Rumford conducted calorimetric experiments in which he estimated the heat generated during boring brass cannons by comparing it with the heat from burning wax candles. Temperature measurement methods have considerably developed since that time. In most machining applications today contacting thermometers such as resistance thermometers and thermocouples as well as radiation thermometers, e.g. pyrometers and infrared cameras are utilized. Other temperature measurement techniques, such as spectroscopic measurements and those that are based on thermophysical processes will not be presented here but are discussed in more detail in [2].

3.1 Contacting Thermometers

3.1.1 Resistance Thermometers

The temperature dependence of the electrical resistance is utilized to determine the temperature with so-called resistance thermometers. With an appropriate measurement setup temperatures can be determined with an accuracy of 10^{-4} K. For resistance thermometers pure metals, in particular platinum and nickel, are preferred due to their well reproducible temperature dependent electrical resistance. For this reason metal alloys are not suitable. Moreover, the advantage of metal resistance thermometers lies in their high accuracy. Platinum is the most common material for resistance thermometers. It has a high chemical resistance and different resistor types such as wire, layer or foil resistors are comparatively easy to produce from platinum. "While they are inexpensive, accurate and relatively easy to use, comparatively speaking, they are relatively large, have long time constants and are limited in temperature range" [3].

3.1.2 Thermocouples

The working principle of a thermocouple (TC) is the Seebeck effect. If an isotropic homogeneous metal, e.g. a copper wire, is subjected to a temperature gradient a diffusive charge carrier transport will generate an electric field that counteracts the diffusion. The electric field is proportional to the temperature difference with the Seebeck coefficient being the constant of proportionality. The Seebeck coefficient is

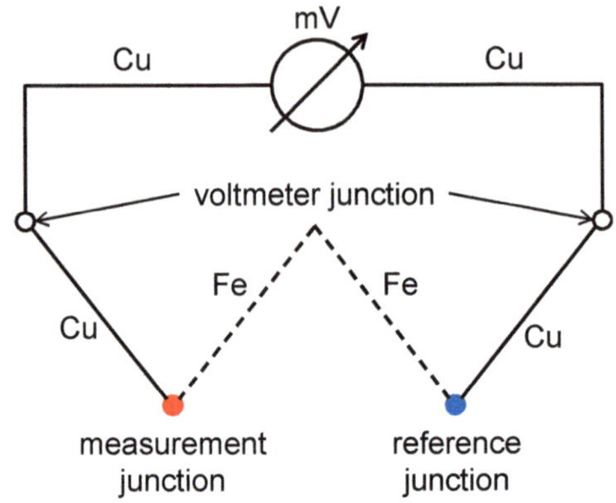

Fig. 1 Circuit diagram of a thermocouple

a material property. If two wires composed of dissimilar metals, e.g. copper and iron, are joined at both ends and one junction is heated (measurement junction), the thermal voltage, i.e. the difference of the electric fields in both wires, causes a current to flow in the thermoelectric circuit. The thermal voltage can be measured if the circuit is broken and a high resistance voltmeter is connected to the leads (Fig. 1). In order to calculate temperatures from the voltage at the measurement junction the second (reference) junction has to be kept at a constant known temperature, e.g. by immersing it into an ice bath (0 °C). Moreover, the measured thermal voltages have to be converted to a corresponding temperature. For this purpose a thermocouple has to be calibrated before the actual temperature measurements (Sect. 4.1).

Each junction between two dissimilar metals in a circuit contributes to the thermal voltage and potentially leads to erroneous temperature measurements. Measurement errors can be avoided either by using the same material for the thermocouple wire and the voltmeter leads as shown in Fig. 1 (Cu). Also dissimilar metals can be used but then the voltmeter junctions have to be at the same temperature.

Depending on the materials used in TCs temperatures from −200 °C to almost 2,000 °C can be measured. Most common in machining applications is the type K (NiCr-Ni) thermocouple with a temperature range from −200 °C to 1,300 °C. TCs are "(1) relatively low cost; (2) rugged; (3) versatile and available for many temperature ranges; (4) reasonably stable and reproducible; (5) subject to relatively low uncertainty when used as designed; and (6) fast responders, depending on size" [3]. Nevertheless uncertainties in TC measurements are associated with cold work of the thermocouple wires, with temperature gradients at the measurement junction and with an inadequate temperature control of the reference junction.

3.2 *Radiation Thermometers*

Planck's law describes the dependence of the spectral emissive power on the temperature of a blackbody. A blackbody absorbs all incident radiation regardless of wavelength and direction. It is an ideal emitter that emits as much as or more energy than any other body at the same temperature and it radiates the energy isotropically (diffuse emitter). When integrated over all wavelengths Planck's law becomes the Stefan-Boltzmann law which states that the total emitted power per unit area of a blackbody is proportional to the fourth power of its temperature. This is the basis for all radiation thermometers: measuring the emitted power can be directly linked to the temperature of a body. However, sources of error in radiation measurements mainly arise from deviations of real bodies from a blackbody. Especially metal surfaces show a strong dependence of the emissivity (= absorptance) on the wavelength and the temperature. Also, the surface properties strongly affect the emissivity. Additionally, the emission characteristics depend on the angle by which the electromagnetic waves are emitted from the surface. From the above mentioned it can be concluded that the emissivity is not solely a surface characteristic, it is the fraction of the total emitted radiation from a technical surface to the radiation emitted by a perfect blackbody at the same temperature. Thus, emissivity values range between zero and one.

Other effects contributing to measurement errors are background radiation and the transmissivity along the optical path from the source, i.e. the surface, to the detector of the thermometer including the lens system. Due to the manifold influences a mathematical description of the emissivity of technical surfaces in closed form is not possible. Thus, a calibration, i.e. an experimental determination of the emissivity, is the central task before conducting temperature measurements with radiation thermometers.

3.2.1 Radiation Detectors

Quantum detectors can be used as a receiver of the incident radiation. These detector types basically count photons and are made of semiconductor materials. The energy of the photons is absorbed and lifts electrons from the valence band to the conduction band of the semiconductor giving rise to a photoelectric current that can be measured.

In another detector type the absorption of the incident radiation leads to a temperature increase of the detector material. The temperature increase causes a change of the electrical resistivity (bolometer) similar to resistance thermometers. Thermopiles are composed of several thermocouples attached to an absorbance material at their "hot" junction. The absorbance material is heated up by the incident radiation and due to the temperature gradient relative to the "cold" junction of the thermocouples a thermoelectric voltage can be measured. The thermocouples are often connected in series aiming at an increase of the thermoelectric voltage, therefore enabling a more sensitive measurement of temperature differences between the "hot" and the "cold" junction. Both bolometers and thermopiles are manufactured by thin-film technology.

3.2.2 Types of Radiation Thermometers

Depending on the wavelength range that is utilized to measure the temperature, radiation thermometers can be classified as follows [3]:

- Total radiation thermometers should collect 90% of the radiation from a surface. This is achieved by maximizing the collection cone and the wavelength band. According to Planck's law the spectral emissive power becomes very small for low temperatures. For this reason total radiation thermometers are especially useful in this temperature range (<50 °C).
- Spectral band thermometers (single color pyrometers) and imagers collect radiation from a small wavelength band of the radiation. From Planck's law it follows that the peak wavelength where the maximum spectral emissive power occurs is shifted to lower wavelengths when the temperature increases. Thus, the detector should be adapted to the wavelength band around the peak wavelength. Since the emissivity of the surface to be measured varies with both, wavelength and temperature, a calibration is inevitable for reliable temperature measurements with single color pyrometers. Spectral thermometers are also band thermometers but with a very limited bandwidth between two and five percent of the wavelength of interest.
- Ratio thermometers, often called two-color pyrometers, collect radiation at two different wavelengths aiming to eliminate the influence of emissivity on the temperature measurement. The main prerequisite for measurements with two-color pyrometers is that the surface to be measured behaves like a gray body, i.e. that the emissivity of both wavelengths is identical. As a consequence the conditions under which ratio thermometers can reliably be applied without being calibrated depend on the a priori knowledge of the emissivity characteristics of the surface. "In fact, for materials where emissivity varies with wavelength, use of the ratio thermometer would be in question" [3].

4 Calibration of Thermometers in the Priority Programme

4.1 Contacting Thermometers

For the calibration of contacting thermometers in the projects of the Priority Programme a setup at the ITT Karlsruhe was developed. It consists of a thermostat in which a copper core is heated up to preset calibration temperatures (Figs. 2 and 3). The "true" temperature is measured with a calibrated precision platinum resistance thermometer PT 25 (25 Ω at 0 °C, Rosemount). A high resistance digital multimeter (DMM) and a constant current source are needed for the measurement with the precision thermometer in a 4-wire sensing setup. It is important to calibrate each temperature sensor, either a resistance thermometer or a thermocouple, as part of the complete measuring chain that will be used in the experiments.

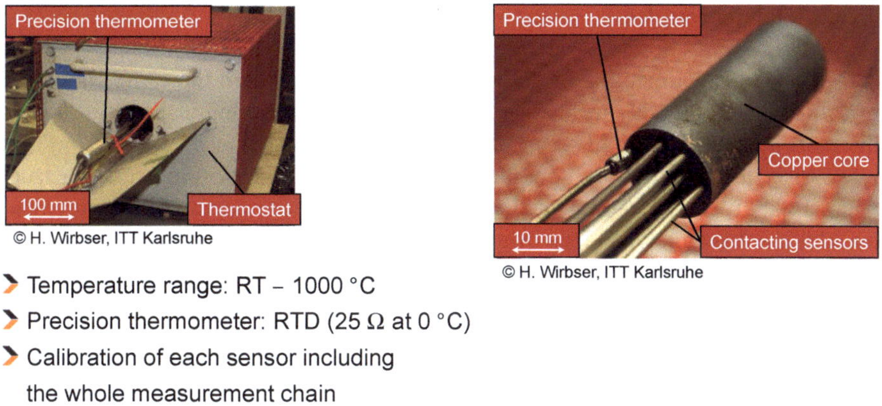

Fig. 2 Calibration setup for contacting thermometers (ITT Karlsruhe)

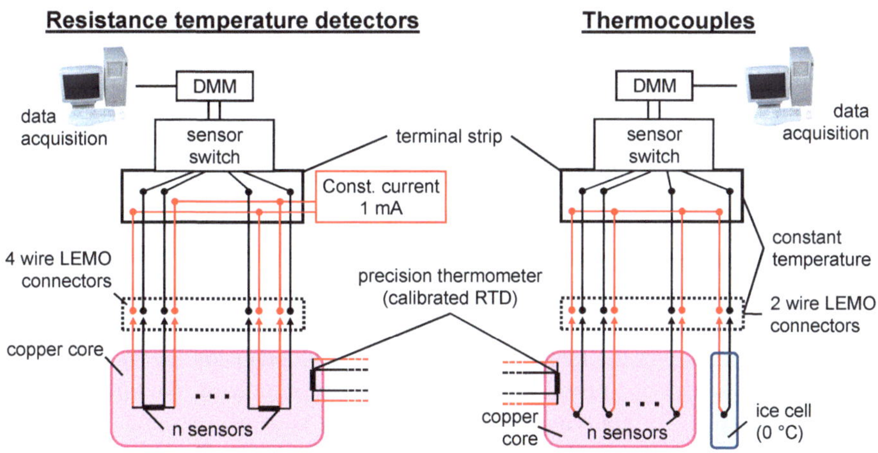

Fig. 3 Circuit diagram of the calibration setup for contacting thermometers (ITT Karlsruhe)

4.1.1 Resistance Thermometers

Resistance thermometers should be connected with 4-wire LEMO connectors in series (not more than ten) to avoid warming by the measuring current (Fig. 3, left). For selecting the different resistance thermometers in the measuring chain a multiplexer is needed. If the measuring chain consists of more than ten sensors a constant current source and a temperature compensated precision resistor should be used. By measuring the voltage drop at the precision resistor, the sensor current can be determined. The resistance (voltage drop) at each sensor is measured in a 4-wire sensing setup. LEMO connectors should also be used to connect the sensors to the multiplexer.

4.1.2 Thermocouples

It is strongly recommended to design the calibration setup in the same way as the measurement setup. This means in particular that the measuring instruments and the thermocouples should be identical and should be arranged to each other in the same way. Also, the surrounding conditions during calibration and the actual measurement should be comparable in order to avoid erroneous temperature measurements by additional thermal voltages originating from junctions between thermocouples and the digital multimeter or voltmeter. Thermocouples and voltmeter should be connected with gold-plated LEMO connectors. Moreover, the connections should be kept on a constant temperature to minimize measurement uncertainties. The use of compensation leads is recommended for connecting reference and measurement junction, e.g. when they are far away from each other. Compensation leads have the same thermo-electrical behavior as the thermocouple wires. This assures that no additional thermal voltage is generated. Since self-heating of the voltmeter might also affect the measurement it should be switched on in advance, so that a stationary temperature of the instrument can be assured before the calibration or measurement starts.

The use of class 1 K-type thermocouples is recommended due to the linearity of the thermal voltage in the temperature range between 0 °C and 1,000 °C (Seebeck coefficient of 41 μV/°C). Additionally, an ice cell can be used as a reference junction (0 °C, Fig. 3, right).

4.1.3 Calibration Procedure

In a comparative calibration as shown in Fig. 3 the sensor signal (resistance or voltage) is plotted over the temperature readings of the precision thermometer. The signal of each sensor at a preset temperature will be read at least five times within a defined time span. By applying the method of least squares a second order polynomial is fitted to the measured output signals of each sensor. The calibration starts with the lowest "true" temperature within the temperature range of interest and ends with the highest "true" temperature. In a second cycle the temperature is decreased to the lowest temperature. This procedure should be conducted two to three times leading to a set of three model parameters for each thermocouple and for each heating or cooling cycle. From these the preliminary model parameters are calculated. In a final step they are used to compute the temperatures from the sensor signals within the temperature range during calibration which are then compared with the true temperature readings from the precision thermometer.

4.2 *Radiation Thermometers*

4.2.1 Calibration Setup

At the IFQ Magdeburg a calibration setup was developed [4] was developed and provided to calibrate the radiation thermometers used in the projects of the Priority Programme (Figs. 4 and 5).

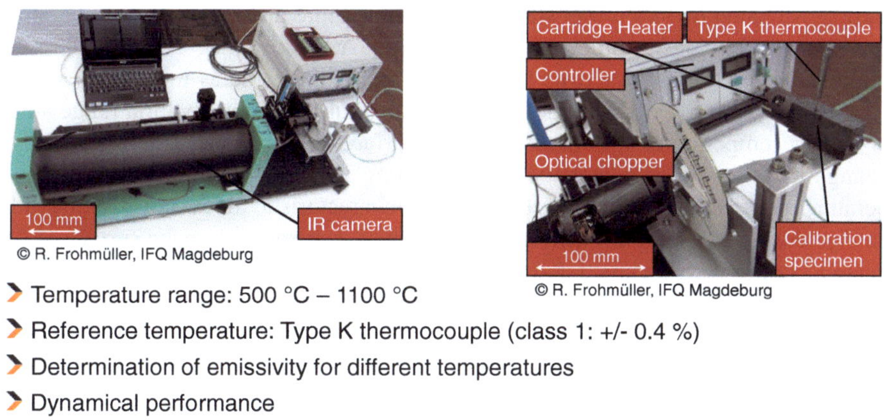

Fig. 4 Calibration setup for radiation thermometers (IFQ Magdeburg)

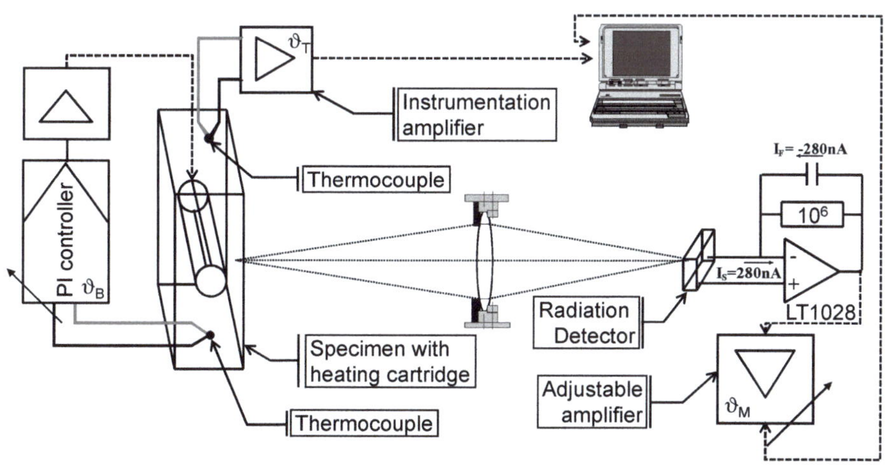

Fig. 5 Circuit diagram of the calibration setup for radiation thermometers (IFQ Magdeburg)

The main component of the calibration setup is a workpiece material sample in which a heating cartridge connected to a PI-controller is inserted. For a reliable calibration it is important that the calibration sample

- is made of the same material as the specimens for the temperature measurements and
- that it is premachined in the same way.

Otherwise the surface characteristics may differ and thus, the calibration might lead to erroneous temperature measurements. The PI-controller assures that the sample is heated up to a preset steady-state-temperature and minimizes temperature gradients within the sample or on its surface. A steady state is reached if the temperature on the surface (true temperature ϑ_T) is the same as in the bulk (ϑ_B).

It is advisable to start with the highest temperature of the calibration range since at higher temperatures oxidization of the surface might occur and permanently change its radiation characteristic. Thus, starting with the highest calibration temperature assures that the surface radiation characteristic does not significantly change during calibration. The true temperature ϑ_T is measured on the surface of the heated sample with a class 1 K-type thermocouple.

4.2.2 Calibration Procedure

When the radiation thermometer is properly aligned to the specimen the measured temperature ϑ_M can be compared to the true temperature. The emissivity of the thermometer is adjusted starting with the highest possible value (1.0) until ϑ_M equals ϑ_T. Then the emissivity for the second highest temperature is calibrated in the same way and so forth. The temperature steps along the calibration curve should adaptively be changed depending on the emissivity change between two temperatures. Two different approaches are feasible to determine emissivities between adjacent calibration temperatures: (1) the complete calibration curve is fitted by an appropriate function over the analyzed temperature range or (2) emissivities are determined by a stepwise linear interpolation between the sampling points of the curve. The actual temperature measurements will be conducted with an emissivity of 1.0. Then the temperature dependent emissivity is utilized to (iteratively) correct the temperature measurements a posteriori. An example for a temperature dependent emissivity of an infrared camera applied in drilling 42CrMo4 samples and the effect of the emissivity correction on the resulting temperature field is shown in Fig. 6.

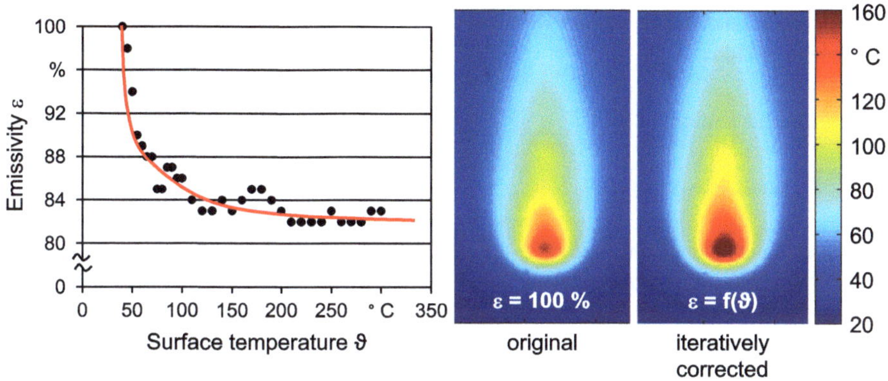

Fig. 6 *Left* temperature dependent emissivity in drilling; *right* corrected temperature field ([5], IWT Bremen)

5 Temperature Measurement Setups and Heat Partitioning

In the projects of the SPP 1480 workpiece temperature measurements were mainly conducted to provide input and validation data for the process simulations. Depending on the project temperatures were measured comparatively far off the chip formation zone making the measurements less error prone. In the following selected measurement setups will be presented for thermocouples and infrared cameras. In the second subsection the results on the heat partition to the workpiece for dry machining processes analyzed in the Priority Programme are presented and discussed.

5.1 Selected Temperature Measurements Setups

5.1.1 Thermocouples

In order to determine the total heat flow to the workpiece during turning of AlSiC metal matrix composites, the temperature increase of the whole workpiece was measured with thermocouples at the FBK Kaiserslautern (Fig. 7). For this purpose a telemetric data acquisition was attached to the workpiece. The energy dissipating to the workpiece can be calculated from the mass of the workpiece, the specific heat of the workpiece material and the measured temperature increase. The total process energy that is needed to finally determine the heat partition to the workpiece can be calculated from the cutting power and the process time.

In a milling setup at the IFW Hannover workpieces were equipped with K-type TCs along the milling path at different distances to the chip formation zone (Fig. 8).

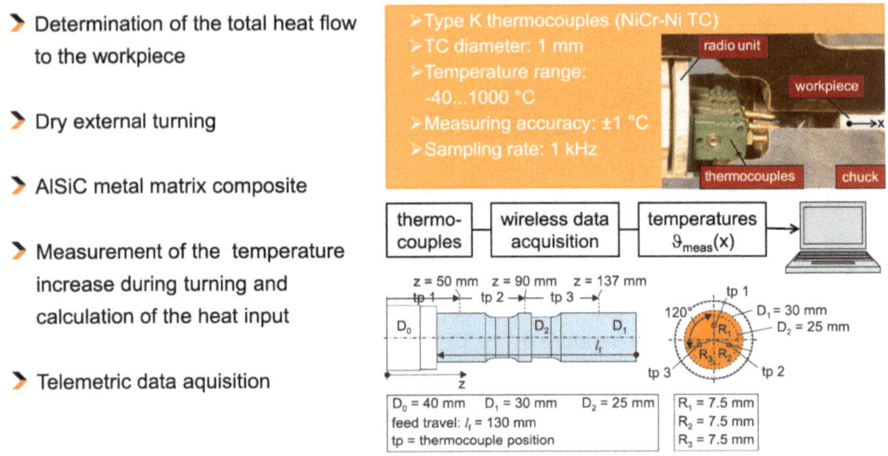

Fig. 7 Telemetric data acquisition for thermocouple measurements in external turning (FBK Kaiserslautern) [6]

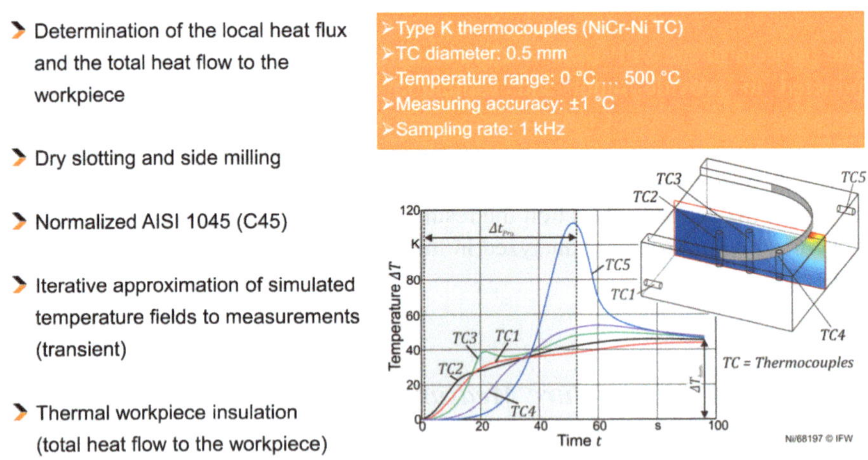

Fig. 8 Thermocouple measurements in milling (IFW Hannover) [7]

Additional TCs were positioned in two corners of the workpiece that was thermally insulated. With this setup the total heat flow to the workpiece could be experimentally determined (mean temperature increase at all TCs). By adjusting milling simulations until simulated and measured transient temperatures at the TCs one, two and three agree, also locally the heat dissipation to the workpiece at the maximum uncut chip thickness can be identified. The heat partition can be calculated by dividing the total heat that dissipated to the workpiece during the milling process by the total milling energy.

5.1.2 Infrared Cameras

The workpiece temperature distribution in milling was measured in a plane orthogonal to the working plane in a setup at the IWT Bremen (Fig. 9). In order to prevent chip collisions workpieces were machined in down-milling and the camera lens was additionally shielded with pressurized air. The aim was to determine the heat partitioning along the cutting arc. This was achieved by varying the width of cut on four levels and by iteratively changing the heat flux to the workpiece in milling simulations until simulated and measured temperature fields were in agreement. The cutting power was approximated by the measured electrical spindle power.

In plunge-turning experiments at WZL and WSA Aachen measured temperatures were utilized as validation data for a multiscale turning model [9]. The temperature field was measured on a workpiece plane orthogonal to the working plane of the process (Fig. 10). A slotted plate prevented chips from colliding with the camera lens and limited the measurement area of the camera to a region close to the chip formation zone.

In a deep-hole drilling setup at the ISF Dortmund workpiece temperatures were measured on the external workpiece surface (Fig. 11). Temperature fields were used to determine the total heat flux to the workpiece and to determine the heat partition to the workpiece for different drilling parameters. This was achieved in an inverse heat transfer analysis in which thermal process simulations were iteratively adjusted until simulated and measured temperature fields agreed.

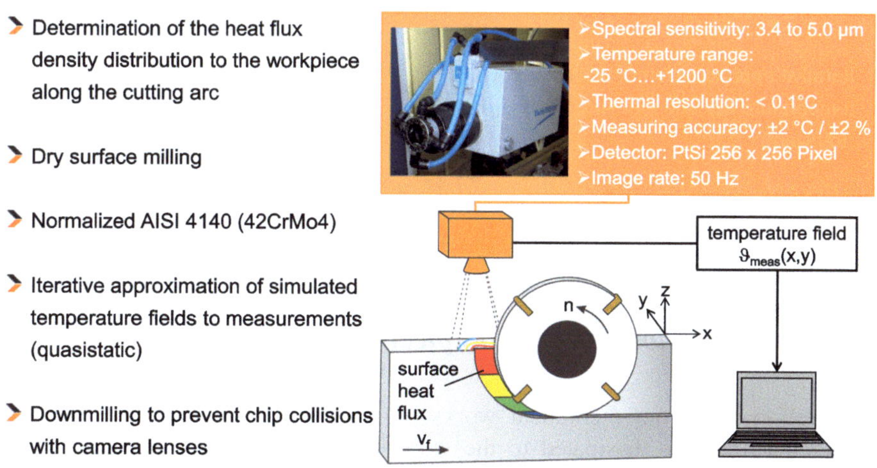

Fig. 9 Infrared camera measurements in milling (IWT Bremen) [8]

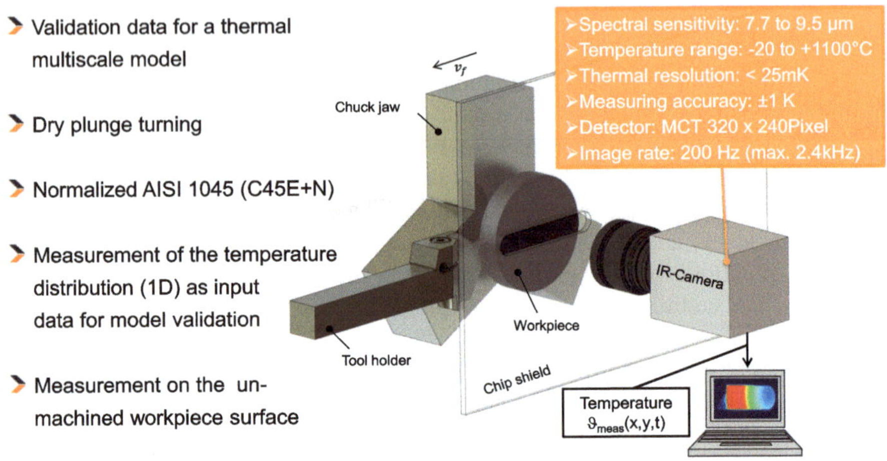

Fig. 10 Infrared camera measurements in plunge turning (WZL Aachen) [9]

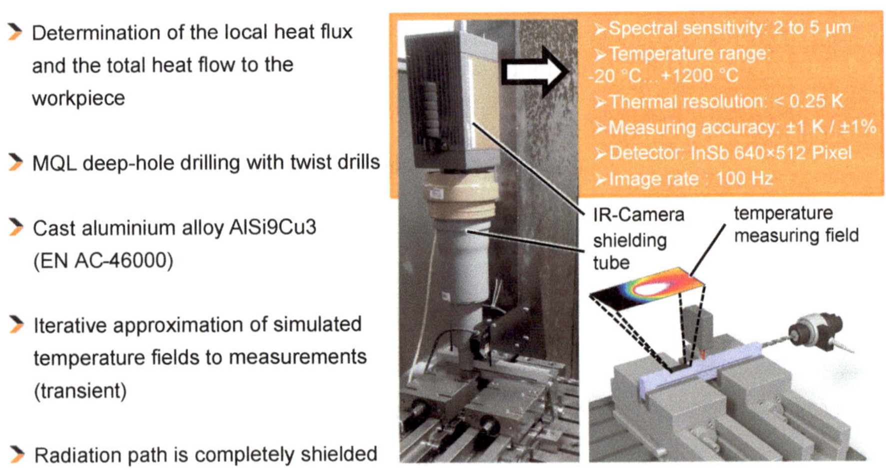

Fig. 11 Infrared camera measurements in deep-hole drilling (ISF Dortmund) [10]

5.2 *Heat Partition to the Workpiece*

It is well-known from theoretical works on the heat partitioning in orthogonal cutting that the heat dissipated to the workpiece predominantly depends on the dimensionless thermal number N_{th} (similar to the Péclet number). The thermal number in orthogonal cutting is the quotient of the cutting velocity times the uncut chip thickness and the thermal diffusivity of the workpiece material. Based on theoretical works of Komanduri and Hou [11] and experimental data from literature, the heat partition to the workpiece was numerically calculated and fitted by a

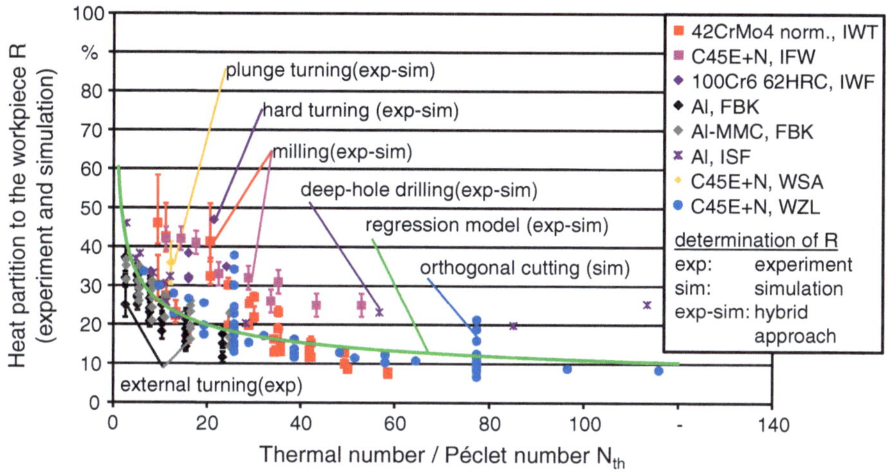

Fig. 12 Heat partition to the workpiece versus the thermal number for different dry machining processes

power law (Fig. 12, green line). It is obvious that finite element simulations of orthogonal cutting from Puls ([9], Fig. 12, blue circles) agree quite well as long as variations in the finite element (FE) model do not differ too much from the simplifying assumptions of [11]. Thermal numbers of 26 and 77 in the FE simulations comprise variations of (1) the width of the flank wear land, (2) the cutting edge radius and (3) the rake angle. Especially (1) and (2) lead to deviations from the analytical model whereas (3) does not significantly affect the FE-simulated heat partition to the workpiece.

The experimentally determined heat partitions follow the theoretical curve. However, depending on the process deviations can become quite high. In external turning the depth of cut was also varied in the experiments but it is not included in the thermal number. Thus, several heat partition values occur for identical thermal numbers. Similarly heat partition values determined in milling of 42CrMo4 increase with decreasing thermal numbers but partially strongly deviate from theoretical calculations. These experiments also included a variation of the depths of cut. Moreover, heat partition values for different uncut chip thicknesses were determined along the cutting arc for a given width of cut [8]. Thus, due to lateral heat flow in the workpiece, the heat partitions for different uncut chip thicknesses along the cutting arc affect each other.

The heat partition to the workpiece determined for milling C45 samples seems to be shifted to higher values compared to the theoretical curve. R values determined in hard turning and in plunge turning are also shifted to higher values. For thermal numbers below 30 heat partition determined in deep-hole drilling agrees reasonably well with theoretical values. However, for comparatively large N_{th} values the heat dissipating to the workpiece is higher in the experiment. Here, the N_{th} values correspond to very high feed rates (2.0, 3.0, 4.0 mm/rev). The deviation to theoretically calculated heat partition values might have been caused by additional friction effects, e.g. at the chisel edge, that become more pronounced if very high feed rates are utilized generating relatively high feed forces of up to 8 kN at a bore diameter of 10 mm [10].

It can be concluded that the theoretical description of the heat partition to the workpiece in orthogonal cutting can be applied to roughly estimate the heat partitioning in industry relevant processes. Especially the dependence on the feed velocity and the cutting velocity is represented by the approach based on orthogonal cutting theory. However, deviations are relevant if the process conditions deviate too much from the assumptions in theory, e.g. influence of tool wear, cutting edge radius as well as the depth of cut. The results indicate that the influence of the workpiece material is adequately described by the thermal diffusivity. The major mechanism that cannot be taken into account in the calculations based on Komanduri and Hou [11] is the removal of preheated workpiece material regions by subsequent cutting edge engagements. This should be analyzed in future research aiming at a more general description of heat partitioning that will then be applicable to industry relevant machining processes.

6 Summary and Conclusions

The working group on temperature metrology in the DFG Priority Programme 1480 was established to support the participating projects in their temperature measurement activities. The main goal of the working group was to assure comparable high quality temperature measurements. For this purpose a calibration setup for (1) contacting thermometers developed at the ITT Karlsruhe and (2) radiation thermometers developed at the IFQ Magdeburg were utilized. Temperature measurement setups from selected laboratories of the Priority Programme are presented and specific features of the setups and the applied approaches are discussed.

Based on the dimensionless thermal number N_{th} the heat partition data for different machining processes determined at the laboratories involved in the Priority Programme were aggregated in a unified way. Although qualitatively theoretical calculations based on orthogonal cutting describe the strong increase of the heat partition to the workpiece with decreasing thermal number, i.e. decreasing cutting velocity and/or decreasing uncut chip thickness correctly, quantitative results from industry relevant processes might be significantly different. This can be explained by the fact that the depth of cut, i.e. the uncut chip width and the feed velocity are not taken into account in the thermal number. Moreover, it is assumed that different process specific mechanisms contributing to the generation and dissipation of heat in machining processes, e.g. friction, are not adequately described by the definition of the thermal number as it is used in this publication. Therefore, it can be concluded that further research on heat partitioning should aim on incorporating effects that occur when industry relevant machining processes are to be described in a theoretical model of heat partitioning.

Acknowledgements The authors thank Mr. Marc Deppermann (WSA Aachen), Mr. Maxim Gulpak (IWT Bremen), Dr. Ivan Iovkov (ISF Dortmund), Mr. Ivan Mitkov Ivanov (IWF Berlin), Dr. Daniel Niederwestberg (IFW Hannover), Dr. Hendrik Puls (WZL Aachen) and Dr. Marco Zimmermann (FBK Kaiserslautern) for providing heat partitioning data and figures on their temperature measurement setups.

The authors greatly appreciate the financial support of the working group on temperature metrology within the Priority Programme 1480 by the Deutsche Forschungsgemeinschaft (DFG).

References

1. Rumford, B.: An inquiry concerning the source of the heat which is excited by friction. Philos. Trans. R. Soc. Lond. **88**, 80–102 (1798)
2. Bernhard, F.: Technische Temperaturmessung. Springer-Verlag, Berlin, Heidelberg, New York (2004)
3. Davies, M.A., Ueda, T., M'Saoubi, R., Mullany, B., Cooke, A.L.: On the measurement of temperature in material removal processes. CIRP Ann.—Manuf. Technol. **56**, 2, 581–604 (2007)
4. Frohmüller, R., Beutner, M., Köchig, M., Karpuschewski, B.: High speed temperature measurement in gear hobbing: Part I—design, concept and physical operation mode of the infrared-camera. Prod. Eng. Res. Devel. **8**, 73–97 (2014)
5. Gulpak, M., Sölter, J.: Thermal modelling in drilling of steel. Adv. Mater. Res. **1140**, 205–212 (2016)
6. Aurich, J.C., Zimmermann, M., Schindler, S., Steinmann, P.: Effect of the cutting condition and the reinforcement phase on the thermal load of the workpiece when dry turning aluminum metal matrix composites. Int. J. Adv. Manuf. Technol. **82**(5–8), 1317–1334 (2016)
7. Niederwestberg, D.: Prognose und Kompensation der temperaturbedingten Werkstückverlagerungen beim Trockenfräsen. Dr.-Ing. Dissertation Universität Hannover, Berichte aus dem IFW, 11/2016, Hrsg.: B. Denkena, PZH Verlag (2016)
8. Sölter, J., Gulpak, M.: Heat partitioning in dry milling of steel. CIRP Ann. Manuf. Technol. **61**(1), 87–90 (2012)
9. Puls, H.: Mehrskalenmodellierung thermo-elastischer Werkstückdeformationen bem Trockendrehen. Dr.-Ing. Dissertation RWTH Aachen, Ergebnisse aus der Produktionstechnik 28/2015, Hrsg.: Brecher, C.; Klocke, F.; Schmitt, R.; Schuh, G., Apprimus Verlag Aachen (2015)
10. Biermann, D., Iovkov, I.: Investigations on the thermal workpiece distortion in MQL deep hole drilling of an aluminium cast alloy. CIRP Ann.—Manuf. Technol. **64**, 1, 85–88 (2015)
11. Komanduri, R., Hou, Z.B.: Thermal modelling of the metal cutting process Part I—Temperature rise distribution due to shear plane heat source. Int. J. Mech. Sci. **42**, 1715–1752 (2000)

Optimization and Compensation Strategies

A. Menzel and A. Schmidt

Abstract The contributions from Priority Program 1480 describe and analyze various applications of machining processes and model the specific influences of process-induced thermal effects. In order to control the resulting deformation and shape errors, specialized compensation and optimization approaches are described. Even though the approaches vary a lot due to the different specialized processes, some general common ideas and concepts can be identified.

1 Introduction

Among the contributions from Priority Program 1480 "Modelling, Simulation and Compensation of Thermal Effects for Complex Machining Processes", all groups work on a joint research field. However, when special processes are being investigated, the activities within the different groups vary. Machining processes vary from drilling over milling and turning up to grinding. Even for the same process, different research groups focus on various special applications such as deep hole drilling or gear hobbing. For every application, special analysis and treatment of the material removal by tool contact and engagement conditions as well as the resulting generation of process forces and production of energy has to be taken into account. For the simulation of the workpiece deformation via thermomechanical models, a special treatment of the material removal and of the boundary conditions has to be implemented in the software (typically based on the Finite Element Method, but new approaches are applied, too [10]). Thus, the mechanisms and possibilities for compensation and optimization also vary from application to application. Nevertheless,

A. Menzel
Institute of Mechanics (IM), TU Dortmund, Leonhard-Euler-Straße 5,
44227 Dortmund, Germany
e-mail: andreas.menzel@udo.edu

A. Schmidt (✉)
Center for Industrial Mathematics, University of Bremen, 28334 Bremen, Germany
e-mail: schmidt@math.uni-bremen.de

D. Biermann and F. Hollmann (eds.), *Thermal Effects in Complex Machining Processes*, Lecture Notes in Production Engineering,
DOI 10.1007/978-3-319-57120-1_3

collective ideas and joint approaches can be identified. Colloquia of the Priority Program 1480 and meetings of the work group "Modelling and Simulation", managed by the authors, were the organized fora to discuss and exchange ideas on related research aspects.

A simulation-based two-step approach was considered in most projects. After an optimization of basic process parameters, a problem-specific compensation strategy was used for reduction of the shape error.

2 Optimization Strategies

The optimization of (aspects of) the machining process can be interpreted as the solution of an *optimal control problem* with an objective function J, a model for the state y, control variables u, and restrictions on the control and, respectively or state. A pair of optimal state and control (y^*, u^*) is a solution of

$$y^* \in Y, u^* \in U: \quad J(y^*, u^*) = \min_{y \in Y, u \in U} J(y, u) \quad \text{under the constraint } y = S(u)$$

and other side conditions (equalities or inequalities) for y or u, where required. Y and U define the admissible sets of state and control, and S gives the relation between control and state, the *state equation*. Various approaches exist for the numerical solution of such optimal control problems.

For machining processes, the state y typically includes parts or all of temperature, deformation, and shape of the workpiece during the process. The state equation typically includes models for material removal and the resulting shape variation, a process model that computes forces and heat flux into the workpiece, as well as heat conduction and thermomechanical state of the workpiece.

Control variables u include at least a minimum of basic process parameters, such as feed velocity and, respectively or rotational velocity of a tool, cutting depth and chip thickness. The approach may be enhanced by control of the tool path with small, continuous variations or discrete variations by change of the underlying machining strategies. Possible variations of the cutting tool (shape) were generally not considered in the related contributions of the Priority Program 1480.

The main goal of a process optimization is the reduction of shape errors (and, possibly, the control of internal stresses) *after* the process. This goal is not easy to achieve directly by an optimal control approach with complex control variations as described above, as the control-to-state equation is complicated and costly to evaluate. Thus, simpler and computationally cheaper control and optimization approaches are applied in most of the projects of the Priority Program 1480.

One standard approach is to select basic process parameters in such a way that the generation of heat and temperature flux are minimized. The idea behind this is that lower temperatures introduce less deformation. Gulpak et al. [6] additionally consider the stresses introduced by face milling of thin plates. Depending on the process considered, parameters are analyzed with respect to their significance for heat generation and then tried to be optimized. Here, mostly scalar process parameters are considered, usually constant over time. Related minimization approaches vary by seeking for the optimal choice in a continuous set of admissible parameter variations, or by simply aiming at the identification of a minimum over a finite number of parameter sets.

For the various processes and applications considered within the Priority Program 1480, examples of parameters considered for optimization are

- drilling: feed rate, cutting speed, machining strategies, and cooling control, [4]
- deep hole drilling: drill velocity and (time dependent) feed, [3]
- gear hobbing: axial feed and cutting speed, [2]
- grinding: grinding wheel and workpiece velocities, axial feed, total radial stock removal, and roughing zone angle, [7]
- milling: greedy approach for optimization of radial and axial engagement and cutting speed during roughing, [5]; different machining strategies (discrete) and (continuous) width of cut, cutting velocity, and feed velocity for each strategy, [6]
- turning: cutting speed, machining strategy, [11]; Peclet number (including feed rate and cutting speed) for coated and non-coated tools, [8]; cutting speed, depth of cut, and feed for composite and non-reinforced workpiece material. [1]

A few groups are presently aiming to establish a "true" optimization framework of the whole process based on the minimization of shape errors. For face milling, Gulpak et al. [6] use a two-part model for source stresses induced by face machining and thermoelastic stresses induced by the process-generated heat. Here, an optimization of the milling strategy (discrete optimization problem) is combined with the continuous optimization of process parameters. A two-phase optimization strategy is used in order to maintain feasibility. Denkena et al. [5] demonstrate the idea of an optimization procedure for a simplified, quasi-2D, milling process. Different objective functions for shape error reduction are discussed in the reports: deviation of shape in the final state or incorrect material removal during the whole process, and the measure of shape deviation can be a global norm or, during the process, a local norm considering over time only the deviations near the current place of tool engagement. The examples show that the full optimization of the thermomechanical deformation during a real cutting process is still a major challenge.

3 Compensation Strategies

Using the optimized process parameters and the standard tool path, a simulation of the process and its thermomechanics is used for the computation of the expected shape deviation by erroneous material removal due to the deformation during the process. The typical compensation strategy tries to reduce the shape error by adjusting the tool path according to the computed deformation. Because a variation of the tool path introduces different conditions of engagement and thus different forces and heat fluxes, a variation of the tool path leads to a variation of deformations during the process. An optimal control approach would take this into account and would iterate the variation of controls until the shape error is minimized. This is mentioned in some reports but has not been implemented for most of the applications. Nevertheless, the typical compensation strategy can be interpreted as a single iteration of a full optimal control strategy. If the shape error after compensation is already below a given tolerance for the process, this compensation approach is successful and sufficient. Otherwise, more iterations might be needed.

Examples of compensation approaches for different processes are

- deep hole drilling: compensation by dynamically changing the drill angle, [3]
- drilling: variation of drilling order (machining strategy), use of 3D (non-)circular milling instead of drilling, and variation of cooling fluids, [4]
- grinding: compensation by inversion of surface profile deviation after cooling, and based on integral material removal, [7]
- milling: variation of NC tool path according to averaged computed deformation, [9]; identification of optimal machining strategy and depth of cut for face milling of thin plates, [6]; local variation of NC tool path according to computed deformation, [5, 12]
- turning: compensation of finishing contour according to simulated deformation. [1, 8]

All compensation approaches result in a substantial reduction of the shape error. Although the respective compensation frameworks show similarities for the different projects, the particularization of the respective approaches and the related implementations for the specific processes is a challenge and requires process-specific concepts and technological approaches, as elaborated in detail in the respective reports.

Acknowledgements This contribution is based on research projects within the DFG Priority Program 1480 "Modeling, Simulation and Compensation of Thermal Effects for Complex Machining Processes". The authors thank the German Research Foundation (DFG) for financial and organizational support.

References

1. Aurich, J.C., Zimmermann, M., Schindler, S., Steinmann, P.: Improvement of the machining accuracy in dry turning of aluminum metal matrix composites via experiments and finite element simulations. In: Biermann, D., Hollmann, F. (eds.) Thermal Effects in Complex Machining Processes, Lecture Notes in Production Engineering, pp 35–62. Springer (2017)
2. Beutner, M., Kadashevich, I., Karpuschewski, B., Halle, T.: Modeling, simulation and compensation of thermal effects in gear hobbing. In: Biermann, D., Hollmann, F. (eds.) Thermal Effects in Complex Machining Processes, Lecture Notes in Production Engineering, pp 181–218. Springer (2017)
3. Biermann, D., Blum, H., Iovkov, I., Rademacher, A., Rosin, K., Suttmeier, F.-T.: Modelling, simulation and compensation of thermomechanically induced deviations in deep-hole drilling with minimum quantity lubrication. In: Biermann, D., Hollmann, F. (eds.) Thermal Effects

in Complex Machining Processes, Lecture Notes in Production Engineering, pp 181–218. Springer (2017)
4. Bollig, P., Michna, J., Faltin, C., Schneider, J., Zanger, F., Schießl, R., Maas, U., Schulze, V.: Experimental and simulative modeling of drilling processes for the compensation of thermal effects. In: Biermann, D., Hollmann, F. (eds.) Thermal Effects in Complex Machining Processes, Lecture Notes in Production Engineering, pp 145–180. Springer (2017)
5. Denkena, B., Maaß, P., Schmidt, A., Niederwestberg, D., Vehmeyer, J., Niebuhr, C., Gralla, P.: Thermomechanical deformation of complex workpieces in milling and drilling processes. In: Biermann, D., Hollmann, F. (eds.) Thermal Effects in Complex Machining Processes, Lecture Notes in Production Engineering, pp 219–250. Springer (2017)
6. Gulpak, M., Wernsing, H., Sölter, J., Büskens, C.: Compensation strategies for thermal effects in dry milling. In: Biermann, D., Hollmann, F. (eds.) Thermal Effects in Complex Machining Processes, Lecture Notes in Production Engineering, pp 251–288. Springer (2017)
7. Holtermann, R., Schumann, S., Menzel, A., Biermann, D.: Modelling and simulation of internal traverse grinding from micro-thermo-mechanical mechanisms to process models. In: Biermann, D., Hollmann, F. (eds.) Thermal Effects in Complex Machining Processes, Lecture Notes in Production Engineering, pp 369–403. Springer (2017)
8. Klocke, F., Kneer, R., Burghold, M., Deppermann, M., Peng, B., Puls, H.: Modelling and compensation of thermoelastic workpiece deformation in dry cutting. In: Biermann, D., Hollmann, F. (eds.) Thermal Effects in Complex Machining Processes, Lecture Notes in Production Engineering, pp 63–94. Springer (2017)
9. Siebrecht, T., Wiederkehr, P., Zabel, A., Schweinoch, M., Byfut, A., Schröder, A.: Modeling, simulation and compensation of thermomechanically induced material deformation in dry NC milling processes. In: Biermann, D., Hollmann, F. (eds.) Thermal Effects in Complex Machining Processes, Lecture Notes in Production Engineering, pp 321–346. Springer (2017)
10. Spreng, F., Eberhard, P.: Modeling of orthogonal metal cutting using adaptive smoothed particle hydrodynamics. In: Biermann, D., Hollmann, F. (eds.) Thermal Effects in Complex Machining Processes, Lecture Notes in Production Engineering, pp 133–144. Springer (2017)
11. Uhlmann, E., Mahnken, R., Ivanov, I.M., Cheng, C.: Thermo-mechanical simulation of hard turning with macroscopic models. In: Biermann, D., Hollmann, F. (eds.) Thermal Effects in Complex Machining Processes, Lecture Notes in Production Engineering, pp 95–132. Springer (2017)
12. Wimmer, S., Loehe, J., Zaeh, M.F.: Coupling analytical and numerical models to simulate thermomechanical interaction during the milling process of thin-walled workpieces. In: Biermann, D., Hollmann, F. (eds.) Thermal Effects in Complex Machining Processes, Lecture Notes in Production Engineering, pp 321–346. Springer (2017)

Material Modelling

V. Schulze

Abstract Within the Priority Program 1480 "Modelling, Simulation and Compensation of Thermal Effects for Complex Machining Processes" different methods to model material behavior in machining operations were developed from experimental tests and implemented into simulations for different materials and workpieces. For the correct description of the thermo-mechanical load, friction tests were performed and compared for AISI4140 QT and AISI1045 QT on a laboratory tribometer and a broaching machine. In addition, extensive experimental and simulative investigations were carried out for the description of phase transformations due to the thermo-mechanical load. In spite of the varying materials and processes, relations between the results could be identified as generally valid statements.

1 Introduction

The Priority Program 1480 aims on the compensation of thermal effects for varying machining processes and materials. Many of the working groups involved use FEM based simulation models to predict forces, temperatures, residual stresses, microstructural changes and shape deformations for turning, drilling, milling and grinding. Therefore, the correct description of the material behavior, friction effects and phase transformations becomes essential and was substantially investigated within the Priority Program.

The field of material laws reaches from standard to advanced Johnson-Cook models regarding e.g. transformation induced plasticity and asymmetric viscoplastic flow behavior up to constitutive models using thermo-physical material properties dependent on the temperature.

V. Schulze (✉)
Institute of Production Science, Karlsruhe Institute of Technology,
Kaiserstraße 12, 76131 Karlsruhe, Germany
e-mail: Volker.Schulze@kit.edu

D. Biermann and F. Hollmann (eds.), *Thermal Effects in Complex Machining Processes*, Lecture Notes in Production Engineering,
DOI 10.1007/978-3-319-57120-1_4

To extend the influences on the material properties due to an adapted thermo-mechanical load, friction coefficients were measured on a laboratory tribometer and a set up close to cutting using a broaching machine. These friction investigations aim on the comparison between the two measurement strategies and a deduction for criteria to become independent from the workpiece-tool interaction.

Last but not least, phase transformations are taken into account to calculate the amount of transformed martensite resulting in workpiece distortions due to microstructural transformations. The investigations include stress states and transformation kinetics modelled from experimental results and material laws as well as a theoretical approach using the free Helmholtz energy densities. All methods are validated in experiments. A final comparison of all three approaches gives a conclusion for the best purpose of each method.

2 Material Laws

For the description and analysis of cutting phenomena material laws need to represent the real material behavior at its best. The researchers from [7] developed a multi-mechanism model based on the work from Johnson and Cook to illustrate strain softening due the influence of the temperature and the SD-effect. The developed asymmetric visco-plasticity uses a dependency on the temperature and the strain rate. With the use of weight functions a correlation to the stress modes and their stress tensors is realized. Investigations showed that the stress modes are independent of the TRIP-effect which is also included in the model according to Leblond's approach.

A second Material law was developed from [1] using constitutive equations which describe the material according to its load and structural parameters in a phenomenological way. The temperature dependent equations were measured for example using laserflash for the heat conductivity, Young's modulus using heated tensile test, dynamic scanning calorimetry for the specific heat or calorimetry measurements for the coefficient of linear thermal expansion.

A comprehensive overview of actual state of the art laws could be published by two of the groups involved in the Priority Program 1480. This is given in [6].

3 Friction

Friction modelling is indispensable for all kinds of FEM machining simulations. However, the field of implementing friction models in machining simulations varies from a constant friction coefficient up to complex friction models based on complex experimental results. Like it is already known from literature, the thermo-mechanical load during the simulation is mainly influenced by the friction modelling and the set up used to measure the friction coefficients. To overcome this problem it is necessary to generalize friction models to reflect for all workpiece-tool interaction conditions which may occur during machining. Therefore, a laboratory tribometer [2] and a broaching machine [4] were used to perform friction tests in dependence of the temperature, contact pressure and sliding speeding for the materials AISI4140 QT and AISI1045 QT. Cutting inserts were used for the laboratory tribometer as well as for the broaching machine. Further, all test methods are able to reach the necessary sliding speeds and to reflect the chip sliding during the cutting process.

The results present a qualitatively similar behavior as the laboratory tribometer and the close to cut set up using a broaching machine. The influence of the sliding speed varies only little between the two test set ups. The identified friction coefficients show a better agreement for medium to high temperature cutting processes for the broaching machine and lower to medium temperatures for the laboratory tribometer. Based on the investigations a coupled friction model is recommended to represent the whole temperature field for a generalized friction model.

4 Phase Transformations

Modelling phase transformations in cutting process with high temperatures are important because of the occurring changes in microstructure, leading workpiece distortions and residual stresses. Within the Priority Program 1480 three different approaches were developed, tested and compared for their ability to predict phase transformations [5]. The first approach mainly focuses on the time-temperature-transformation using a relation between the austenitization temperature and thermal activation [2]. The second approach uses a multi-mechanism model [7] based on the Clausius-Clapeyron equation. The third approach uses free Helmholtz energy densities to consider austenite and martensite phases in a so called “phase landscape” [3].

A comparison of the three approaches shows special benefits and disadvantages for every method. The first approach is able to calculate every phase state like austenite, bainite, ferrite, pearlite and martensite but demands extensive experimental investigations. The second approach solves the stress states using mechanical laws and the transformation heat which makes it easy to implement and experimentally less elaborate. The mechanically motivated third approach can be adapted to experimental results without any phenomenological material parameter and considers the crystal orientation as well. Thermal rate dependency and cooling effects will be modelled in future work.

As a conclusion, the amount of austenite and martensite can be calculated by every approach in very good agreement to experimental results. So, the implementation of each approach becomes dependent on the aim of the simulation and the required experimental data [5].

Acknowledgements The author gratefully thank the "Deutsche Forschungsgemeinschaft DFG" for the support of the priority program 1480 "Modelling, Simulation and Compensation of Thermal Effects for Complex Machining Processes".

References

1. Beutner, M., Kadashevich, I., Karpuschewski, B., Halle, T.: Modeling, simulation and compensation of thermal effects in gear hobbing. In: Biermann, D., Hollmann, F. (eds.) Thermal Effects in Complex Machining Processes, Lecture Notes in Production Engineering, pp. 347–367. Springer (2017)
2. Bollig, P., Michna, J., Faltin, C., Schneider, J., Zanger, F., Schießl, R., Maas, U., Schulze, V.: Experimental and simulative modeling of drilling processes for the compensation of thermal effects. In: Biermann, D., Hollmann, F. (eds.) Thermal Effects in Complex Machining Processes, Lecture Notes in Production Engineering, pp. 145–180. Springer (2017)
3. Holtermann, R., Schumann, S., Menzel, A., Biermann, D.: Modelling and simulation of internal traverse grinding from micro-thermo-mechanical mechanisms to orocess models. In: Biermann, D., Hollmann, F. (eds.) Thermal Effects in Complex Machining Processes, Lecture Notes in Production Engineering, pp. 369–403. Springer (2017)
4. Klocke, F., Kneer, R., Burghold, M., Deppermann, M., Peng, B., Puls, H.: Modelling and compensation of thermoelastic workpiece deformation in dry cutting. In: Biermann, D., Hollmann, F. (eds.) Thermal Effects in Complex Machining Processes, Lecture Notes in Production Engineering, pp. 63–94. Springer (2017)
5. Schulze, V., Uhlmann, E., Mahnken, R., Menzel, A., Biermann, D., Zabel, A., Bollig, P., Ivanov, I.M., Cheng, C., Holtermann, R., Barthel, T.: Evaluation of different approaches for modeling phase transformations in machining simulation. Prod. Eng. **9**(4), 437–449 (2015)
6. Schulze, V., Vöhringer, O., Halle, T.: Plastic Deformation: Constitutive Description, Reference Module in Materials Science and Materials Engineering (2016)
7. Uhlmann, E., Mahnken, R., Ivanov, I.M., Cheng, C.: Thermo-mechanical simulation of hard turning with macroscopic models. In: Biermann, D., Hollmann, F. (eds.) Thermal Effects in Complex Machining Processes, Lecture Notes in Production Engineering, pp. 95–132. Springer (2017)

Part II
Final Reports of the Research Projects

Improvement of the Machining Accuracy in Dry Turning of Aluminum Metal Matrix Composites via Experiments and Finite Element Simulations

M. Zimmermann, S. Schindler, J.C. Aurich and P. Steinmann

Abstract This chapter presents experimental investigations and a numerical approach for the achievement of a better machining accuracy when dry turning aluminum metal matrix composites (Al-MMCs) and aluminum alloys. Three differently reinforced Al-MMCs regarding the volume percentage (17 and 30%) and the average particle size (0.6 and 3 μm) were considered. Experimentally validated finite element (FE) models were developed to determine cutting conditions, which reduce the thermal load of the workpiece and the tool. Three approaches were investigated for the determination of the heat flow into the workpiece and the tool: a local FE model of chip formation, an experimental approach, and a fitting algorithm. Global FE models of the workpiece and the tool use this thermal boundary condition in order to calculate the temperature evolution and the thermal expansion. Moreover, a computer-aided design/computer-aided manufacturing (CAD-CAM) approach for the compensation of the workpiece and the tool deformations via an accordingly adapted depth of cut is outlined. As a result of the decreased thermal load on the workpiece and tool in combination with the compensation of the deformations, a significant increase of the machining accuracy was observed.

M. Zimmermann (✉) · J.C. Aurich
Institute for Manufacturing Technology and Production Systems, University of Kaiserslautern, Kaiserslautern, Rhineland-Palatinate, Germany
e-mail: marco.zimmermann@mv.uni-kl.de

J.C. Aurich
e-mail: jan.aurich@mv.uni-kl.de

S. Schindler · P. Steinmann
Chair of Applied Mechanics, Friedrich-Alexander-Universität Erlangen-Nürnberg, Erlangen, Germany
e-mail: stefan.schindler@ltm.uni-erlangen.de

P. Steinmann
e-mail: paul.steinmann@ltm.uni-erlangen.de

D. Biermann and F. Hollmann (eds.), *Thermal Effects in Complex Machining Processes*, Lecture Notes in Production Engineering,
DOI 10.1007/978-3-319-57120-1_5

1 Introduction

During turning, most of the mechanical work is dissipated into heat via plastic deformation of the workpiece material and friction. The generated heat causes a thermal expansion of the workpiece and the tool [10]. As a result, the actual depth of cut differs from the nominal depth of cut and the accuracy of machining decreases. Recently, the deformation of the workpiece attained particularly significance due to the increased use of dry cutting and higher demands on the accuracy of machining in general. Aluminum metal matrix composites (Al-MMCs) are two-phase high-performance materials. The reinforcement of aluminum via silicon carbide particulates enhances the material properties. Such composites are therefore increasingly applied in the aerospace and automobile industry in order to be able to fulfill higher demands on engineering applications [22]. The heterogeneous structure (ductile matrix and brittle reinforcement), however, leads to a poor machinability [20]. In particular for such composites, considerable thermal workpiece and tool loads are expected [24].

The use of an adequate cutting condition reduces the thermal load, and consequently the thermal expansion of the workpiece and the tool. A deviation from the intended workpiece geometry, however, remains because workpiece and tool are still deformed [2]. Moreover, further demands on machining (e.g. in terms of surface roughness or process time) need to be considered in industrial applications when determining the cutting condition. The cutting condition cannot be defined solely with regard to the machining accuracy. However, the correction of the depth of cut according to the remaining deformation of the workpiece and the tool allows for the compensation of these deformations.

In order to determine adequate cutting conditions and to compensate the deformations, we developed three experimentally validated finite element (FE) models. With the use of a local model of chip formation, the thermal and mechanical load acting on the workpiece and the tool can be determined with regard to the cutting condition and the workpiece material used. These results serve as boundary conditions for a global FE model of the workpiece and a global FE model of the tool. These models are suitable to determine the deformations. The approach used for the compensation of the workpiece and tool deformations, the experimental investigations, and the finite element models are outlined in this chapter.

2 Experimental Investigations

Dry turning investigations were carried out on a computerized numerical control (CNC) lathe (Fig. 1a). The workpieces were clamped via a three-jaw chuck. A tailstock was used to support the workpiece in order to reduce its deflection due to the forces. Two different workpiece geometries were considered (Fig. 1b). Workpiece geometry 2 allows for the measurement of the actual workpiece diameter after

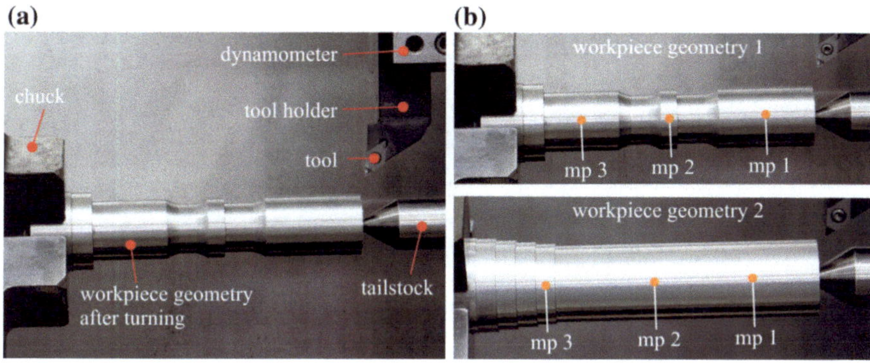

Fig. 1 **a** Experimental setup. **b** Workpiece geometries and axial measurement positions (mp) of the thermocouples

Table 1 Tool geometry (DCMT 11T304)

Clearance angle α	7°	Tool cutting edge inclination λ	0°
Rake angle γ	0°	Corner radius r_ϵ (mm)	0.4
Tool cutting edge angle κ	93°	Cutting edge radius r_β (μm)	12

each tool engagement, and thus for the evaluation of the thermal expansions of the workpiece and the tool.

The cutting material used was polycrystalline diamond (PCD; 90% diamond, grain size: 10 μm, 10% cobalt). A single workpiece can thus be turned without an appreciable wear progress. However, after the turning of multiple workpieces using a single tool, significant wear occurs. Wear increases the forces during turning. As a result, greater quantities of heat are generated. In order to minimize such undesirable influences on the results, new tools were used prior to a width of flank wear land of 100 μm. The tool geometry is listed in Table 1.

Four workpiece materials were investigated: one non-reinforced aluminum matrix alloy and three differently reinforced Al-MMCs (Fig. 2). The three considered composites consist of the following volume percentages of the particulate silicon carbide (SiC) reinforcement and average particle sizes: 17% and 0.6 μm (17/0.6), 17% and 3 μm (17/3), and 30% and 3 μm (30/3). The thermo-elastic material properties are listed in Table 3 for the aluminum alloy Al 2024 and in Table 4 for the Al-MMCs.

The Al-MMCs were produced via a powder metallurgical processing route. The particles and the aluminum powder were mixed, pre-compressed, hot-isostatic pressed, and extruded. The Al-MMCs and the non-reinforced aluminum alloy were subjected to a T6 heat treatment: solution annealing, cold water quenching, and artificial aging. The powder metallurgical processing route and the heat treatment ensure

Fig. 2 Metallographic sections of the aluminum metal matrix composites

that the results of the investigations are not affected (e.g. through an in-homogeneous particle distribution or voids) by the processing route of the materials (Fig. 2).

The cutting force, feed force, and passive force were measured by means of a three-component dynamometer. The workpiece temperature was measured using three thermocouples (type K; wire diameter: 1 mm). The axial positions of the thermocouples are illustrated in Fig. 1. All thermocouples were positioned centered regarding the radius of the workpiece after turning. A part of the generated heat during turning flows into the workpiece and the tool; as a result, the temperature in the two parts increases. The temperature of the workpiece at a measurement position rises during a tool engagement until the tool passes the respective measurement position. Afterwards, the workpiece temperature decreases due to the heat flow to the environment, particularly during the final stages of turning, cf. Fig. 10. The temperature distribution and temperature evolution at the workpiece results from the sum of all heat flows into and out of the workpiece. The heat flow to the environment takes place in terms of heat convection with the ambient air and heat conduction into the chuck. Heat radiation can be neglected at the present temperatures (< 100 °C). The magnitude of the heat flow to the environment can be described by the coefficients of heat convection and heat conduction, the time of turning, and the temperature difference between the workpiece and the environment. The remaining quantity of heat at the workpiece after tool engagement i

$$Q_i^{\mathrm{w}} = m_i c \, [\vartheta_i - \vartheta_0] \tag{1}$$

can be calculated from the current mass m_i, the specific heat capacity c, the initial temperature ϑ_0 and averaged current temperature ϑ_i of the workpiece after tool engagement i. At the first tool engagement, the temperature difference between workpiece and environment and thus also the heat flow to the environment is small (contrary to the final stages of turning). Neglecting this small heat loss, the heat flow into the workpiece

$$\mathfrak{Q}_{\mathrm{w}} = m_1 c \frac{\vartheta_1 - \vartheta_0}{t_1}, \tag{2}$$

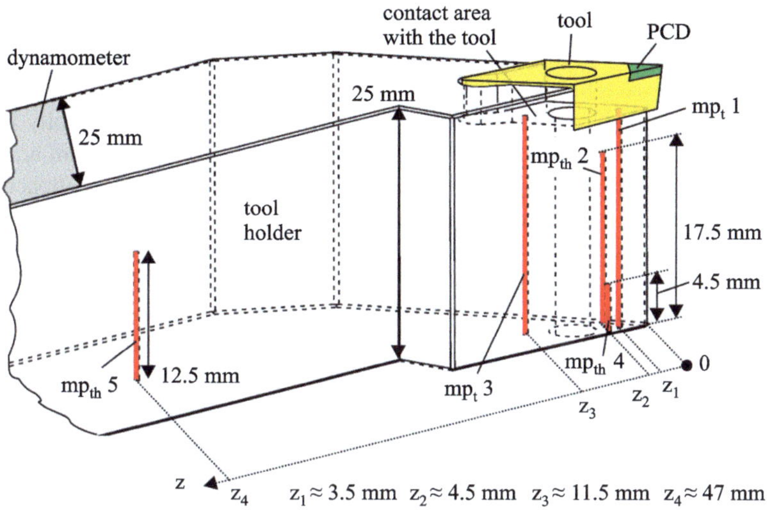

Fig. 3 Thermocouple positions in the tool (mp_t) and tool holder (mp_{th}) [17]

can be approximated from the remaining quantity of heat at the workpiece after the first tool engagement and the time t_1 elapsed during the first tool engagement. In order to determine the average workpiece temperature ϑ_i after tool engagement i, the temperature peek at each of the three measurement positions, i.e. the temperature when the tool passes the respective position, is considered and the average of the three temperature peeks is calculated. The average temperatures ϑ_0 and ϑ_i do not exactly represent the workpiece temperature at each point along the feed travel. At the final stages of turning, the workpiece temperature near the chuck (mp3, Fig. 1b) is lower compared to the temperature in the remaining workpiece, i.e. the workpiece exhibits a temperature gradient in axial direction. Therefore, the temperature at the measurement position near the chuck is, contrary to the first tool engagement, not considered when determining ϑ_i after the last tool engagement. The difference in the workpiece temperature at the measurement positions used to calculate ϑ_i is up to 8%.

Tool and tool holder temperatures were measured by means of thermocouples (type K; wire diameter: 0.5 mm). The temperature of the tool holder was determined at three positions: directly below the dynamometer (mp_{th} 5) and in two distances (mp_{th} 2 and mp_{th} 4) behind the contact area with the tool (Fig. 3). Two thermocouples measured the tool temperature at the contact area between the tool holder and the tool (mp_t 1 and mp_t 3) (Fig. 3).

The actual diameter of the workpiece after turning D_i^{act} was measured at multiple positions along the feed travel via a coordinate measuring machine. At each measurement position, 9 points were homogeneously distributed along the circumference in order to evaluate the actual diameter by the least square method. The difference $D_i^{act} - D_i$ between the nominal diameter D_i and the actual diameter D_i^{act}

represents the diameter deviation d_i. This deviation includes all sources for the deviation from the nominal diameter, such as the thermal expansion of workpiece and tool, the surface roughness, and kinematic errors of the machine tool. The profile peaks and profile valleys of the workpiece roughness were mathematically approximately smoothed when evaluating d_i. Therefore, the average maximum height Rz was subtracted from the respectively measured actual diameter. The impact of the cutting parameters (and particularly of the feed) on the diameter deviation can hence be evaluated approximately unaffected through the workpiece roughness.

In order to detect the thermal effects on the depth of cut at each tool engagement, the workpiece geometry 2 (Fig. 1b) was turned. At this workpiece geometry, a shaft shoulder remains after each tool engagement. Most of the factors influencing the machining accuracy are approximately constant for a single cutting condition and consecutive tool engagements. However, the thermal expansion of the workpiece and the tool (including the tool holder) increases with each tool engagement due to continuous heating of both components during the tool engagements. Differences in terms of d_i between consecutive tool engagements thus are primarily the result of the change in the thermal workpiece and tool expansion. To evaluate the magnitude of these thermal expansions experimentally, the diameter deviation was normalized. The normalized diameter deviation d_i^{norm} results from the difference of the diameter deviation after tool engagement i and 1 [17].

$$d_i^{\mathrm{norm}} = d_i - d_1 \tag{3}$$

3 Finite Element Modeling

The modeling of the whole turning process is considered at two scales. At the small scale, the formation of a single chip is investigated in detail on a small section of the whole workpiece during a short time sequence. The chip formation is actually a process of plastic deformation, whereby mechanical work is dissipated into thermal energy. The most important results of this local model of chip formation are the arising process forces and the heat flows into the workpiece and the tool. At the large scale, the whole geometry of the workpiece and the tool are considered each in a separate model and the appearing mechanical and thermal effects are investigated during the whole process time. In these global models, the results of the local model are used as boundary conditions in order to calculate the temperature evolution and temperature distribution, the subsequent thermal expansion, and the deflection as a result of the acting heat flows and forces. This time, the material removal is not considered in terms of chip formation but via the elimination of material. The goal of the global models and thus of the whole modeling approach is the optimization of the machining accuracy via the compensation of the predicted deformations of the workpiece and the tool.

3.1 Local Model of Chip Formation

During chip formation, mechanical work is dissipated into thermal energy via plastic deformation in particular in the primary and secondary shear zones and via friction between the workpiece material and the tool, compare Fig. 4. These effects are captured in a thermo-mechanically coupled 2D plane strain model, originally documented in [16]. In order to consider the arising large plastic deformations, nonlinear continuum mechanics are considered in a nonlinear finite element analysis provided by the commercial FE program MSC MARC. This software provides the advancing front quad remeshing technique, with which the severe deformation can be considered during the analysis. It is sufficient to consider only a small section of the workpiece, where the chip formation takes place, see Fig. 5. Convergence in terms of cutting force has been revealed for approximately ten elements across the chip thickness and about 3,000 linear quadrilateral elements overall. The mesh density was prescribed with large element edges of 0.1 mm length in regions of low interest and small element edges of 0.01 mm length in regions of high interest, compare Fig. 5.

The thermo-viscoplastic behavior of the workpiece material when machining is influenced by high temperatures, high strain rates, and finite plastic strains. To characterize this material behavior, a novel isotropic hardening model, in the following abbreviated as S^2AZ model, has been developed by the present authors. A detailed description, including a derivative from the basics of nonlinear continuum mechanics embedded in a thermodynamical consistent frame, is presented in [13]. In the following, only the definition of the yield stress

$$\begin{aligned}\sigma_{\mathrm{y}} = {} & A_{\mathrm{c}}\left[1 - T^{A_{\mathrm{T}}}\right]\left[1 + A_{\mathrm{r}}\exp\left(A_{\mathrm{re}}\dot{\varepsilon}_{\mathrm{p}}\right)\right] + \\ & B_{\mathrm{c}}\left[1 - T^{B_{\mathrm{T}}}\right]\left[1 + \ln(1 + B_{\mathrm{r}}\dot{\varepsilon}_{\mathrm{p}}) + \ln\left(B_{\mathrm{rs}} + B_{\mathrm{rl}}\dot{\varepsilon}_{\mathrm{p}}\right)\varepsilon_{\mathrm{p}}\right] \\ & \left[1 - \exp\left(B_1\left[1 + B_2 T^{B_3}\right]\left[1 + B_4\exp\left(B_5\dot{\varepsilon}_{\mathrm{p}}\right)\right]\varepsilon_{\mathrm{p}}\right)\right] \end{aligned} \tag{4}$$

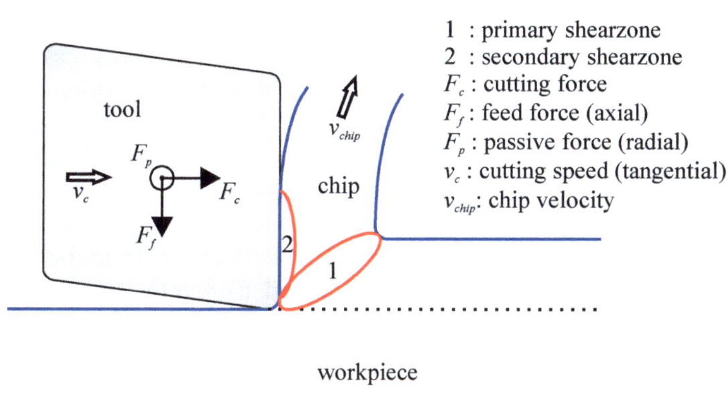

Fig. 4 Principle of chip formation [16]

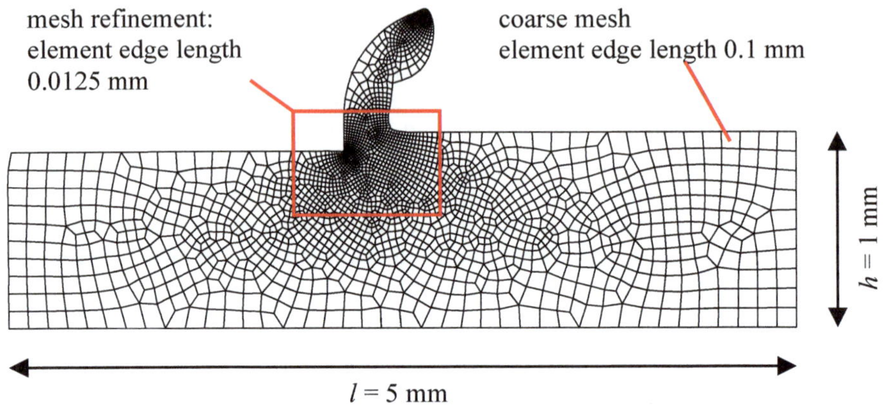

Fig. 5 Meshed section of the workpiece for chip formation [16]

is presented. The yield stress depends on the effective plastic strain $\varepsilon_{\mathrm{pl}}$, its rate $\dot{\varepsilon}_{\mathrm{pl}}$, the temperature (in terms of a normalized temperature $T = \frac{\vartheta - \vartheta_0}{\vartheta_{\mathrm{m}} - \vartheta_0}$, where $\vartheta_0 = 0\,^{\circ}\mathrm{C}$ is a reference temperature and ϑ_{m} is the melting temperature), and on 14 parameters, that adjust the strain hardening (A_{c}, B_{c}, B_{rs}, and B_1), the thermal softening (A_{T}, B_{T}, B_2, and B_3), and the strain rate hardening (A_{r}, A_{re}, B_{r}, B_{rl}, B_4, and B_5). This model is an extension of the von Mises flow law. A parameter identification (cf. Table 2) and a validation via experimental data has been performed in [13] for the aluminum alloy Al 2024. Experimental stress-strain measurements under high strain rates ($\dot{\varepsilon}_{\mathrm{p}} > 10,000\ \mathrm{s}^{-1}$) require specialized equipment and are therefore rare in the literature. Respective parameters for the S^2AZ model regarding the Al-MMCs are not available due to missing experimental data required for the identification. The heat dissipation has not to be constant, but can depend on the plastic strain. Hodowany [8] has investigated this effect experimentally for Al 2024 under high strain rates. Based on these data, we defined a phenomenological constitutive law for the fraction

$$\eta(\varepsilon_{\mathrm{p}}) = \begin{cases} 0.3 & \text{for} \quad \varepsilon_{\mathrm{p}} \leq 0.1 \\ \sin\left(\frac{\varepsilon_{\mathrm{p}} - 0.1}{0.6 - 0.1}\pi - \frac{\pi}{2}\right) 0.35 + 0.65 & \text{for} \quad 0.1 < \varepsilon_{\mathrm{p}} < 0.6 \\ 1 & \text{for} \quad \varepsilon_{\mathrm{p}} \geq 0.6 \end{cases} \tag{5}$$

of mechanical work dissipated into thermal energy. For a detailed documentation, it is referred to [13]. Thermo-elastic material properties have also been considered and are listed in Table 3.

The shear friction model was used to describe the friction at the tool-chip interface. Numerical and experimental results were compared in order to determine the friction coefficient of $\mu = 0.6$. The main object of this contribution within the model of chip formation is the development of a new hardening model for isotropic plasticity.

Table 2 Parameters for the S^2AZ model with $\eta = \eta(\varepsilon_p)$ [13]

Parameters for strain hardening corrected by the effect of the strain rate parameters					
A_c (MPa)	B_c (MPa)	B_1 [−]	B_{rs} [−]		
74.5847	203.6439	−13.6082	2.1885		
Parameters for thermal softening					
A_T [−]	B_T [−]	B_2 [−]	B_3 [−]		
1.3163	1.2570	37.4254	2.2865		
Parameters for strain rate hardening					
A_r [−]	A_{re} [s]	B_r [s]	B_{rl} [s]	B_4 [−]	B_5 [s]
3.9474	1.6212×10^{-5}	3.5664×10^{-7}	6.3709×10^{-6}	1.1562×10^{-3}	7.8617×10^{-4}

Table 3 Thermal properties for Al 2024 according to [1, 6, 12, 21]

density ρ (kg/m^3)	2780
Young's modulus $E(\vartheta)$ (GPa)	$74.0 - 0.0462\,\vartheta$
Poisson's ratio $\nu(\vartheta)$ [-]	$0.3282 + 9.0\,10^{-5}\vartheta$
thermal expansion α (µm/m°C)	$22.2 + 0.00891\,\vartheta$
specific heat c (J/(kg°C))	$858.0 + 0.85\,\vartheta$
thermal conductivity $\kappa(\vartheta)$ (W/m°C)	$162.6 + 0.118\,\vartheta$
melting point ϑ_m (°C)	502

For detailed investigations of frictional effects during chip formation including the influence of temperature and relative velocity, it is referred e.g. to [11, 18].

A part of the generated heat is dissipated by the chips, while the remaining heat flows into the tool and the workpiece [7]. The latter causes heating and consequently unwanted deformations of the tool and the workpiece. Neglecting the heat convection with the ambient air and the heat conduction into the not considered material, which are quite small at the regarded model dimensions and the considered time sequence, the heat flow (through the boundary $\partial\Omega$ of the body Ω) into the workpiece and the tool due to chip formation can be calculated from the rate of change of internal energy inside the respective body according to

$$\mathfrak{Q}_{w/t} = \int_{\partial\Omega_{w/t}} q_n \mathrm{d}S = \int_{\Omega_{w/t}} \rho c(\theta) \frac{\mathrm{D}\theta}{\mathrm{D}t} \mathrm{d}V, \tag{6}$$

with the density ρ, the heat capacity c, and the rate of change of the temperature $\frac{\mathrm{D}\theta}{\mathrm{D}t}$ integrated over the respective volume V. Note that the absolute temperature θ is measured in K, while ϑ denotes the Celsius-temperature in °C. For a detailed derivation from the first law of thermodynamics, it is referred to [16]. The integration of all not cut off elements of the workpiece in every time step reveals the heat flow $\mathfrak{Q}_w$ into the workpiece, which can be used as boundary condition for the global model. In analogy, the integration of all elements of the tool in every time step reveals the heat flow $\dot{Q}_t$ into the tool. Machining the aluminum alloy Al 2024 results in continuous chips [3]. The amount of generated heat per time as well as the heat flow into the workpiece hence is approximately constant once chip formation reached a steady state although the heat conduction is transient because of the transient location of the heat source.

Note that the described 2D plane strain model is primarily suitable for cutting processes, which feature a much greater depth of cut than the corner radius of the tool, i.e. processes that can be assumed as orthogonal cutting. In this case, the undeformed chip thickness is approximately constant along the major cutting edge. For cutting processes with a small depth of cut and a large corner radius, however, the undeformed chip thickness varies along the major cutting edge. The shape of the cross-section of undeformed chip thus differs from a rectangular area, which is assumed in the 2D plane strain model used. Despite this simplification, the surface area of the cross-section of undeformed chip and consequently also the corresponding volume is well represented in the present case. In [16], the suitability of our 2D model for not ideal orthogonal cutting processes in terms of determining the heat flow into the workpiece is investigated and validated. 3D models promise a more accurate geometrically depiction of the process and thus of the arising effects, but they usually suffer under greater inaccuracies due to the much more challenging 3D remeshing. The improvement of 3D models of chip formation is suggested for future work.

3.2 Global Model of the Workpiece

In a global model of the workpiece (originally proposed in [16] and enhanced in [15, 17]), the whole geometry of the workpiece is considered, whereby the detail of chip formation is omitted. Therefore, the results of the chip formation, the heat flow into the workpiece, and the forces acting on the workpiece are considered as boundary conditions. The respective quantities of the forces (cutting F_c, feed F_f, and passive F_p) and the heat flow $\dot{Q}_w$ have to be determined by experimental investigations, by a separate model of chip formation, or by an inverse identification method as described later on in this section. Figure 6 depicts the global model of the workpiece and the applied boundary conditions. Besides the heat flow into the workpiece also the heat flow out of the workpiece to the environment (convective $\dot{Q}_{cvw}$ and conductive $\dot{Q}_{cdw}$) have to be considered. Therefore, the respective coefficients of heat convection with the ambient air α_{cvw} and heat conduction into the chuck α_{cdw} were determined experimentally.

The mechanical boundary condition in terms of the chuck is assumed as a rigid restraint. The center of the tailstock is geometrically considered due to its noteworthy elastic behavior, which affects the deflection of the workpiece. The right hand side of the center of the tailstock is rigidly restrained. Since the temperature at the right hand side does not significantly increase during turning, it can be assumed isother-

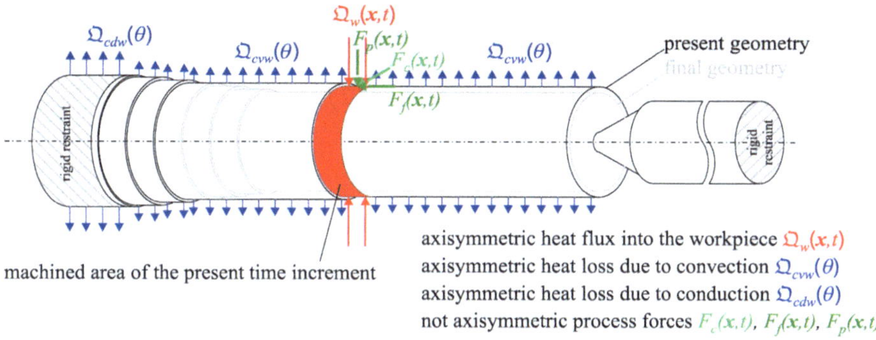

Fig. 6 Boundary conditions at the model of the workpiece [17]

mal or adiabatic. Experimental investigations in [2] showed solely axisymmetric temperature distributions in the workpiece in terms of relevant time discretization. Consequently, the thermal boundary conditions of the workpiece can be assumed axisymmetric. Rotational body forces and heat radiation have been neglected, since they are quite small compared to the process forces and heat convection. The forces F_c, F_f, and F_p do not act axisymmetrically on the workpiece and cause a kind of pure elastic beam bending. It is therefore not necessary to consider the rotation of the workpiece in the present model.

Beside the mechanical and thermal boundary conditions, the removal of material has to be considered. The removal of material, the heat flow into the workpiece, and the forces occur in the area of chip formation and thus follow the tool path. An accurate time and local discretization is required in order to consider these three effects. Therefore, a preprocessor has been developed that calculates first the time-dependent position of the tool from the numerical control (NC) code of the lathe. In a second step, the preprocessor defines all elements according to the nominal tool path, whereby all element edges are parallel or perpendicular to the closest tool path, compare Fig. 7.

The workpiece material undergoes thermal expansion and elastic deflection during turning; in the FE analysis the nodes undergo a displacement and the elements a deformation. Consequently, the element edges will not fit the nominal tool path. Therefore, the elements are arranged so that the tool path approximately halves each element, as depicted in the 2D illustration of Fig. 7. With the use of an h-adaptive mesh refinement each hexahedral element is subdivided into eight new hexahedral elements. The new nodes are defined exactly on the nominal tool path. The new element edges consider the deformations due to thermal expansion and elastic deflection and thus represent the actual (deformed) workpiece geometry accurately. The removal of material is performed via deactivation of all elements above the nominal tool path. For an accurate time discretization of the element refinement and element elimination, again, the NC code is evaluated. Using a fixed time incrementation for the simulation, the position of the tool at the end of a time increment and thus the

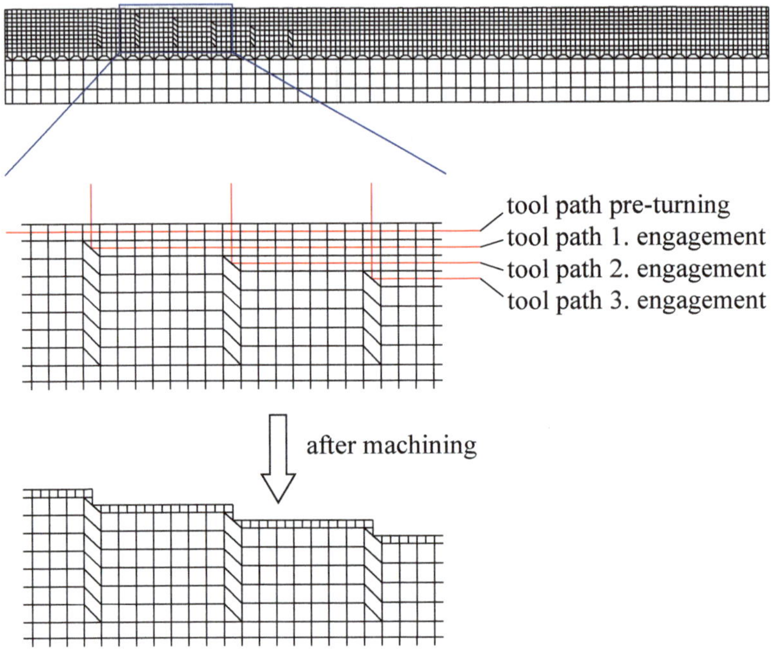

Fig. 7 Mesh design and h-adaptive mesh refinement for the workpiece [17]

elements to be subdivided and eliminated are known already prior to each time increment. As a result, subdivision and elimination can be performed accordingly. The element refinement is carried out prior to each time increment and the element deactivation at the beginning of each time increment. During each time increment the heat flow into the workpiece and the forces are applied to the newly defined elements. In the following time increment, the now already machined surface is considered for heat convection with the ambient air.

A tool path halves each element only once. This restricts the maximum element edge length in radial direction in the machined part of the workpiece to the depth of cut. The element edge length in axial direction is chosen according to the radial direction in order to guaranty well shaped elements. Along the circumference of the workpiece a sufficient number of elements should be defined for an adequate representation of a circle. Here 24 elements have been used. In the not machined part of the workpiece, three times larger elements (in radial and axial direction) can be used without loss of numerical accuracy. The applied h-adaptive mesh refinement can handle solely linear elements. Convergence studies revealed that the used discretization is sufficient.

Since the effect of chip formation is omitted, i.e. no plastic deformation occurs, thermo-elastic material properties are sufficient to describe the material behavior. For the aluminum alloy Al 2024, the respective temperature-dependent properties are listed in Table 3. For the Al-MMCs, we have performed a two-scale material

Table 4 Material properties of the two Al-MMC types for the temperature range $\vartheta = 0\,°C$ to $502\,°C$

MMC Al 2024-SiC, 17% SiC at 3 and 0.6 μm	
Density ρ (kg/m^3)	2844
Thermal expansion $\alpha(\vartheta)$ (μm/m°C)	$19.8 + 5.919 \times 10^{-3}\,\vartheta - 2.223\,e^{-2.239\times10^{-3}\,\vartheta}$
Young's modulus $E(\vartheta)$ (GPa)	$96.75 - 0.05397\,\vartheta$
Poisson's ratio $\nu(\vartheta)$ [−]	$0.3067 + 8.6656 \times 10^{-5}\,\vartheta$
Heat capacity $c_{\varepsilon m}(\vartheta)$ (J/kg°C)	$905.6 + 0.66\,\vartheta - 80.24\,e^{-0.003\,\vartheta}$
Thermal conductivity $\kappa(\vartheta)$ (W/m°C)	$122.7 + 0.0876\,\vartheta + 21780\,e^{-2.846\times10^{-6}\,\vartheta}[\vartheta + 682.3]^{-1}$
Melting point ϑ_m (°C)	502
MMC Al 2024-SiC 30% SiC at 3 μm	
Density ρ (kg/m^3)	2894
Thermal expansion $\alpha(\vartheta)$ (μm/m°C):	$17.5 + 4.190 \times 10^{-3}\,\vartheta - 3.194\,e^{-2.538\times10^{-3}\vartheta}$
Young's modulus $E(\vartheta)$ (GPa)	$119.81 - 0.05950\,\vartheta$
Poisson's ratio $\nu(\vartheta)$ [−]	$0.2881 + 8.1134 \times 10^{-5}\,\vartheta$
Heat capacity $c_{\varepsilon m}(\vartheta)$ (J/kg°C)	$940.56 + 0.52\,\vartheta - 139.23\,e^{-0.003\,\vartheta}$
Thermal conductivity $\kappa(\vartheta)$ (W/m°C)	$96.5 + 0.0675\,\vartheta + 32839\,e^{-0.136\times10^{-6}\vartheta}[\vartheta + 630.5]^{-1}$
Melting point ϑ_m (°C)	502

modeling in order to determine the temperature-dependent thermo-elastic material properties from the respective properties of the single constituent parts, i.e. matrix and reinforcement, of the composite. A detailed documentation, including an illumination of the usually violated thermodynamical consistency for non-constant temperature fields, is provided in [12]. The thermodynamical consistency is not conserved through the scale transition, but the error is negligible for sufficiently small temperature differences inside the considered microscopic model. Note that the thermo-elastic material behavior depends primarily on the volume percentage of the two constituents and not on the size of the particles. Therefore, the Al-MMC with 17% SiC features identical material properties for 3 μm and 0.6 μm particle size. The respective thermo-elastic material properties of the Al-MMCs are listed in Table 4.

There are various methods to determine the required values of the thermal boundary conditions for the heat flow into the workpiece $\mathfrak{Q}_w$, coefficient of heat convection α_{cvw}, and coefficient of heat conduction α_{cdw}. Generally, the respective values have to be determined once per cutting condition for $\mathfrak{Q}_w$, once per cutting speed for α_{cvw}, and once per workpiece and chuck material for α_{cdw}. Afterwards, they can be used for arbitrary workpiece geometries. The heat flow into the workpiece due to chip formation can be approximated from experimental investigations, see Sect. 2, or calculated with the use of a suitable FE model of chip formation, see Sect. 3.1. A possibility to determine the heat flow to the environment, that requires separate experimental investigations, is described in [16]. An alternative technique according to [17] is outlined in the following, which allows also for the determination of the heat flow into the workpiece.

The values of the thermal boundary conditions can be determined via an inverse parameter identification using a least square curve-fit algorithm, e.g. available in the program MATLAB, and temperature measurements from experimental investigations, e.g. those outlined in Sect. 2. The mathematical formulation of the least square curve-fitting problem is given by

$$\min_{y_j} \|G\left(y_j, t\right) - \vartheta\left(t\right)\|_2^2 = \min_{y_j} \sum_i \left[G\left(y_j, t_i\right) - \vartheta\left(t_i\right)\right]^2. \tag{7}$$

The time t is the given input data (in the present case the process time) and the reference data $\vartheta(t)$ is the measured temperature evolution at selected positions, e.g. the positions depicted in Fig. 1b. The function $G(y_j, t)$, in the present case the FE model of the workpiece, calculates the temperature evolution at the same positions as experimentally used to measure the temperature from the applied heat input and heat output. In the present case, the vector y_j has three entries, each for the magnitude of one of the three thermal boundary conditions $\mathfrak{Q}_w$, α_{cvw}, and α_{cdw}. All thermal boundary conditions act perpendicular to the surface of the workpiece. The time and local discretization for the application of the thermal boundary conditions is determined by the evaluation of the NC code.

The minimization problem tries to find a vector y_j in order to minimize the residuum of the square of the difference of $G(y_j, t)$ and $\vartheta(t)$. In an iterative procedure, the optimizer executes the FE simulation several times until the pre-defined accuracy is reached. Since the optimization problem is a solely thermal problem, the described, thermo-mechanical coupled 3D FE model can be reduced to a 2D axisymmetric FE model, in order to save computational costs. The exclusive experimental determination of the heat flow into the workpiece from Eq. 2 is used as initial value for the optimization algorithm. The identification of the three unknowns was performed using geometry 1 and afterwards the resulting parameters have been applied to simulate the turning of geometry 2 in order to ensure the transferability of the identified values.

3.3 Global Model of the Tool

In a global model of the tool, originally documented in [15], all parts shown in Fig. 8 are considered. These are the PCD insert, the cemented carbide substrate, and the tool

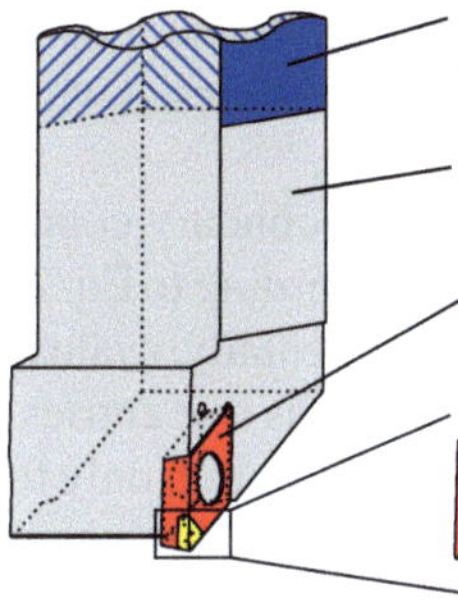

Fig. 8 Boundary conditions of the model of the tool according to [15]

Table 5 Material properties of the tool

	PCD	K10 (cemented carbide)	AISI 4140
Density ρ (g/cm^3)	3.52	14.9	7.85
Young's modulus E (GPa)	410	650	210
Thermal conductivity κ (W/(mK))	520	100	48
Specific heat capacity c (J/(kgK))	600	200	480
Thermal expansion α (10^{-6}/K)	2.6	5	11.5

holder. All of them undergo a significant temperature increase during turning, while the dynamometer, in which the tool holder is fixed, is almost isothermal. The whole model undergoes solely thermo-elastic deformations so that thermo-elastic material properties are sufficient to describe the material behavior. The respective parameters are listed in Table 5. The temperature dependency of the thermo-elastic material properties of the three tool materials (see e.g. [9, 19]) is quite small at the present temperatures and is therefore neglected. The discretization of the tool is chosen according to the temperature gradients occurring. The PCD insert is discretized with the smallest elements, the average element edge length is approximately 0.4 mm. At the cemented carbide substrate, an average element edge length of approximately 1 mm is used. In the tool holder, the element edge length increases from 1 mm in the vicinity of the tool to almost 10 mm near the dynamometer.

In analogy to the model of the workpiece, also the model of the tool inputs the heat flow into the tool Ω_t, the heat convection with the ambient air according to the coefficients of heat convection α_{cvth}, α_{cvt}, and α_{cvPCD}, the heat conduction into the dynamometer according to the coefficient of heat conduction α_{cdt}, and the forces F_c, F_f, and F_p as boundary conditions, compare Fig. 8. The local discretization of the heat flow into the tool and the forces is, contrary to the model of the workpiece, not time-dependent, because the contact area between the PCD insert and the chip does not change. The contact area is approximated as the cross-section of undeformed chip at the tool tip. The time discretization of the heat input and the forces is performed in analogy to the model of the workpiece. The preprocessor described in Sect. 3.2 indicates the times at which the tool is in engagement, i.e. the times at which the heat flow and the forces have to be applied.

For the determination of the magnitude of the thermal boundary conditions ($\dot{Q}_t$, α_{cvth}, α_{cvt}, α_{cvPCD}, and α_{cdt}) the least square curve-fitting according to Eq. 7 is applied. This time, the temperature ϑ is numerically and experimentally evaluated at five measurement positions (cf. Fig. 3). The vector y_j therefore has five entries. The temperature evolution inside the tool shows, at the beginning of turning, only for the thermocouple at $\mathrm{mp_t}$ 1 a temperature increase. All other thermocouples stay at constant temperature. Afterwards, the thermocouples at mp_{th} 2, mp_{t} 3, and $\mathrm{mp_{th}}$ 4 undergo a significant temperature increase. The last thermocouple mp_{th} 5 is still almost isothermal. In a third time sequence, also $\mathrm{mp_{th}}$ 5 depicts a temperature increase, compare Fig. 11. Due to this staggered temperature evolution at the evaluation positions, a staggered calculation during the least square curve-fitting is recommendable in order to save computational time.

Three independent calculations according to the three described time intervals are consecutively performed in order to determine in each calculation only one or three unknowns instead of five. In the first time interval, only the heat flow into the tool is considered. The heat loss to the environment can be neglected due to the almost constant temperature at the measurement positions 2–5. In the second time interval, the heat flow into the tool is known and the heat conduction into the fixture is still negligible due to the almost constant temperature at $\mathrm{mp_{th}}$ 5. The magnitudes of α_{cvth}, α_{cvt}, and α_{cvPCD} remain as the values to be determined. In the last time interval, α_{cdt} has to be identified. In analogy to the workpiece, it is sufficient to determine the coefficient of heat conduction into the dynamometer once per tool, the coefficients of heat convection for each material of the tool once per cutting speed, and the amount of heat flow into the tool once per cutting condition in order to use them afterwards for arbitrary workpiece geometries.

4 Approach for the Compensation of the Workpiece and Tool Deformation

A CAD-CAM approach is applied for the compensation of the process-induced thermo-mechanical deformations of the workpiece and the tool. Therefore, the grid points of the CAD-model of the workpiece are displaced according to the sum of the calculated workpiece and tool deformations (Fig. 9). The distance between the grid points is 1 mm. Linear interpolation is used in order to design the contour of the workpiece between adjacent grid points. The thermal expansion of the workpiece and the tool increases the depth of cut; their deflection due to the forces reduces the depth of cut. The direction of the grid point displacement is determined by the superior effect. Greater thermal effects than mechanical effects (as in the present

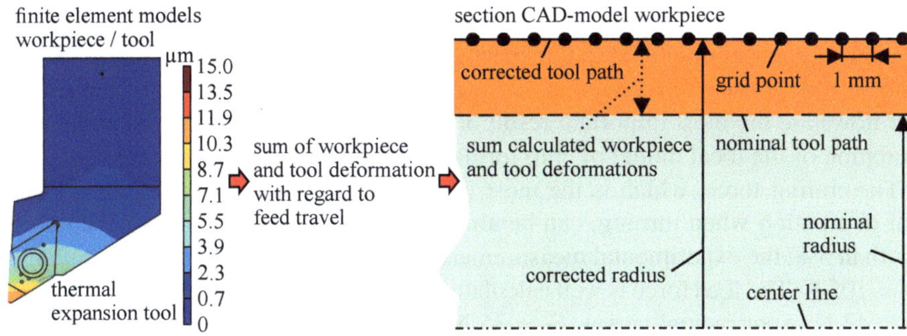

Fig. 9 Correction of the workpiece CAD-model according to the calculated deformations

case) require an increase of the nominal workpiece radius in the CAD-model in order to compensate the deformations (Fig. 9) [25].

The deformation of the workpiece and the tool is not uniform along the feed travel. This can be attributed to the varying stiffness of the system machine tool-workpiece, the rising tool temperature with increasing feed travel, and the axial temperature gradient in the workpiece due to the heat flow into the chuck. The calculated deformation needs hence to be evaluated in terms of the feed travel. The evaluation is carried out on the element nodes. The position of the element nodes is identical to the position of the grid points in the CAD-model of the workpiece. Thus, the axial element edge length in the FE model of the workpiece is 1 mm. The element edge length in radial direction matches the depth of cut. The time-dependent tool deformation is related to the evaluation positions via the NC-code of the lathe. The positions for the evaluation of the deformation of the workpiece and the tool are the same [25].

Each grid point of the workpiece CAD-model is displaced by the CAD-CAM software according to the sum of the calculated workpiece and tool deformations. The CAD-CAM software inputs the corrected radial positions of the grid points as tabular data. Computer-aided manufacturing is used in order to generate the NC-code for the lathe. For finish turning, the NC-code is corrected by the sum of the calculated deformations (via the global finite element models) of the workpiece and the tool. The depth of cut used for finish turning thus varies along the feed travel [25].

5 Validation of the FE Models

5.1 Local Model

The results of the local model of chip formation are exemplarily validated for the cutting condition $v_c = 100$ m/min, $f = 0.1$ mm/rev, and $a_p = 0.9$ mm using the non-reinforced aluminum alloy as the workpiece material. The criteria of validation are the forces, which are the most popular validation criteria due to the easy experimental determination, the chip thickness, and the heat flow into the workpiece and the tool. The present investigations focus on the thermal effects. Thus, the arising heat flows are the most important result of the local model. A more comprehensive validation of the local model of chip formation is given in [16].

The cutting force, which is the most important force component in view of the heat dissipation when turning, can be almost exactly predicted with a deviation of less than 5%; the experimental measurement reveals $F_{\mathrm{c}} = 100$ N and the simulation $F_{\mathrm{c}} = 105$ N. The feed force is well calculated with a deviation of approximately 17%; $F_{\mathrm{f}} = 42$ N experimental versus $F_{\mathrm{f}} = 35$ N numerical. The passive force cannot be captured with a 2D model. Since there is no movement in the direction of the passive force, it performs no work and thus it does not contribute to the heat dissipation. However, the passive force is significant for the deflection of the workpiece. In the present investigations, the deflection of the workpiece is small compared to the thermal expansion because of the small passive force and the support of the workpiece via a tailstock.

The experimentally measured chip thicknesses of 0.19 mm is slightly underestimated by the simulation, which renders a chip thickness of approximately 0.17 mm. Considering the slightly varying chip thickness of a single chip along the chip length, the experimental measuring inaccuracy (in particular the determination of a perpendicular cross-section), and the numerical errors due to continuous remeshing, numerical and experimental results are in good agreement. Also the chip type, a continuous chip, is well predicted.

The iterative procedure, explained in the Sects. 3.2 and 3.3, for the determination of the heat flow into the workpiece and the tool serves as reference for the respective results of the local model. The iterative procedure determines a heat flow of $\dot{Q}_{\mathrm{w}} = 70$ W into the workpiece and $\dot{Q}_{\mathrm{t}} = 22$ W into the tool. The local model of chip formation features a deviation of approximately 20% for the workpiece, $\dot{Q}_{\mathrm{w}} = 55$ W. The heat flow into the tool is almost exactly predicted, $\dot{Q}_{\mathrm{t}} = 22$ W.

A detailed analysis of error sources is given e.g. in [16] and the references therein. The most important ones are the numerical inaccuracies during remeshing, the extrapolation of the yield stress for strain rates above the validated range, the friction model, and the simplification due to the 2D plane strain assumption.

5.2 *Global Models*

The thermal boundary conditions have been identified with the use of the iterative procedure, which is described in the Sects. 3.2 and 3.3. The thermal boundary conditions determined from the investigations using workpiece geometry 1 were applied for the calculation of the temperature evolution during the turning of geometry 2, compare Fig. 1b. The most important validation criteria are the temperature distribution and temperature evolution, which can be measured accurately with the method described in Sect. 2. In Figs. 10 and 11, the calculated temperature evolution at the

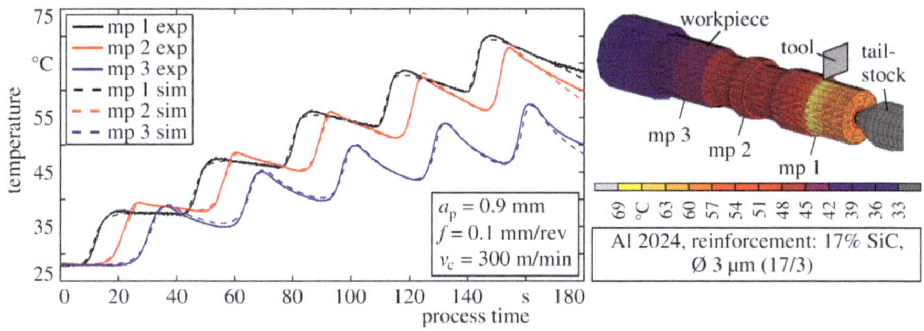

Fig. 10 Calculated and measured workpiece temperature [5]

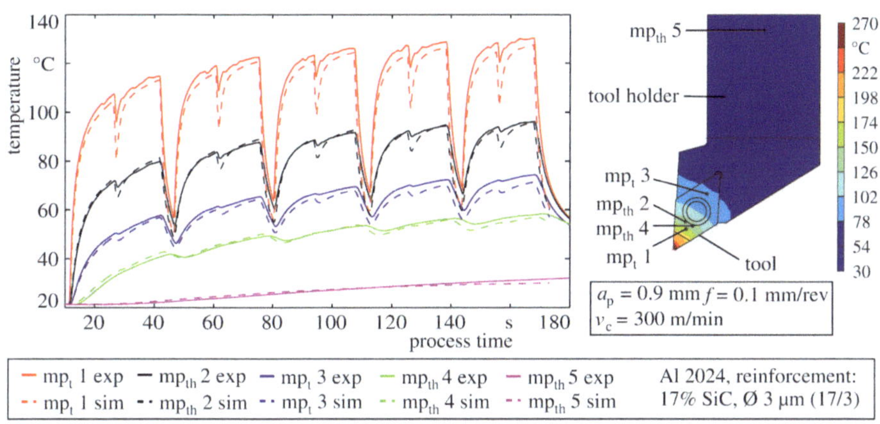

Fig. 11 Calculated and measured tool temperature and tool holder temperature [5]

three measurement positions in the workpiece and the five measurement positions in the tool is validated exemplarily for a single cutting condition. Both models provide an accurate prediction of the temperature evolution during the whole process.

In general, the physical phenomena of thermal expansion is well known and mathematically described. The mesh refinement and the technique for the elimination of elements at the workpiece (see Sect. 3.2) are accurate in terms of predicting the thermal expansion. The global model of the tool, which gets along without any refinement technique, is anyway accurate in that terms. Therefore, it can be assumed that the thermal expansion is well calculated as long as the temperature distribution is well predicted. An additional validation of the predicted diameter deviation due to thermal expansion and mechanical deflection is provided in [17]. Comprehensive validations of the global models using the non-reinforced aluminum alloy and the Al-MMCs, including also a validation of the calculated tool temperature on the rake face via thermographic measurements, are given in [2, 14–17].

6 Thermal Load of the Workpiece

The remaining quantities of heat at the workpiece (geometry 1, Fig. 1b) directly after the last tool engagement are depicted in Fig. 12 with regard to the cutting condition and the workpiece material. The nominal material removal ($V_w = 61,200\ \mathrm{mm}^3$) was the same for each cutting condition. An increase of feed or cutting speed reduces the thermal load of the workpiece. However, for the least considered depth of cut of $a_p = 0.5$ mm, an increase of the remaining quantity of heat at the workpiece was observed when only slightly raising feed or cutting speed. Greater depths of cut cause a higher thermal load of the workpiece except for the cutting conditions, which exhibit the largest considered cutting speed and feed. The simultaneous use of high cutting speeds and feeds thus allows for the use of high depths of cut without increasing the thermal workpiece load [3].

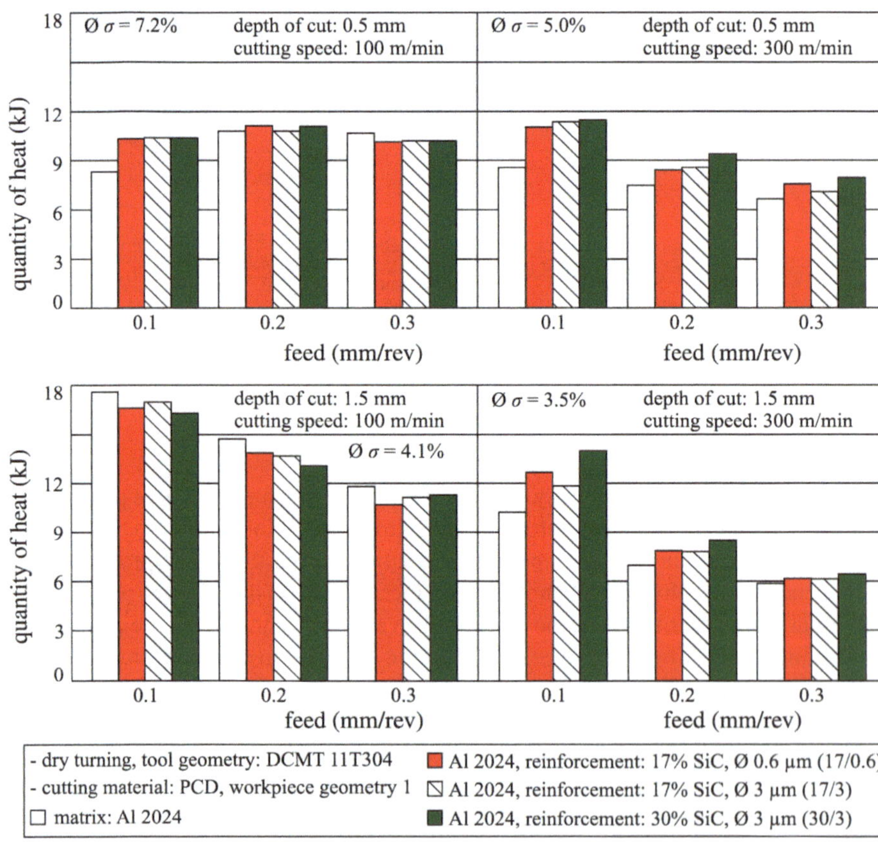

Fig. 12 Thermal load Q_W of the workpiece directly after the last tool engagement regarding the workpiece material and the cutting condition [3]

The composite workpieces are subjected to a greater thermal load in comparison to the workpieces of the non-reinforced aluminum alloy. Only at low cutting speeds ($v_c = 100\,\mathrm{m/min}$) and high depths of cuts ($a_p = 1.5$ mm), a larger thermal load was found on the workpieces of the non-reinforced aluminum alloy. The feed is the most important cutting parameter for the thermal load of the composite workpieces. On the contrary, the cutting speed primarily determines the quantity of heat flowing into the workpieces of the non-reinforced aluminum alloy. The aluminum matrix composite 30/3 undergoes compared to the composites with a particle volume percentage of 17 on average over all considered cutting conditions a 4.1% greater thermal load. The non-reinforced aluminum alloy exhibits a 12.8% lower quantity of heat at the workpieces than the Al-MMC 30/3. The comparison of the thermal workpiece load between the composites with a particle volume percentage of 17 and average particle sizes of 0.6 and 3 μm reveals differences of up to 9.5% regarding particular cutting conditions. However, this difference is not affected by a process parameter. The average particle size of 0.6 μm causes on average over all investigated cutting conditions a 0.3% larger quantity of heat at the workpiece. For a detailed description of the reasons for the differences in the thermal workpiece load, we refer to [23].

7 Machining Accuracy and Surface Integrity

Figure 13 shows the deviation d_i from the nominal diameter with regard to the workpiece material and the cutting condition. Each deviation depicts the average over sixteen measurement positions along the feed travel (workpiece geometry 1, Fig. 1b) and the three repetitions of each investigation. Thermal expansions of the workpiece and the tool increase the depth of cut. The mechanically induced deflection of the workpiece and the tool decreases the depth of cut. Both effects thus compensate each other partially. The actual workpiece diameters D_i^{act} are undersized. At the present workpiece geometry, the thermal expansions hence are dominant in comparison to the deflections.

The composite workpieces exhibit better machining accuracies (on average over all cutting conditions: 14%) than the workpieces of the non-reinforced aluminum alloy [5]. The significantly smaller coefficients of thermal expansion of the Al-MMCs in comparison to the non-reinforced aluminum alloy thus overcompensate the greater thermal workpiece and tool loads, i.e. the thermal expansion of the composite workpieces during turning, and consequently the rise of the depth of cut is significantly lower. The difference in terms of the coefficients of thermal expansion between the particle volume percentages of 17 and 30 is much lower than the difference between the composites and the non-reinforced aluminum alloy. The larger thermal workpiece (4.1%) and tool (5.0%) load during the turning of the Al-MMC 30/3 compared to the Al-MMCs 17/0.6 and 17/3 thus cannot be overcompensated. The workpieces with a particle volume percentage of 17 therefore exhibit on average over all cutting conditions a 5.3% better machining accuracy. The turning of the Al-MMC 17/3 is accompanied with the least diameter deviation. In comparison to the Al-MMC 17/0.6,

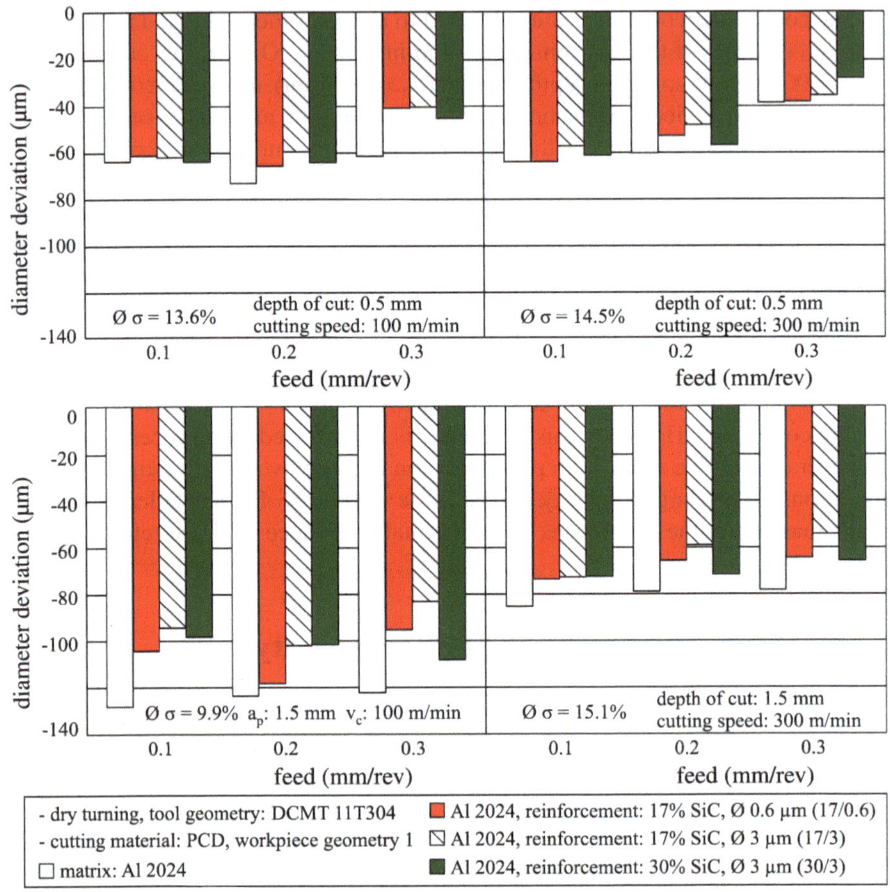

Fig. 13 Deviation d_i from the nominal workpiece diameter in terms of the workpiece material and the cutting condition [3]

this might be due to the 5.9% lower tool temperatures when turning the Al-MMC 17/3 [3].

The impact of the cutting parameters on the machining accuracy is in accordance to their effect on the thermal load of the workpiece and the tool. The lower the thermal workpiece and tool load (and thus the more heat is removed by the chips), the better the machining accuracy. The best machining accuracy was achieved using the highest cutting speed ($v_c = 300\,\text{m/min}$), greatest feed ($f = 0.3\,\text{mm/rev}$), and smallest depth of cut ($a_p = 0.5\,\text{mm}$) for turning. This cutting condition did not show the least thermal workpiece load. However, the comparatively low thermal load of both the workpiece and the tool results in the best machining accuracy [24]. The cutting condition causing the least thermal workpiece load ($v_c = 300\,\text{m/min}$, $f = 0.3\,\text{mm/rev}$, $a_p = 1.5\,\text{mm}$) exhibits in comparison to the cutting condition ($v_c = 300\,\text{m/min}$, $f = 0.3\,\text{mm/rev}$, $a_p = 0.5\,\text{mm}$, best machining accuracy) an approximately 40 μm greater deviation from the nominal diameter of the workpieces of the non-reinforced aluminum alloy. This can be attributed to the significantly greater tool temperature (approximately 200 °C on the rake face). The use of an adequate cutting condition already allows for a considerable improvement of the machining accuracy. d_i varies between the cutting conditions considered up to 283% for the Al-MMC 17/3, 306% for the Al-MMC 17/0.6, 327% for the non-reinforced aluminum alloy, and 378% for the Al-MMC 30/3 [23].

The cutting condition and the reinforcement affect moreover the roughness of the workpiece and the residual stress in the workpiece surface. The roughness of the composite workpieces is at low feeds, contrary to high feeds, larger than the roughness of the workpieces of the non-reinforced aluminum alloy. A higher particle volume percentage reduces within the Al-MMCs investigated the surface roughness. The effect of the average particle size on the roughness needs to be evaluated in terms of the cutting condition used. A general trend cannot be concluded. Tensile residual stresses were revealed in the surface of the Al-MMCs and the non-reinforced aluminum alloy. The tensile residual stress in the workpiece surface of the composite with a particle volume percentage of 30 was on average over all investigations approximately 89% lower in comparison to the Al-MMCs with a particle volume percentage of 17. The ratio between the composites 17/0.6 and 17/3 and the non-reinforced aluminum alloy needs to be evaluated with respect to the individual cutting conditions. Contrary to that, the Al-MMC 30/3 exhibits for all cutting conditions smaller tensile residual stresses than the non-reinforced aluminum alloy. The feed is the most important cutting parameter for all workpiece materials regarding the workpiece roughness and the residual stress. An increase of feed reduces the tensile stress in the surface, but deteriorates the surface quality of the workpiece considerably. For a detailed description, we refer to [4].

8 Results of the Compensation of the Workpiece and Tool Deformation

Workpiece geometry 2 (Fig. 1b) was used in order to perform the compensation of the deformations of the workpiece and the tool during finish turning. Figure 14 illustrates the normalized diameter deviation d_i^{norm} when compensating and not-compensating the deformations. In Fig. 14a, the results when turning the non-reinforced aluminum alloy using an identical cutting condition for each tool engagement are shown. The results of the composite 17/0.6 are depicted in Fig. 14b. In the case of the composite, we used two different cutting conditions for turning. The tool engagements 1–4 represent rough turning and the last tool engagement 5 is performed using a cutting condition for finish turning.

The compensation of the deformations enhances the machining accuracy significantly. The magnitude of the reduction of the normalized diameter deviation in microns depends on the magnitude of the thermal workpiece and tool load. The greater the thermal load, the greater the enhancement of the machining accuracy.

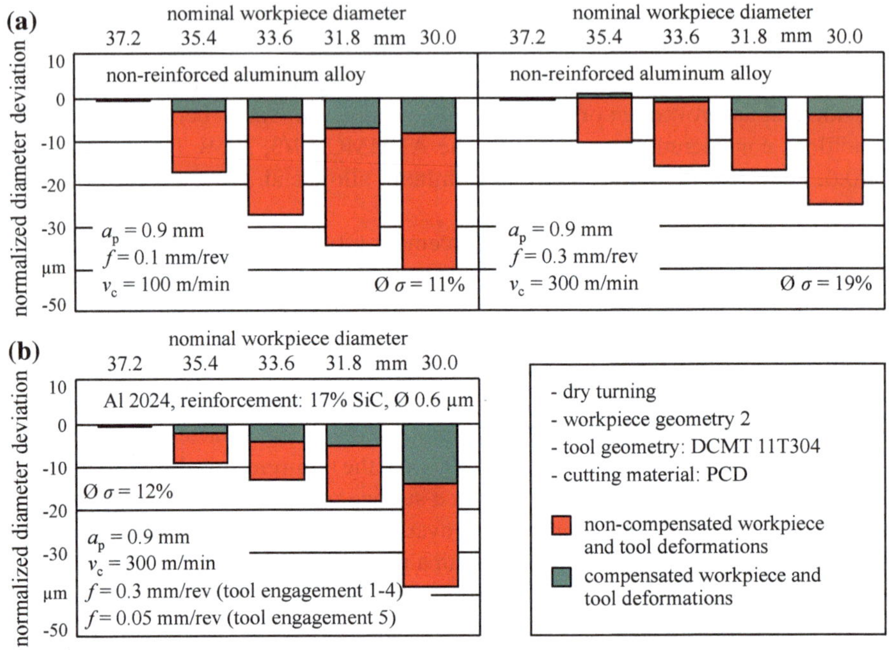

Fig. 14 Normalized diameter deviation d_i^{norm} for dry turning workpiece geometry 2 when compensating (*green*) and not-compensating (*red*) the deformations of the workpiece and the tool according to the results of the FE models [25]

Thus, particularly at low cutting speeds or feeds a considerable improvement was observed. Nevertheless, the percentage reduction of the normalized diameter deviation is the same for all considered cutting conditions. Concerning the non-reinforced aluminum alloy, d_i^{norm} is at the first tool engagement (D_1 = 37.2 mm) reduced by approximately 85% through the compensation approach. This percentage reduction declines to approximately 80% at the last tool engagement (D_5 = 30.0 mm). The remaining normalized diameter deviation after the first tool engagement when compensating the deformations is close to zero microns. The reduction of 80% at the least turned nominal diameter leads at the considered cutting conditions to a remaining d_i^{norm} in the range from 4 to 8 µm (Fig. 14a) [25]. The enhancement of the machining accuracy is in the case of the composite approximately 5% smaller in comparison to the non-reinforced aluminum alloy.

The capability of the presented compensation approach is affected by numerical and experimental uncertainties. Concerning the experiments, multiple factors of influence impede the achievement of the intended correction of the depth of cut. Such factors of influence are: the formation of built-up edges (particularly at low cutting speeds), the characteristics of the machine tool (e.g. the positioning accuracy or the vibration behavior), tool wear, and measurement uncertainties. Such experimental uncertainties are not considered in the present FE models, but separate FE models (e.g. of the machine tool) are required. However, the experimental uncertainties are indicated by the standard deviation σ [25]. The corrected depth of cut used for turning might influence the thermal and mechanical load of the workpiece and the tool, and thus their deformation. The temperatures and forces measured revealed, however, that the change in depth of cut does not significantly influence the workpiece and tool load.

The thermal boundary conditions required for the FE models were determined via a least square curve-fitting algorithm. Nevertheless, the primary numerical uncertainty is the heat flow to the environment. In particular for the tool, it is difficult to exactly determine the magnitude of the heat flow to the environment due to heat convection and heat conduction (into the dynamometer). The environmental conditions, such as the temperature or the speed of the ambient air, seem to be not constant as a result of the varying tool position. Radiation is neglected. Concerning the tool, radiation, however, might slightly affect the heat flow to the environment due to the temperatures ($< 500\,°C$) occurring on the rake face. Generally, slight inaccuracies in the magnitude of the boundary conditions influence the calculated tool temperature more strongly than the workpiece temperature. This is due to the significantly more elevated tool temperature compared to the workpiece temperature. Each tool engagement raises the numerical uncertainty because the numerical uncertainty at a tool engagement includes the uncertainties of each prior tool engagement. The uncertainty of the calculated temperature and therefore the remaining normalized diameter deviation hence slightly increases with the number of tool engagements. The workpiece and tool temperature increases with each tool engagement. As a result, also their thermal expansion rises. An increase of d_i^{norm} was therefore also observed when not-compensating the deformations of the workpiece and the tool [25].

9 Conclusion

In this chapter, experimental investigations and finite element models for the calculation of the workpiece and tool deformation when turning were outlined. Experimental investigations were carried out in order to generate fundamental knowledge with respect to the thermal workpiece load and the machining accuracy in dry turning of aluminum metal matrix composites (Al-MMCs). The results showed that the thermal load of the workpiece is significantly affected by the cutting condition and the reinforcement. As a result, also the thermal expansion of the workpiece and thus the machining accuracy varies accordingly. The effect of the cutting condition on the thermal workpiece load is in accordance to the impact of the cutting condition on the machining accuracy. An increase of feed or cutting speed, or a decrease of the depth of cut reduces the thermal load of the workpiece. The composites undergo a greater thermal load than the non-reinforced aluminum alloy. The larger the particle volume percentage, the larger the thermal workpiece load. However, the turning of the composites is accompanied with better machining accuracies in comparison to the non-reinforced aluminum alloy due to the smaller coefficient of thermal expansion. The use of an adequate cutting condition enhances the machining accuracy considerably. We revealed differences in the machining accuracy of up to approximately 380%. Nevertheless, despite the use of an adequate cutting condition, workpiece and tool are still deformed during turning leading to a remaining machining inaccuracy.

Three finite element models were developed in order to determine the deformations of the workpiece and the tool: a local model of chip formation, a global model of the workpiece, and a global model of the tool. Three alternatives, the local model, an experimental based analysis, and an inverse identification method, were proposed for the determination of the thermal and mechanical loads on the workpiece and the tool with respect to the cutting condition. These data are required as boundary conditions for the global models in order to calculate the deformations of the workpiece and the tool during turning. For the non-reinforced aluminum alloy considered, we developed a novel isotropic plastic hardening model to enhance the description of the material behavior under high strain rates and high temperatures and thus the prediction of the chip formation. As a result, the thermal and mechanical workpiece and tool load can be calculated more precisely. A parameter identification for the composites is addressed to future work because of missing experimental data. All three finite element models require thermo-elastic material properties. For the aluminum alloy, the silicon carbide reinforcements, and the tool materials, respective data is available in the literature. For the Al-MMCs on the contrary, the respective data has been derived from a two-scale material modeling. The comparison of the measured and calculated temperature evolution in the workpiece and the tool revealed good agreement. The finite element models thus are suitable to calculate the deformations of the workpiece and the tool.

We developed a computer-aided design/computer-aided manufacturing (CAD-CAM) approach for the compensation of the workpiece and tool deformations. For this purpose, the CAD-model of the workpiece is corrected according to the sum of the calculated deformations. The corrected CAD-model is used to generate a corrected numerical control code for finish turning via CAM, i.e. the depth of cut is adapted according to the calculated deformations. With the use of the compensation approach the machining accuracy has been increased by 85% in case of the non-reinforced aluminum alloy and by 80% in case of the composites. The compensation approach thus allows for a significant enhancement of the machining accuracy.

Acknowledgements The authors would like to thank the German research foundation (DFG) for funding the project "Thermal effects when turning Al-MMC—experiments and simulations AU 185/26, STE 544/42" within the priority program SPP 1480.

References

1. Asad, M., Girardin, F., Mabrouki, T., Rigal, J.F.: Dry cutting study of an aluminium alloy (A2024–T351): a numerical and experimental approach. Int. J. Mater. Form. **1**, 499–502 (2008)
2. Aurich, J.C., Zimmermann, M., Schindler, S., Steinmann, P.: Analysis of the machining accuracy when dry turning via experiments and finite element simulations. Prod. Eng. Res. Dev.

8(1–2), 41–50 (2014)

3. Aurich, J.C., Zimmermann, M., Schindler, S., Steinmann, P.: Effect of the cutting condition and the reinforcement phase on the thermal load of the workpiece when dry turning aluminum metal matrix composites. Int. J. Adv. Manuf. Technol. **82**(5), 1317–1334 (2016)
4. Aurich, J.C., Zimmermann, M., Schindler, S., Steinmann, P.: Turning of aluminum metal matrix composites: Influence of the reinforcement and the cutting condition on the surface layer of the workpiece. Adv. Manuf. **4**(3), 225–236 (2016)
5. Aurich, J.C., Zimmermann, M., Steinmann, P., Schindler, S.: Fertigungsgenauigkeit beim Drehen steigern - Einfluss der Prozessparameter auf die Maßhaltigkeit bei der Trockenbearbeitung. VDI-Z Integrierte Produktion **157**(4), 42–45 (2015)
6. Brammer, J.A., Percival, C.M.: Elevated-temperature elastic moduli of 2024 aluminum obtained by a laser-pulse technique. Exp. Mech. **10**(6), 245–250 (1970)
7. Denkena, B., Tönshof, H.K.: Spanen - Grundlagen. Springer, Heidelberg (2011)
8. Hodowany, J., Ravichandran, G., Rosakis, A., Rosakis, P.: Partition of plastic work into heat and stored energy in metals. Exp. Mech. **40**(2), 113–123 (2000)
9. Leichtfried, G., Sauthoff, G., Spriggs, G.: Refractory, hard and intermetallic materials. In: Advanced Materials and Technologies. Springer, Berlin (2002)
10. Moriwaki, T., Horiuchi, A., Okuda, K.: Effect of cutting heat on machining accuracy in ultra-precision diamond turning. CIRP Ann. Manuf. Technol. **39**(1), 81–84 (1990)
11. Puls, H.: Mehrskalenmodellierung thermo-elastischer Werkstückdeformationen beim Trockendrehen. Dissertation, RWTH Aachen, Aachen (2015)
12. Schindler, S., Mergheim, J., Zimmermann, M., Aurich, J.C., Steinmann, P.: Numerical homogenization of elastic and thermal material properties for metal matrix composites (mmc). Continuum Mech. Thermodyn. **29**(1), 51–75 (2017)
13. Schindler, S., Steinmann, P., Aurich, J.C., Zimmermann, M.: A thermo-viscoplastic constitutive law for isotropic hardening of metals. Arch. Appl. Mech. **87**(1), 129–157 (2017)
14. Schindler, S., Zimmermann, M., Aurich, J.C., Steinmann, P.: Modeling deformations of the workpiece and removal of material when turning. Procedia CIRP **8**, 39–44 (2013)
15. Schindler, S., Zimmermann, M., Aurich, J.C., Steinmann, P.: Finite element model to calculate the thermal expansion of the tool and the workpiece in dry turning. Procedia CIRP **14**, 535–540 (2014)
16. Schindler, S., Zimmermann, M., Aurich, J.C., Steinmann, P.: Thermo-elastic deformations of the workpiece when dry turning aluminum alloys—a finite element model to predict thermal effects in the workpiece. CIRP J. Manuf. Sci. Technol. **7**(3), 233–245 (2014)
17. Schindler, S., Zimmermann, M., Aurich, J.C., Steinmann, P.: Identification of thermal effects on the diameter deviation of inhomogeneous aluminum metal matrix composite workpieces when dry turning. Prod. Eng. **9**(4), 473–485 (2015)
18. Schulze, V., Zanger, F., Boev, N., Michna, J., Maas, U., Faltin, C., Schneider, J., Bollig, P.: FE-simulation of machining induced phase transformations considering friction as a function of temperature and sliding speed and detailed modeling of the heat transport. Adv. Eng. Mater. **16**(2), 137–141 (2014)
19. Spittel, M., Spittel, T.: Metal forming data—ferrousalloys. In: Advanced Materials and Technologies. Springer, Berlin (2002)
20. Tomac, N., Tannessen, K., Rasch, F.: Machinability of particulate aluminium matrix composites. CIRP Ann. Manuf. Technol. **41**(1), 55–58 (1992)
21. Zhao, N., Yang, Y.Q., Han, M., Luo, X., Feng, G.H., Zhang, R.J.: Finite element analysis of pressure on 2024 aluminum alloy created during restricting expansion-deformation heat-treatment. Trans. Nonferrous Met. Soc. China **22**, 2226–2232 (2012)
22. Zhou, L., Wang, Y., Ma, Z., Yu, X.: Finite element and experimental studies of the formation mechanism of edge defects during machining of SiCp/Al composites. Int. J. Mach. Tool Manuf. **84**, 9–16 (2014)

23. Zimmermann, M.: Drehen von Aluminium-Matrix-Verbundwerkstoffen - Analyse thermischer und mechanischer Einflüsse auf die Fertigungsgenauigkeit bei der Trockenbearbeitung. Dissertation, TU Kaiserslautern, Kaiserslautern (2015)
24. Zimmermann, M., Schindler, S., Steinmann, P., Aurich, J.C.: Drehen metallischer Verbundwerkstoffe - Einfluss von Prozessparametern und Verstärkungsphase auf die Fertigungsgenauigkeit. wt Werkstattstechnik online **104**(1–2), 33–39 (2014)
25. Zimmermann, M., Schindler, S., Steinmann, P., Aurich, J.C.: Compensation of thermomechanically induced workpiece and tool deformations during dry turning. Procedia CIRP **46**, 31–34 (2016)

Modelling and Compensation of Thermoelastic Workpiece Deformation in Dry Cutting

F. Klocke, R. Kneer, M. Burghold, M. Deppermann, B. Peng and H. Puls

Abstract Dry cutting ranks among the most significant developments within manufacturing technology. Compared to wet cutting, a major problem of dry machining is a stronger heat generation and thus, workpiece warming. This leads to thermoelastic workpiece deformation. Therefore, within this work a model is developed to predict and compensate the thermoelastic workpiece deformation. At first, friction behavior and heat transfer at the tool-chip interface in the orthogonal cutting process are experimentally investigated. Based on the fundamental investigations, a multi-scale model for the dry turning process is developed. It contains two submodels, a mesoscopic FE-model for the chip formation and a macroscopic FE-model for the turning process. To validate the mesoscopic FE-model, experiments of orthogonal turning are performed and the temperature fields are measured. Hereby, the occurring heat flow into the workpiece is calculated by solving the inverse heat conduction problem. The macroscopic FE-model calculates the thermoelastic workpiece deformation based on heat inputs of the mesoscopic model. By means of the developed approach, minimization and compensation strategies are developed, applied and evaluated based on complex processing examples.

1 Introduction

Machining with minimum quantity lubrication (MQL) or dry machining in combination with high performance cutting (HPC) ranks among one of the most significant developments in manufacturing technology. More and more companies recognize the enormous economic and ecological advantages of this technology and convert

F. Klocke (✉) · B. Peng · H. Puls
Laboratory for Machine Tools and Production Engineering, RWTH Aachen University, Aachen, Germany
e-mail: F.Klocke@wzl.rwth-aachen.de

R. Kneer · M. Burghold · M. Deppermann
Institute of Heat and Mass Transfer, RWTH Aachen University, Aachen, Germany
e-mail: kneer@wsa.rwth-aachen.de

D. Biermann and F. Hollmann (eds.), *Thermal Effects in Complex Machining Processes*, Lecture Notes in Production Engineering,
DOI 10.1007/978-3-319-57120-1_6

their production to dry cutting or MQL-technology [1, 2]. Drivers for the implementation of dry cutting are in particular companies with series production in the field of automobile manufacturers and their suppliers [3, 4]. But the conversion from wet to dry cutting comes along with major problems, a strong heating of the workpiece during dry cutting and finally, thermally caused dimension and form deviations [2]. Until now, based on the blank contour and depending on cutting sequence and conditions, it is not possible to consistently calculate the expectable dimension and form deviations during dry cutting.

The amount of heat and heat partition were mostly estimated by experiments and are sometimes extended by analytical and numerical models according to the recent research [5]. For example, Pabst et al. investigated the heat distribution inside the workpiece in dry milling with a calorimetric methodology [6]. Abukhshim et al. analyzed the heat partition in the dry cutting process to identify the thermal field in the cutting zone [7]. To build knowledge of the mechanics of the dry cutting process, cutting forces were also determined in addition to temperature measurements. Besides, Sölter and Gulpak used an inverse method to analyze the heat flux into the workpiece in dry milling based on experiments and simulations [8]. The approach of using experiments to determine the amount of heat and heat partition is common. However, a direct, precise measurement of the workpiece's temperature distribution is still a big challenge, due to high temperature gradients and a broad temperature range. Optical temperature measurement devices offer a sufficient spatial and temporal resolution, but are sensitive to the emissivity of the workpiece material. Further, the emissivity of metals varies over temperature, angle of view, surface roughness and surface oxidation [9].

Therefore, this work presents a numerical multiscale approach to calculate the heat generation during the chip formation process on a mesoscopic scale and the thermoelastic workpiece deformation on a macroscopic scale. After that, compensation strategies are outlined to avoid thermally-induced dimension and form deviations.

2 Fundamental Investigations

Dry cutting is a complex process of mechanical deformation and heat transfer. Heat is generated due to deformation of workpiece material and friction at tool-chip interface (see Fig. 1). Further, a partition of heat is transferred to the tool by surface contact.

To account for friction and contact conductance at the deformation zone, input parameters are needed for the mesoscopic FE-model: friction, contact conductance and chip formation. Thus, these input parameters are determined experimentally. The measurement of forces during chip formation is presented in Sect. 2.1 and compared to forces of the mesoscopic FE-model in Sect. 3.1. Afterwards in Sect. 2.2, boundary conditions of friction and in Sect. 2.3, boundary conditions of contact conductance for the mesoscopic FE-model are determined.

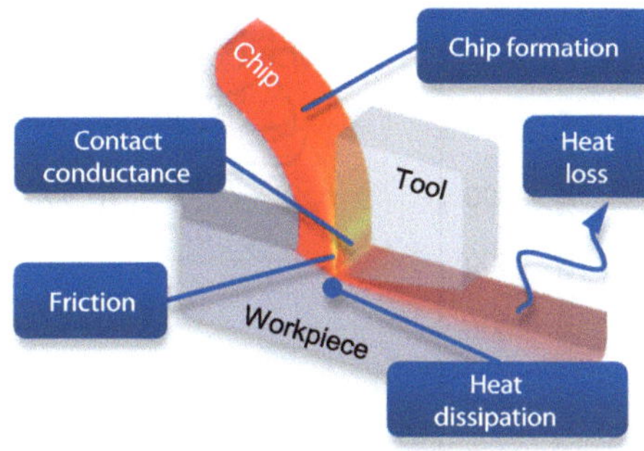

Fig. 1 Deformation zone of the orthogonal cut

2.1 In-situ Recording of Chip Formation in Orthogonal Cutting

To validate the chip formation simulation in Sect. 3.1, a test bench has been developed for the orthogonal cutting process on a vertical broaching machine by the manufacturer Forst Technologie of type RASX 8 × 2200 × 600 M/CNC, as shown on the left hand side of Fig. 2. The maximum cutting speed is $v_c = 150\,\text{m/min}$. The tool holder was mounted on a custom-made Kistler dynamometer 9257B with a custom-made cage. The workpiece with the dimensions 200 mm × 40 mm × 3.5 mm was clamped with the standard tool clamping holder of the machine tool, which was located on the linear machine tool drive. The linear drive provides the relative velocity to execute the cutting operation. The cutting process was captured with a

Fig. 2 Work bench design for the orthogonal cutting processes

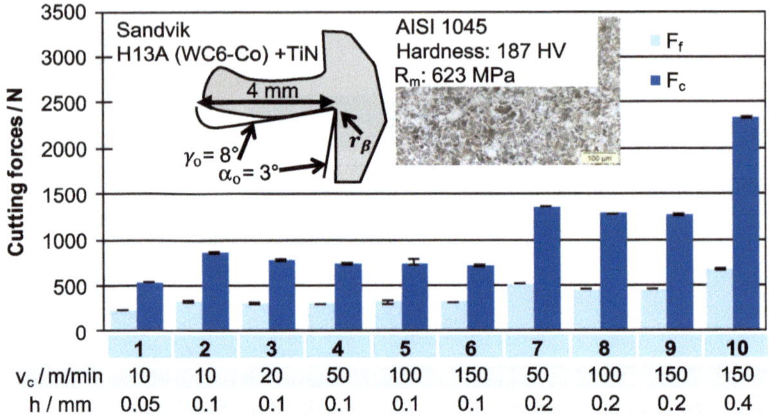

Fig. 3 Orthogonal cutting force components F_c and F_f of AISI 1045 with different cutting speeds v_c and uncut chip thicknesses h

high speed video camera of type Vision Research Phantom v7.3, which can capture frame rates from 6,688 fps by 800 × 600 pixels to 500,000 fps by 32 × 16 pixels.

The workpiece material used for these experiments is normalized carbon steel AISI 1045, with the tensile strength $R_m = 623$ MPa. The substrate of the cutting inserts is WC6-Co (Sandvik H13A), which is coated with TiN. The cutting inserts have a rake angle of $\gamma = 8°$ and a clearance angle of $\alpha = 3°$. The cutting edge radius of the cutting inserts is $r_\beta = 33.3$ µm. The orthogonal cutting tests were performed at different cutting speeds v_c (10–150 m/min) and uncut chip thicknesses h (0.05–0.4 mm). The cutting force F_c and the feed force F_f were measured by dynamometer and shown in Fig. 3. Regarding the experimental results in Fig. 3, the process forces decrease slightly with increasing the cutting speed. This phenomenon can be explained by thermal softening with increasing cutting speed. Compared to the cutting speed v_c, the uncut chip thickness h has stronger influence on the cutting force components.

2.2 *Friction Experiment Under Cutting Conditions*

The thermomechanical load at the tool-chip interface influences the plastic deformation and friction in primary and secondary shear zone and thus, influences the mechanical stresses. Due to this reason, the thermomechanical load at the tool-chip interface is vitally important for the design of metal cutting processes. In combination with the sliding velocities at the tool-chip interface, these stresses generate very high temperatures up to the melting temperature of the workpiece material [10]. The determination of friction at the tool-chip interface is still a challenging research topic and not fully understood [11].

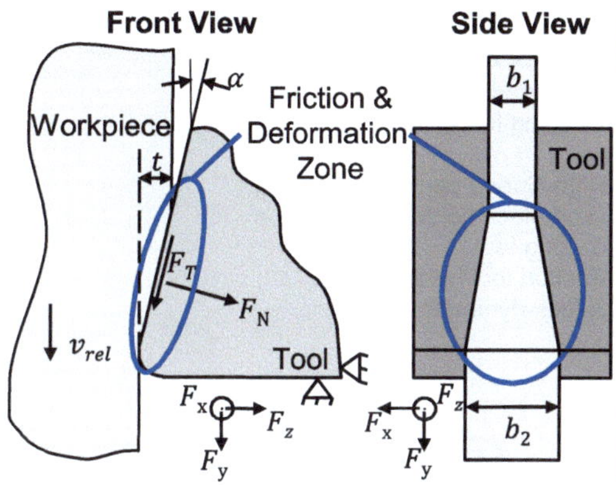

α:	Tool inclination angle
t:	Depth of engagement
F_N:	Normal contact force
F_T:	Tangential contact force
F_x:	Measured force x-direction
F_y:	Measured force y-direction
F_z:	Measured force z-direction
b_1:	Undeformed workpiece width
b_2:	Deformed workpiece width
v_{rel}:	Relative velocity between tool and workpiece

Fig. 4 Kinematic concept of the friction test

To analyze the local contact behavior, there are two common approaches: tribometer experiments and the direct analysis of the metal cutting process. In order to determine the friction at the tool-chip interface in cutting conditions, an advanced friction test [12] was designed to obtain contact forces with different temperature levels and sliding velocities in the contact zone. Figure 4 shows the kinematic concept of the friction test. The tests are derived from the orthogonal cutting process on a broaching machine. In contrast to cutting, an extremely negative rake angle α is utilized to suppress chip formation which results in a simple high speed forming process and plastic metal flow over the inclined plane of the tool. Thus, the normal contact force F_N and tangential contact force F_T can be calculated by geometrical transformation of the measured forces F_z and F_y. The undeformed width of the workpieces b_1 is significantly smaller than the tool width. The deformation zone has approximately a trapezoidal shape and results in a deformed workpiece width b_2.

In summary, local strain and strain rate levels can be primarily controlled by parameters t, α and v_{rel}. The test concept is a promising approach to realize the metal cutting conditions and to investigate different parameters separately by a variation of the process parameters. An experimental setup in Fig. 5 was designed on a vertical broaching machine with the maximum cutting velocity of 100 m/min. Workpieces in sheet geometry were clamped by a holder which was mounted on the linear machine tool drive. In this case the drive provides the relative velocity v_{rel}. Surface temperatures were measured by a two color fibre pyrometer and the measurement spot was located 1.5 mm below the cutting insert. An advantage of the two color ratio pyrometer is that the measured temperature is independent of emissivity variations [13]. Furthermore, a high speed video camera was installed to analyze the process with a maximum frame rate of 6688 fps at a resolution of 800 × 600 pixels.

With respect to different mechanical and thermal properties, different workpiece materials have been selected for the investigation (see Table 1). Normalized carbon steel AISI 1045, quenched and tempered steel AISI 4140 and the aged nickel based

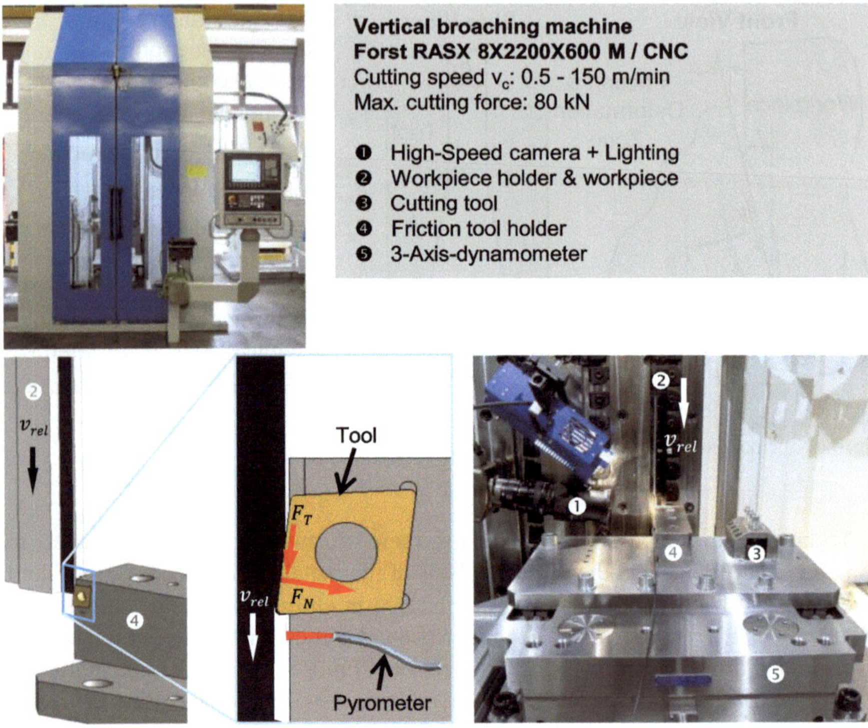

Fig. 5 Experiment setup for friction test [14]

Table 1 Workpiece and tool samples

	AISI 1045	Inconel 718	AISI 4140
Heat treatment	Normalized	Aged	Tempered
Hardness	200 HV 30	450 HV 30	318 HV 30
Thermal conductivity at 20°C/W/(mK)	48	11	40
Tool material	Carbide WC-6Co (Sandvik H13A)		
Tool geometry	CNMG120408	CNMG120408	CNMG160608

alloy Inconel 718 have been used. The undeformed workpiece width b_1 is 1.5 mm. The tools are uncoated cemented carbides (Sandvik H13A) with a composition of 94% WC and 6% Co [15].

Figure 6 shows a sample plot of the measured forces and temperatures for a relative velocity of 100 m/min and parameters $\alpha = 10°$, $t = 1$ mm. Furthermore, it shows the coefficient of friction which represents the tangential contact force divided by the normal contact force. To analyze the relationship between different process parameters and temperature levels, the analysis of the steady state conditions was performed.

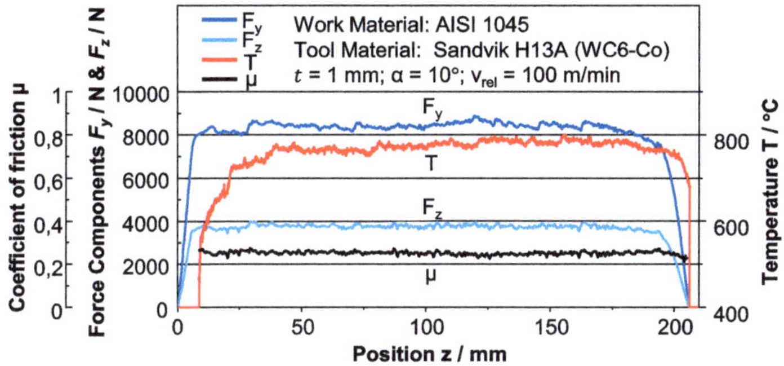

Fig. 6 Sample plot of the measured forces in friction test

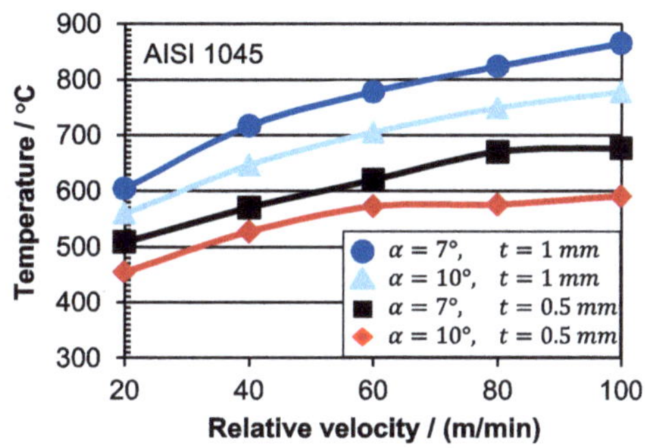

Fig. 7 Temperature with different relative velocities: AISI 1045

Figure 7 shows the temperature evolution for the different process parameters at various relative velocities for the AISI 1045 steel. With increasing contact length and relative velocity, the temperature below the contact zone rises from 450 to 850 °C. Due to the distance to the end of the contact zone, the temperature can be significantly higher and therefore causes strong softening of the material. In metal cutting, contact pressures are very high. Thus, the frictional stress should be limited to the shear flow stress of the workpiece material. Despite of that, the effective friction coefficient is determined by dividing the tangential force by the normal force.

Figure 8 shows the evolution of friction coefficients with different relative velocities for different process parameters. The friction coefficient decreases with the increase of relative velocity. The non-constant friction coefficients for various velocities were observed by other authors [16, 17] as well. This fact emphasizes that a model using a constant coefficient of friction could not describe metal cutting and the friction process for all conditions. As concluded by the experimental results, the softening effect should be considered in the friction model, which leads to a decreasing coefficient of friction.

The following Eq. 1 has been used to model the phenomena [12]. A softening term is analogously formulated to the well established constitutive Johnson-Cook material model. Regarding the shear stress limit it should be noted that this limit is automatically considered by common finite element software. If the shear stress

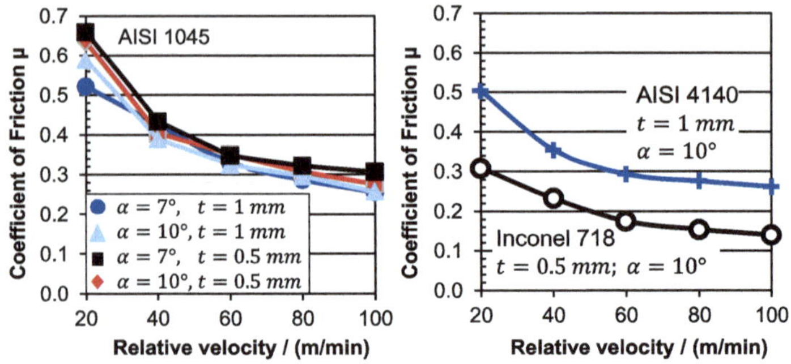

Fig. 8 Friction coefficient for different relative velocities: AISI 1045, AISI 4140 and Inconel 718

reaches the shear flow stress of the material, it causes sticking and internal shearing. For contact temperature lower than T_0 it is reasonable to use the initial friction coefficient μ_0. The μ_0, T_0 and the softening term m_r are determined by an inverse calibration [12] of the proposed equation with finite element simulations.

$$\mu = \mu_0 \cdot \left(1 - \left(\frac{T - T_0}{T_m - T_0}\right)^{m_r}\right) \tag{1}$$

2.3 *Shear Zone Contact Conductance Model*

One of the required boundary conditions for modelling of thermo-elastic workpiece deformation, as described later in Sect. 3, is the contact heat transfer between the workpiece and the tool, see Fig. 9. The most common way to incorporate the heat transfer between two solid bodies is by means of the contact heat transfer coefficient h_c, which is formally defined as the ratio of the occurring heat flux density $\dot{q}''$ and the temperature difference ΔT_c across the contacting interfaces, thus

Fig. 9 Contact heat transfer in contact between tool and chip/workpiece

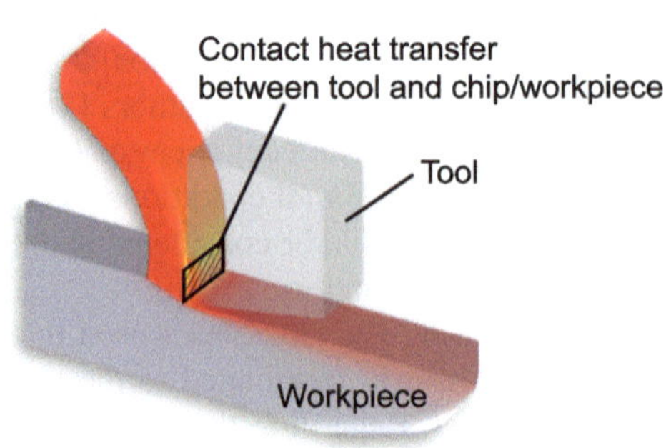

$$h_c = \frac{\dot{q}''}{\Delta T_c}. \quad (2)$$

Thus, the coefficient accounts for any deficiencies in the contact of the involved components. These shortcomings arise due to the technical nature of the tool's and workpiece's surfaces (as opposed to ideal, perfectly smooth surfaces), which limit the real area in contact to below 10% of the nominal one. The heat transfer coefficient is a macroscopic way to account for the microscopic processes at the involved surfaces during temperature equalization and therefore it has to be known in advance for all occurring scenarios, to allow for a precise thermal modelling of workpiece deformation.

The following subsection presents experimental results to quantify the contact heat transfer during the chip-tool interaction in the secondary shear zone, see Fig. 9. In the experiment, the tool is represented by a generic specimen made out of cemented carbide (CC) with an optional surface coating. It is brought into contact with a specimen made out of AISI 1045 (C45E+N), representing the workpiece and chip, respectively. In addition to the coating, the applied contact load and the specimen's initial temperatures have been varied to account for different possible contact situations.

The experimental approach has been described in literature before [18] and is an advancement of the setup presented in [19, 20], see Fig. 10. The basic principle is to record the temperature equalization of two previously heated specimens on different temperature levels made of AISI 1045 and H10F, respectively, as they are brought into contact, using high-speed infrared (IR) thermography. The specimens' geometry is given in Fig. 11.

The material testing system employed is custom-made and can apply forces up to 100 kN, the load being recorded by a strain gauge and controlled by a PID controller.

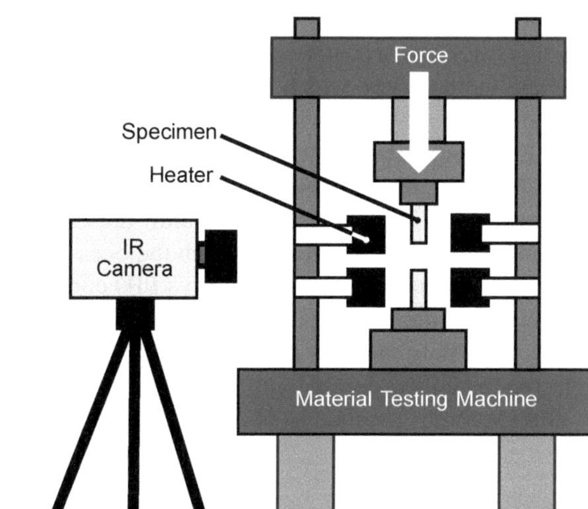

Fig. 10 Experimental setup used for the contact heat transfer experiments [18]

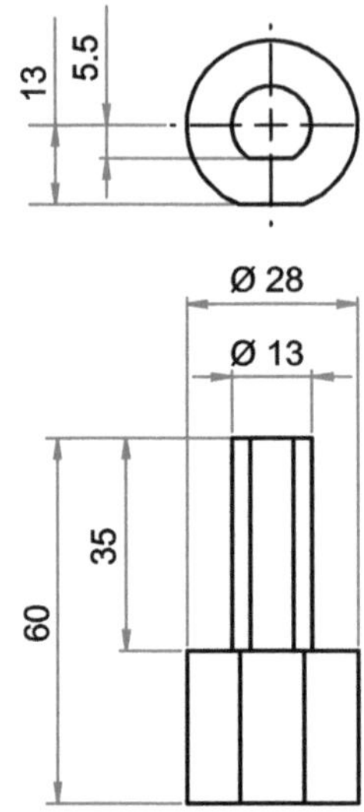

Fig. 11 Geometry of the specimen for the contact heat transfer experiments

The IR camera is an InfraTec ImageIR 5300 with an overall noise equivalent temperature difference (NETD) of less than 25 mK at 30 °C. The built-in detector is a focal-plane-array made out of Mercury-Cadmium-Tellurium (MgCdTe), the optical resolution of the objective used is 25 µm/pixel. In order to minimize ambient influences on the measurement and thus, increase accuracy, the specimens' surfaces facing the IR camera have been painted black using a special paint to achieve an emissivity of 97.2%.

After the preheated specimens have been brought together, the contact heat transfer reduces the initial temperature difference, here 50 K. According to Eq. 2, a contact heat transfer coefficient can be determined for this equalization process, if the rate of heat flow is known. To calculate the contact heat transfer coefficient, an inverse heat conduction problem (IHCP) has to be solved. This is done by means of the conjugate gradient method [21], which has been used before to reconstruct contact heat transfer coefficients [19] and is able to resolve temporal changes of the contact coefficient [18]. For this, a temporal temperature profile is extracted from the IR recording for each specimen involved. Next, an initially assumed contact heat transfer coefficient is altered in an iterative procedure, until the calculated temperatures based on the current heat transfer coefficient match the ones extracted from the IR video.

Due to the material testing system's PID controller, some oscillations during the load build-up are inevitable. To resemble the contact between tool and chip as far as possible, only the data of the first load cycle during each experiment has been selected to be incorporated into the modelling, see Fig. 12.

Subsequently, the data has been fitted with a regression model of the form

$$h_c = a \cdot \ln (p + 1) + b \tag{3}$$

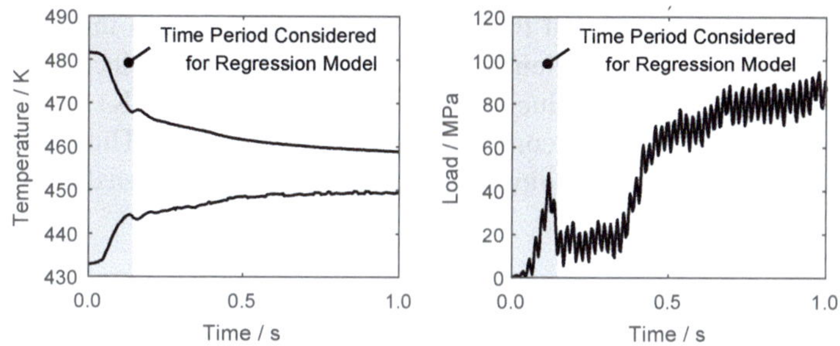

Fig. 12 Exemplary temperature and applied load for one experimental run

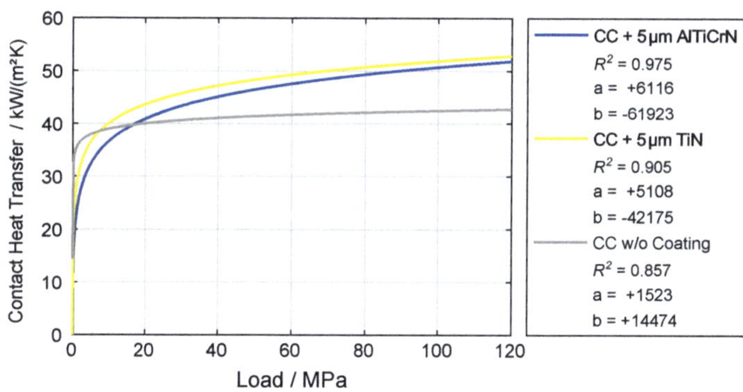

Fig. 13 Regression curves of contact conductance of the shear zone for cemented carbide tools (CC) with an optional AlTiCrN or TiN coating. *a* and *b* are the fitting coefficients according to Eq. 3 and R^2 the coefficient of determination

to provide an equation to use in any further modelling.

Figure 13 illustrates the results of the fitted coefficients (*a* and *b*) and the coefficient of determination R^2 of the regression model.

In all experiments, the upper specimen is made out of AISI 1045, while the lower one is made out of cemented carbide (Sandvik H10F, the substrate consists of chromium carbide and cobalt), with an optional 5 µm coating of TiN and AlTiCrN, respectively, as stated in the legend.

All data show a logarithmic trend, with the boundary value of the contact heat transfer being reached at approximately 100 MPa, which is in line with literature [22]. In addition, there is no significant difference in the final heat transfer coefficient for the experiments with the coated H10F specimen. But then, the coated specimens show a higher heat transfer coefficient compared to the uncoated specimens. This might be due to the coating covering some of the base materials' surface defects and thus, leading to a larger surface percentage in contact.

An additional temperature dependency of the contact heat transfer could not be noticed. This is probably due to the high loads investigated, wherefore a maximum of contact heat transfer was reached.

All in all, it can be seen that the common assumption of an infinite high thermal contact heat transfer coefficient in the shear zone is wrong, since the coefficient approaches a finite boundary value of approximately 50 kW/(m^2 K) for a coated and 42 kW/(m^2 K) for an uncoated contact between chip and tool. Thus, the proposed shear zone model could help to improve the accuracy of heat transfer modelling in chip removal.

3 Modelling and Compensation of Thermoelastic Workpiece Deformation in Dry Turning

The development of simulation models and compensation strategies within this work can be divided in two major steps. At first, the heat flux into the workpiece due to dry turning is investigated by chip formation simulation on a mesoscopic scale and by experiments. The experimental thermal analysis is compared to the simulation results. Secondly, the thermoelastic workpiece deformation in dry turning is modeled based on the determined heat flux into the workpiece and compensation strategies are outlined. The overall procedure is illustrated in Fig. 14.

3.1 Mesoscopic FE-Modelling

The first step in the application of the model is to compute the heat flux into the workpiece by using the mesoscopic model. Figure 15 briefly shows the concept

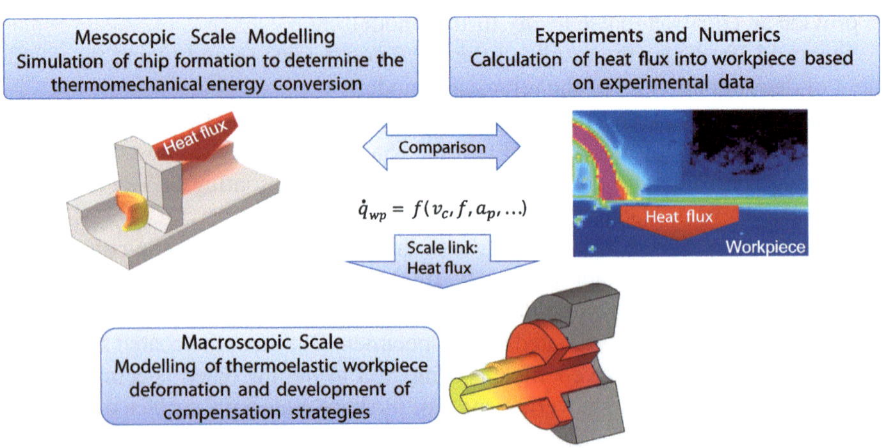

Fig. 14 Overview of the general approach to predict and compensate thermoelastic workpiece deformation by use of different simulation scales and experimental thermal analyses

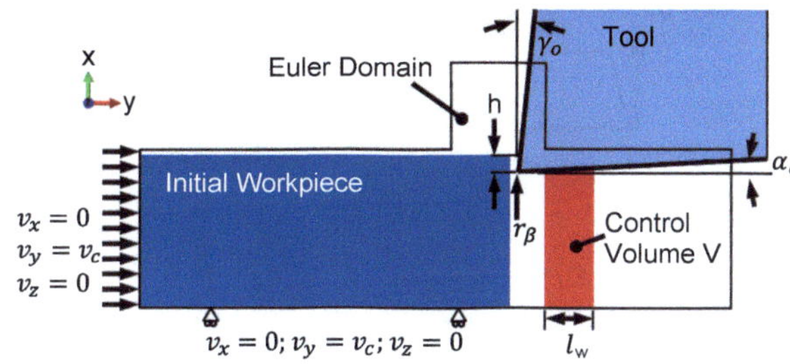

Fig. 15 Modelling of orthogonal cutting with coupled Eulerian-Lagrangian method

Table 2 Parameters of Johnson-Cook material model for AISI 1045

A (MPa)	B (MPa)	n	m	C	T_m (°C)	T_0 (°C)	ϵ_0 (s^{-1})
546	487	0.25	0.631	0.027	1500	20	0.001

of this finite element model. To model the chip formation, the coupled Eulerian-Lagrangian finite element method was employed using the commercial software package ABAQUS/Explicit. With this model, the chip formation is simulated similar to a fluid flow model. The fluid in this case is the workpiece material which continuously enters the Eulerian domain at the top and then flows against the tool which is fixed in space. The main advantage of this approach is the robust handling of large deformations inside the Eulerian mesh. In addition, the mesh is fixed in space which facilitates the evaluation of state variables at defined locations of the cutting zone. Convection and radiation are not considered in the mesoscopic model.

To model the visco-elastoplastic material behavior of AISI 1045, the Johnson-Cook material model with the parameters in Table 2 has been used, which was obtained by inverse parameter identification [23]. The thermal properties for the workpiece material AISI 1045 [24] and tool substrate WC6-Co [25] were obtained from literature. Thermal properties for the specific coating are not available. Instead, values for a coating with similar chemical composition AlTiN were used [26]. To model the thermal contact without significant resistance, a heat transfer coefficient of $h_c = 10,000\ \text{kW/(m}^2\text{K)}$ was utilized. Furthermore, a temperature dependent friction model has been implemented as described in Sect. 2.2.

The method to determine the rate of heat flow into the workpiece $\dot{Q}_w$ is illustrated in Fig. 16 [14]. For this purpose, the internal heat energy in a moving control volume is evaluated. In the cutting process, a certain rate of heat flow $\dot{Q}_w$ enters the control volume V_{1,t_0} inside the workpiece at a certain time t_0. Later, at the time t, the control volume has moved by the distance $l_w = v_c(t - t_0)$, if the tool is assumed to be fixed. Between t_0 and $t = t_0 + l_w/v_c$, the rate of heat flow $\dot{Q}_w$ has changed the internal heat energy in the volume $V_{1,t}$. If the temperature field reaches a steady state, the internal energy in the control volume V_{1,t_0} and $V_{1,t}$ will reach the steady state as well. The difference of the internal energies in both control volumes can be used to calculate the rate of heat flow (see in Eq. 4). More details about the methodology to calculate

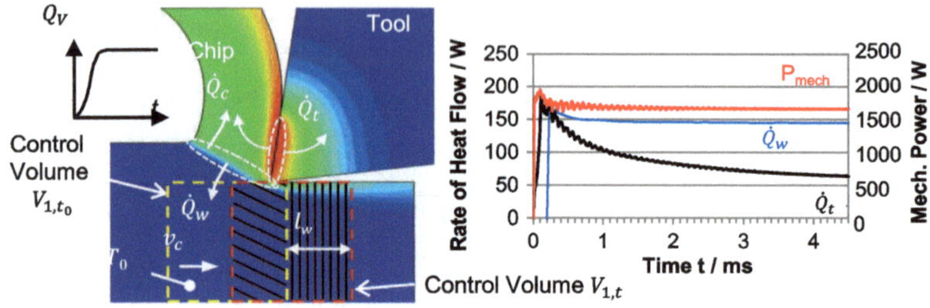

Fig. 16 Control volume to calculate rate of heat flow into workpiece $\dot{Q}_w$, tool $\dot{Q}_t$ and mechanical power P_{mech} [14]

the rate of heat flow into the workpiece, the CEL method, boundary conditions and model parameters are explained in [14, 27].

$$\dot{Q}_w(t) \cong \frac{\Delta Q_w}{t - t_0} = \int_V c_{p_w} \cdot \rho_w \cdot (T - T_0) dV \cdot \frac{v_c}{l_w} \tag{4}$$

Figure 17 shows the effects of uncut chip thickness and cutting velocity on the rate of heat flow into the workpiece $\dot{Q}_w$ and into the tool $\dot{Q}_t$. It shows an increasing rate of heat flow $\dot{Q}_w$ with rising uncut chip thickness and cutting velocity. The fraction β_w decreases by 7% up to 34% with rising uncut chip thickness and cutting velocity. The uncut chip thickness increases the distance of the heat to flow from the primary shear zone to the workpiece and the cutting speed reduces the time that is available for this heat transport. Besides, these technological parameters also modify the process mechanics. An increase in both uncut chip thickness and cutting velocity results in thermal softening in the primary and secondary shear zone, which reduces the total heat generation of the process. In contrast to that, the heat flux into the tool $\dot{Q}_t$ mainly differs with the uncut chip thickness h (Fig. 17). The cutting speed only has a minor influence on this particular rate of heat flow, although the fraction β_t decreases significantly.

In Fig. 18 the effect of cutting edge radius r_β on the rate of heat flow is shown. The cutting edge radius r_β also has a strong effect on the rate of heat flow into the workpiece but only a minor effect on the mechanical process power P_{mech}. Severe plastic deformation resulting from a larger cutting edge radius generates heat close by the generated surface. However, it only slightly increases the cutting force and therefore the mechanical process power.

To calculate the rate of heat flow into the workpiece in the turning processes, 3D CEL chip formation simulations were investigated and the change of internal heat energy inside a control volume was evaluated below the cutting edges. This is done for the main cutting edge $\dot{Q}_{(w,m)}$ and the corner radius $\dot{Q}_{(w,r)}$ separately. More details about the rates of heat flow at varied main cutting edge and corner radius are discussed in [28]. To validate the 3D CEL chip formation simulation, orthogonal

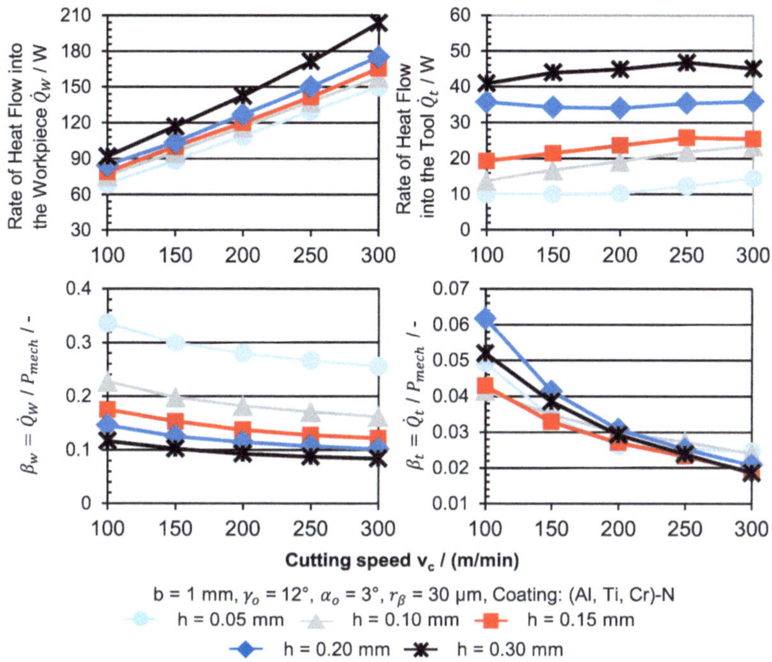

Fig. 17 Effect of uncut chip thickness h and cutting speed v_c on rate of heat flow into workpiece and tool in orthogonal cutting of AISI 1045

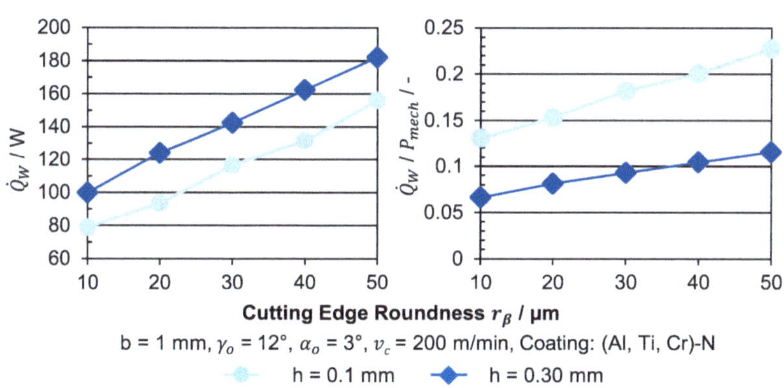

Fig. 18 Influence of cutting edge radius on rate of heat flow $\dot{Q}_w$ and β_w

cutting trials were performed at the experiment setup in Sect. 2.1. The cutting forces have been measured for selected cutting parameters. The comparison of cutting force components between simulations and experiments is shown in Fig. 19.

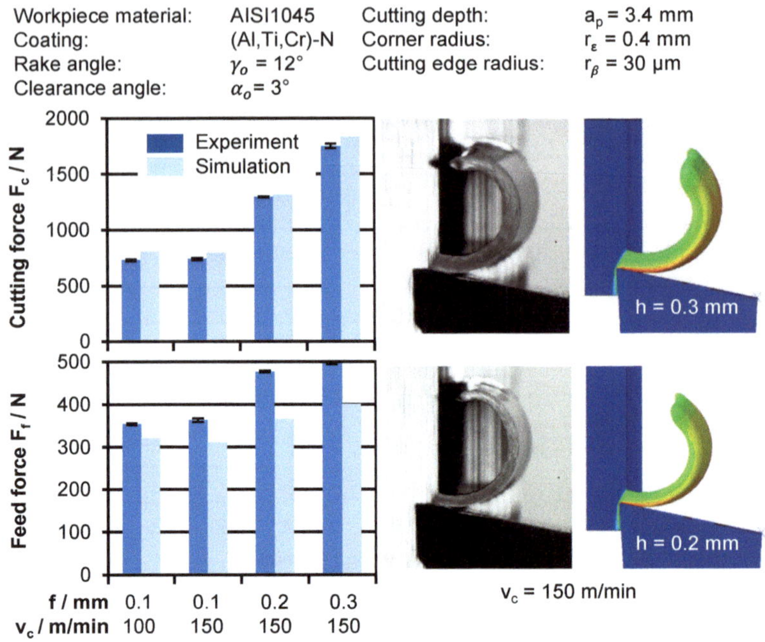

Fig. 19 Comparison of the cutting forces between 3D simulations and experiments

3.2 Heat Flux into the Workpiece During Orthogonal Turning

The orthogonal turning process without any cooling lubricants has been investigated and transient temperature profiles of the workpiece were measured. Detailed results can be found in Deppermann et al. [29]. The thermal analysis performed is aimed at heat flux histories into the workpiece during the cutting process. The reconstruction of heat flux histories at inaccessible outer boundaries based on temperature measurements requires solving an ill-posed inverse heat conduction problem (IHCP). This instance will be discussed later in Sect. 3.2.2.

The main steps of the proposed approach can be structured as follows:

1. Measuring temperature distribution with high spatial and temporal resolution
2. Identification of dominant heat transfer mechanism
3. Formulation of an analogous heat transfer model
4. Implementation of a stable, accurate, and fast solution algorithm

In the next subsections the experimental procedure and set-up are summarized for an understanding of the orthogonal turning process and the measured data. Afterwards, the conversion of the dominant heat transfer mechanism into the analogous heat transfer model is described. In the last subsection, the implemented solution algorithm is introduced and finally, the determined heat fluxes are presented.

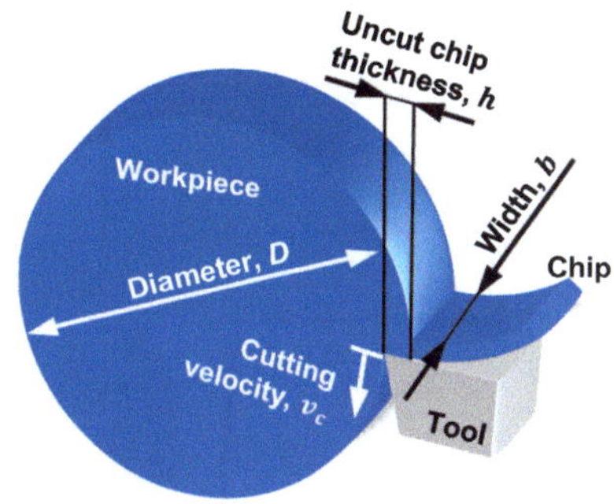

Fig. 20 The dry orthogonal turning process

3.2.1 Experimental Investigation of Dry Orthogonal Turning

Orthogonal cutting can be considered two dimensional and is geometrically simple, as shown in Fig. 20. Within one rotation of the workpiece, the diameter is reduced twice the feed f, which happens to be equal to the uncut chip thickness h for the present process. The uncut chip width and the workpiece width b are equal as well. The cutting speed v_c is equivalent to the tangential velocity of the workpiece at the outer diameter D. Thus, the heat generation due to deformation and friction is mainly influenced by uncut chip width, cutting speed, feed rate and materials of tool and workpiece.

Therefore, the influence of cutting velocities and feed rates on the workpiece temperatures have been investigated for uncoated cemented carbide inserts (Sandvik N151.2-540-40-3B H13A) and coated inserts (TiN-coating layer of 4 μm). The tools had a rake angle of 0°, a relief angle of 10° and the substrate consisted of 94 m% WC and 6 m% Co. The cutting edge radius $\boldsymbol{r}_\beta$ was 25 μm. Metal disks made of carbon steel AISI 1045 in normalised state (C45E+N, Mat.No. 1.1191) were cut with constant cutting velocity and feed rate, compare to Fig. 20. The initial diameter of each circular workpiece was 180 mm, with a width of 3 mm.

The experimental set-up was installed on a lathe with a fixed tool and a rotating workpiece, the lathe Frontor M1 by Weisser. Thus, this set-up enabled a free and fixed measurement area on the front face of the workpiece (see Fig. 21).

A high-speed infrared (IR) camera was fixed perpendicular to the workpiece's front surface, providing a temporal resolution of 1250 Hz and a temperature resolution of 25 mK at 25 °C. The optical resolution was 195.5 μ/pixel. The camera uses a MCT-detector (mercury cadmium telluride) sensitive to infrared radiation in the wavelength range of 3.7–4.8 μmm.

The workpiece has been coated with paint without any metallic substances, which is temperature resistant up to roughly 600 °C. In this manner, the surface of the workpiece was considered as a grey emitter with high emissivity of $\varepsilon = 0.972$.

Within the camera's field of view, a measurement area has been evaluated 0.394 mm apart from the cutting edge of the tool (see Fig. 22).

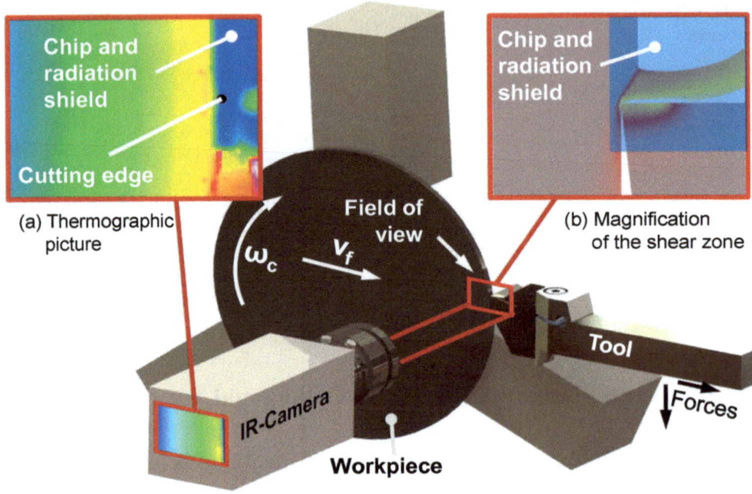

Fig. 21 Set-up of the orthogonal turning process

Fig. 22 Position of the temperature measurement area on the workpieces' front face with distance $L = 0.394$ mm

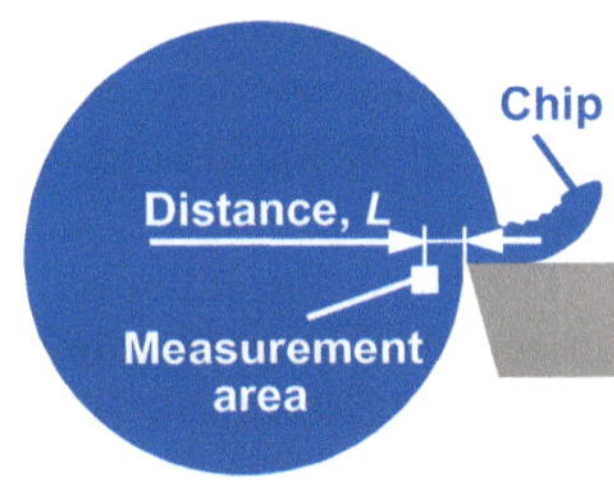

The distance L was read off the thermographic video with an accuracy of $\pm 97.7\ \mu$m. The position of the measurement area was chosen closest to the cutting edge with marginal disruptions during the machining.

Experimentation
The temperature evolution was recorded over the first 5 s for each parameter variation of cutting velocity (100, 125 and 150 m/min), feed (0.05, 0.10 and 0.20 mm), and the two different tools (un-/coated). The experiments were repeated three times with constant parameters.

3.2.2 Numeric Determination of the Heat Flux into the Workpiece

An inverse heat conduction problem is solved for each of the 54 data sets recorded to determine the induced transient heat flux at the cutting edge. Therefore, a stable, accurate, and fast algorithm with little computational effort is desirable, although not all criteria can be fulfilled completely at the same time. A complexity reduction of the real machining process and conversion into an analogous heat transfer model of lower

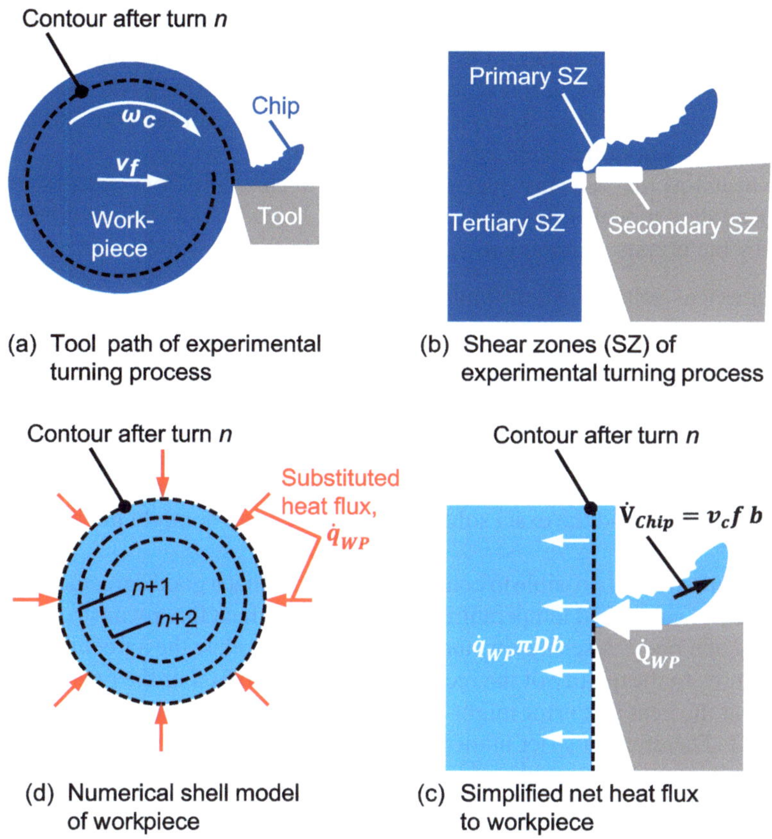

Fig. 23 Heat transfer model of the dry orthogonal turning process

dimensions offers a compromise. This thermal model and the numerical procedure necessary for heat flux determination are explained in the next two paragraphs.

Analogous Thermal Model

The real orthogonal cutting process (see Fig. 23a and b) has been transferred into an analogous thermal model for the numerical investigation, see Fig. 23c and d. The tool path of the real process can be described by an Archimedean spiral (a). The heat sources of the real process are located at the three shear zones, compare to Fig. 23b. From these shear zones, the heat flows to the chip, the tool, and the workpiece, if heat losses to the environment are neglected. The heat load on the workpiece results from the combined dissipated heat flow rates of the primary, the secondary, and marginally of the tertiary shear zone. By measuring the workpiece temperatures in the area illustrated in Fig. 22 the total workpiece heat load can be calculated.

The analogous thermal model, used to calculate the heat flux to the workpiece, is related to the ideal circular outer surface of the disk for each revolution, illustrated in Fig. 23c. The outer diameter of the disk is reduced stepwise twice the feed for each revolution. In this manner, the continuous tool path is transferred into a stepwise cutting process. The heat flux to the workpiece is assumed to be uniform over the idealized outer diameter in radial direction for each step. This is the net heat flow rate to the workpiece, $\dot{Q}_{WP}$, without the enthalpy flow rate of the chip and heat loss to the environment of the shear zones, see Fig. 23d.

The heat flow rate can vary over time in terms of changing from step to step in the introduced shell model. The resulting transient 1-D heat conduction problem can be solved by the proposed numerical approach in the next subsection.

The Numerical Solution Algorithm

The reconstruction of heat flux histories from transient temperatures inside a domain, in this case the measurement area of a workpiece, requires to calculate from an "effect" back to the "cause" and thus, pose an inverse problem [30]. Any inverse heat conduction problem with unknown boundary conditions is very sensitive to temperature measurement errors and unavoidable noise, resulting in an ill-posed problem [21, 31]. Thus, heat flux histories calculated from measured temperatures are mathematical estimates and the exact solution cannot be found or might not be physically feasible.

As a result, it is not possible to conclude in general, that a good agreement between measured and calculated temperatures proves an exact heat flux reconstruction at the boundary of a body. A sensitivity analysis is needed to relate the agreement between temperatures to the quality of the reconstructed heat fluxes. For the present case it is found, that the heat flux error might be about 200 times higher than the temperature error [32]. Therefore, a strict numerical approach is utilized, that does not rely on any manual or visual comparison to judge a "good" agreement between measured and calculated temperatures.

The applied solution approach is based on the sequential function specification algorithm with future times regularization as described by Beck et al. [31]. The core of the solution algorithm is to minimize least squares of measured minus calculated temperatures and to provide an analytic solution for the heat equation. The algorithm works without any iteration and determines heat flux histories according to 6250 time steps in 0.73 s in Matlab R2015a running on a Windows desktop computer with a clock rate of 3.4 GHz. Further details on the algorithm can be found in [32].

An exemplary set of measured temperatures over time and calculated heat fluxes is shown in Fig. 24. The experiment was run with a cutting velocity of 100 m/min and feed of 0.05 mm. The temperature evolution marked as "calculated" is congruent with the measured temperatures in the plot. The general sensitivity of an inverse heat conduction problem can be seen by the amplified fluctuations in the heat flux profile. Therefore, a representative average is determined for each investigated experimental run, considering the quasi-steady profile of the heat fluxes. The average heat flux of the run presented is $\dot{q}_{WP} = 259.1\ \mathrm{mW/mm^2}$.

Results

The averaged heat flux into the workpiece of each parameter variation is plotted in Fig. 25 for the different tools, cutting velocities, and uncut chip thickness or feed respectively. The heat flux rises proportionally with the feed and cutting velocity.

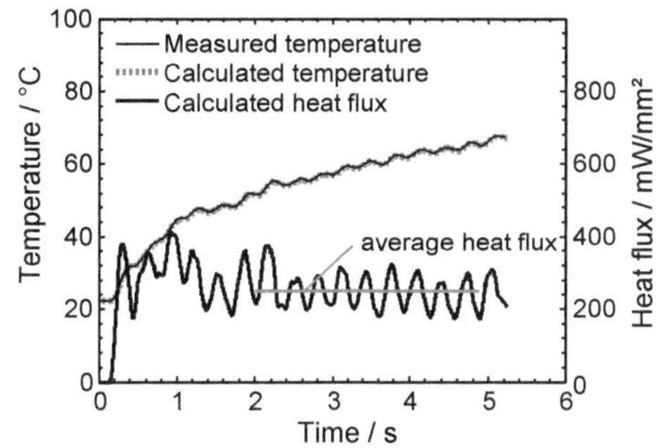

Fig. 24 Exemplary set of measured and calculated temperatures and the reconstructed heat fluxes plotted over the time of an experimental run with a cutting velocity of 100 m/min and feed of 0.05 mm for an uncoated carbon carbide tool. The average heat flux is 259.1 mW/mm^2

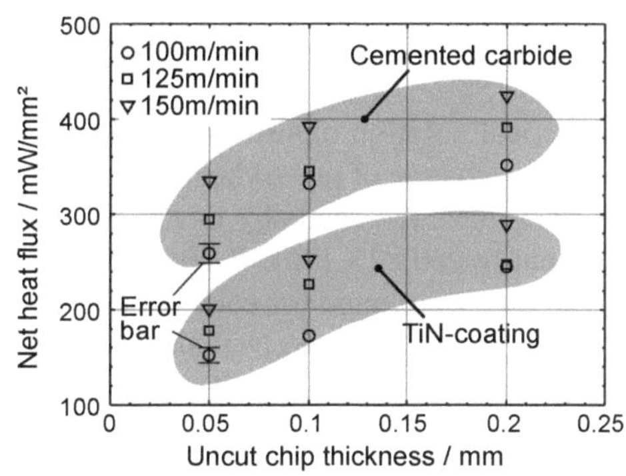

Fig. 25 Net heat flux into workpiece for different cutting velocities as a function of uncut chip thickness or feed respectively; *grey* areas mark same tool type

The evolution of heat flux over the feed shows an overall asymptotic trend for each cutting velocity. The TiN-coated cemented carbide tools decrease the heat load on the workpiece by −37.2% in average.

3.3 *Comparison of the Heat Flux Between IHCP and Mesoscopic FE-Model*

In the previous Sects. 3.1 and 3.2 the rate of heat flow and heat flux into the workpiece have been determined for different cutting parameters, see Fig. 26. The experimentally based results serve as validation data for the simulation model.

An error analysis is conducted for the validation data. Sources of errors are measurement uncertainties and numerical errors for solving the inverse heat conduction problem. An overall temperature measurement error of ±1 K is estimated. The experimental uncertainty includes the temperature measurement error of the IR-camera and reflections of thermal radiation of other hot bodies near to the workpiece, for instance the chip. Additionally, a localization uncertainty of ±1 Pixels (±97.7 μm) for measuring the distance from the workpiece edge to the location of temperature measurement is assumed (compare to Fig. 22). The error propagation due to the

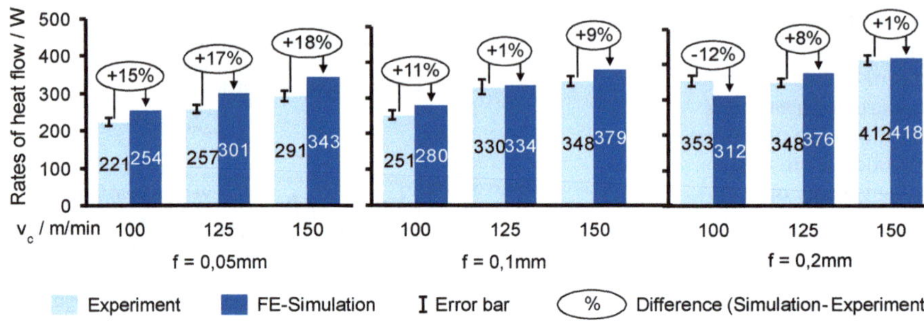

Fig. 26 Comparison between rate of heat determined by chip formation simulation and calculated heat rate based on experiments of orthogonal turning

numerical solution algorithm was investigated by numerical experiments and the resulting change of rate of heat flow is determined for each experiment individually. Furthermore, the accuracy of the numerical solution approach itself is taken into account by ±0.82% [32]. Finally, the sum of all individual errors is calculated and error bars are plotted for each parameter variation in Fig. 26. The average error for the calculated heat flows over all parameter variations is between +4.63 and −4.73%.

The results are feasible to each other, the heat flow rises with increasing cutting velocity and feed. The differences between simulation and experiment are highlighted by bubbles above the bar chart. The discrepancy ranges from +18 to −12%, which is a satisfactory agreement for a complex process like dry turning.

3.4 *Minimization of Workpiece Warming*

The major difference between the use and the absence of cooling lubricants is the significant workpiece warming during machining. Thus, by preventing any temperature rise, or at least reducing the temperature increase, thermally-induced workpiece deviations are reduced at the same time. The idea of the compensation strategy *minimization of warming* is to reduce the thermal difference between wet and dry machining.

Uncontrolled workpiece warming has additionally negative side effects like an unintended heat treatment and a resulting change of microstructure [33]. The presented approach has a positive impact on the workpiece's quality, because an unintended heat treatment and the resultant change of the microstructure are reduced along with thermal deformations.

An orthogonal turning process with constant cutting parameters over a multitude of revolutions constitutes a complete *process step*, by which the workpiece diameter is reduced from an initial diameter D_0 to its desired final diameter D_{end}, see Fig. 27.

Heat fluxes and rates of heat flow have been determined in Sects. 3.1 and 3.2, but the workpiece warming during a complete process step is not well characterized.

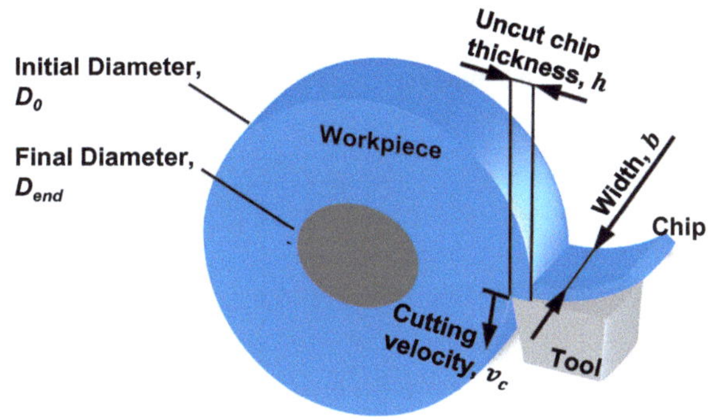

Fig. 27 Process step for orthogonal turning

The heat flux, for instance, does not account for the duration of a process step, but depends on cutting velocity and feed rate. These cutting parameters determine the duration of a process step. Thus, a lower heat flux due to slower cutting velocities may generate more heat in total over time, as the workpiece is exposed to the heat flux for a longer period of time.

3.4.1 Heat Load on Workpiece

A measure is needed that accounts for the working duration and the amount of heat flowing towards the workpiece as a result of dry machining. By normalizing the rate of heat flow towards the workpiece, $\dot{Q}_{wp}$, using the volumetric flow rate of the chip formed, $\dot{V}_{chip}$, a volumetric heat source can be determined, see Fig. 23d.

The volumetric heat source is a measure to characterize the heat load ϕ_{wp} of the workpiece due to machining as suggested by Deppermann and Kneer [32]. The heat load gives the amount of heat energy remaining in the workpiece after removing a specified amount of material and can be determined as follows,

$$\phi_{wp} = \frac{\dot{Q}_{wp}}{\dot{V}_{chip}} = \frac{\dot{q}_{wp}\,\pi D}{v_c\,h}. \tag{5}$$

The heat load per unit volume removed depends only on the rate of heat flow and cutting parameters and provides the following properties:

- measure for workpiece warming
- constant for constant cutting parameters
- accounts for duration of machining
- scalable by total chip volume

The heat load will be used as a process criterion to identify the minimal workpiece warming. The influence of different cutting parameters on the heat load, in particular the cutting velocity v_c and the feed rate f, which is equal to the uncut chip thickness h, is investigated over the Péclet number. This number is a dimensionless quantity of

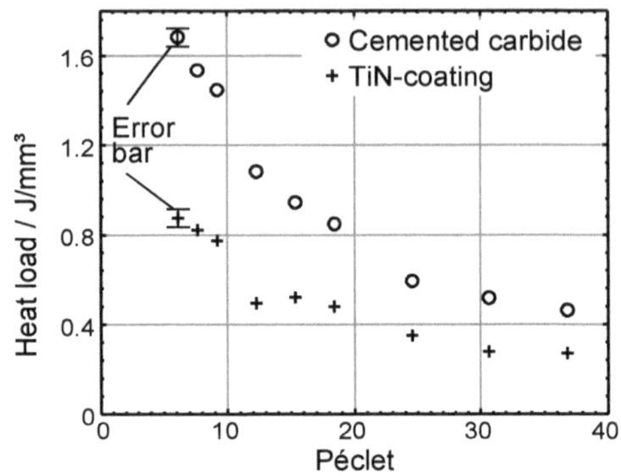

Fig. 28 Heat load on the workpiece for different Péclet numbers and different tools

heat transfer and therefore, conclusions are transferable to other working materials and cutting parameters with same Péclet numbers. The Péclet number is defined as

$$\mathrm{Pe} = \frac{h\, v_c}{\alpha}, \tag{6}$$

where α is the thermal diffusivity of workpiece material.

In Fig. 28, heat load is plotted over Péclet number. The data points for the different tools investigated tend to an asymptotic minimum at Péclet numbers greater than 30. At higher cutting parameters heat dissipation at the cutting edge of the tool has less time to flow into the workpiece. Instead, the heat moves along with the chip in form of an enthalpy flow, compare to Ref. [32]. Further, the duration of the cutting process step is reduced, as well as the time to induce heat into the workpiece. Therefore, the heat load on the workpiece decreases with rising cutting velocities and feed rates.

The graph for the uncoated tool reaches a minimum of approximately $0.46\,\mathrm{J/mm^3}$ and of $0.27\,\mathrm{J/mm^3}$ for the TiN-coated tool. The comparison between both tools indicates, that the friction between workpiece and tool is reduced by the TiN-coating.

The results for cemented carbide tools show, that raising the Péclet number from $\mathrm{Pe} = 6.1$ to 36.8 reduces the volumetric heat source of the workpiece by 72.6%. Using the same tool with a TiN-coating for the same Péclet number of $\mathrm{Pe} = 36.8$ reduces the heat source for an orthogonal turning process by 30.0%. Combining both measures, the amount of heat induced into the workpiece can be reduced by 78.2%.

3.4.2 Thermally-Induced Geometry Deviations

Minimization of workpiece warming can be linked to thermally-induced geometry deviations by use of the equation for thermal expansion,

$$\frac{\Delta D_{th}}{D_{end}} = \overline{\beta}\, \Delta T_{WP}, \tag{7}$$

where ΔD_{th} is the change in diameter due to the temperature change ΔT_{WP} of the workpiece from its initial temperature to its final temperature after the process step, compare to Fig. 27. Further, $\overline{\beta}$ is the thermal expansion coefficient, which is well documented for a wide range of metals [34].

In an adiabatic case, the total temperature rise of the workpiece, ΔT_{WP}, is equal to the heat load multiplied by the total chip volume of the process step, V_{chip}, thus

$$m_{WP,end}\, c_p\, \Delta T_{WP} = \phi_{wp}\, V_{chip}. \tag{8}$$

where $m_{WP,end}$ is the mass of the workpiece in its final dimension and c_p is the heat capacity.

For a desired process step, the thermally-induced geometry deviation depends only on the magnitude of the heat source,

$$\frac{\Delta D_{th}}{D_{end}} = \underbrace{\frac{\overline{\beta}\, V_{chip}}{m_{WP,end}\, c_p}}_{constant} \phi_{wp}. \tag{9}$$

The fraction $\frac{\Delta D_{th}}{D_{end}}$ can be defined as a dimensionless thermal error, ε_{th}, due to dry machining, which is proportional to the heat load on the workpiece,

$$\varepsilon_{th} \propto \phi_{wp}. \tag{10}$$

To sum up, in the present case increasing the cutting parameters and the use of a TiN-coated tool would reduce the dimensionless thermal error by 78.2%.

A positive side effect of the proposed compensation strategy is, that unintended heat treatment of the workpiece is reduced along with geometry deviations, due to lower temperatures. The limits of the proposed compensation strategy are, that some heat load on the workpiece caused by dry turning is inevitable and will induce thermal deviations to some extent. This error could potentially be compensated by an active engagement on the cutting path during the process.

3.5 *Macroscopic FE-Modelling for Thermoelastic Workpiece Deformation*

After the calculation of the rates of heat flow, the corresponding heat sources can be transferred into the macroscopic FE-model that was realized in the commercial software SFTC DEFORM. The transient, Lagrangian implicit 2D macroscopic model includes the whole workpiece and uses the tool path, the rates of heat flow at the

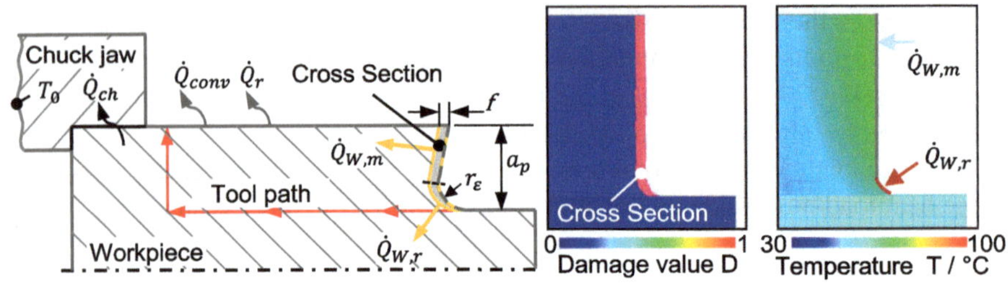

Fig. 29 Concept of macroscopic FE-modelling

main cutting edge and corner radius to model the heating and resulting thermoelastic deformation of the workpiece. The macroscopic model concept is shown in Fig. 29.

The removal of material was modeled by element deletion. Based on tool path and geometry, the position of the cutting edge was calculated for each revolution of the workpiece. The rates of heat flow into the workpiece were defined at the main cutting edge $\dot{Q}_{(w,m)}$ and corner radius $\dot{Q}_{(w,r)}$ separately. The difference of the position after one revolution represents the cross section of uncut chip. The elements inside this cross section were then deleted by a user subroutine. Therefore, after each revolution the workpiece geometry changes resulting in several thousand calculation steps for a machining cycle. In addition, the heat transfer mechanisms convection ($h = 22.6\,\mathrm{W/(m^2K)}$), conduction to the chuck ($h = 240\,\mathrm{W/(m^2K)}$), and radiation (emissivity = 0.3) were considered by the simulation model [28].

To reduce the computation time an adaptive remeshing algorithm provided by the software has been employed. The element size ranges between 10 µm in the cross section of uncut chip and 5 mm inside the workpiece resulting in a total number of 10,000–20,000 elements. This is also the reason for the choice of an implicit simulation method for the macroscopic model. An explicit time integration would be inefficient regarding the computation time for the necessary minimum element size in combination with long time periods to simulate. The application of the macroscopic model was demonstrated by means of a case study which represents external turning of a gear shaft, Fig. 30. The machining sequence was performed by experiments and multiscale modelling based FE-simulations.

Five sets of cutting parameters have been used to define the process alternatives (Table 3). Case 1 to case 4 are external longitudinal turning processes. Case 5 is an external radial turning process. In all cases, CNMG120408 coated carbide cutting tool inserts were used to machine the work material AISI 1045 in normalized state. The rates of heat flow at the main cutting edge and corner radius have been calculated with the mesoscopic FE-model in Sect. 3.1 with consideration of flank wear VB. Details about the part geometry, process kinematics and parameters can be found in [28]. In the following section only case 3 and case 5 will be discussed. More case studies can be found in [28].

Figure 31 illustrates the evolution of the workpiece geometry and temperature field for case 3 at three different moments. Furthermore, the figure shows a com-

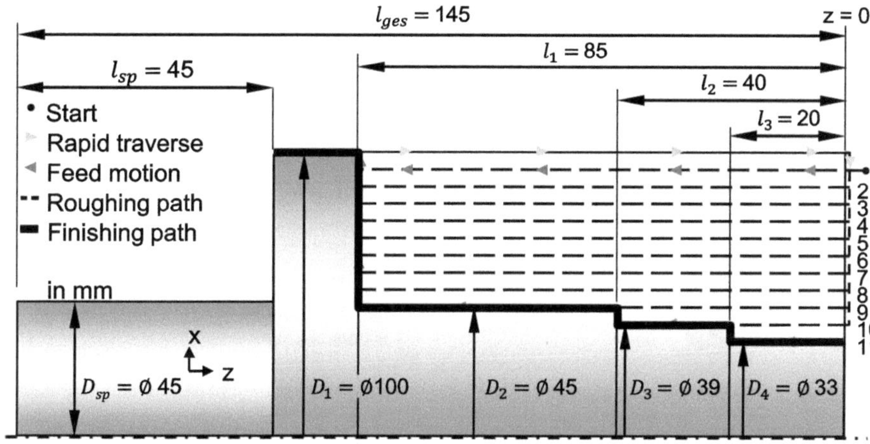

Fig. 30 Case study of external turning of a gear shaft [28]

Table 3 Rates of heat flow for different process parameters in external turning

Nr.	v_c (m/min)	f (mm)	a_p (mm)	$\overline{VB}$ (μm)	External turning	$\dot{Q}_{(w,m)} + \Delta\dot{Q}_{(VB,m)}$ (W)	$\dot{Q}_{(r,m)} + \Delta\dot{Q}_{(VB,r)}$ (W)
1	150	0.25	3	14	Longitudinal	322 + 27	255 + 19
2	150	0.18	3	16	Longitudinal	304 + 31	210 + 20
3	75	0.25	3	14	Longitudinal	212 + 17	165 + 11
4	150	0.25	1.5	13	Longitudinal	101 + 8	238 + 17
5	150	0.25	3	15	Radial	322 + 29	255 + 20

parison of the final temperature on the front face between simulation and thermal imaging 5 s after the process. In addition, the temperature evolution at this measurement spot was recorded for 30 min, and plotted in the diagram for the simulation and experiment.

In this way the temperature evolution in the measurement point is also evaluated in Fig. 32 for external radial turning. The results of other experiment sets can be found in [28]. The deviation of the temperature evolution between experiments and simulations is less than 13%. Regarding the temperature evolution on the front surface it can be concluded that the multiscale simulation model can predict the temperature field of the workpiece sufficiently precise.

After computation of the transient workpiece temperature immediately after roughing, the deformed workpiece geometry at this moment was used to compensate the thermoelastic workpiece deformation. Figure 33 illustrates the calculated longitudinal and radial thermoelastic displacement after rough turning.

It can be observed that the displacement is significantly different between the cases due to the different cutting strategy, total cutting time and different rate of

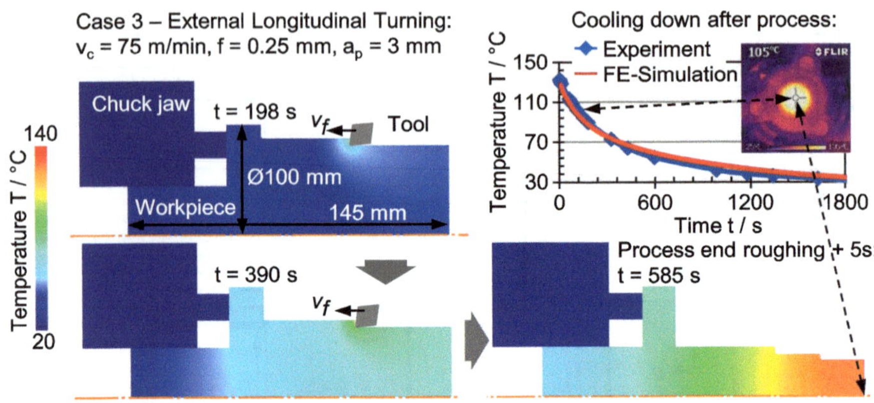

Fig. 31 Sample result of the workpiece temperature evolution calculated with the macroscopic model for case 3

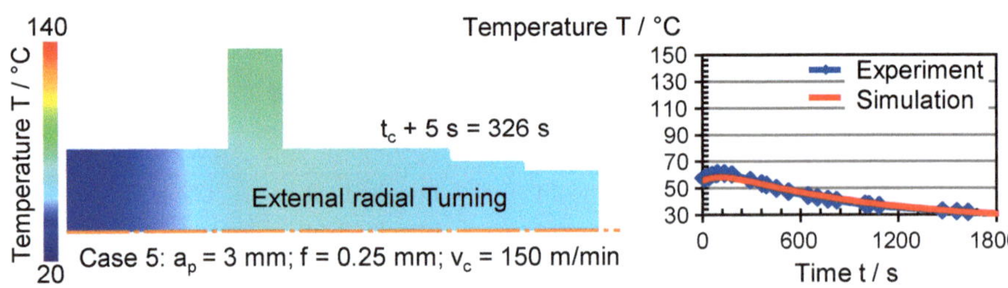

Fig. 32 Comparison of temperature evolution after the cutting process between simulation and experiment for case 5

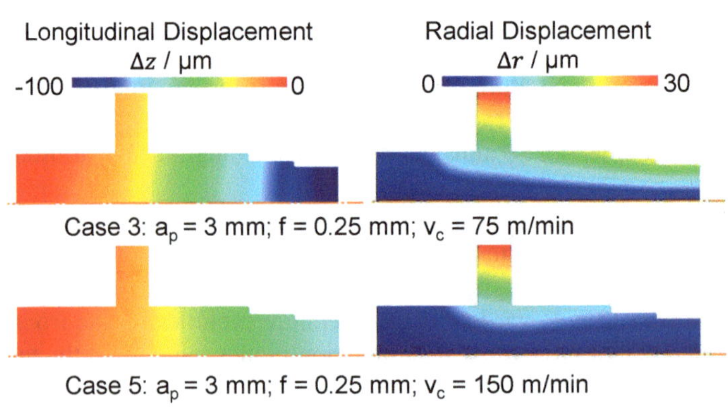

Fig. 33 Calculated thermoelastic displacement after rough turning for case 3 and case 5

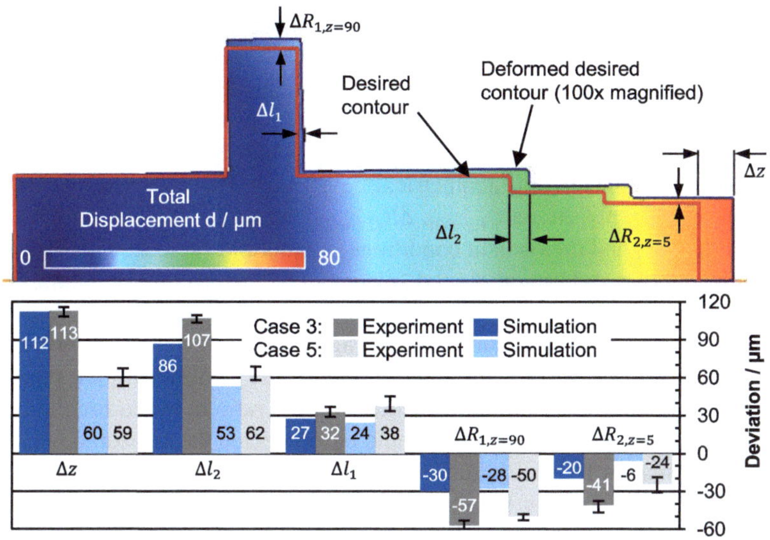

Fig. 34 Compensation strategy of thermoelastic deformation in experiments and simulations

heat flow into the workpiece. The simulated radial displacement is up to 30 µm for these cases resulting in a diameter deviation of 60 µm. In longitudinal direction deviations are higher than 100 µm due to the workpiece geometry and temperature distribution. In most of the cases these deviations would be unacceptable. Due to the expansion before finishing, the final part dimensions would be smaller than desired. The deformed desired contour is 100× magnified illustrated in Fig. 34. One possible strategy to compensate the thermoelastic deformations would be the usage of this contour as the new finishing contour. In this case, the final part would shrink or rather deform to the originally desired contour after cooling down. However, using this strategy, several aspects have to be considered. First, a finish process of critical part features is required and these features must be larger than the thermoelastic deformation. Otherwise, the contour of the feature would be eventually cut during roughing which results in dimensional errors. Second, the user has to decide if the additional heat input or a possible modification of the temperature field occurs during finishing. In these cases it will further increase the precision if the finishing process is also included in the simulation of the deformation field. However, this approach needs additional algorithms to follow the deformed desired contour during finishing. This strategy would be also suitable for processes without finishing processes and will be integrated in the future.

To validate the simulation model, a comparison between simulated and measured dimensional deviations is done according to Fig. 34. As reference experiment, a turning roughing cycle has been performed with coolant and additional cooling time before finishing. Therefore, no thermoelastic deformations occurred in this case. In the second step, turning experiments under dry conditions were performed resulting in thermal expansion and geometrical deviations. The measured difference for the dimensions Δz, Δl_1, Δl_2, $\Delta R_{(1,z=90)}$, $\Delta R_{(2,z=5)}$ between the dry conditions (thermal expansions occur) and the coolant conditions (no thermal expansion) is compared to the simulated radial and longitudinal displacements. The average deviation between experiments and simulations for both cases for the longitudinal dimensions is 8.5 μm. The comparison shows sufficiently precise agreement. However, larger deviations (average 22 μm) are found for the radial dimensions. Considering the relationship between temperature field and deformation and the ratio between longitudinal and radial deformation, a systematic error could have affected the measured radial deformations. One reason for this is the relative thermal displacement of the tool with respect to the workpiece axis. For the longitudinal direction this effect does not occur, because of a facing process right before roughing. The front face of the workpiece is the measurement reference for the longitudinal dimensions. Thus, in future works an additional model considering thermoelastic tool deformation should be integrated into the model to further increase the precision.

4 Conclusion and Future Work

This contribution presents fundamental investigation on chip formation, contact conductance and friction behavior at tool-chip interface. The significant heating in dry turning can result in unacceptable dimensional and geometrical deviations of the workpiece. The heat generation and heat partition into the workpiece were calculated by a mesoscopic FE-model for the orthogonal cutting processes. Besides, the rate of heat flow was experimentally investigated for orthogonal turning and the induced heat was calculated by a numerical algorithm.

Subsequently, based on the relation of rates of heat flow and cutting parameters, an evaluation criterion of workpiece warming was defined. A compensation strategy for thermoelastic workpiece deformation was introduced and recommendations for cutting parameters were given to minimize workpiece warming. Hereby, the heat load on the workpiece could be reduced up to 78% and accordingly the thermally-induced deformation. Furthermore, a macroscopic simulation model of the workpiece was utilized to simulate transient temperature fields and thermoelastic deformations for a case study, which represents two different turning cycles to manufacture a gear shaft. Two cutting strategies were used to compute the deformation of the desired workpiece contour. During the cutting process, this contour can be used as a modified tool path. In this case, the workpiece geometry will nearly approach the desired workpiece geometry after cooling down. For the presented case study, the multiscale model shows promising prediction of temperature and longitudinal thermoplastic deformation by dry turning. However, the precision could be increased, if the machine tool and tool deformations were also considered in the model. Finally, it would be also necessary to include a broad range of work materials, cutting tools and cutting operations to account for complex processing cycles in industrial practice.

Acknowledgements The authors would like to thank the German Research Foundation (Deutsche Forschungsgemeinschaft, DFG) for the funding of the depicted research within the priority programme 1480 (SPP 1480).

References

1. Weinert, K., Inasaki, I., Sutherland, J.W., Wakabayashi, T.: Dry machining and minimum quantity lubrication. CIRP Ann. Manuf. Technol. **53**(2), 511–537 (2004)
2. Dyck, M.: Beitrag zur analyse thermisch bedingter werkstückdeformationen in trockenbearbeitungsprozessen. Dissertation, Universität Karlsruhe (2007)
3. Eisenblätter, G.: Spanende Fertigung. Hrsg.: K. Weinert, 4. Auflage, Vulkan Verlag, chap Ganzheitlicher Ansatz zur Gestaltung der Trockenbearbeitung-Erfahrungen bei der Bearbeitung von Bauteilen aus Grauguss und Aluminium (2005)
4. Zielasko, W., Schirsch, R.: Trockenbearbeitung in der großserienfertigung. Werkstatttechnik **90**(6), 259–260 (2000)
5. Arrazola, P.J., Özel, T., Umbrello, D., Davies, M., Jawahir, I.S.: Recent advances in modelling of metal machining processes. CIRP Ann. Manuf. Technol. **62**(2), 695–718 (2013)
6. Pabst, R., Fleischer, J., Michna, J.: Modelling of the heat input for face-milling processes. CIRP Ann. Manuf. Technol. **59**(1), 121–124 (2010)
7. Abukhshim, N.A., Mativenga, P.T., Sheikh, M.A.: Investigation of heat partition in high speed turning of high strength alloy steel. Int. J. Mach. Tools Manuf. **45**(15), 1687–1695 (2005)
8. Sölter, J., Gulpak, M.: Heat partitioning in dry milling of steel. CIRP Ann. Manuf. Technol. **61**(1), 87–90 (2012)
9. Vollmer, M., Möllmann, K.P.: Infrared Thermal Imaging: Fundamentals, Research and Applications. Wiley (2010)
10. Hoppe, S.: Experimental and numerical analysis of chip formation in metal cutting. Dissertation, RWTH Aachen (2003)
11. Shi, B., Attia, H.: Current status and future direction in the numerical modeling and simulation of machining processes: a critical literature review. Mach. Sci. Technol. Int. J. **14**(2), 149–188 (2010)
12. Puls, H., Klocke, F., Lung, D.: Experimental investigation on friction under metal cutting conditions. Wear **310**(1–2), 63–71 (2014)
13. Müller, B., Renz, U.: Development of a fast fiber-optic two-color pyrometer for the temperature measurement of surfaces with varying emissivities. Rev. Sci. Instrum. **72**(8), 3366 (2001)
14. Puls, H., Klocke, F., Veselovac, D.: Fem-based prediction of heat partition in dry metal cutting of aisi 1045. Int. J. Adv. Manuf. Technol. **86**(1), 737–745 (2015)
15. Brookes, K.J.A.: World Directory and Handbook of Hardmetals and Hard Materials, 6th edn. International Carbide Data, East Barnet, Herts, U.K. (1996)
16. Arrazola, P.J., Ugarte, D., Domínguez, X.: A new approach for the friction identification during machining through the use of finite element modeling. Int. J. Mach. Tools Manuf. **48**(2), 173–183 (2008)
17. Rech, J., Arrazola, P.J., Claudin, C., Courbon, C., Pusavec, F., Kopac, J.: Characterisation of friction and heat partition coefficients at the tool-work material interface in cutting. CIRP Ann. Manuf. Technol. **62**(1), 79–82 (2013)

18. Burghold, E., Frekers, Y., Kneer, R.: Determination of time-dependent thermal contact conductance through IR-thermography. Int. J. Therm. Sci. **98**, 148–155 (2015)
19. Fieberg, C., Kneer, R.: Determination of thermal contact resistance from transient temperature measurements. Int. J. Heat Mass Transf. **51**(5–6), 1017–1023 (2008)
20. Ustinov, V., Al-Sibai, F., Kneer, R., Schulz, S., El-Magd, E.: Influence of surface roughness on contact heat transfer. In: 14th International Heat Transfer Conference, vol. 3, pp. 305–312 (2010)
21. Bergles, A.E., Mayinger, F. (eds.): Inverse Heat Transfer Problems. Springer, New York (1994)
22. Mesnyankin, S., Vikulov, A., Vikulov, D.: Solid-solid thermal contact problems: current understanding. Phys. Usp. **52**(9), 891–914 (2009)
23. Klocke, F., Lung, D., Buchkremer, S.: Inverse identification of the constitutive equation of inconel 718 and aisi 1045 from fe machining simulations. Procedia CIRP **8**, 212–217 (2013)
24. Spittel, M., Spittel, T.: Steel symbol/number: C45/1.0503. In: Warlimont, H. (ed.) Metal Forming Data—Ferrous Alloys—Deformation Behaviour: SpringerMaterials—The Landolt-Börnstein Database, Landolt-Börnstein—Group VIII Advanced Materials and Technologies. Springer, Berlin (2009)
25. Beiss, P., Ruthardt, R., Warlimont, H. (eds.): Powder metallurgy data. In: Refractory, Hard and Intermetallic Materials. Landolt-Börnstein—Group VIII Advanced Materials and Technologies. Springer, Berlin (2002)
26. Martan, J., Beneš, P.: Thermal properties of cutting tool coatings at high temperatures. Thermochim. Acta **539**, 51–55 (2012)
27. Puls, H.: Coupled Eulerian-Lagrangian modelling of high speed metal cutting processes. In: Advances in Manufacturing Technology: 11th International Conference High Speed Machining, Prague (2014)
28. Puls, H.: Mehrskalenmodellierung thermo-elastischer werkstückdeformationen beim trockendrehen. Dissertation, RWTH Aachen University (2015)
29. Deppermann, M., Puls, H., Burghold, M.E., Kneer, R., Klocke, F.: Experimental investigation of the work piece temperatures in dry orthogonal metal turning. In: Proceedings of the 15th International Heat Transfer Conference, Begell House, Kyoto, Japan, no. 9223 in IHTC15 (2014)
30. Maillet, D.: Experimental inverse problems: potentials and limitations. In: Proceedings of the 15th International Heat Transfer Conference, Begell House, Kyoto, Japan, no. KN10 in IHTC15 (2014)
31. Beck, J.V., Blackwell, B., Clair, C.R.S.: Inverse Heat Conduction. Wiley, New York (1985)
32. Deppermann, M., Kneer, R.: Determination of the heat flux to the workpiece during dry turning by inverse methods. Prod. Eng. Res. Dev. **1**(9), 465–471 (2015)
33. Brinksmeier, E., Brockhoff, T.: Utilization of grinding heat as a new heat treatment process. CIRP Ann. Manuf. Technol. **45**(1), 283–286 (1996)
34. VDI eV: VDI-Wärmeatlas, 11th edn. Springer, Berlin, Heidelberg, Wiesbaden (2013)

Thermo-Mechanical Simulation of Hard Turning with Macroscopic Models

E. Uhlmann, R. Mahnken, I.M. Ivanov and C. Cheng

Abstract Based on the concept of generalized stresses, a multi-mechanism model has been developed concerning asymmetric visco-plasticity, asymmetric hardness dependency and transformation induced plasticity (TRIP), which are decisive phenomena affecting the process-related ductility in hard turning. The asymmetric effects are taken into account using the concept of weighting functions related to stress modes such as tension, compression and shear. The model also considers the phase transformations between martensite and austenite during the material heating as well as austenite and white layer formation due to the so-called reverse martensite transformation. Hardness modifications as a result of the white layer formation are considered. Moreover, the model is specialized for chrome bearing steel AISI 52100 and applied in cutting simulations using the commercial Finite-Element-Method (FEM) software systems ABAQUS and Deform. The evaluation of the simulation results enables the analysis of the influence of crucial actuating variables on the machining accuracy.

1 Introduction

Hard turning aims at substituting grinding as a high-precision finishing production process of workpieces characterized by a hardness higher than 45 HRC [38]. The primary advantages of hard turning in comparison to grinding are productivity enhancement, higher flexibility during machining of geometrically intricate workpieces as well as resource efficiency [7, 29, 33]. In order to create a basis for realistic hard turning simulations providing the groundwork for computer-aided thermal process

E. Uhlmann (✉) · I.M. Ivanov
Institute for Machine Tools and Factory Management (IWF),
Technische Universität Berlin, Pascalstraße 7-8, 10587 Berlin, Germany
e-mail: eckart.uhlmann@iwf.tu-berlin.de

R. Mahnken · C. Cheng
Chair of Engineering Mechanics (LTM), University of Paderborn,
Warburgerstr. 100, 33098 Paderborn, Germany
e-mail: mahnken@ltm.upb.de

D. Biermann and F. Hollmann (eds.), *Thermal Effects in Complex Machining Processes*, Lecture Notes in Production Engineering,
DOI 10.1007/978-3-319-57120-1_7

optimization, the potential of in-depth modelling techniques needs to be researched. Moreover, the analysis of the influence of the actuating variables relevant for the chip formation on the thermomechanical load enables a better comprehension of surface damage sources as well as thermally induced machining inaccuracies. For the purpose of deriving convenient measures for thermomechanical workpiece load reduction, a comprehensive hard turning simulation model considering phase transformations caused by temperature gradients as well as white layer formation ought to be developed.

In hard turning the workpiece is machined under high speed up to $v_{c,max} = 200$ m/min causing high strain rates and temperatures, which can change the material microstructure in workpiece surfaces affecting physical effects such as elasticity, plasticity, phase transformation and transformation induced plasticity (TRIP). The consideration of microstructure within material behavior is a big challenge for material modelling. *Extended continuum models* or (*generalized continua*, [27]) have been developed in the last century to account for size effects on the materials response. Based on the so-called concept of generalized stresses initially introduced by Gurtin [13] and extended by Forest et al. [3, 10], we formulate a multi-mechanism model in a thermodynamic framework within a large-strain theory. The following effects are taken into account for the modelling.

Asymmetric visco-plasticity: High strength steels exhibit different behaviors under different loading types such as tension, compression and shear. Stouffer and Dame [31] show that for a superalloy René 95 creep rates in tension are much greater than the corresponding rates in compression. Further examples can be seen e.g. in Spitzig et al. [32] for a martensitic steel, in Iwamoto et al. [18] and Miller for an austenitc stainess steel. This observation is labelled *asymmetric effect*. In this work we use the concept of *stress mode dependent weighting functions* (firstly introduced in [23]) to model the asymmetric effect.

Asymmetric hardness dependency: Umbrello et al. [39] prove that white layer formation causes hardness modifications and propose an empirical function to describe this effect. The function takes the austenitization and the martensitic retransformation and the maximal reached temperature into account. Experimental tests show that hardness has a strong influence on material behavior, see e.g. Umbrello et al. [37]. The harder the material, the higher is the yield stress and tensile strength. A linear hardness-dependent function proposed in [37], which allows itself to be integrated into a flow function, will be used to describe this characteristic. Furthermore, the asymmetric effect of this characteristic will be considered using the concept of weighting functions and studied with the help of experimental tests under different loading types such as tension, compression and shear.

Transformation induced plasticity: White layer is formed due to a austenitization and a reverse martensite transformation with martensite as initial phase. The austenitization is described by an approach due to Leblond and Devaux [21], and the reverse martensite transformation is taken into account by the classical Koistinen-Marburger relation [20]. Both approaches are extended with the first gradient of the austenite phase fraction using the concept of generalized stresses. Both transformations induce plastic strain under mechanical loadings, see e.g. Leblond et al. [22] for a martensitic transformation and Bökenheide et al. [6] for a austenitic transformation. Both TRIP-strains will be described using the approach proposed in [22].

This work is organized as follows. In Sects. 2 and 3, the test conditions for cutting and thermal experiments, the applied measurement methods and the measurement results are described. In Sect. 4 we develop a multi-mechanism model for cutting simulations and specialize it for the steel AISI 52100. The related parameter identifications will be shown in Sect. 5. In order to test the model, cutting simulations are investigated using the program Abaqus in Sect. 6. The simulation results generated using the Finite-Element-Method software Deform-3D developed by SFTC, Columbus, USA, and the derived model-based optimization strategies are summarized in Sect. 7 and Sect. 8, respectively.

2 Test Conditions and Measurement Methods

2.1 Cutting Experiments

In order to validate and calibrate the developed hard turning model, cutting tests using bearing rings IR 70X80X30 consisting of a chrome bearing steel AISI 52100 with a surface hardness of 62 HRC were performed, see Fig. 1. In this regard, the cutting speed v_c, the feed f and the depth of cut a_p represent suitable machining parameters to be varied for the purpose of evaluating a hard turning simulation model [5]. The indexable inserts used of the specification CNMA 120408 are composed of cemented carbide metal base plates and soldered polycrystalline cubic boron nitride

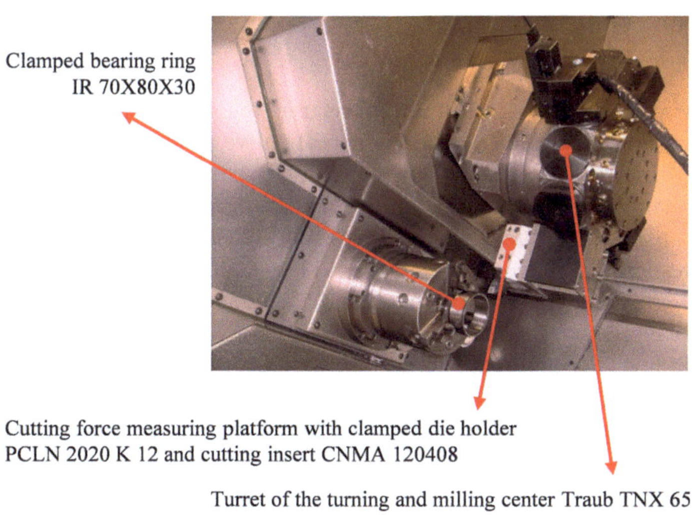

Fig. 1 Experimental setup for cutting trials

inserts. Due to the fact that the cutting insert's microgeometry exerts an essential influence on the thermomechanical workpiece load, its characteristics were experimentally examined. Thus, the width of the cutting edge chamfer and the angle of chamfer were proven to equal $b_f = 0.15$ mm and $\gamma_f = 53°$ respectively. In addition, the tool holder clearance angle of $\alpha = 6°$ and the effective rake angle $\gamma_{eff} = -59°$ were determined. Furthermore, the cutting edge features a radius between rake face and chamfer of $r_\gamma = 39$ µm and a radius between flank face and chamfer of $r_\alpha = 41$ µm.

The cutting experiments were performed on a CNC turning and milling center TRAUB TNX 65 manufactured by INDEX-Werke GmbH & Co. KG, Esslingen. The external surface of the bearing rings was dry machined and the resultant force components were measured using a dynamometer of type 9121 manufactured by Kistler Instrumente GmbH, Sindelfingen. The cutting speed was varied between $v_{c,min}$ = 100 m/min and $v_{c,max}$ = 200 m/min. Moreover, the feed and depth of cut were varied between 0.5 mm $\leq f \leq$ 0.2 mm and 0.1 mm $\leq a_p \leq$ 0.3 mm, respectively. The aforementioned process parameters and the corresponding ranges are suitable in the case of machining hardened steel with hardness value above 45 HRC. Due to the considerable cutting forces and temperature levels, an effective hard turning process requires machine tools of high rigidity. Using the depicted experimental setup, the measurement results show that the resultant force components do not depend on the cutting speed v_c, which can be explained by the neutralization of higher material removal rates and higher cutting forces at rising cutting speed v_c, temperature increase, material softening and concurrent decrease of the rake face friction coefficient [4]. Furthermore, the recorded measurement data was used for verification of the hard turning model.

2.2 Thermal Experiments

Prior to the cutting trials, the test setup was used to establish the specimen surface temperature. For this purpose, an initial representative temperature value of $\theta_r = 700$ °C was determined. Under laboratory conditions, specimens were heated up to the previously defined decisive temperature θ_r and the corresponding emissivity ε_r was ascertained using an infrared camera of type Jade Mwir II manufactured by InfraTec GmbH, Dresden. Using the special test rig in Fig. 2, the specimen's surface temperature was varied and concurrently measured at a constant emissivity ε_r. Due to the fact that the emissivity ε_r is valid merely for the initially determined representative temperature level θ_r, incorrect temperature values were measured. Hence, a correction factor k_{th} was derived, which, by multiplication with the incorrect temperature values yields the real workpiece surface temperature. The described procedure was

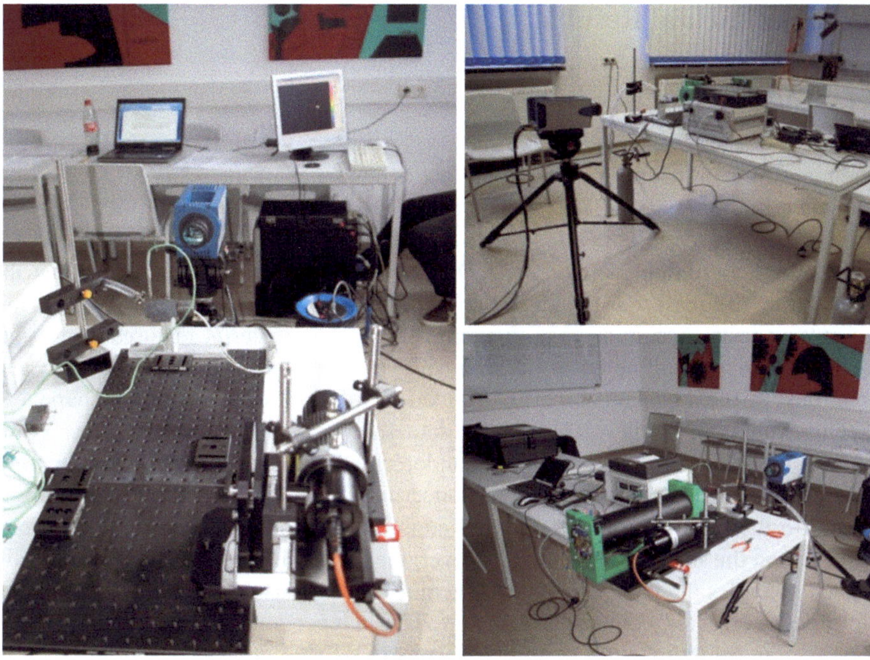

Fig. 2 Test rig for the calibration of the radiation temperature measurement

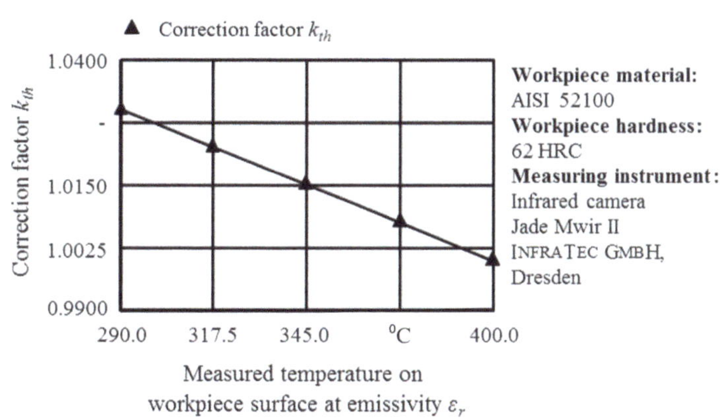

Fig. 3 Correction factors as function of measured surface temperatures

applied in order to establish the correction factors for the cases of oxidized as well as partially oxidized specimen surfaces. Figure 3 shows the correction factors of the partially oxidized specimen surface.

3 Experimental Results

3.1 Cutting Forces

The measurement results do not reveal any dependence of the resultant force components on cutting speed. This can be explained by the temperature increase with rising cutting speed v_c which leads to a material softening and concurrent decrease of the rake face friction coefficient [4]. Due to the consistent alloying element distribution within the analyzed specimen, continuous chip formation was observed during the cutting tests. Assuming that an exemplary cutting material features lower tool wear than polycrystalline cubic boron nitride, the cutting speed v_c can be increased regardless of the mechanical workpiece load as well as the resulting chip shape.

The evaluation of the measurement data showed that the resultant force components related to the cross section of undeformed chip decrease with higher feed. Hence, higher feeds are advantageous concerning the reduction of the passive force related to the cutting cross section and consequently the reduction of the thermal influence on the surface integrity. Moreover, the experimental analysis of the cutting capacity enables the identification of machine control variables capable of reducing the process-oriented power consumption. The independence of the resultant force components from the cutting speed v_c results in a linear dependency of the cutting rate at varying cutting velocity [36]. As shown in Fig. 4, the cutting rate $P_{c,A}$ related to the cutting cross section decreases at increasing feed, which can be explained by a hydrostatic compressive stress state augmentation in the shear plane and subsequent material softening [40]. Concerning the depth of cut variation, it can be concluded that the cutting power increases faster than the cutting cross section at lower depth of cut and vice versa at higher cutting depth. Hence, a process optimization regarding a desirable power requirement reduction can be achieved by applying either low or high depth of cut at the highest acceptable feed.

In subsequent experiments, the front face and lateral area of rolling bearing rings composed of bearing steel AISI 52100 were machined. The resultant force components as well as the cutting power related to the cutting cross section during face turning exhibit lower values compared to lateral area cylindrical turning. In comparison with outer longitudinal cylindrical turning, the heat zone advancing the cutting insert during face turning covers a shorter propagation path, leading to a higher extent of heat accumulation and material softening in the cutting area [36]. Moreover, the specific cutting power decreases at higher feeds and depths of cut. The lower specific passive forces detected during face turning of rolling bearing rings demonstrate that a reduction of the steep temperature rise in the workpiece surface layer at the beginning of the cutting process can be achieved by an adjustment of the process control.

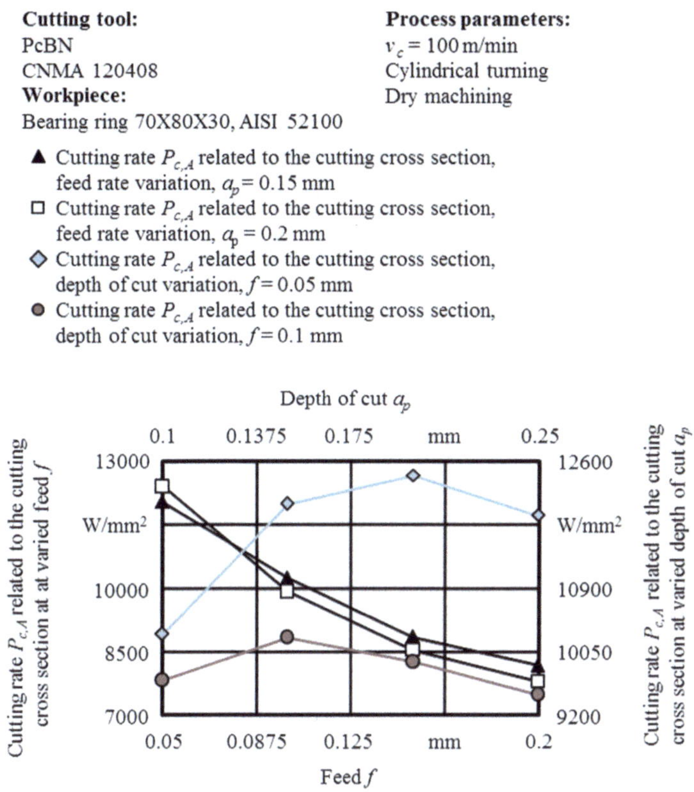

Fig. 4 Cutting rate related to the cutting cross section at varied feed and depth of cut

3.2 *Workpiece Surface Temperature*

Based on the measuring procedure described in Sect. 2.2, the process-oriented workpiece surface temperature immediately before chip separation was quantified. For this purpose, cutting trials on bearing rings composed of bearing steel AISI 52100 under variation of cutting speed, depth of cut and feed rate were thermographically recorded by an infrared camera of type Jade Mwir II manufactured by InfraTec GmbH, Dresden. As a result, the maximum specimen surface temperature prior to the chip separation was determined. During the evaluation of the thermographic images, a high value was set on the choice of snapshots free of chip segments that could falsify the temperature calculation by collateral thermal radiation.

The measurement results reveal no dependency of the maximal thermographically recorded workpiece surface temperature on the tool cutting edge angle κ at a cutting speed of $v_c = 100$ m/min. At higher cutting speeds, a slight temperature decrease at a higher tool cutting edge angle was measured be established. Due to the relationship between cutting speed v_c, cutting performance P_z and surface temperature level, the maximum specimen surface temperature values rise with increasing cutting speed. Moreover, no perceptible dependence of the temperature values on the tool cutting edge angle κ at a constant depth of cut a_p was determined [36]. Similarly, the maximal surface temperature levels do not depend on the tool cutting edge angle κ at constant feed f as well as on the feed f at steady tool cutting edge angle κ. Accordingly, the feed f does not represent an appropriate control variable with regard to a desirable thermal load collective diminishment in the specimen fringe area. In conclusion, at higher cutting speeds, a comparatively slight temperature reduction can be achieved by increasing the tool cutting edge angle. The measurement results and the identified maximum temperature levels represent a crucial basis for the evaluation and calibration of hard machining simulation models.

3.3 Heat-Up and Cooling Rates

The thermographic snapshots generated during hard turning of bearing rings enable the calculation of the heat-up rate as the temperature increase divided by the elapsed span before chip separation starts. Discrete measurement points were tracked along their trajectory considering the corresponding temperature shift. The evaluation of the measurement results discloses no perceptible dependence of the maximum heat-up rate of exemplary measurement points on the specimen surface on the depth of cut a_p. On the other hand, a slight heat-up rate increase could be established at rising feed rates. The largest heat-up rate increase was determined at the highest cutting speed v_c = 200 m/min. The magnitude of the heat-up rate at the cutting speed v_c = 200 m/min during the performed research equals 10^6 °C/s.

In order to ascertain the cooling rate during hard turning of the bearing steel AISI 52100, the test rig was modified in such a way that the applied infrared camera could record the machined workpiece surface temperature. For this purpose, the tool holder was clamped into the revolver of the lathe TRAUB TNX 65 with the rake face directed to the machine bed. Thus, the infrared camera positioned on the opposite side of the interior was capable of recording the cooling process on the machined specimen surface. In this configuration, the cooling rate was calculated analogously to the heat-up rate. The experimental analysis reveals no dependency of the cooling rate on depth of cut or feed. A significant increase of the cooling rate could be observed at rising cutting speeds. It could be established that the maximal cooling rate lies in the order of magnitude of $\sim 10^5$ °C/s and exhibits significantly lower values than the maximal heat-up rate.

4 A Multi-mechanism Model for Cutting Simulations

According to the experimental results in Sect. 3, the workpiece is machined under high speeds with high heat-up and cooling rates. Therefore, we developed a multi-mechanism model to describe the material behavior of the workpiece taking

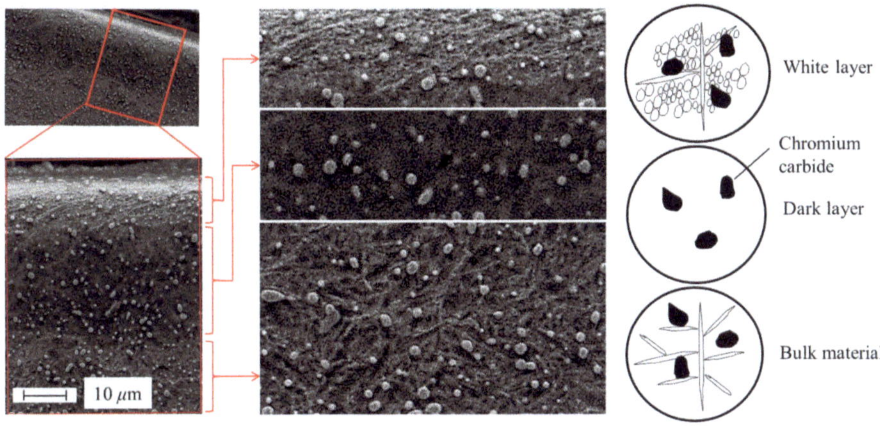

Fig. 5 Microstructure of steel AISI 52100 after hard turning with formation of white layer and dark layer (LWK, University of Paderborn)

temperature and strain rate dependency, hardness dependency, phase transformations and TRIP into account, which are typical phenomenon occurring in hard turning. To this end, the classical Johnson-Cook model is extended by an asymmetric effect and hardness dependency. The classical models for phase transformations, i.e. Leblond and Devaux [21] as well as Koistinen-Marburger [20] are extended by a gradient term based on the concept of generalized stresses.

To physically motivate the gradient term, we take a look at the microstructure of the steel AISI 52100 after hard turning in Fig. 5. White layer and dark layer are clearly formed, which have different microstructures than the bulk material. As mentioned in Sect. 1 the white layer forms due to the so-called reverse martensite transformation, which is schematically illustrated in Fig. 6. After reaching the austenitizing temperature A_{c1} the microstructure transforms due to the rapid cooling down to martensite. In comparison, the microstructure of the new formed martensite is circular, which is much smaller than the initial martensite with an acicular microstructure. Due to this we have two type of interfaces: I. Interface between initial martensite and austenite; II. Interface between new formed martensite and austenite. Thus we have two type of surface energies, which leads to different macroscopic free energies. The size effect and the difference of the free energies shall be considered using the gradient term.

In this section we introduce firstly a thermodynamic framework within a large-strain theory. Then we propose a prototype model.

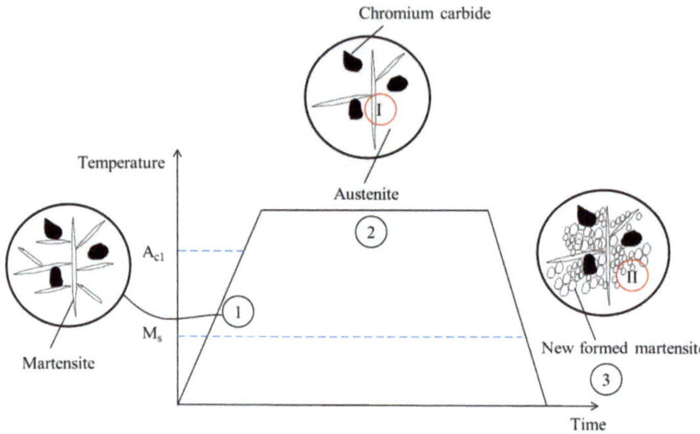

Fig. 6 Schematic illustration of phase transformations with different interface types: *I*. Interface between initial martensite and austenite; *II*. Interface between new formed martensite and austenite

4.1 A Thermodynamic Framework for Visco-Plasticity and Phase Transformations Based on the Concept of Generalized Stresses

4.1.1 The Concept of Generalized Stresses—General Setting

We consider an open set $\mathscr{D}_0$ as a subdomain of the reference configuration $\mathscr{B}_0$ for a continuum body within a large-strain theory. Then we introduce the placement of the related particle at the current configuration **x** associated with the mapping $\boldsymbol{\varphi}$, the deformation gradient **F** and the Jacobian of **F**

$$1.\ \mathbf{x} = \boldsymbol{\varphi}(\mathbf{X}, t), \qquad 2.\ \mathbf{F} = \nabla\boldsymbol{\varphi}(\mathbf{X}, t), \qquad 3.\ J = \det \mathbf{F}. \tag{1}$$

Here, the vector **X** labels material particles in $\mathscr{D}_0$. t represents the time. The gradient operator ∇ is introduced in relation of the reference configuration.

Under consideration of elasticity, plasticity, thermal effects and phase transformations, the deformation gradient **F** and the Jacobian J are split as

$$1.\ \mathbf{F} = \mathbf{F}^{\text{vol}} \cdot \mathbf{F}_e^{\text{iso}}, \qquad 2.\ \mathbf{F}^{\text{vol}} = J^{1/3}\mathbf{1} \qquad 3.\ \tilde{F} = \mathbf{F}_e^{\text{iso}} \cdot \mathbf{F}_e^{\text{iso}} \qquad 4.\ J = J_e \cdot J_\theta \cdot J_z \tag{2}$$

where $\mathbf{F}_e^{\text{iso}}$ and $\mathbf{F}^{\text{vol}}$ represent isochoric and volumetric deformations, respectively. J_e, J_θ and J_z represent the elastic volumetric, the thermal and the tranformational deformations, respectively. The elastic isochoric deformation is defined as $\mathbf{F}_e^{\text{iso}}$, and the plastic deformation $\mathbf{F}_i$ is assumed to be purely isochoric.

According to [26] the following additive decomposition for a velocity gradient with respect to an intermediate configuration is obtained

$$
\begin{aligned}
&1.\ \bar{\mathbf{L}} := \bar{\mathbf{L}}_e + \bar{\mathbf{L}}_i + \bar{\mathbf{L}}_\theta + \bar{\mathbf{L}}_z, \quad \text{where} \quad 2.\ \bar{\mathbf{L}}_e := \mathbf{F}_e^{-1} \cdot \dot{\mathbf{F}}_e, \\
&3.\ \bar{\mathbf{L}}_i := \dot{\mathbf{F}}_i \cdot \mathbf{F}_i^{-1} = -\mathbf{F}_i \cdot \dot{\mathbf{F}}_i^{-1}, \quad 4.\ \bar{\mathbf{L}}_\theta := \frac{1}{3}\frac{d}{dt}(\ln J_\theta)\mathbf{1}, \quad 5.\ \bar{\mathbf{L}}_z := \frac{1}{3}\frac{d}{dt}(\ln J_z)\mathbf{1}.
\end{aligned} \tag{3}
$$

Consequently $\bar{\mathbf{L}}_e, \bar{\mathbf{L}}_i, \bar{\mathbf{L}}_\theta, \bar{\mathbf{L}}_z$, represent respectively the elastic, inelastic, thermal and transformation part. $\mathbf{F}_e = J_e^{1/3}\mathbf{F}_e^{\text{iso}}$ represents the elastic deformation. On details of the terms $\ln J_\theta$ and $\ln J_z$ we refer to Mahnken et al. [26].

Based on the concept of generalized stresses of Gurtin [13] with consideration of phase transformations, the placement $\mathbf{x}$ is complemented by an *additional degree of freedom* z_A—the austenite phase fraction. Thus, a first gradient theory is build as

$$
1.\ \text{DOF} = \{\mathbf{x},\ z_A\}, \qquad 2.\ \text{STRAIN} = \{\mathbf{F},\ z_A,\ \nabla z_A\}. \tag{4}
$$

Following the principle of virtual power due to Ammar et al. [3] and the derivations in Cheng et al. [8], the strong formulations are given as

$$
\begin{aligned}
&1.\ \nabla \cdot \mathbf{P} + \mathbf{f} = \mathbf{0} \ \text{ in } \mathscr{B}_0, && 2.\ \mathbf{t} = \mathbf{P} \cdot \mathbf{N} \ \text{ on } \partial\mathscr{B}_0^t, \\
&3.\ \nabla \cdot (\boldsymbol{\xi} - \boldsymbol{\Gamma}) + \pi + \gamma = 0 \ \text{ in } \mathscr{B}_0, && 4.\ \zeta = (\boldsymbol{\xi} - \boldsymbol{\Gamma}) \cdot \mathbf{n} \ \text{ on } \partial\mathscr{B}_0^\zeta,
\end{aligned} \tag{5}
$$

where $\mathbf{P}$ is the first Piola-stress tensor, $\mathbf{f}$ the volumetric density of force, $\mathbf{t}$ is the macro traction vector on the Neumann boundary $\partial\mathscr{B}_0^t$, π is the scalar internal microstress, $\boldsymbol{\xi}$ is the internal vector microstress. π and $\boldsymbol{\xi}$ work conjugate to z_A and ∇z_A, respectively. γ and $\boldsymbol{\Gamma}$ are the external scalar and vector microstresses, work conjugate to z_A and ∇z_A, respectively. ζ is the generalized microtration on the boundary $\mathscr{B}_0^\zeta$.

According to Forest et al. [11] and analogously to Cheng et al. [8], the local balance of energy and the entropy principle in the reference configuration $\mathscr{B}_0$ are given by

$$
\begin{aligned}
&1. && \rho_0\, \dot{\epsilon} + \nabla \mathbf{q}_0 = p^{(i)} + \rho_0\, r_\theta \quad \text{in } V, \\
&2. && -\rho_0 \dot{\Psi} + \rho_0\, \theta\, \dot{\eta} + p^{(i)} - \frac{1}{\theta}\mathbf{q}_0 \cdot \nabla\theta \geq 0 \quad \text{in } V.
\end{aligned} \tag{6}
$$

Here, ρ_0 represents the density in the reference configuration, ϵ is the specific internal energy. $\mathbf{q}_0$ and r_θ represent the heat-flux density vector and the mass density of heat supply, respectively. The density of internal power is introduced as

$$
p^{(i)}[\dot{z}_A, \dot{\mathbf{u}}] = \mathbf{P} : \dot{\mathbf{F}} - \pi \dot{z}_A + \boldsymbol{\xi} \cdot \nabla \dot{z}_A, \tag{7}
$$

which is an extension of a standard formulation, see e.g. Haupt [15].

4.1.2 Constitutive Framework

Based on the STRAIN definition in Eq. (4.2), we select the state variables as

$$\text{STATE} = \{\overline{\mathbf{C}}_e, \underline{q}, z_A, \nabla z_A, \theta, \nabla\theta\}, \tag{8}$$

with the elastic right Cauchy–Green tensor $\overline{\mathbf{C}}_e = \mathbf{F}_e^T \cdot \mathbf{F}_e$ and internal variables for strain hardening $\underline{q} = [q_i, \dots, q_{n_q}]$. Therefore, the Helmholtz energy Ψ is assumed as

$$\Psi = \Psi[\overline{\mathbf{C}}_e, \underline{q}, z_A, \nabla z_A, \theta, \nabla\theta]. \tag{9}$$

The stress-power $\mathbf{P} : \dot{\mathbf{F}}$ in Eq. (7) can also be formulated as a dual pairing of the Mandel stress tensor $\overline{\mathbf{M}} = \mathbf{F}_e^T(\mathbf{P}\mathbf{F}^T)\mathbf{F}_e^{-T}$ and the velocity gradient $\bar{\mathbf{L}}$ (Eq. 3.1) with respect to the intermediate configuation $\bar{\mathscr{B}}$

$$\mathbf{P} : \dot{\mathbf{F}} = \overline{\mathbf{M}} : \bar{\mathbf{L}}. \tag{10}$$

Using the additive decomposition of the velocity gradient $\bar{\mathbf{L}}$ in Eq. (3.1) for the stress power in Eq. (6.2) and the identity $\partial\Psi/\partial\overline{\mathbf{C}}_e : \dot{\overline{\mathbf{C}}}_e = 2(\overline{\mathbf{C}}_e \cdot \partial\Psi/\partial\overline{\mathbf{C}}_e) : \bar{\mathbf{L}}_e$ the Clausius-Duhem inequality (6.2)—with quantities with respect to the intermediate configuration—reads as

$$\begin{aligned} &\left(\overline{\mathbf{M}} - \rho_0 2\overline{\mathbf{C}}_e \cdot \frac{\partial\Psi}{\partial\overline{\mathbf{C}}_e}\right) : \bar{\mathbf{L}} + \left(\xi - \rho_0\frac{\partial\Psi}{\partial\nabla z_A}\right) \cdot \nabla\dot{z}_A - \left(\eta + \frac{\partial\Psi}{\partial\theta}\right)\dot{\theta} \\ &+ \rho_0 2\overline{\mathbf{C}}_e \cdot \frac{\partial\Psi}{\partial\overline{\mathbf{C}}_e} : \bar{\mathbf{L}}_i - \rho_0\frac{\partial\Psi}{\partial\underline{q}}\dot{\underline{q}} - \left(\pi + \rho_0\frac{\partial\Psi}{\partial z_A}\right)\dot{z}_A - \frac{1}{\theta}\mathbf{q}_0 \cdot \operatorname{Grad}\theta \geq 0. \end{aligned} \tag{11}$$

Next, we define the Mandel stress tensor $\overline{\mathbf{M}}$, the vector microstress $\boldsymbol{\xi}$, the entropy η, thermodynamic forces $\underline{Q} = [Q_1, Q_2, \dots, Q_{n_q}]^T$ and a chemical force Z_A

$$\begin{aligned} &1.\ \overline{\mathbf{M}} = \rho_0 2\overline{\mathbf{C}}_e\frac{\partial\Psi}{\partial\overline{\mathbf{C}}_e}, \quad 2.\ \xi = \rho_0\frac{\partial\Psi}{\partial\nabla z_A}, \quad 3.\ \eta = -\frac{\partial\Psi}{\partial\theta}, \\ &4.\ \underline{Q} = \rho_0\frac{\partial\Psi}{\partial\underline{q}}, \quad 5.\ Z_A = \rho_0\frac{\partial\Psi}{\partial z_A} + \pi. \end{aligned} \tag{12}$$

where Q_{n_q} are also called hardening stresses. The following inequalities are sufficient for validity of the Clausius-Duhem inequality (6.2) to make the model thermodynamic consistent:

$$1.\ \mathscr{D}^i = \overline{\mathbf{M}} : \bar{\mathbf{L}}_i - \underline{Q}\dot{\underline{q}} \geq 0, \qquad 2.\ \mathscr{D}^z = -Z_A\dot{z}_A \geq 0, \qquad 3.\ \mathscr{D}^\theta = -\frac{1}{\theta}\mathbf{q}_0 \cdot \nabla\theta \geq 0. \tag{13}$$

Remark 1 4.1.1

1. In order to make the model thermodynamically consistent it is necessary to formulate the following evolution equations to fulfil the inequalities (13.1) and (13.2):

$$1.\ \bar{\mathbf{L}}_i = \bar{\mathbf{L}}_i\,[\overline{\mathbf{M}}, \underline{Q}, Z_A, \underline{q}, z_A, \nabla z_A, \theta], \quad 2.\ \dot{\underline{q}} = \dot{\underline{q}}[\overline{\mathbf{M}}, \underline{Q}, Z_A, \underline{q}, z_A, \nabla z_A, \theta],$$
$$3.\ \dot{z}_A = \dot{z}_A\,[\overline{\mathbf{M}}, \underline{Q}, Z_A, \underline{q}, z_A, \nabla z_A, \theta], \tag{14}$$

   which are given in Sect. 4.2.
2. Assuming a non-negative chemical dissipation $\mathscr{D}^z = \beta \dot{z}_A^2$ with a positive constant β, the following constraint describing the relation between the free energy Ψ and the evolution of austenite phase transformation $\dot{z}_A$ is derived as

$$\rho_0 \frac{\partial \Psi}{\partial z_A} - \rho_0 \nabla \cdot \frac{\partial \Psi}{\partial \nabla z_A} = -\beta \dot{z}_A. \tag{15}$$

   This constraint must be fulfilled in the formulation of the evolution equation of $\dot{z}_A$.
3. The inequality (13.3) is fulfilled due to a common approach $\mathbf{q}_0 = -\lambda_\theta (\det \mathbf{F})\mathbf{C}^{-1} \cdot \nabla\theta$.

4.1.3 Heat-Conduction Equation

The heat-conduction equation can be derived in a standard way from the energy equation (6.1) (e.g. Haupt [15]) with $\Psi = \epsilon - \theta\eta$. Taking the relations (3.1), (7) and (12) into account, one gets from Eq. (6.1):

$$\rho_0\,\theta\,\dot{\eta} + \mathrm{div}\mathbf{q}_0 = \overline{\mathbf{M}} : (\bar{\mathbf{L}}_i + \bar{\mathbf{L}}_\theta + \bar{\mathbf{L}}_z) - \underline{\underline{Q}\dot{q}} - Z_A\,\dot{z}_A + \rho_0 r_\theta. \tag{16}$$

4.2 *A Prototype Model for Hard Turning Simulation*

In this subsection we present a multi-mechanism model for the simulation of hard turning. The model is specialized for the metal AISI 52100. To this end, the well-known Johnson-Cook model [19] is extended with inelastic asymmetric effect. To get a better description on strain rate dependency, the original Johnson-Cook rate form is replaced by the Huh-Kang [16] rate form. Furthermore, asymmetric hardness dependency and hardness modification due to white layer formation are taken into account for the inelasticity. Moreover, phase transformations, which are very important phenomenons in metal working, are taken into account. Martensite as the initial phase transforms into austenite and retransforms back to martensite. To this end, we have the phase fractions for

1. austenite (z_A) and
2. martensite (z_M).

Since only two phases occur, the martensite fraction can be expressed as

$$z_M = 1 - z_A. \tag{17}$$

At the end of this section, the transformations are combined with transformation induced plasticity (TRIP).

4.2.1 Helmholz Energy

The Helmholtz energy Ψ of Eq. (9) is decomposed as

$$\Psi\left[\overline{\mathbf{C}}_e, \underline{q}, z_A, \nabla z_A, \theta\right] = \Psi^{el}\left[\overline{\mathbf{C}}_e, z_A, \theta\right] + \Psi^{pl}\left[\underline{q}, z_A, \theta\right] + \Psi^{\theta}\left[\theta\right] + \Psi^{ch}\left[z_A, \nabla z_A, \theta\right]. \tag{18}$$

The terms Ψ^{el}, Ψ^{pl} and Ψ^{θ} are specialized in [8]. The chemical part Ψ^{ch} is defined as

$$\Psi^{ch}\left[z_A, \nabla z_A, \theta\right] = \begin{cases} \dfrac{\beta}{\rho_0}\left(-\dfrac{1}{2}\alpha\nabla^2 z_A \cdot z_A + \dfrac{3}{4}\alpha\nabla^2 z_A + \dfrac{1}{2}z_A^2 - z_A\right)\mu_{MA}H(\theta - A_{c1}) \text{ for } \dot{\theta} \geq 0 \\ \dfrac{\beta}{\rho_0}\left(-\dfrac{1}{2}\alpha\nabla^2 z_A \cdot z_A + \dfrac{1}{4}\alpha\nabla^2 z_A + \dfrac{1}{2}z_A^2\right)\left\langle\dfrac{-\dot{\theta}}{k_\theta}\right\rangle H(\theta_{MS} - \theta) \text{ for } \dot{\theta} < 0. \end{cases}, \tag{19}$$

β, μ_{MA} and k_θ are positive constants. A_{c1} and θ_{MS} are the austenite and martensite start temperature, respectively. The Heaviside function H plays the role as a switcher with $H(s) = 1$ for $s > 0$ and $H(s) = 0$ for $s \leq 0$. The gradient terms including ∇z_A with the related parameter α (named as *gradient factor*) are regarded as a macroscopic contribution for Ψ^{ch} reflecting microscopic surface energies, which are used to consider the size effect and the difference of the marcoscopic free energies as mentioned at the begin of Sect. 4. The specific formulation for Ψ^{ch} has been obtained with the motivation to extend the classical Leblond-Devaux and Koistinen-Marburger approaches for phase transformations with a gradient term by inserting Eq. (19) into Eq. (15), see Sect. 4.2.3. Under consideration of convexity properties, the partial derivations $\partial^2\Psi^{ch}/\nabla^2 z_A$ and $\partial^2\Psi^{ch}/z_A^2$ must be positive. These are automatically fulfilled for $\dot{\theta} \geq 0$ and only under the condition that $z_A < 0.5$ for $\dot{\theta} \geq 0$. Since the cutting process is very fast, the duration of the austenitization is very short. Therefore, the austenite fraction is far smaller than 0.5 (see the cutting simulation in Sect. 6), which leads to a convex function of Ψ^{ch}.

4.2.2 Evolution Equations of the Visco-Plasticity and the TRIP

The inelastic velocity gradient $\bar{\mathbf{L}}_i$ is introduced and decomposed additively as

$$\bar{\mathbf{L}}_i = \bar{\mathbf{L}}_p + \bar{\mathbf{L}}_t, \tag{20}$$

where $\bar{\mathbf{L}}_p$ and $\bar{\mathbf{L}}_t$ represent the visco-plastic and the TRIP part, respectively.

We assume the following evolution equations for the visco-plastic part $\bar{\mathbf{L}}_p$ of the velocity gradient in Eq. (20) and the internal variables q_1 and q_2 in the relation (12.4)

$$
\begin{aligned}
&1.\ \bar{\mathbf{L}}_p = \dot{\lambda}\sqrt{\frac{3}{2}}\mathbf{N}, \quad \text{where} \quad \mathbf{N} = \frac{\overline{\mathbf{M}}^{dev}}{||\overline{\mathbf{M}}^{dev}||},\\
&2.\ \dot{q}_1 = \dot{\lambda}\mathscr{J}\left(1 - c\frac{Q_1}{H_1}\right),\\
&3.\ \dot{q}_2 = \dot{\lambda}\mathscr{J},
\end{aligned}
\tag{21}
$$

where $\mathscr{J}$ is the Johnson-Cook coefficient introduced in [8] representing temperature and strain rate dependency. Furthermore, the plastic multiplier $\dot{\lambda}$ is obtained from the loading/unloading conditions, see e.g. [30],

$$
1.\ \dot{\lambda} \geq 0, \quad 2.\ \phi \leq 0, \quad 3.\ \dot{\lambda}\phi = 0. \tag{22}
$$

Here, we have also introduced the rate of equivalent plastic strain $\dot{e}_v$, defined as $\dot{e}_v^2 = 2/3\, \bar{\mathbf{L}}_p : \bar{\mathbf{L}}_p$, such that by use of Eq. (21.1) one obtains the relation

$$
\dot{e}_v = \sqrt{\frac{2}{3}\bar{\mathbf{L}}_i : \bar{\mathbf{L}}_i} = \dot{\lambda}\sqrt{\frac{2}{3}\frac{3}{2}\mathbf{N} : \mathbf{N}} = \dot{\lambda}. \tag{23}
$$

As shown in [8, 26] the hardening stresses Q_1 and Q_2 are obtained from the Helmholtz energy Ψ^{pl} in Eq. (9), to render the following evolution equations:

$$
\begin{aligned}
&1.\ \dot{Q}_1 = H_1\dot{q}_1 = \dot{\lambda}\mathscr{J}\left(H_1 - cQ_1\right)\\
&2.\ \dot{Q}_2 = H_2\dot{q}_2 = \dot{\lambda}H_2\mathscr{J}.
\end{aligned}
\tag{24}
$$

Based on the Johnson-Cook model, we formulate a yield function

$$
\begin{aligned}
&1.\ \phi &&= \sigma_v - (Y_0 + F + Ge_v + Q_1 + Q_2)\mathscr{J}, \quad 2.\ \mathscr{J} = \mathscr{J}^\theta \mathscr{J}^R_{HK},\\
&3.\ \mathscr{J}^\theta &&= (1 - (\theta^*)^m), \quad 4.\ \mathscr{J}^R_{HK} = 1 + C_1 \ln\left(\left\langle\frac{\dot{e}_v}{\dot{\varepsilon}_0}\right\rangle_1\right) + C_2\left(\ln\left(\left\langle\frac{\dot{e}_v}{\dot{\varepsilon}_0}\right\rangle_1\right)\right)^2,\\
&5.\ F[\mathscr{H}] &&= F_1 \cdot \mathscr{H} - F_2, \quad 6.\ G[\mathscr{H}] = G_1 \cdot \mathscr{H} - G_2,\\
&7.\ Y_0 &&= \sum_{i=1}^{S} w_i Y_{0i}, \ c = \sum_{i=1}^{S} w_i c_i, \ C_1 = \sum_{i=1}^{S} w_i C_{1i}, \ C_2 = \sum_{i=1}^{S} w_i C_{2i}, \ \dot{\varepsilon}_0 = \sum_{i=1}^{S} w_i \dot{\varepsilon}_{0i},\\
&\quad m &&= \sum_{i=1}^{S} w_i m_i, F_1 = \sum_{i=1}^{S} w_i, F_{1i}, \ F_2 = \sum_{i=1}^{S} w_i F_{2i}, \ G_1 = \sum_{i=1}^{S} w_i G_{1i}, \ G_2 = \sum_{i=1}^{S} w_i G_{2i},\\
&8.\ \boldsymbol{\kappa}_{vp} &&= [Y_{0i}, H_{1i}, H_{2i}, c_i, C_{1i}, C_{2i}, \dot{\varepsilon}_{0i}, m_i, F_{1i}, F_{2i}, G_{1i}, G_{2i}]^T
\end{aligned}
\tag{25}
$$

where the von Mises stress is written in terms of the deviatoric part of the Mandel stress tensor as $\sigma_v = \sqrt{3/2}||\overline{\mathbf{M}}^{dev}||$.

The term $F + Ge_v$ is a linear function proposed by Umbrello et al. [37] for consideration of hardness dependency. F and G are functions of the hardness $\mathscr{H}$. F_1, F_2, G_1, and G_2 are parameters. Subsequently, the coefficient $\mathscr{J}$ in Eq. (25.2) shall be refered to as the Johnson-Cook coefficient, where $\mathscr{J}^{\theta}$ and $\mathscr{J}^{R}_{HK}$ reflect the dependences of temperature and strain rate, respectively. The rate form $\mathscr{J}^{R}_{HK}$ is called Huh-Kang rate form, which is quadratic in the logarithm of the strain rate as a replacement of the original form with linear dependence proving better material description. $Y_0, c, C_1, C_2, \dot{\varepsilon}_0, m$ are parameters.

In order to consider the asymmetric effect, we weight the parameters in Eq. (25) using weighting functions w_i with the properties

$$1.\ \sum_{i=1}^{S} w_i[\overline{\mathbf{M}}] = 1, \quad 2.\ w_i[\overline{\mathbf{M}}_j] = \delta_{ij}, \quad 3.\ w_i[\overline{\mathbf{M}}] \geq 0, \tag{26}$$

The weighting functions w_i are associated to different independent characteristic *stress modes* characterised by stress tensors $\overline{\mathbf{M}}_j$, $j = 1, 2, \ldots, S$. For more details on the weighting functions we refer to Mahnken [23]. Under consideration of three typical loadings (tension, compression and shear), we set $S = 3$. To this end, the new parameters are summarized as $\boldsymbol{\kappa}_{vp}$ in Eq. (25.8).

Due to potential white layer formation, the hardness $\mathscr{H}$ in Eqs. (25.5) and (25.6) can be varied. This is proved by Umbrello et al. [39]. They also propose an empirical model to describe the hardness modification

$$\begin{aligned} &1.\ \mathscr{H} &&= \mathscr{H}_{init} + \Delta\mathscr{H} \\ &2.\ \Delta\mathscr{H} &&= F_{\mathscr{H}} \frac{\mathscr{H}_r - \mathscr{H}_{init}}{\theta_r - A_{c1}} (\theta_{max} - A_{c1}) \cdot H(\theta_{max} - A_{c1}) \cdot H(\theta_{MS} - \theta). \end{aligned} \tag{27}$$

Here $\mathscr{H}_{init}$ is the hardness of the bulk material. $\Delta\mathscr{H}$ represent the change of the hardness. $\mathscr{H}_r$ and θ_r are parameters representing the reference hardness and the reference temperature. θ_{max} is the maximal temperature. The parameters for the hardness variation is summarized as $\boldsymbol{\kappa}_{\mathscr{H}} = [\mathscr{H}_r, \theta_r, F_{\mathscr{H}}]$. $\mathscr{H}_r$ and θ_r are empirically chosen as $\mathscr{H}_r = 67$ HRC and $\theta_r = 1030$ °C by Umbrello et al. [39].

According to the approach of Leblond [22] among others for a small strain theory, the TRIP for austenitization and martensitic transformation are supposed to be

$$\bar{\mathbf{L}}_t = \begin{cases} \dot{z}_A \frac{3}{2} f'_A K_{tpA} (\overline{\mathbf{M}}^{dev})^t & \text{for } \dot{z}_A \geq 0 \\ \dot{z}_M \frac{3}{2} f'_M K_{tpM} (\overline{\mathbf{M}}^{dev})^t & \text{for } \dot{z}_A < 0 \end{cases} . \tag{28}$$

Here, K_{tpA} and K_{tpM} are the Greenwood-Johnson parameters with respect to austenitization and martensitic transformation. The saturation functions f_A and f_M are heuristic functions satisfying

$$1.\, f[0] = 0, \qquad 2.\, f[1] = 1, \qquad 3.\, f'[z] = \frac{df}{dz} \geq 0. \tag{29}$$

A possible formulation is given by Denis [9] as

$$1.\, f[z_A] = \left(2 - z_A\right) z_A \quad 2.\, f[z_M] = \left(2 - z_M\right) z_M. \tag{30}$$

To this end, the unknown parameters are summarized as $\boldsymbol{\kappa}_{tp} = [K_{tpA},\ K_{tpM}]$.

4.2.3 Evolution Equations of Phase Transformations

Inserting Eq. (19) into Eq. (15) and using Eq. (17), we obtain the evolution equations of austenite phase fraction and the retransformation:

$$1.\, \dot{z}_A = \begin{cases} \mu_{MA}(1 + \alpha\nabla^2 z_A)(1 - z_A)H(\theta - A_{c1}) & \text{for } \dot{\theta} \geq 0 \\ -\left\langle \dfrac{-\dot{\theta}}{k_\theta} \right\rangle (1 + \alpha\nabla^2 z_A) z_A H(\theta_{MS} - \theta) & \text{for } \dot{\theta} < 0 \end{cases}, \quad 2.\, \dot{z}_M = -\dot{z}_A. \tag{31}$$

Here, α, μ_{MA}, k_θ are parameters summarized as $\boldsymbol{\kappa}_{pt} = [\alpha,\ \mu_{MA},\ A_{c1},\ k_\theta,\ M_s]$. For details on the derivations we refer to Cheng et al. [8].

5 Parameter Identification

In this section we identify the parameters of the multi-mechanism model. Two type of experiments are investigated for parameter identifications. In the first experiment mechanical tests are carried out at different temperatures and strain rates under different loading types such as tension, compression and shear with different hardened specimens. In the second example dilatometer tests are investigated under different stress loadings for identifying the parameters related to TRIP and phase transformations.

5.1 Mechanical Tests for Visco-Plasticity and Hardness Dependency

This section exhibits identification of the parameters $\boldsymbol{\kappa}_{pl}$ in Sect. 4.2.2 related to the asymmetric visco-plastic material behavior and the asymmetric hardness dependency for the steel AISI 52100. Two parts of experiments are investigated under tension, compression and shear. Part I concentrates on the visco-plasticity. Therefore, the hardness of all specimens is chosen to be constant (62 HRC). Temperature and strain rate are varied. In part II the tests are carried out at room temperature (20 °C)

Table 1 Summary of mechanical tests for AISI 52100

	Notation	Strain rate (s^{-1})	Compression (C)/Tension (T) /Shear (S)	Temperature (°C)	Hardness HRC
Part I	*T-R01-T20*	0.1	*Tension*	20	62
	T-R1000-T20	1000	*Tension*	20	62
	T-R100-T200	100	*Tension*	200	62
	T-R100-T400	100	*Tension*	400	62
	T-R100-T600	100	*Tension*	600	62
	C-R0001-T20	0.001	*Compression*	20	62
	C-R01-T20	0.1	*Compression*	20	62
	C-R184-T20	184	*Compression*	20	62
	C-R1890-T20	1890	*Compression*	20	62
	C-R183-T400	183	*Compression*	400	62
	C-R185-T600	185	*Compression*	600	62
	S-R01-T20	0.1	*Shear*	20	62
	S-R100-T200	100	*Shear*	200	62
	S-R100-T400	100	*Shear*	400	62
Part II	*T53*	0.1	*Tension*	20	53
	T57	0.1	*Tension*	20	57
	T60	0.1	*Tension*	20	60
	T62	0.1	*Tension*	20	62
	T64	0.1	*Tension*	20	64
	C53	0.1	*Compression*	20	53
	C57	0.1	*Compression*	20	57
	C60	0.1	*Compression*	20	60
	C62	0.1	*Compression*	20	62
	C64	0.1	*Compression*	20	64
	S53	0.1	*Shear*	20	53
	S57	0.1	*Shear*	20	57
	S60	0.1	*Shear*	20	60
	S62	0.1	*Shear*	20	62
	S64	0.1	*Shear*	20	64

with the same strain rate but different hardened specimens, in order to concentrate on the hardness dependency. Concerning specimen geometry and performance of the experiments we refer to Halle [14]. Furthermore, the devices for performing the dynamic tests under tension, compression and torsion are described in [26]. All specimens used in the experiments were made and all tests were performed by the company Nordmetall, Adorf, Germany.

The experiments are summarized according to the nomenclature in Table 1. Note, $T-R01-T20$, $C-R01-T20$ and $S-R01-T20$ in part I are the same tests as $T62$, $C62$ and $S62$ in part II, respectively.

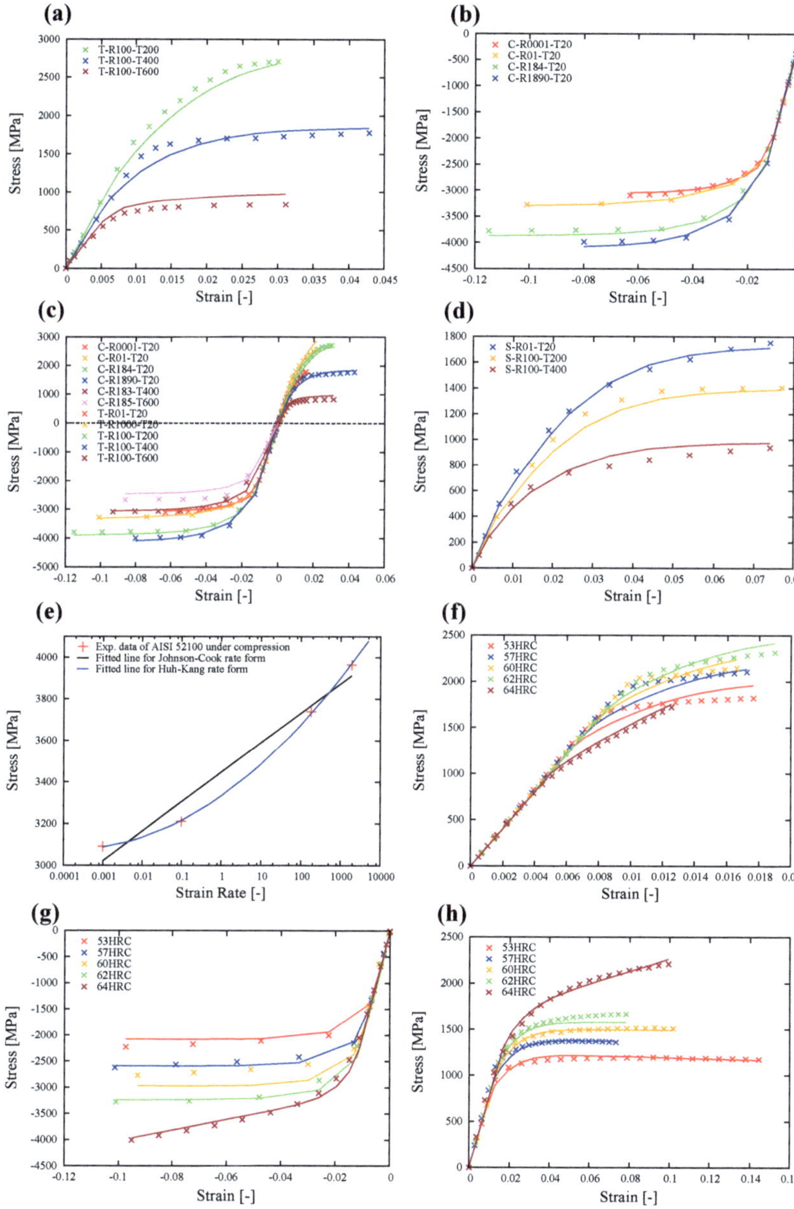

Fig. 7 Steel AISI 52100: **a** Temperature dependency for tension, **b** Strain rate dependency for compression, **c** Asymmetric visco-plastic behavior under tension and compression, **d** Stress-strain curves under shear, **e** Comparison of Huh-Kang and Johnson-Cook rate form, **f**, **g**, **h** Hardness dependency under tension, compression and shear. Symbols refer to experiment, solid lines to simulation

Figure 7 illustrates the resulting stress strain curves of the experiments and the simulations. The results in Fig. 7a show an identical elastic behavior and a different inelastic behavior for different strain rates in tension. Figure 7b shows the strain rate dependency in compression. Figure 7c and d show the asymmetric visco-plastic behavior in tension, compression and shear. Figure 7e shows the advantage of the Huh-Kang rate form in comparison with the original Johnson-Cook rate form, where the points represent the effective stresses at different strain rates (from C-R0001-T20, C-R01-T20, C-R184-T20 and C-R1890-T20) with 5% effective plastic strain. Figures 7f–h show the hardness dependency in tension, compression and shear, respectively. The curves exhibit an asymmetric behavior. Generally, the harder the specimen, the higher the stress level, except for the hardened state 64 HRC in tension. Since the hardened state 64 HRC is very brittle such that the specimen starts to fail already at elastic loading only with very small plastic deformation for pure tension. Also under compression and shear the material behavior of 64 HRC doesn't match with other hardened states as the stress level is higher. To this end, a distinction of cases for the parameter identification is made, where the hardened state 62 HRC is treated as the threshold.

Using a least-squares functional as an identification criterion, the parameters of $\boldsymbol{\kappa}_{pl}$ are identified by minimizing the difference between experiments and simulations. For more details on the functional and the procedure we refer to [24, 25]. The identified parameters are given in Table 2.

Table 2 Material parameters related to visco-plasticity concerning hardness dependency for steel AISI 52100

$\boldsymbol{\kappa}_{pl}$:	F_{11} $[\frac{\text{MPa}}{\text{HRC}}]$	F_{12} $[\frac{\text{MPa}}{\text{HRC}}]$	F_{13} $[\frac{\text{MPa}}{\text{HRC}}]$	G_{11} $[\frac{\text{MPa}}{\text{HRC}}]$	G_{12} $[\frac{\text{MPa}}{\text{HRC}}]$	G_{13} $[\frac{\text{MPa}}{\text{HRC}}]$
$\mathscr{H} \geq$ 62HRC:	−3.43E + 02	2.04E + 01	1.27E + 02	2.73E + 04	4.48E + 03	6.707E + 03
$\mathscr{H} <$ 62HRC:	3.78E + 01	1.049E + 02	6.43E + 01	4.48E + 02	4.012E + 00	2.504E + 02
	F_{21} [MPa]	F_{22} [MPa]	F_{23} [MPa]	G_{21} [MPa]	G_{22} [MPa]	G_{23} [MPa]
$\mathscr{H} \geq$ 62HRC:	−2.126E + 04	1.265E + 03	7.87E + 03	1.69E + 06	2.77E + 05	4.158E + 05
$\mathscr{H} <$ 62HRC:	2.34E + 03	6.503E + 03	3.987E + 03	2.778E + 04	2.487E + 02	1.55E + 04
	Y_{01} [MPa]	Y_{02} [MPa]	Y_{03} [MPa]	H_{11} [MPa]	H_{12} [MPa]	H_{13} [MPa]
	1.475E + 03	2.0953E + 03	1.7066E + 03	1.4875E + 05	1.4875E + 05	1.4875E + 05
	H_{21} [MPa]	H_{22} [MPa]	H_{23} [MPa]	c_1 [−]	c_2 [−]	c_3 [−]
	1.97E-02	1.97E-02	1.97E-02	2.61E + 02	2.61E + 02	2.61E + 02
	C_{11} [−]	C_{12} [−]	C_{13} [−]	C_{21} [−]	C_{22} [−]	C_{23} [−]
	2.91E-02	2.55E-02	2.39E-02	1.1E-03	8.1E-04	5.1E-04
	$\dot{\varepsilon}_{01}$ [−]	$\dot{\varepsilon}_{02}$ [−]	$\dot{\varepsilon}_{03}$ [−]	m_1 [−]	m_2 [−]	m_3 [−]
	1.0E-05	1.0E-05	1.0E-05	7.23E-01	1.1E + 00	0.7E-01

5.2 Dilatometer Tests for TRIP-strains

In order to identify the parameters of $\boldsymbol{\kappa}_{tp}$ in Sect. 4.2.2 related to TRIP-strains and $\boldsymbol{\kappa}_{pt}$ in Sect. 4.2.3 related to transformations of the material AISI 52100, so called dilatometer tests are investigated. The tests were performed by Stiftung Institut für Werkstofftechnik (IWT), University of Bremen. Details on the geometry and preparation of the hollow cylindrical specimens are given in Irretier [17]. Note, the gradient factor is assumed as $\alpha = 0$ for the identifications. Its influence will be studied in Sect. 6.

Two parts of the dilatometer tests are investigated, which are summarized in Table 3. Part I is related to TRIP for martensitic transformation (TRIP-M). Part II is related to TRIP for austenitization (TRIP-A). The program for the specimens of temperature and stress in part I is illustrated in Fig. 8a: The specimens are heated with constant heating speed 6.2 K/s to over the austenitization start temperature holding for 10 min for the homogenization of the specimens in the measured length. Then the specimens are cooled with quenching gas (−170 °C) in order to obtain martensite. During cooling stresses (5, 50, 72, 100 and −50 MPa) are applied for analysing the effects of external stresses on the phase transformation behavior and the TRIP effect.

The resulting simulations for the longitudinal strain-temperature curves under tension and compression are shown in Fig. 8b. A satisfying agreement with experimental data is obtained.

The specimens of part II fully consist of martensite (62 HRC), and their geometry is shown in Fig. 9a. They were heated up to 1000 °C with a constant heating rate 20 K/s, where the longitudinal and radial strain are measured. A schematic illustration of the temperature and the loaded stresses in time is shown in Fig. 9a.

Figure 9b shows the longitudinal strains under different stress loadings over the temperature. According to the phase transformations the curves are decomposed into five phases illustrated in Fig. 10a. In order to express the difference between loaded and unloaded cases we compare the two curves (A0 and A5) in this diagram. In phase I, the specimens are heated from room temperature (ca. 15 °C) to 420 °C. The strain curve without stress loading (A0) is linear due to a pure thermal loading. In

Table 3 Summary of dilatometer tests

TRIP-M		TRIP-A	
Notation	Stress (MPa)	Notation	Stress (MPa)
M–50	–50	A –5	–5
M5	5	A0	0
M50	50	A5	5
M72	72	A10	10
M100	100	A15	15
		A20	20
		A25	25

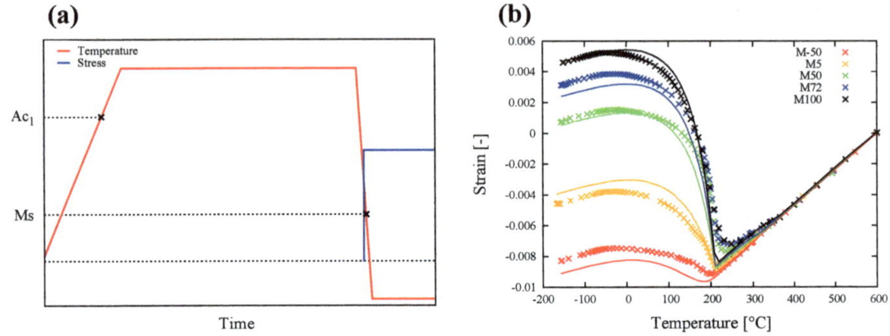

Fig. 8 TRIP-M: **a** Schematic of temperature and stress loading with respect to time; **b** Longitudinal strain-temperature curves (Comparison of experimental and computed results, *symbols* refer to experiment, *solid lines* refer to simulation)

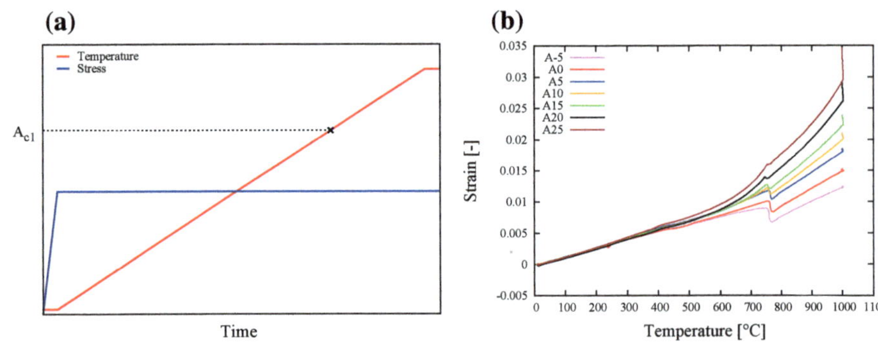

Fig. 9 TRIP-A: **a** Schematic of temperature and stress loading with respect to time; **b** longitudinal strain-temperature curves

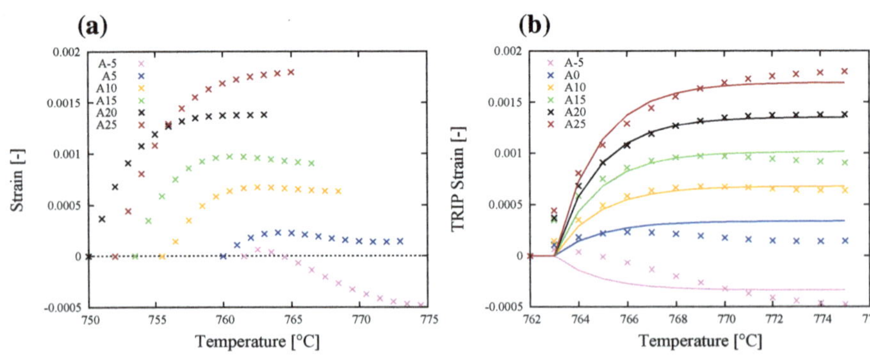

Fig. 10 TRIP-A: **a** Decomposition of the transformation in five phases; **b** TRIP-strains (Comparison of experimental and computed results, *symbols refer* to experiment, *solid lines* refer to simulation)

comparison the curve with stress loading (A5) has not only thermal strain but also an additional creep strain because of 5 MPa stress loading. Phase II begins with tempering at ca. 420 °C and ends at ca. 460 °C, the slopes of both curves are smaller due to the volume reduction of the tempering. In phase III, analogously to phase I, the additional creep strain on A5 makes that the curves A0 and A5 separating further from each other. In phase IV, the huge volume reduction indicates that the martensite transforms into austenite (from ca. 760 to 770 °C). In phase V, transformation is finished and the specimens are heated further (from ca. 770 to 1000 °C).

To this end, the strain change of A5 in phase IV $\Delta\varepsilon_5^{IV}$ is expressed as

$$\Delta\varepsilon_5^{IV} = \Delta\varepsilon_5^{\theta} + \Delta\varepsilon_5^{cr} + \Delta\varepsilon_5^{tv} + \Delta\varepsilon_5^{tp}, \tag{32}$$

where $\Delta\varepsilon_5^{\theta}$, $\Delta\varepsilon_5^{cr}$, $\Delta\varepsilon_5^{tv}$ and $\Delta\varepsilon_5^{tp}$ represent the strain change due to heating, creep, volume change of phase transformation and TRIP under 5 MPa stress loading, respectively. Note, the expressions of Eq. (32) are also valid for other curves A−5, A10, A15, A20 and A25.

According to Eq. (28) TRIP-strain can't be induced without stress loading, therefore, the strain change of D0 in phase IV $\Delta\varepsilon_0^{IV}$ reads

$$\Delta\varepsilon_0^{IV} = \Delta\varepsilon_0^{\theta} + \Delta\varepsilon_0^{tv}, \tag{33}$$

where $\Delta\varepsilon_0^{\theta}$ and $\Delta\varepsilon_0^{tv}$ represent the strain change due to heating and volume change of phase transformation. Since the stress loading doesn't affect the thermal strain and volume change of phase transformation, the following relations hold

$$1.\ \Delta\varepsilon_5^{\theta} = \Delta\varepsilon_0^{\theta}, \qquad 2.\ \Delta\varepsilon_5^{tv} = \Delta\varepsilon_0^{tv}. \tag{34}$$

Inserting Eq. (33) and Eq. (34) into Eq. (32) solving for $\Delta\varepsilon_5^{tp}$ we obtain

$$\Delta\varepsilon_5^{tp} = \Delta\varepsilon_5^{IV} - \Delta\varepsilon_0^{IV} - \Delta\varepsilon_5^{cr}, \tag{35}$$

where the creep strain $\Delta\varepsilon_5^{cr}$ can be approximated as

$$\Delta\varepsilon_5^{cr} \approx \Delta\theta * (K_5 - K_0). \tag{36}$$

Here $\Delta\theta$ is the temperature increase in phase IV, K_5 and K_0 are the slopes of the curves D5 and D0 at the beginning of the phase V. Note, that Eq. (35) holds also for the other curves with stress loadings.

According to the above relations the TRIP strains are calculated and illustrated in Fig. 10b (points). Moreover, the simulated results are also shown in this figure (lines). A satisfying agreement with experimental data is obtained. The procedure for identifying the parameters is the same as in the previous subsection and shall not be discussed here.

Table 4 Material parameters for steel 100Cr6

κ_{tp} :	K_{tpA} $[\frac{1}{\text{MPa}}]$	K_{tpM} $[\frac{1}{\text{MPa}}]$		
	1.37E-03	1.03E-04		
κ_{pt} :	μ_{MA} $[-]$	A_{c1} [°C]	k_θ $[\frac{1}{°\text{C}}]$	M_s [°C]
	4.857	504	0.013	211

According to Fig. 10b the austenitization start temperature A_{c1} is identified to be 762 °C, however, the value is actually much lower due to the intense mechanical loading in hard turning. Following the so-called Clausius-Clayperon equation proposed by Griffiths [12] and used by Ramesh et al. [28]

$$\Delta A_{c1} = \frac{\sigma_v A_{c1}^{nom} \Delta V_{tr}}{\Delta H_{tr}} \quad \text{with} \quad \Delta V_{tr} = \frac{M}{\rho_A} - \frac{M}{\rho_M} \tag{37}$$

we calculate a new value for A_{c1}. Here ΔA_{c1} is the variation of A_{c1} due to stress loading. ΔV_{tr} is the volume change per mole due to transformation and M is the molar weight. ΔH_{tr} is the heat of transformation involved in the austenitization. The following values are given by Ramesh et al. [28]

$$\begin{aligned} &1.\ \Delta H_{tr} = 215\ \text{cal} = 215 \cdot 41.3\ \text{cm}^3\text{atm} = 8879.5\ \text{cm}^3\text{atm} \\ &2.\ M = 55.85\ \text{g} \\ &3.\ \sigma_v = 1300\ \text{MPa} = 1.28 \times 10^4\ \text{atm} \end{aligned} \tag{38}$$

The densities of the austenite and the martensite for AISI 52100 are determined by Acht et al. [1] with $\rho_A = 8.03\text{g/cm}^3$, $\rho_M = 7.765\text{g/cm}^3$. ΔA_{c1} is calculated as

$$\Delta A_{c1} = \frac{1.28 \times 10^4\text{atm} \cdot 762\ °\text{C} \cdot 55.85\ \text{g}(\frac{1}{8.03\ \text{g/cm}^3} - \frac{1}{7.765\ \text{g/cm}^3})}{8879.5\ \text{cm}^3\text{atm}} = -258\ °\text{C} \tag{39}$$

The austenitization temperature A_{c1} for cutting with martensite as the initial phase is reduced to 504 °C.

The values of the parameters are summarized in Table 4.

6 Cutting Simulations in ABAQUS for Testing the Model

Some cutting simulations were investigated in order to test our model related to the extensions. Figure 11 shows the geometry and the finite-element discretization. The tool has a negative rake angle $\gamma = -6°$ as given in Sect. 2.1 and an clearance angle $\alpha = 6°$. It is treated as a rigid body and therefore meshed coarsely. It moves into the workpiece with a constant velocity $v_c = 200$ m/min as the experiment in Sect. 2.1.

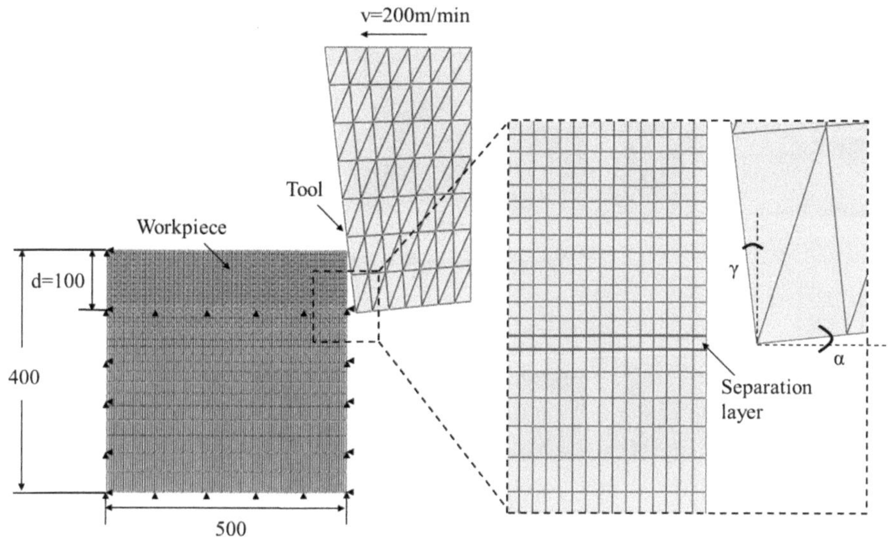

Fig. 11 Cutting simulation: geometry and finite-element discretization

The workpiece is defined to be 500 μm long and 400 μm high; the depth of cut used was $h = 100\,\mu\text{m}$. In order to get a realistic chip formation we define a separation layer. Above the separation layer the meshing is very fine and uniform. In comparison, a biased meshing is used for the elements below the separation layer with a fine mesh for upper area and coarse mesh for lower area, since the elements at the bottom area are not deformed. The boundary conditions on the workpiece are applied as shown in Fig. 11. The initial conditions assume room temperature, and for cooling a conventional temperature condition is applied over the surfaces of the workpiece. Note that, friction between the tool and workpiece is not considered for simplicity. A 4-node plane strain thermally coupled quadrilateral element (CPE4RT) is used for the workpiece. A 3-node plane strain thermally coupled triangle element (CPE3T) is used for the tool.

The model is implemented as a user-defined subroutine (VUEL) on the element level and linked to ABAQUS v6.14. Based on the parameter identification for the material AISI 52100 in Sect. 5.1, the model is applied to the material behavior of the workpiece. To the separation layer we assign a shear failure criterion $e_v \leq e_v^f$ with a constant e_v^f for separating the chip from the workpiece.

The model testing concentrates on the following characteristics:

1. Asymmetric effect and influence of hardness on the material behavior,
2. Influence of the gradient term,
3. Hardness modification due to the white layer formation,
4. TRIP of austenitization.

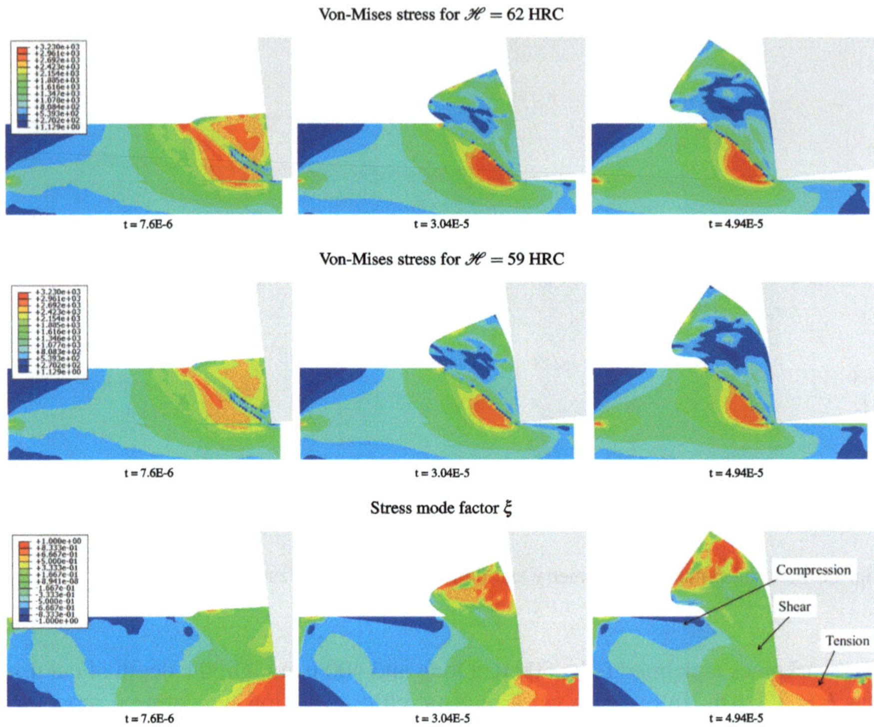

Fig. 12 Cutting simulation: contours of von-Mises stress for different initial hardness states and related stress mode factor

In order to study the influence of hardness on the material behavior, two simulations with $\mathscr{H}_{init} = 62$ HRC and $\mathscr{H}_{init} = 59$ HRC are investigated. Figure 12 shows the von-Mises stresses of both simulations at time instants 7.6E-6, 3.04E-5, and 4.94E-5 s. It can be seen that the chips are formed continuously and have the same geometry. The von-Mises stress in the cutting zone for both simulations are different, the one with $\mathscr{H}_{init} = 62$ HRC is about 250 MPa higher than the one with $\mathscr{H}_{init} = 59$ HRC. Furthermore, with the help of the stress mode factor we are able to indicate the different stress modes of the machined workpiece such as tension, compression and shear.

Figure 13 shows contour plots of the temperature. The maximum value is approximately 1000 °C and concentrates on the cutting area due to the intense inelastic dissipation. Within an area, where the temperature is greater than A_{c1}, austenite forms over the time (see Fig. 14). This can be seen at the time point t = 4.94E-5 s, the duration of the temperature over A_{c1} for the red area is longer than that for the green area. Furthermore, we study the influence of the gradient term by setting the gradient factor α to be 0, 50 and 100. We can find, that the austenite only differs along the chip edge next to the cutting tool, where the temperature and phase gradient are much

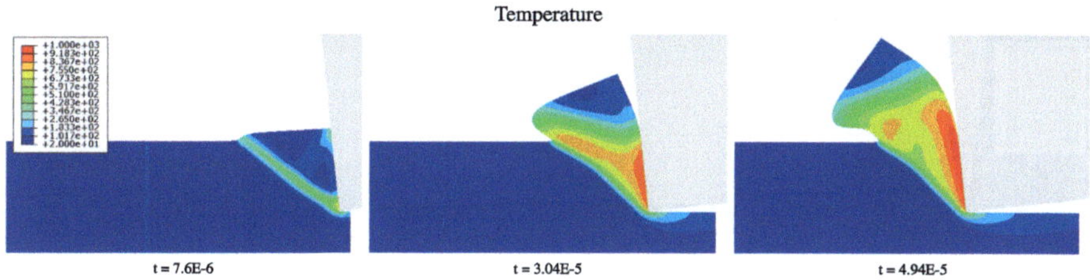

Fig. 13 Cutting simulation: contours of temperature for cutting depths 100 μm

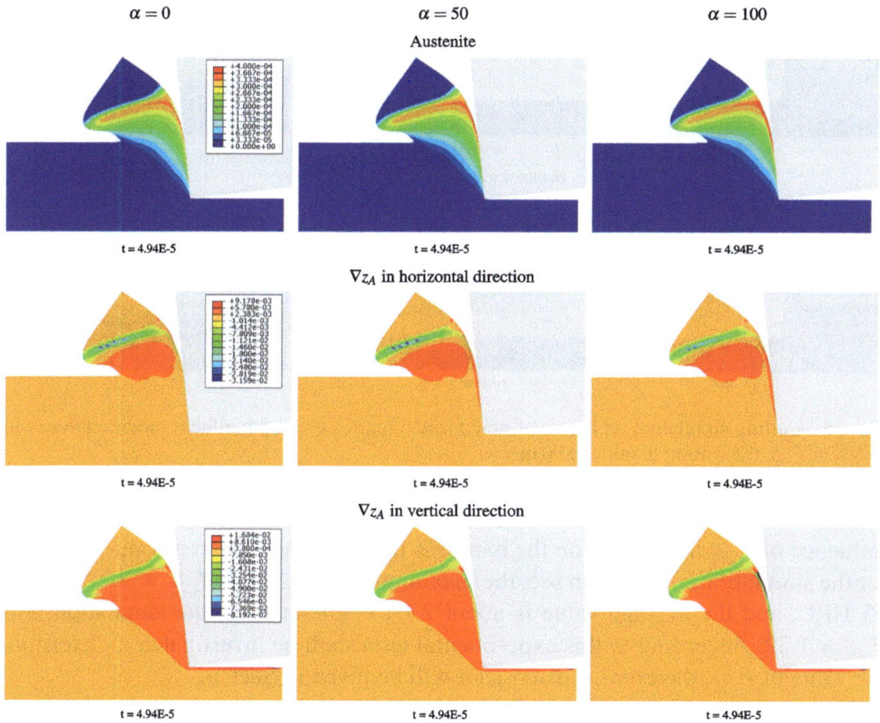

Fig. 14 Cutting simulation: austenite and its gradient with different gradient factor α

higher than in other area. The difference can also be seen at the components of the gradient. About identification of the gradient term we give a discussion in Sect. 9.

As proposed in Sect. 4.2.2, Eq. (27) describes the hardness modification due to the white layer formation. According to this equation the white layer is induced if the maximal temperature of a cutting area is greater than the austenitization temperature and the current temperature is lower than the martensite start temperature. Due to rapid cooling the austenite retransforms back to martensite and as a result the white layer is formed on the newly generated surface of the workpiece (see Fig. 15 "new formed martensite"), which leads to a hardness modification. In order to study the

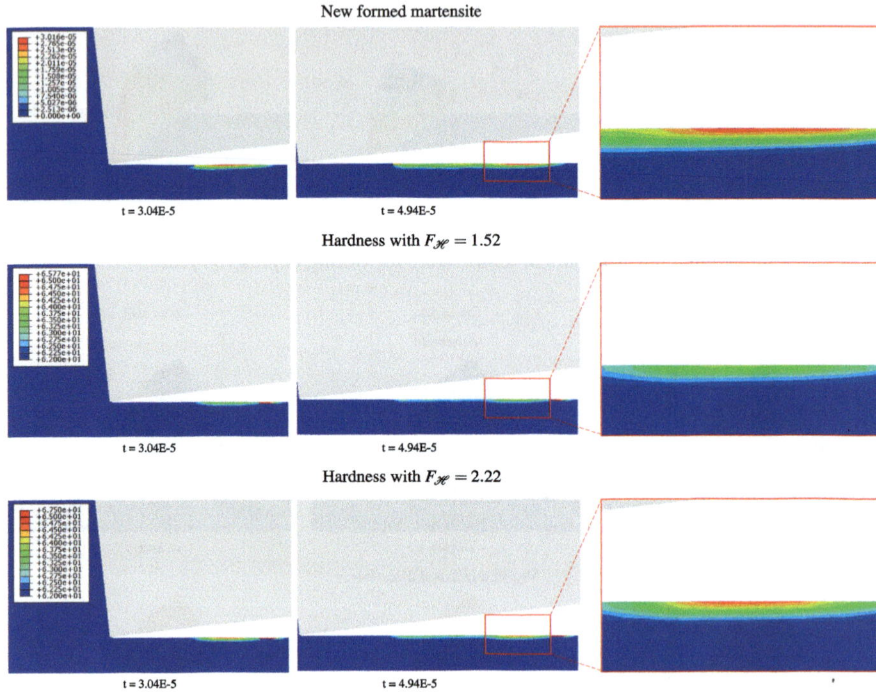

Fig. 15 Cutting simulation: contours of new formed martensite and hardness modifications with different $F_{\mathcal{H}}$ for cutting depths 100 μm

influence of $F_{\mathcal{H}}$ in Eq. (27) on the hardness modification, different values are used for the simulations. As we can see, the maximum hardness with $F_{\mathcal{H}} = 2.22$ is about 65 HRC, and the average value is about 2 HRC greater than the simulation with $F_{\mathcal{H}} = 1.52$. According to this experimental tests shall be investigated to determine the value of $F_{\mathcal{H}}$, therefore, a discussion will be given in Sect. 9.

Next, we intend to study the influence of the TRIP for austenitization (TRIP-A). We define a scalar TRIP-factor

$$w_A = \dot{z}_A \frac{3}{2} f'_A K_{tpA} \tag{40}$$

from Eq. (28) in order to represent the quantity of TRIP-A. Then we plot this value in Fig. 16. The value is very homogeneous at the area where the austenitization occurs, and a bit smaller at the boundaries. Furthermore, we show the von-Mises stress with and without the influence of TRIP-A. In comparison we find that the TRIP-A has a strong effect on the stress at the austenitization boundaries, although w_A is very small. At the homogeneous area the TRIP-A has almost no influence on the stress.

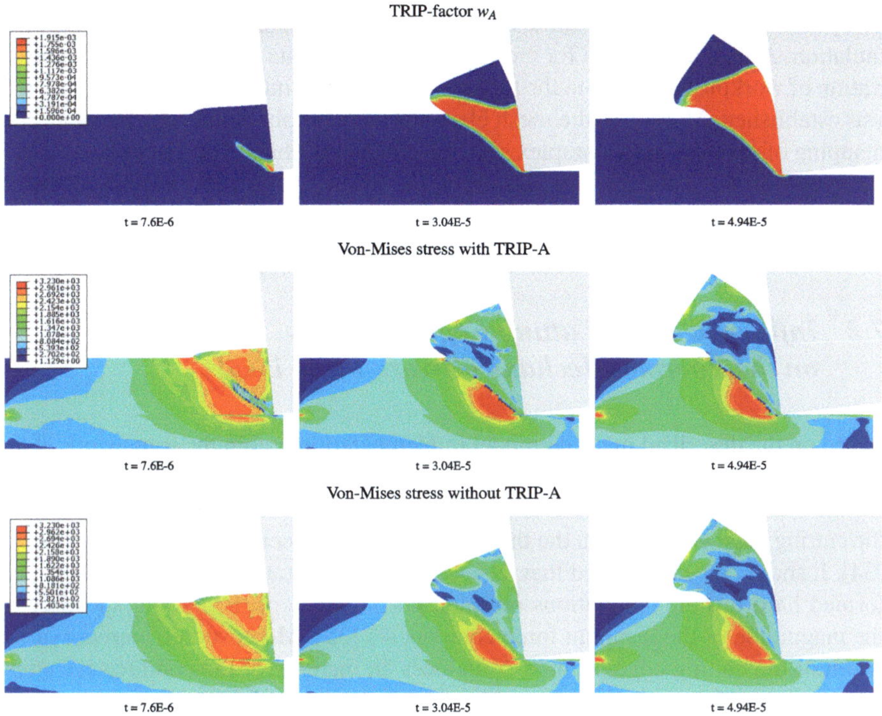

Fig. 16 Cutting simulation: contours of austenite, TRIP-A and von-Mises stress with influence of TRIP-A for cutting depths 100 μm

7 Simulation in Deform

7.1 Modelling in Deform

In order to validate the thermo-mechanical workpiece load during hard turning, a material routine with the purpose to calculate the material flow stress considering the theoretical fundamentals described in the previous sections was developed using the programming language Fortran. In this context, the constitutive model equations and parameters were provided by the Chair of Technical Mechanics of the University of Paderborn, whereas the subsequent implementation in an executable machining simulation was conducted at the Institute for Machine Tools and Factory Management of the Technical University Berlin. The hard turning simulations were executed using the commercial Finite Element Method software Deform-3D manufactured by SFTC, Columbus, USA. Moreover, a cutting insert model was generated by means of the CAD-software SolidWorks manufactured from Dassault Systémes, Vélizy-Villacoublay, France. The tool holder applied within the machining simulations was defined through its tool cutting edge angle κ, rake angle γ and angle of inclination λ.

The performed hard turning simulations were based on the implicit Lagrangian formulation. A further criterion for mesh generation and data interpolation is the penetration of workpiece mesh into the tool mesh. Hence, an interference depth of 25 μm was established. Moreover, the mesh elements were tetrahedrons, which enables the mapping of geometrically complex structures like flow chips. The viscoplastic material behavior was influenced by the strain, the strain rate as well as the constitutive volume element temperature.

7.2 Influence of the Cutting Edge Geometry on the Thermo-Mechanical Work Piece Load

Based on the described material model implementation within the framework of the concept of weighting functions considering the asymmetric viscoplastic flow behavior the bearing steel AISI 52100, fundamental conclusions about the influence of the cutting edge geometry on the thermo-mechanical specimen load could be drawn [34]. It should be emphasized that, at each investigated tool edge geometry, the performed hard turning simulations revealed no influence of the cutting speed v_c on the magnitude of the resultant force components [35]. Moreover, an increase of the width of the cutting edge chamfer b_f induces an increase of the compressive stress areas in the cutting zone and the remaining workpiece as well as a concurrent temperature reduction, which can be attributed to the augmented hydrostatic compressive stress fraction. Further numerical results reveal that a higher chamfer angle arouses an extension of the thermally loaded areas in the primary shear zone as well as a workpiece surface temperature increase. In addition, the temperature at the interface between the cutting tool's flank face and the machined workpiece surface rises at a higher corner radius r_ε.

7.3 Impact of Dissimilar Stress States on Work Piece Heat Balance

Based on the theoretically developed procedure to identify areas in the cutting zone subjected to dissimilar stress conditions using the stress mode factor ξ, the generated chrome bearing steel material routine is capable of determining high compressive stress regions in the chip root as well as low tensile stress spots beneath the cutting edge influencing the reached temperature level and the machined surface quality [36]. Hence, an incentive to derive an approach which appraises the impact of the cutting zone stress distribution motivation on the specimen's heat balance is provided. By means of the algebraic sign of the stress mode factor ξ attributed to each workpiece volume element, the associated stress state can be established, the correspondent yield stress magnitude in the material routine can be purposely

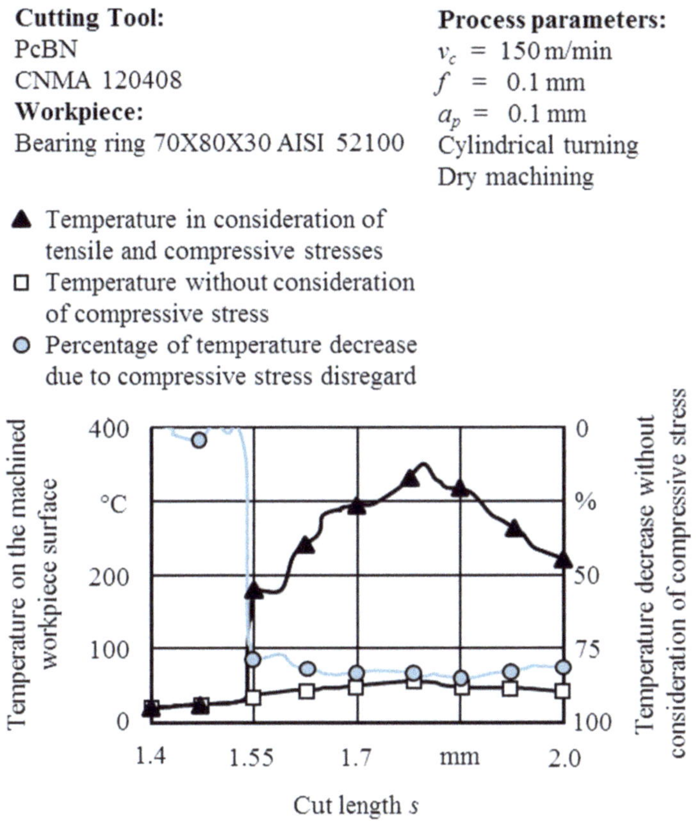

Fig. 17 Influence of the compressive stress in the cutting zone on the temperature development in the workpiece surface layer

decreased and the subsequent influence on the machined surface temperature level can be analyzed. In this regard, Fig. 17 depicts the average temperature profiles of an exemplary point on the machined surface both with and without consideration of the compressive stress in the cutting zone. Thus, the tensile stress acting underneath the cutting edge has no discernible influence on the specimen's heat balance. Furthermore, ignoring the compressive stress in the material routine yields a computation error of approximately 80%. In principle, this approach can be applied for the simulation of any chip-removing manufacturing process with the aim of gaining a better understanding of the stress allocation in the specimen's cutting zone as well as investigating the impact of dissimilar stress conditions on the temperature development in the surface layer.

7.4 Method for Analysis of the Influence of Physical Phenomena on the Workpiece Stress Distribution During Hard Turning

In order to achieve a high simulation accuracy when modeling the material behavior in consideration of the thermo-mechanical cutting conditions during hard turning, several physical phenomena like asymmetric visco-plasticity, temperature dependent phase transformations and transformation induced plasticity ought to be incorporated. It should be recalled that the aforementioned physical phenomena exert a significant influence on the temperature and stress onset in the machined surface layer. The question that inevitably arises is to what extent such a model extension could induce a considerable variation of the evaluated state variables. For instance, the comparison of numerically generated temperature profiles with and without consideration of transformation induced plasticity exhibits a specimen surface temperature reduction, which can be attributed to a ductility increase during phase transitions. Accordingly, the implementation of the Leblond-approach in order to consider the transformation induced plasticity effect exerts a substantial influence on the temperature and stress levels within the workpiece surface layer. This poses the question whether physical phenomena capable of being modeled by means of suitable numerical approaches affect the qualitative distribution of stress acting in the cutting area. A convenient tool for the purpose of analyzing the impact of numerical approaches on the aforesaid stress allocation within the cutting zone is the frequency distribution vertical-bar graph of the described stress mode factor ξ. According to the simulation results, the implementation of transformation induced plasticity within the material model induces no frequency distribution variation of the stress mode factor ξ. Hence, the transformation induced plasticity effect influences the stress level appearing in the cutting area, however not the qualitative allocation of the stress types occurring in the workpiece surface layer.

8 Model-Based Optimization of the Hard Turning Process

8.1 Process Parameters

An increase of the cutting speed v_c induces an augmentation of the workpiece areas subjected to compressive stress. The volume enlargement of the workpiece elements characterized by a negative stress mode factor and consequently by preponderance of compressive stress is more pronounced during the simulation of a continuous cut. The hard turning simulations using a tool trajectory along the workpiece's contour border indicate a less perceptible dependence of the frequency distribution of the workpiece volume elements subjected to compression stress on cutting speed v_c. Furthermore, the comparison of the frequency distribution of volume elements subjected to tensile stress exhibits no significant alteration at varied cutting speed v_c.

The simulation results indicate a workpiece surface temperature increase at higher cutting speeds. Assuming that the surface temperature does not exceed the austenitization temperature, which is guaranteed by applying unworn cutting tools, a hard turning process optimization can be achieved by adjusting the cutting speed v_c. Considering the wear of the cutting insert, the cutting speed v_c has to be chosen as high as possible in order to increase the fraction of volume elements subjected to compressive stress, which leads to a material softening and a reduction of the required cutting performance. In order to exploit the effect of cutting speed v_c dependent volume fraction enlargement of compressive stress areas, the cutting speed v_c should be maximized primarily during continuous machining. In case of a tool trajectory along the workpiece's contour border, a lower cutting speed v_c is recommendable due to the reduced dependency of the frequency distribution of the workpiece volume elements exhibiting a negative stress mode factor on cutting speed during edge finishing as well as the subsequent tool wear rate decrease.

According to the simulation results, an increased depth of cut arouses a volume fraction growth of both compressive and tensile stress areas within the machined workpiece surface layer. Moreover, the temperature analysis of the machined workpiece surface indicates that higher depths of cut cause a temperature rise in the cutting area. With regard to a reduction the thermal impact on the machined surface integrity, a decrease of the depth of cut appears to be an advantageous measure in the context of production planning.

The evaluation of the simulation results indicates a maximum volume fraction of workpiece areas subjected to both compressive and tensile stress at an average feed rate of $f = 0.1$ mm. The volume rise of the aforesaid specimen areas is particularly evident in the case of a continuous turning process simulation. Accordingly, a maximum feed rate is beneficial concerning a desirable mechanical load and machining time reduction. Due to the feed rate's influence on the machined surface quality, the requested surface roughness should also be considered while determining an optimum feed rate value. Moreover, the simulation results reveal no workpiece surface temperature dependency on a feed rate variation. Hence, the spatial distribution and volume fraction of stress acting within the workpiece surface layer represent suitable criteria in the context of feed rate optimization.

The evaluation of the stress mode factor frequency distribution at various rake and inclination angles indicates a tensile stress frequency reduction applying cutting inserts with rake and inclination angles less than zero. The frequency distribution of negative stress mode factor values in front of the cutting edge reveals no explicit tendency. Accordingly, cutting inserts with negative rake and inclination angle prove to be beneficial with regard to a tensile stress decrease acting beneath the clearance surface. Furthermore, the performed simulations of a continuous turning process indicate a slight workpiece temperature reduction at decreasing rake and inclination angles. This observation correlates with the fact that hydrostatic compressive stress is acting within the machined workpiece surface at increasing negative rake and inclination angles arouses material softening as well as temperature reduction. Thus, the reduction of the aforesaid angles is a substantial requirement in order to ensure optimum stress and thermal conditions during hard turning.

The analysis of the stress mode factor frequency distribution at various tool cutting edge angles reveals no respective dependency of stress allocation within the cutting area. Similarly, no influence of the tool cutting edge angle on the calculated workpiece surface temperature could be determined.

8.2 Machining Strategies

The experimentally determined reduction of the passive force related to the product of feed f and depth of cut a_p during face turning of bearing rings, compared to cylindrical turning, indicates that an adjustment of the process is capable of reducing the steep specimen surface temperature rise at the beginning of the hard turning operation. Accordingly, a feed increase from $f = 0.1$ mm to $f = 0.5$ mm induces a maximum temperature rise within the work piece surface layer during the first cut, whereas a feed decrease to $f = 0.001$ mm arouses lower temperature peaks up to a cut length of $s = 500$ mm. A thermally optimized hard turning process can be achieved by choosing a low feed of $f = 0.001$ mm up to a maximal cut length of $s = 500$ mm as well as a subsequent increase to $f = 0.1$ mm until the end of the machining operation. The described tool kinematic adjustment induces a temperature profile slope reduction within the specimen surface layer, which corresponds to a smooth thermal load increase until reaching a stationary surface temperature level. Since the aforesaid cut length of $s = 500$ mm equals about two workpiece revolutions, no significant processing time extension should be expected as a consequence of the described feed decrease. A further machining strategy aiming at reducing the calculated temperature profile slope during the first cut consists in front surface face turning of bearing rings at a low depth of cut of $a_p = 0.05$ mm until a cut length in feed direction crosswise to the workpiece rotation axis of $s = 0.1$ mm is reached. The described strategy is shown in Fig. 18. Due to the low depth of cut and the resulting mean ther-

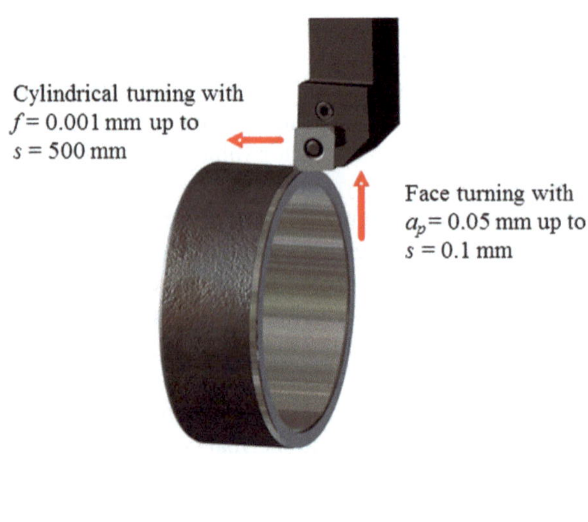

Fig. 18 Machining strategy for lowering the temperature in the first cut

momechanical workpiece load, the heat zone moving previous to the cutting insert increases gradually. Afterwards, the feed direction is changed by 90° and the bearing ring outer surface is machined by outer longitudinal cylindrical turning. The feed path initially covered by the cutting insert crosswise to the specimen rotation axis corresponds to the depth of cut during the second cutting procedure. The simulation of the described machining strategy indicates a less steep temperature profile within the workpiece surface layer as well as a temperature decrease at the beginning of the hard turning operation.

9 Summary and Outlook

Concerning material modelling we developed a multi-mechanism model based on the concept of generalized stresses. The model considers asymmetric visco-plasticity, asymmetric hardness dependency, phase transformations and TRIP-strains. To this end, the classic Johnson-Cook model is extended with the concept of weighting functions for the asymmetric visco-plasticity. The consideration of the asymmetric hardness dependency is based on a study of Umbrello et al. [37] for the hardness-dependent material AISI 52100. Phase transformations are taken into account by extending the classical Leblond-Devaux [21] and Koistinen-Marburger [20] with a gradient term, where the austenite phase fraction is regarded as an additional degree of freedom in the frame of the generalized concept. According to Leblond [22] TRIP-strains are considered for martensitic transformation as well as for austenitization. Furthermore, we identified all related parameters by comparing the simulated and experimented data while assuming $\alpha = 0$. Moreover, cutting simulations are investigated for testing the model, where the characteristics were studied as well as the influence of the gradient factor α and the hardness modification related parameter $F_{\mathscr{H}}$. The identification of these two parameters are an interesting challenge for future work. Since the original martensite and newly formed martensite have different microstrutures, as shown at the begin of Sect. 4, austenitization along the cutting chips from experiments can be determined and compared with the results of the simulations in order to identify α. According to the characteristic that $F_{\mathscr{H}}$ affect the hardness modification of the white layer, experiments and simulations can be compared to determine $F_{\mathscr{H}}$.

The material routine based on the theoretical findings serves as a basis for analyzing the interdependency between the process parameters, the cutting insert geometry and the resulting workpiece's surface temperature. By means of the obtained simulation results, the correlations between the available hard turning process parameters and the thermomechanical workpiece load can be explained. Moreover, the described simulation model enables the identification of optimized tool trajectories, which induce a thermomechanical load reduction as well as an increase of the machining accuracy. The described rolling bearing steel routine provides a fundamental basis for achieving high accordance between simulation results and measurement data by including further physical phenomena. For example, the process-induced residual stress formation within the workpiece's surface layer can be estimated using the available state variables as well as suitable analytical methods. Moreover, the validation of the intermediately oxidized workpiece surface during hard turning, unfeasible with currently available measurement equipment, can lead to a significant accuracy increase of the described non-contact temperature measurement procedure. The generation of the experimentally determined intermediately oxidized specimen surface layer under laboratory conditions could enable the precise derivation of the corresponding correction factor and an accurate non-contact surface temperature measurement in case of rotating workpieces.

Acknowledgements This paper is based on investigations of SPP 1480 which is kindly supported by the Deutsche Forschungsgemeinschaft (DFG). Furthermore, we gratefully acknowledge the support of the company Nordmetall, Adorf, Germany, for performing the mechanical tests related to the asymmetric visco-plasticity and hardness dependency and Stiftung Institut für Werkstofftechnik (IWT), University of Bremen, for performing the dilatometer tests related to the TRIP-strains.

References

1. Acht, C., Dalgiç, M., Frerichs, F., Hunkel, M., Irretier, A., Lübben, T., Surm, H.: Ermittlung der materialdaten zur simulation des Durchhärtens von komponenten aus 100Cr6—Teil 1. HTM J. Heat Treat. Mater. **63**, 234–244 (2008)
2. Ahrens, U.: Beanspruchungsabhängiges Umwandlungsverhalten und Umwandlungsplastizität niedrig legierter Stähle mit unterschiedlich hohen Kohlenstoffgehalten, Dissertation, University of Paderborn, Germany (2003)
3. Ammar, K., Appolaire, B., Cailletaud, G., Feyel, F., Forest, S.: Finite element formulation of a phase field model based on the concept of generalized stresses. Comput. Mater. Sci. **45**, 800–805 (2009)
4. Biermann, D., Liedschulte, M.: Plasmaunterstttztes drehen von hartlegierungen auf eisenbasis mit PKB. Ind. Diamanten Rundschau **28**(2), 71–77 (1994)
5. Biermann, D., Höhne, F., Sieben, B., Zabel, A.: Finite element modeling and three-dimensional simulation of the turning process incorporating the material hardness. Int. J. Mater. Form. **3–1**, 459–462 (2010)
6. Bökenheide, S., Wolff, M., Dalgic, M., Lammers, D., Linke, D.: Creep, phase transformations and transformation-induced plasticity of 100Cr6 steel during heating. In: Zoch, H.W., Lübben, T. (eds.) Proceedings of 3rd International Conference on Distortion Engineering, Bremen, Germany, 14–16, September (IWT Bremen), pp 411–418 (2012)
7. Byrne, G., Dornfeld, D., Denkena, B.: Advancing cutting technology. CIRP Annals **52**(2), 483–507 (2003)
8. Cheng, C., Mahnken, R.: A multi-mechanism model for cutting simulations based on the concept of generalized stresses. Comput. Mater. Sci. **100B**, 144–158 (2015)
9. Denis, S., Simon, A., Beck, G.: Analysis of the thermomechanical behaviour of steel during martensitic quenching and calculation of internal stresses, Hrsg.: Macherauch, E., Hauk, V., pp. 211–238 (1983)
10. Forest, S., Ammar, K., Appolaire, B.: Micromorphic vs. phase-field approaches for gradient viscoplasticity and phase transformations. Lecture Notes in Applied and Computational Mechanics, vol. 59, pp. 69–88 (2011)
11. Forest, S., Aifantis, E.C.: Some links between recent gradient thermo-elasto-plasticity theories and the thermomechanics of generalized continua. Int. J. Solids Struct. **47**, 3367–3376 (2011)

12. Griffiths, B.J.: Mechanisms of white layer generation with reference to machining and deformation processes. Trans. ASME J. Tribol. **109**, 525–530 (1987)
13. Gurtin, M.: Generalized Ginzburg-Landau and Cahn-Hilliard equations based on a microforce balance. Physica D **92**, 178–192 (1996)
14. Halle, T.: Zusammenhänge zwischen Spanvorgängen und dem mechanischen Werkstoffverhalten bei hohen Dehnungsgeschwindigkeiten, Dissertation, Technical University of Chemnitz, Germany (2005)
15. Haupt, P.: Continuum Mechanics and Theory of Materials. Springer, Berlin (2002)
16. Huh, H., Kang, W.J.: Crash-worthiness assessment of thin-walled structures with the high-strength steel sheet. Int. J. of Veh. Des. **30**(1/2) (2002)
17. Irretier, A.: Abschlussbericht Projekt C1 "Stoffwertebestimmung", Sonderforschungsbereichs 570 "Distorsion Engineering". University of Bremen, Germany (2008)
18. Iwamoto, T., Tsuta, T., Tomita, T.: Investigation on deformation mode dependence of strain-induced martensitic transformation in TRIP steels and modeling of transformation kinetics. Int. J. Mech. Sci. **40**(2–3), 173–182 (1998)
19. Johnson, G.R., Cook, W.H.: A constitutive model and data for metals subjected to large strain rates and high temperatures. In: Proceedings 7th International Symposium on Ballistics, The Hague, pp. 541–547 (1983)
20. Koistinen, D.P., Marburger, R.E.: A general equation prescribing the extent of the austenite-martensite transformation in pure iron-carbon alloys and plain carbon steels. Acta Metallica **7**, 59–60 (1959)
21. Leblond, J.B., Devaux, J.: A new kinetic model for anisothermal metallurgical transformations in steels including effect of austenite grain size. Acta Metall. **32**, 137–146 (1984)
22. Leblond, J.B.: Mathematical modelling of transformation plasticity in steels II: coupling with strain hardening phenomena. Int. J. of Plast. **5**, 537–591 (1989)
23. Mahnken, R.: Creep simulation of asymmetric effects by use of stress mode dependent weighting functions. Int. J. Solids Struct. **40**, 6189–6209 (2003)
24. Mahnken, R., Johansson, M., Runesson, K.: Parameter estimation for a viscoplastic damage model using a gradient-based optimization algorithm. Eng. Comput. **15**(7), 925–955 (1998)
25. Mahnken, R.: Identification of material parameters for constitutive equations, In: Stein, E., de Borst, R., Hughes, T.J.R. (eds.), Encyclopedia of Computational Mechanics, vol. 2, Chapter 19. Wiley Ltd, Chichester (2004)
26. Mahnken, R., Wolff, M., Cheng, C.: A multi-mechanism model for cutting simulations combining visco-plastic asymmetry and phase transformation. Int. J. Solids Struct. **50**, 3045–3066 (2013)
27. Maugin, G.A., Metrikine, A.V.: Mechanics of Generalized Continua, One Hundred Years After the Cosserats. Springer, Berlin, Heidelberg (2010)
28. Ramesh, A., Melkote, S.N.: Modeling of white layer formaion under thermally dominant conditions in orthogonal machining of hardened AISI 52100 steel. Int. J. Mach. Tools Manuf. **48**, 402–414 (2007)
29. Rech, J., Moisan, A.: Surface integrity in finish hard turning of case hardened steels. Int. J. Mach. Tools Manuf. **43**, 543–550 (2003)
30. Simo, J.C., Hughes, T.J.R.: Computational Inelasticity, Interdisciplinary Applied Mathematics, vol. 7. Mechanics and Materials, Springer, Berlin (1998)
31. Stouffer, D.C., Dame, L.T.: Inelastic Deformation of Metals. Wiley, New York (1996)
32. Spitzig, W.A., Sober, R.J., Richmond, O.: Pressure dependence of yielding and associated volume expansion in tempered martensite. Acta Metall. **23**, 885–893 (1975)
33. Tönshoff, H.K., Arendt, C., Ben, A.R.: Cutting of hardened steel. CIRP Annals **49**(2), 547–566 (2000)
34. Uhlmann, E., Ivanov, I.M.: Zerspankraftsimulation beim Hartdrehen Dreidimensionale Modellierung des Hartdrehens zur Zerspankraftberechnung und Werkzeugoptimierung, wt Werkstattstechnik. **1**(2), 10–15 (2012)
35. Uhlmann, E., Mahnken, R., Ivanov, I.M., Cheng, C.: FEM modeling of hard turning with consideration of viscoplastic asymmetry and phase transformation. In: Proceedings of the XXIV

CIRP Sponsored Conference on Supervising and Diagnostics of Machining Systems, 11–14 March 2013, Karpacz, Poland (2013)
36. Uhlmann, E., Mahnken, R., Ivanov, I.M., Cheng, C.: A novel finite element approach to modeling hard turning in due consideration of the viscoplastic asymmetry effect. In: Proceedings of the 15th CIRP Conference on Modelling of Machining Operations, 11–12 June 2015, Karlsruhe, Germany (2015)
37. Umbrello, D., Hua, J., Shivpuri, R.: Hardness-based flow stress and fracture models for numerical simulation of hard machining AISI 52100 bearing steel. Mater. Sci. Eng. A **374**, 90–100 (2004)
38. Umbrello, D., Ambrogio, G., Filice, L., Shivpuri, R.: An ANN approach for predicting subsurface residual stresses and the desired cutting conditions during hard turning. J. Mater. Process. Technol. **189**, 143–152 (2007)
39. Umbrello, D., Filice, L.: Improving surface integrity in orthogonal machining of hardened AISI 52100 steel by modeling white and dark layers formation. Annals CIRP **58**(1), 73–76 (2009)
40. Zwirlein, O.: Die Beeinflussung der mechanischen Eigenschaften von gehrtetem Wlzlagerstahl 100Cr6 durch hydrostatischen Druck. Materialwissenschaft und Werkstofftechnik **8**(10), 344–353 (1977)

Modeling of Orthogonal Metal Cutting Using Adaptive Smoothed Particle Hydrodynamics

F. Spreng and P. Eberhard

Abstract This contribution provides an overview over different aspects of the numerical modeling of orthogonal metal cutting using adaptive Smoothed Particle Hydrodynamics. First, the basic principle of the employed spatial discretization technique and its application to the equations of solid continuum mechanics are introduced. Furthermore, a brief description is given of the most important extensions to this meshless Lagrangian simulation method that are necessary to correctly model the process of metal cutting, e.g. a variable resolution scheme. Here, references to in-depth literature are provided where necessary. Second, the applicability of the introduced enhanced discretization technique to model cutting processes is shown. To that end, the results of orthogonal cutting simulations for a steel of type AISI 1045 are analyzed in terms of chip formation, stress distribution as well as the cutting force, and they are compared to experimentally obtained data.

1 Introduction

Machining, as the most important manufacturing process, has been in the research focus of both academia and industry for several decades. Substantial progress has been made in raising its efficiency, but most of these innovations were results of expensive and time-consuming experimental investigations. For a further enhancement, comprehensive knowledge on the interaction between the cutting tool and the workpiece, as well as on the material-dependent behavior of the workpiece itself is required. Numerical simulations can help to boost this process of innovation. However, mesh-based spatial discretization schemes such as the Finite Element Method (FEM) [1] show some drawbacks when being applied to cutting

F. Spreng · P. Eberhard (✉)
Institute of Engineering and Computational Mechanics, University of Stuttgart,
Pfaffenwaldring 9, 70569 Stuttgart, Germany
e-mail: peter.eberhard@itm.uni-stuttgart.de

F. Spreng
e-mail: fabian.spreng@itm.uni-stuttgart.de

D. Biermann and F. Hollmann (eds.), *Thermal Effects in Complex Machining Processes*, Lecture Notes in Production Engineering,
DOI 10.1007/978-3-319-57120-1_8

processes, e.g. the need for frequent remeshing, due to the challenging setup including, among other things, severe changes in the system topology and a continuous generation of new surfaces. As an alternative, meshless methods like Smoothed Particle Hydrodynamics (SPH) [2] naturally allow to describe material separation and large deformations. For this reason, the promising approach of modeling the process of metal cutting with the meshless Lagrangian discretization technique SPH is both presented and thoroughly discussed in this paper.

2 Smoothed Particle Hydrodynamics for Solid Continua

In this section, both the basic discretization procedure of SPH and its application to the Euler equations are discussed. Afterwards, the extensions to the original SPH solid scheme that are necessary to model the process of orthogonal metal cutting in a correct way are presented. All numerical results shown in this paper have been obtained using the particle simulation package Pasimodo [3].

2.1 Basic Equations of Continuum Mechanics

The well-known Euler equations [4], i.e. the acceleration equation

$$\frac{\mathrm{d}\mathbf{v}}{\mathrm{d}t} = \frac{1}{\rho}\nabla \cdot \boldsymbol{\sigma} + \mathbf{g} \tag{1}$$

and the continuity equation

$$\frac{\mathrm{d}\rho}{\mathrm{d}t} = -\rho \nabla \cdot \mathbf{v}, \tag{2}$$

allow to numerically describe the behavior of a solid continuum. Here, $\mathbf{v}$ denotes the velocity vector, t the time, ρ the density, $\boldsymbol{\sigma}$ the Cauchy stress tensor, and $\mathbf{g}$ the vector of volume forces per unit mass. For solving this system of partial differential equations (PDEs) in time and space, the meshless Lagrangian spatial discretization technique SPH [5, 6] can be used. Originally developed for the purpose of investigating astrophysical phenomena, the SPH simulation method also enables a conversion of the PDEs of solid continuum mechanics (1) and (2) into ordinary differential equations (ODEs) in time [7]. The resulting ODEs then can be numerically integrated with usual time integration schemes.

For the transformation process from PDEs to ODEs, a cluster of abstract particles is introduced. Each SPH particle represents a certain part of the modeled solid domain and, accordingly, carries all of its properties, such as its density and mass. Using these SPH particles and a so-called kernel function W, which defines to what extent a neighboring particle j contributes to the properties of the particle of interest named i, the acceleration equation (1) and the continuity equation (2) read

$$\frac{d\mathbf{v}_i}{dt} = \sum_j m_j \left(\frac{\boldsymbol{\sigma}_i}{\rho_i^2} + \frac{\boldsymbol{\sigma}_j}{\rho_j^2} \right) \cdot \nabla_i W(\mathbf{r}_i - \mathbf{r}_j, h) + \mathbf{g} \tag{3}$$

and

$$\frac{d\rho_i}{dt} = \sum_j m_j (\mathbf{v}_i + \mathbf{v}_j) \cdot \nabla_i W(\mathbf{r}_i - \mathbf{r}_j, h), \tag{4}$$

where m represents the mass, $\mathbf{r}$ the position vector, and h the smoothing length. The smoothing length determines at what distance from the reference particle's position the kernel function is cut off in order to reduce the computational effort.

To complete the system of governing equations of continuum mechanics, the Mie-Grüneisen equation of state [8] is employed. Furthermore, a Gaussian distribution with a smoothing length of 1.7 times the initial particle spacing serves as a kernel function. In addition, both the artificial viscosity [2] and the artificial stress [9] correction technique are applied to the SPH Euler equations (3) and (4); the chosen parameter values are $\alpha = 1.0$, $\beta = 2.0$, and $\varepsilon = 0.15$.

When intending to also investigate the thermal aspects of the orthogonal cutting problem, the energy equation, which describes the temporal evolution of a particle's specific internal energy, as well as Fourier's law, which allows to reproduce the flux of thermal energy over time for solid continua in a simulation, are to be considered, too [10].

2.2 *Plasticity and Fracture Behavior*

During the cutting of metal, both severe plastic deformation and failure of the processed material are seen, and they are of paramount importance to the overall dynamics of the cutting system. To properly model these phenomena in numerical simulations, the basic equations of solid continuum mechanics are to be extended by more or less sophisticated plasticity and fracture models.

One material model that is commonly applied in the context of cutting simulations is the Johnson-Cook (JC) flow stress model [11] in combination with the associated damage model [12]. This enhanced plasticity and damage model takes into account the influences of the current levels of the actual plastic strain, of the strain rate, and of the temperature on the response of the material. For this purpose, the elastically calculated deviatoric part of the stress tensor is to be reduced in order to bring it back onto the yield surface being provided by the employed plasticity model.

Besides, a possible decrease in the existing stress level due to material damage is considered by replacing the original, i.e. undamaged, stress tensor of an SPH particle by its damaged equivalent. The related damage parameter is allowed to grow under both tensile and compressive loading until total damage is observed and the simulated domain of material is no longer able to take any load. Further details concerning the implementation of the JC plasticity and fracture model into an SPH-based simulation environment are to be found in [13].

2.3 Boundary Contact Force

A common way to model rigid boundaries in SPH is to replace them by triangular surface meshes, which are used to prevent particles from passing them by applying penalty forces. For the cutting simulations presented in this paper, the modified Lennard-Jones penalty force function introduced in [14] is employed. The stiffness scaling factor of the model has a value of $k = 2.1 \times 10^7$, the maximum distance up to which a contact between an SPH particle and a triangle occurs is equal to the initial particle spacing, and only the repulsive terms of the force law are applied. Moreover, this penalty approach provides the possibility to model friction based on the difference in the interaction partners' tangential velocities. Here, a value of $\mu = 0.5$ is used for the boundary viscosity control parameter of the model.

2.4 Variable Smoothing Length

To achieve an exclusively fracture-model-driven damage behavior of the simulated solid matter even in case of large deformations of the modeled structure, the smoothing length of any SPH particle must be allowed to vary with increasing simulation time. Otherwise, when using a fixed smoothing length h, an increasing distance between adjacent particles might result in a loss of contact and, thus, a numerically induced material failure. In case of solids, the so-called ASPH algorithm [15] provides a proper remedy for this issue. Here, the average distance of any particle to all of its neighbors is calculated and, subsequently, used to correct its smoothing length. This is done by comparing the current value of average distance with a user-defined target. In contrast to the original algorithm proposed in [15], we do not treat the target value as a user-specified one and apply the same single value to all particles of an SPH simulation, but the initial average distance is determined separately for each particle in a preprocessing operation. That way, the ASPH extension becomes more flexible and convenient.

When allowing for a time-dependent variation in the smoothing length, one has to think about how to deal with the interaction between SPH particles of different size. The strategy that proved to provide the most stable results among the several

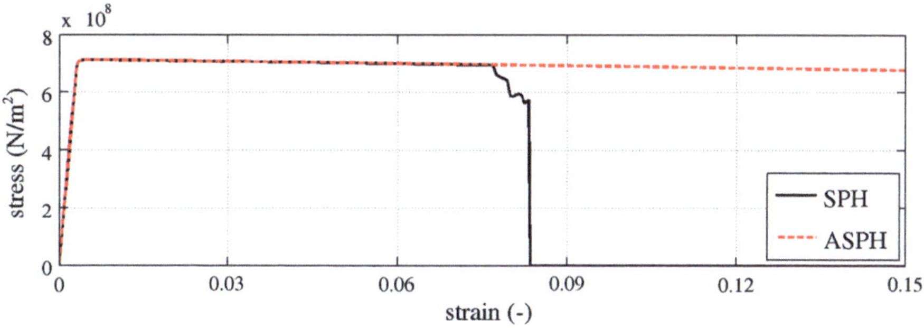

Fig. 1 Stress-strain curves for the simulated tensile specimen [17]

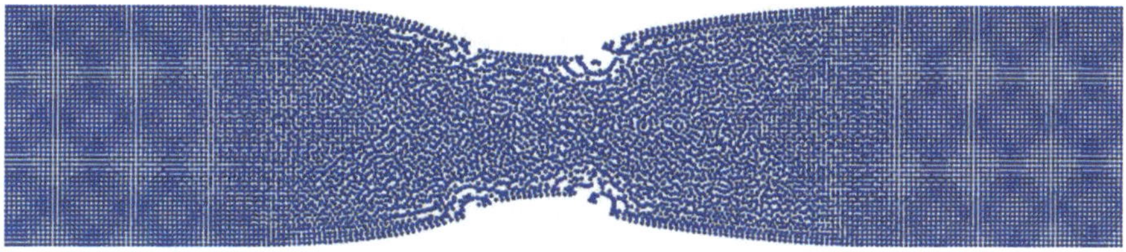

Fig. 2 Particle distribution for the tensile specimen without ASPH extension

available approaches, is to use the averaged smoothing length $h_{ij} = 1/2\,(h_i + h_j)$ for the evaluation of both the kernel function and its derivatives [16].

The benefit gained from the ASPH algorithm becomes obvious when modeling the standardized test setup of a uniaxial tensile test with and without this extension. While the severe drop in the relationship between stress and strain that is found for the SPH solid formulation with a fixed smoothing length (solid line) reveals the occurrence of a loss of contact between neighboring particles and, thus, a numerical fracture in case a critical strain is exceeded, this effect is not observed for the curve obtained with the modified method (dashed line) shown in Fig. 1. When the critical strain value is reached, the particles experiencing the highest strain fall short in terms of adjacent interpolation points for the original formulation and, hence, become decoupled from each other as can be seen from Fig. 2.

2.5 Local Adaptive Particle Refinement

One major disadvantage of the SPH particle method in comparison with mesh-based techniques such as the FEM, is its higher computational effort. In order to mitigate this problem, adaptive resolution schemes for SPH have been developed, see e.g. [18]. They allow for a reduction of the computational cost of an SPH simulation by locally adapting the spatial resolution in all parts of a model to the required level. Accordingly, a finer discretization by particle splitting is provided in the critical areas of the simulation domain. To that end, the particle refinement strategy proposed in [13] consists of three steps.

First of all, the decision whether to refine a particle or not is taken based upon a recurring evaluation of a user-defined adaptivity criterion. In case that the checked criterion is met, the SPH particle is split into several smaller ones according to a prescribed refinement pattern. Finally, the state variables of the additional, smaller particles are to be calculated and any history-dependent data is to be transferred from the old to the new ones. A detailed description of both the developed variable resolution technique as well as its applicability to different kinds of simulation setups covering fluid dynamics and solid mechanics is provided in [13].

3 Orthogonal Metal Cutting Simulation

In the second half of this chapter, the capability of the introduced enhanced SPH solid discretization technique to reproduce the behavior of a real machined workpiece in simulations regarding chip formation, obtained stress distribution, as well as resulting cutting force is discussed. First, the model setup used for the orthogonal machining simulation in Pasimodo is presented. Second, the obtained results are analyzed in detail and compared with experimental data.

3.1 Model Setup

If the width of cut is by far larger than the depth of cut for the considered cutting process, it can be assumed that a two-dimensional state of stress exists in the workpiece. For this reason, the machining setup becomes a planar problem and, hence, it is sufficient to employ a two-dimensional model of the cutting specimen [19]. In both the conducted experiments and the performed simulations, the workpiece to be machined has a length of 1.15×10^{-2} m and a height of 3×10^{-3} m. The applied depth of cut is 6×10^{-4} m while the cutting speed is 1.6 m/s. Due to its wide application in mechanical engineering, the workpiece material is steel AISI 1045. One proper set of JC parameter values for this type of steel is listed in Table 1. Here, A, B, C, n, and D_1 to D_4 are empirically determined model coefficients, and $\dot{\varepsilon}_{\mathrm{pl},0}$ denotes the reference equivalent plastic strain rate.

The SPH particles used for discretizing the metal workpiece are initially positioned on a grid consisting of square cells with an edge length of 5.58×10^{-5} m. In addition, five layers of particles being completely fixed in space are attached at the lower edge of the model, representing the real clamp. As a result of this, the overall particle number is approximately 12,000 in the initial configuration.

Table 1 Parameters for the JC model

Parameter	Value	Unit
A	5.53×10^{8}	N/m^2
B	6.00×10^{8}	N/m^2
C	1.34×10^{-2}	–
n	2.34×10^{-1}	–
$\dot{\varepsilon}_{pl,0}$	1.00×10^{-3}	1/s
D_1	0.06	–
D_2	3.31	–
D_3	−1.96	–
D_4	1.80×10^{-3}	–

Fig. 3 Cutting tool geometry

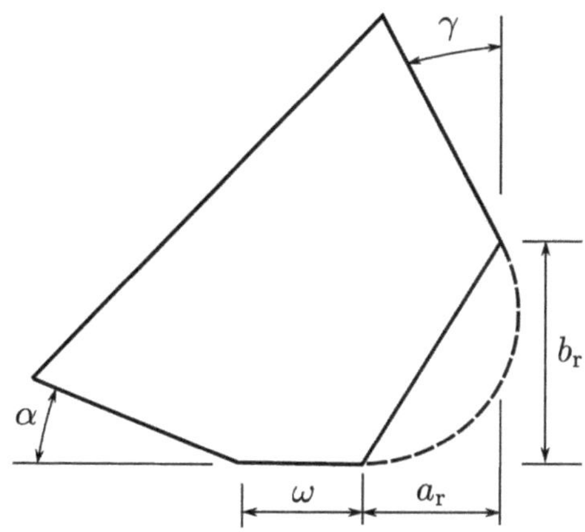

As the shape of the cutting tool edge directly influences the obtained results of both experiment and numerical simulation, it is of prime importance to use a tool model which enables to reproduce the conditions of a real tool up to a satisfying level. Due to wear and limitations in the manufacturing process, real tools do not show the ideal geometry of a wedge shape with a sharp edge [20]. One model that takes into account the imperfections of a slightly worn tool is provided in [21]. The proposed tool geometry is depicted in Fig. 3. For the simulations presented in the following section, the rake angle is $\gamma = 8°$ and the clearance angle is $\alpha = 4°$, respectively. The remaining model parameter values are $\omega = 1 \times 10^{-4}$ m and $a_r = b_r = 5 \times 10^{-5}$ m.

3.2 Results

First, the SPH simulation of an orthogonal cutting process with a fixed particle size, i.e. an invariant smoothing length h of the particles, is considered. As can be seen

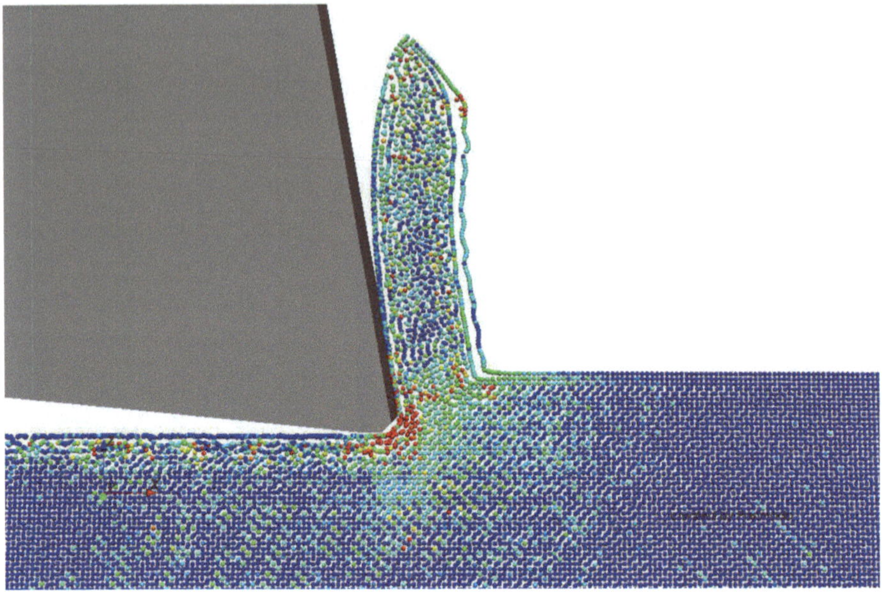

Fig. 4 Distribution of the von Mises equivalent stress in the workpiece (*red* high stress; *blue* low stress) [22]

from Fig. 4, a continuous chip is formed, and the highest values of von Mises equivalent stress are found next to the cutting tool edge and in the primary shear zone, which is indicated by the red color of the particles. Around these zones, one sees SPH particles of yellow and green color, i.e. of mid-level equivalent stress, whereas the far-off particles show a low stress. For this reason, high gradients in stress exist in the vicinity of the cutting edge of the tool, requiring a higher spatial resolution of these particular parts of the simulation model in order to be able to resolve the abovementioned high gradients in stress in a sufficient manner. The previously introduced local adaptive refinement algorithm allows for such an adaptation in particle size at runtime.

If a stress-based refinement criterion together with a square particle positioning pattern is applied, the distribution of SPH particles for the investigated orthogonal machining scenario that is shown in Fig. 5 is found. As expected, the load-carrying particles located in the different shear zones are adaptively refined, i.e. replaced by smaller ones (red color), when necessary, while the particles located outside the critical areas of the model keep their initial size throughout the entire simulation (blue color). A thorough discussion on the potential of machining simulations using adaptive SPH is provided in [22].

In Fig. 6, the cutting forces obtained from both experiment and numerical simulation are contrasted. As can be seen, the experimental curve (solid line) is characterized by a steep increase following the impact between cutting tool and workpiece. With increasing simulation time, the fluctuations in the cutting force

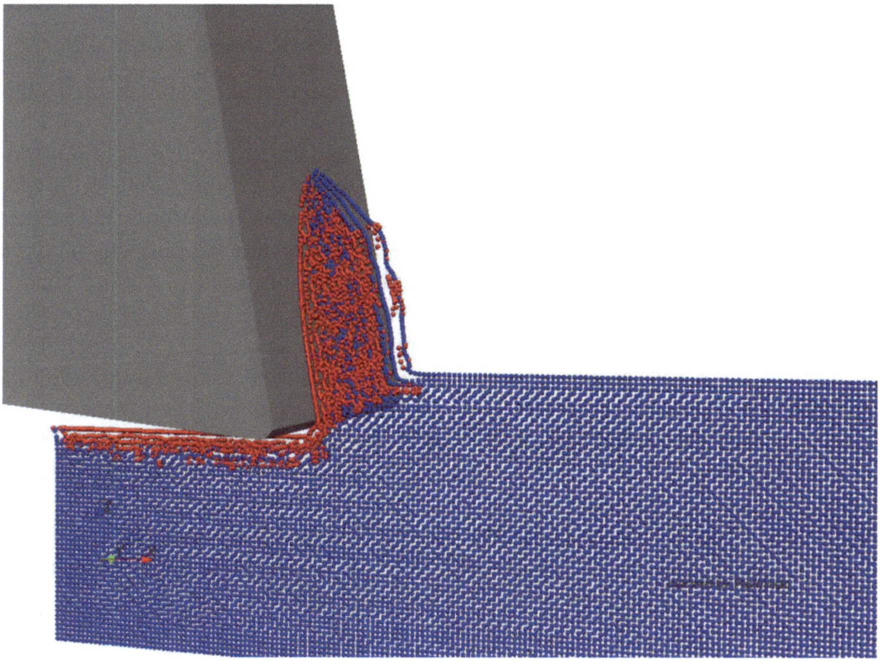

Fig. 5 Particle distribution for the adaptively refined workpiece (*red* small particle; *blue* large particle) [13]

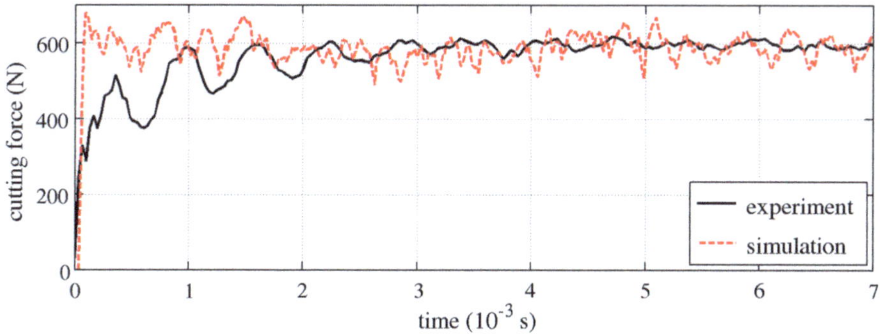

Fig. 6 Comparison of the experimental and the simulated cutting force

more and more decrease and a quasi-steady-state at a force level of approximately 600 N is reached. Similar to these experimental results, also the simulated curve (dashed line) shows both a steep rise and a subsequent oscillation around a force value of nearly 600 N. Despite the fact that the simulated cutting force provides a slightly wider range of variation compared with the experimentally measured one, they are still in good agreement.

4 Conclusion

It was shown in this contribution that the meshless Lagrangian spatial discretization technique SPH allows for a numerical modeling of real machining processes. First, the SPH discretization procedure in its original formulation as well as its application to the basic equations of solid continuum mechanics were discussed. Following this, its most important extensions being necessary to correctly describe the process of metal cutting in simulations were presented. Here, the JC plasticity and the corresponding fracture model, which are both commonly used in cutting simulations, were introduced. They take into account the influences of the current values of plastic strain, strain rate, and temperature on the response of the material. To describe the interaction between the cutting tool discretized using triangles and the workpiece represented by SPH particles, a modified Lennard-Jones penalty force function, which also allows to model the phenomenon of friction, was employed. In order to avoid the appearance of a numerically induced failure and, thus, ensure a exclusively damage-model-driven fracture behavior of the simulated solid matter also in case of large deformations, the ASPH modification allowing for a variation of the smoothing length of any SPH particle at runtime was further introduced. Finally, a local adaptive refinement scheme was presented, which can be used to improve the accuracy of the obtained results and/or to reduce the required computational cost by dynamically adjusting the particle size.

Then, the ability of the enhanced SPH solid formulation to capture the essential characteristics of real machining processes and, hence, its applicability to the modeling of metal cutting was shown. By taking the example of orthogonal cutting for steel of type AISI 1045, it was demonstrated that the meshless SPH method naturally allows to describe large deformations, the process of material separation, and the generation of new surfaces. So, a continuous chip is formed, and the highest values of equivalent stress appear next to the cutting tool edge and in the primary shear zone. In accordance with the high gradients in stress that are seen in the different shear zones, the load-carrying particles are adaptively refined to enable an improved results accuracy. To further assess the capability of the presented model to reproduce a real machining process in a simulation, also the quantitative values for the cutting force obtained from both experiment and numerical simulation were compared, and the identified level of agreement emphasizes the high quality of the proposed SPH cutting model.

References

1. Zienkiewicz, O.C., Taylor, R.L., Zhu, J.Z.: The Finite Element Method: Its Basis and Fundamentals. Butterworth-Heinemann, Oxford (2013)
2. Monaghan, J.J.: Smoothed Particle Hydrodynamics. Rep. Prog. Phys. **68**, 1703–1759 (2005)
3. Pasimodo. http://www.itm.uni-stuttgart.de/research/pasimodo/pasimodo_en.php

4. Batchelor, G.K.: An Introduction to Fluid Dynamics. Cambridge University Press, Cambridge (2000)
5. Lucy, L.B.: A numerical approach to the testing of the fission hypothesis. Astron. J. **82**, 1013–1024 (1977)
6. Monaghan, J.J.: Simulating free surface flows with SPH. J. Comput. Phys. **110**, 399–406 (1994)
7. Liu, M., Liu, G.: Smoothed Particle Hydrodynamics (SPH): an overview and recent developments. Arch. Comput. Method E **17**, 25–76 (2010)
8. Burshtein, A.I.: Introduction to Thermodynamics and Kinetic Theory of Matter. Wiley-VCH, Berlin (2005)
9. Monaghan, J.J.: SPH without a tensile instability. J. Comput. Phys. **159**, 290–311 (2000)
10. Cleary, P.W., Monaghan, J.J.: Conduction modelling using Smoothed Particle Hydrodynamics. J. Comput. Phys. **148**, 227–264 (1999)
11. Johnson, G.R., Cook, W.H.: A constitutive model and data for metals subjected to large strains, high strain rates and high temperatures. In: Proceedings of the 7th International Symposium on Ballistics, vol. 547, pp. 541–547 (1983)
12. Johnson, G.R., Cook, W.H.: Fracture characteristics of three metals subjected to various strains, strain rates, temperatures and pressures. Eng. Fract. Mech. **21**, 31–48 (1985)
13. Spreng, F., Schnabel, D., Mueller, A., Eberhard, P.: A local adaptive discretization algorithm for Smoothed Particle Hydrodynamics. Comput. Part. Mech. **1**, 131–145 (2014)
14. Müller, M., Schirm, S., Teschner, M., Heidelberger, B., Gross, M.: Interaction of fluids with deformable solids. Comput. Anim. Virtual Worlds **15**, 159–171 (2004)
15. Owen, J.M.: ASPH modeling of material damage and failure. In: Proceedings of the 5th International SPHERIC SPH Workshop, pp. 297–304 (2010)
16. Mueller, A.: Dynamic refinement and coarsening for the Smoothed Particle Hydrodynamics method. Dissertation, University of Stuttgart (2015)
17. Spreng, F., Eberhard, P.: The way to an enhanced Smoothed Particle Hydrodynamics formulation suitable for machining process simulations. In: Proceedings of the 8th International SPHERIC SPH Workshop, pp. 255–262 (2013)
18. Vacondio, R., Rogers, B.D., Stansby, P.K., Mignosa, P., Feldman, J.: Variable resolution for SPH: a dynamic particle coalescing and splitting scheme. Comput. Methods Appl. Mech. Eng. **256**, 132–148 (2013)
19. Jaspers, S.P.F.C.: Metal cutting mechanics and material behaviour. Dissertation, Eindhoven University of Technology (1999)
20. Gaugele, T., Eberhard, P.: Simulation of cutting processes using mesh-free Lagrangian particle methods. Comput. Mech. **51**, 261–278 (2013)
21. Armarego, E.J.A., Brown, R.H.: The Machining of Metals. Prentice-Hall, Englewood Cliffs (1969)
22. Spreng, F., Schnabel, D., Mueller, A., Eberhard, P.: Smoothed Particle Hydrodynamics with adaptive discretization. In: Proceedings of the 9th International SPHERIC SPH Workshop, pp. 192–199 (2014)

Experimental and Simulative Modeling of Drilling Processes for the Compensation of Thermal Effects

P. Bollig, J. Michna, C. Faltin, J. Schneider, F. Zanger, R. Schießl, U. Maas and V. Schulze

Abstract This work focuses on the prediction of phase transformations and shape deviations for drilling of demonstrator workpieces. In a first step, a 2D chip formation simulation was developed with all physical effects. Based on this simulation a simplified workpiece geometry with only one drilling hole was 3D modeled and tested for its predictive capability of phase transformations and shape deviations. Therefore, the feed rate, the rotation of the drilling tool and the material removal were considered for the process kinematics. Using these results the modeling approach was optimized and transferred into a simulation model with a demonstrator workpiece to minimize shape deviations and control phase transformations using different compensation strategies. The simulations were validated by experiments.

P. Bollig · J. Michna · F. Zanger (✉) · V. Schulze
Institute of Production Science, Karlsruhe Institute of Technology, Kaiserstraße 12, 76131 Karlsruhe, Germany
e-mail: Frederik.Zanger@kit.edu

P. Bollig
e-mail: Patrick.Bollig@kit.edu

J. Michna
e-mail: Juergen.Michna@kit.edu

V. Schulze
e-mail: Volker.Schulze@kit.edu

C. Faltin · R. Schießl · U. Maas
Institute of Technical Thermodynamics, Karlsruhe Institute of Technology, Kaiserstraße 12, 76131 Karlsruhe, Germany
e-mail: Carsten.Faltin@kit.edu

R. Schießl
e-mail: Robert.Schiessl@kit.edu

U. Maas
e-mail: Ulrich.Maas@kit.edu

J. Schneider
Institute for Applied Materials – Computational Materials Science, Karlsruhe Institute of Technology, Kaiserstraße 12, Karlsruhe, Germany
e-mail: Johannes.Schneider@kit.edu

D. Biermann and F. Hollmann (eds.), *Thermal Effects in Complex Machining Processes*, Lecture Notes in Production Engineering,
DOI 10.1007/978-3-319-57120-1_9

1 Introduction

Nowadays, the demand for pollution free and fast manufacturing processes is continuously increasing and makes dry machining a promising alternative [6]. The problems with dry machining are the high surface temperatures causing phase transformations or shape deviations due to the lack of coolant. To overcome these issues compensation strategies to minimize the temperatures by optimized machining strategies [1], cooling methods [18] and tool path correction [7] have been taken into account. These strategies were applied in turning experiments that were carried out and a 2D chip formation simulation as a first step towards modeling thermal effects during drilling was developed. Additional drilling experiments have been used in combination with turning experiments and 2D simulation in order to model the thermal and mechanical load in a simplified 3D drilling and finally in a demonstrator workpiece simulation model. At the end of the investigations, the mentioned compensation strategies were defined in the simulation to reduce shape deviations and phase transformations.

2 Experimental Investigations

In the following sections, the experimental set-ups and testing methods for the development and validation of the simulation models are introduced.

2.1 Orthogonal Turning Experiments

In this work turning experiments were used pursuing two different goals. The first goal was to validate the 2D simulation model and the other objective included the identification of thermal and mechanical loadings caused by different cutting and tool parameters.

The turning experiments were performed on an INDEX V100 vertical lathe with parallel geometry. For force measurements a Kistler type 9256 dynamometer was used. Temperature measurements were performed with a two color pyrometer (type FIRE II). To measure the temperature close to the cutting zone, the turning inserts were drilled from the backside to the rake face with a drilling hole diameter of $d = 300$ μm using electrical discharge machining (EDM). Using the small EDM drilling hole a glass fiber was placed in such a way, that the temperatures could be measured directly on the surface of the workpiece. These uncoated rhombical carbide turning inserts had a clearance angle of $\alpha = 7°$ and wedge angle of $\beta = 83°$ with a cutting edge radius of $r_\beta = 30$ μm.

In order to try different rake angles, a special tool holder was used to adjust the angle range from $\gamma = -5°$ to $+5°$. In addition, the tool holder was positioned

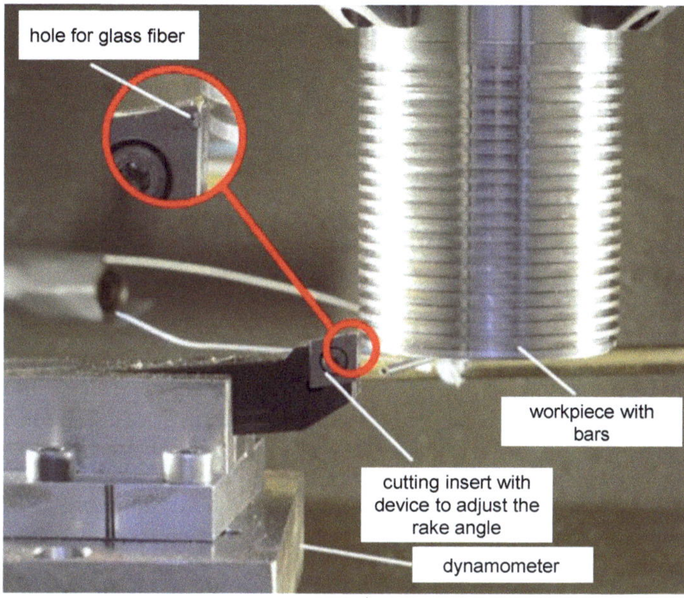

Fig. 1 Experimental set-up for the turning experiments

orthogonal to the workpiece. For the comparison with the 2D chip formation simulation a special type of workpiece was needed to be as close as possible to an orthogonal cutting process. Due to that reason cylindrical workpieces of the quenched and tempered material AISI4140 QT (42CrMo4) were primed with 1.5 mm thick webs, see Fig. 1. The experimental set up for the turning experiments and more detailed information can be found in [9, 13].

Results of Orthogonal Turning Experiments

During the turning experiments the parameters cutting speed (v_c = 75 m/min up to 400 m/min), feed (f = 0.05 mm to 0.15 mm) and rake angle (γ = −5° to +5°) were varied. These parameters were selected based on the parameters chosen for the drilling experiments (Sect. 2.2) to model the thermal and mechanical load on the main cutting edge for the 3D drilling simulation.

Force Measurement

The specific cutting forces for varying cutting speeds are shown in Fig. 2. This figure demonstrates a small decrease for the specific cutting force for higher cutting speeds. Such behavior can be observed for all feed rates and rake angles. A rise in feed rate leads to a lower specific cutting force level. For every feed, a minimum of the specific cutting force can be found for a rake angle of γ = +2°.

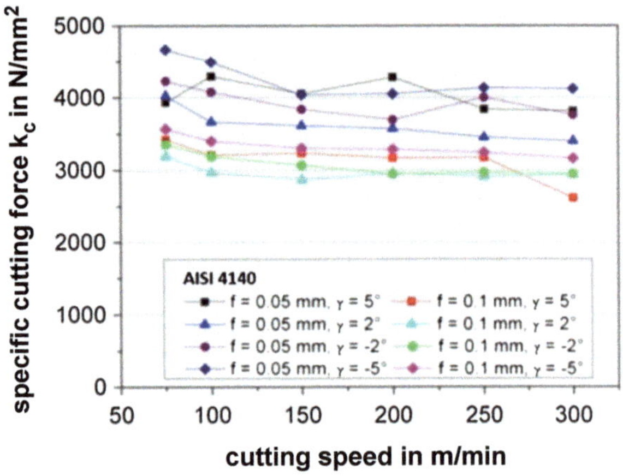

Fig. 2 Specific cutting force in dependence of the cutting speed for varying feed rates and rake angles [9]

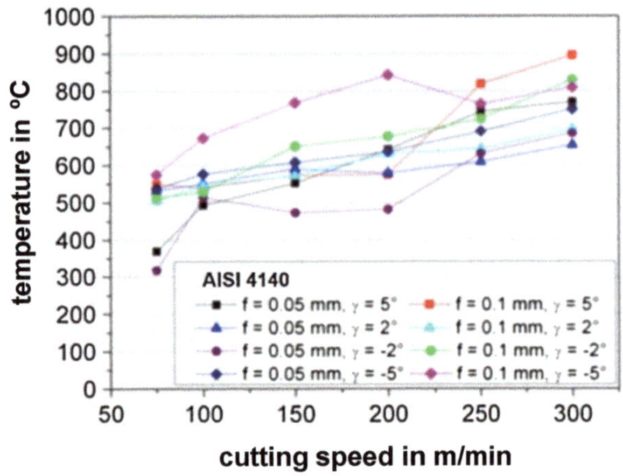

Fig. 3 Temperature in dependence of the cutting speed for varying feed rates and rake angles [9]

Temperature Measurement

In Fig. 3, the temperature measurements are plotted over the cutting speed. The results reveal a significant increase in temperature for the cutting speeds from $v_c = 75$ m/min up to 300 m/min. The influence of the feed shows a higher temperature level for the feed rate of $f = 0.1$ mm compared to $f = 0.05$ mm. For the rake angles, a minimum is found for $\gamma = 2°$ at a feed of $f = 0.1$ mm and for $\gamma = -2°$ at a feed of $f = 0.05$ mm.

Metallurgical Investigations—Characteristics of the Surface

During the austenitization of austenitizing steels, such as AISI4140 QT, high temperatures can lead to microstructural changes in the surface layer of the workpiece. When the transformation temperature to austenite is exceeded, critical cooling rates enable the formation of white layers. Within the turning experiments

Fig. 4 Surface at a cutting speed of 300 m/min, feed rate 0.05 mm and rake angle of 5° [9]

Fig. 5 Vickers hardness change in the microstructure for the cutting speeds 300 m/min and 400 m/min at a feed of 0.05 mm and a rake angle of 5° [9]

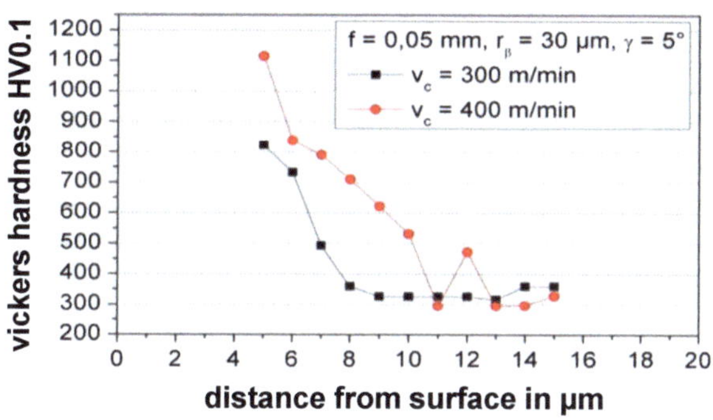

the results show a distinct influence of the cutting speed for the formation of white layers. Figure 4 illustrates an etched surface for a cutting speed of $v_c = 300$ m/min at a feed of $f = 0.05$ mm and a rake angle of $\gamma = 5°$. The whole surface demonstrates a white layer formation with a thickness of about 5 µm.

For a lower cutting speed of $v_c = 75$ m/min, no white layer formation appeared. The reason for that constitutes in the reduced temperatures during the turning process. Only for some samples a small change in microstructure was observed for the cutting speed of $v_c = 75$ m/min. A reason for this could be the increased cutting forces at low cutting speeds or the bad relation between the cutting edge radius and the cutting depth [9].

Successive measurements of Vickers hardness were carried out on the microstructural transformed surface. Figure 5 depicts the hardness for exemplary surfaces when using a cutting speed of $v_c = 300$ m/min and $v_c = 400$ m/min. It shows a clearly higher hardness 5 µm apart from the surface for both cutting speeds of $v_c = 400$ m/min (1150 HV) and $v_c = 300$ m/min (822 HV) as a sign of a martensitic microstructure.

2.2 Drilling Experiments

For modeling the thermal and mechanical load of a drilling process, drilling and turning experiments as well as 2D simulations (Sect. 3.1.4) were performed. The drilling experiments in this work were separated into two categories and carried out on a machining center made by HELLER of the type MC16.

The first category included drilling of ring shaped workpieces to determine the thermal and mechanical load along the main cutting edge, see Fig. 6. A small drilling hole with a diameter of 1 mm was drilled into this ring to measure defined points along the main cutting edge by inserting a thin glass fiber into the EDM drilled hole. By varying the size of the ring diameter as well as the adapted position of the hole, the measured point on the cutting edge was determined. In total three points along the main cutting edge were chosen to specify the load of a drilling tool with a diameter of 12 mm. The chisel edge of the drilling tool was not taken into account. According to these three points along the main cutting edge, the thermal load could be mathematically described. During these drilling experiments the cutting forces were also measured to approximate a derived mechanical load equation for the simulation model.

The second category used solid cylindrical test samples with one skewed side for the identification of the thermal load along the minor cutting edge, see Fig. 7. In total, four holes on the circular face of the workpiece have been manufactured, one below the other orthogonal to the cutting direction. Hence, the temperatures can be

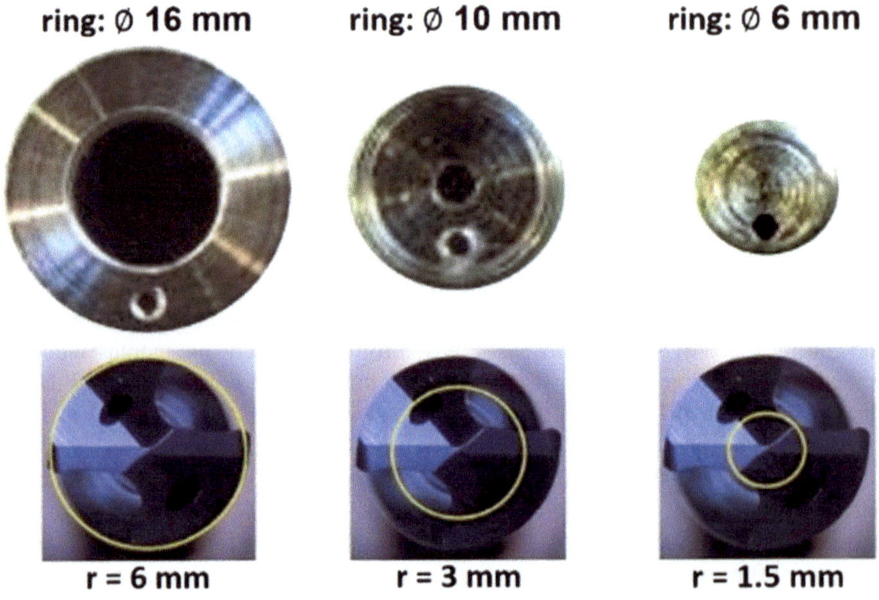

Fig. 6 Test samples of rings and their assignment to the main cutting edge [9]

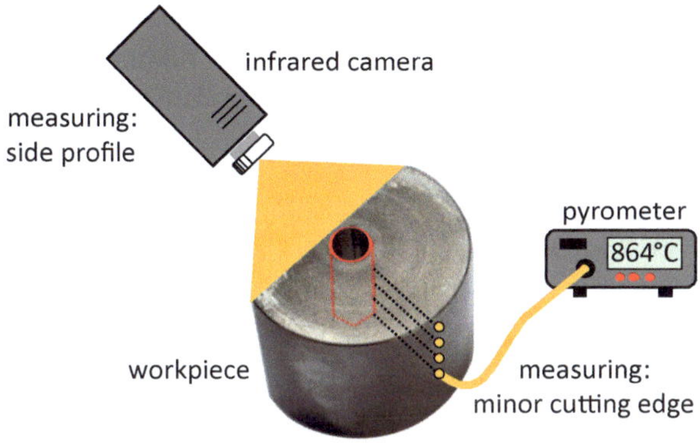

Fig. 7 Infrared and pyrometer measurement of the skewed workpiece

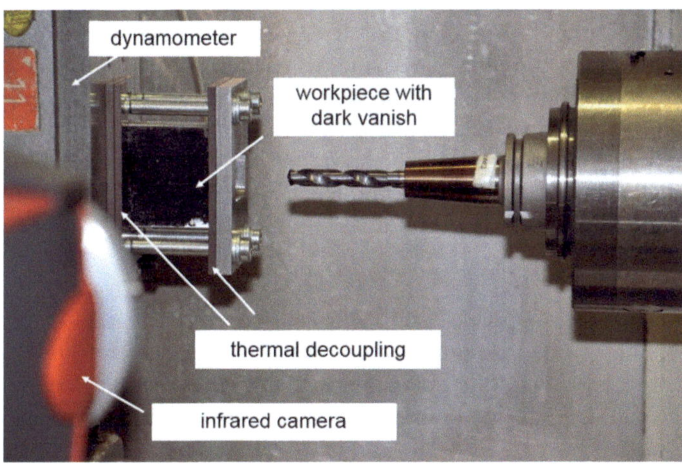

Fig. 8 Experimental set-up for the infrared camera and force measurement [9]

measured in 4 drilling depths (6, 12, 18 and 24 mm) for different cutting speeds and feed rates [9, 15].

For the dry drilling tests, the EDM-holes end at the later wall surface of the drilling hole. For the MQL tests, however, it has been ensured that the EDM-holes end close to the surface wall with an interspace of 15 μm to protect the glass fiber from the MQL. It was assumed that the temperature change between the real temperatures on the minor cutting edge is small compared to the measure points with an interspace of 15 μm [5].

The skewed side was used to measure the temperature distribution in the workpiece close to the wall surface of the drilling hole with a high speed infrared camera (Type ImageIR3300 from Infratec). To minimize the perturbation of the infrared images the skewed side was darkened with black varnish and the test samples were thermally decoupled from the clamping system, see Fig. 8.

Drilling torque and feed force were determined by force measurements of a Kistler multicomponent dynamometer type 9215A for the minimum-quantity-lubrication (MQL) and with a Kistler type 9255 for the dry experiments.

Results of the Drilling Experiments

The investigated parameter field for the drilling experiments included the cutting speeds of $v_c = 30, 75, 100, 150, 200$ and 300 m/min at a constant feed of 0.1 mm. This feed was chosen to compare the turning experiments and 2D chip formation simulations. All experiments were performed with the aim of developing a loading spectrum for the 3D drilling simulation.

Force Measurement

Defined points along the main cutting edge were taken into account to measure drilling forces. Figure 9 shows the results of the measured feed forces at the radius of the points r = 1.5 mm, r = 3.0 mm and r = 6.0 mm of the tool. The feed force at the position of r = 1.5 mm demonstrates a decreasing behavior from the cutting speed 75 m/min to 300 m/min. The same tendency can be seen for the position of r = 3.0 mm. At the cutting edge corner at r = 6.0 mm an increasing behavior from slow to higher cutting speeds is recorded. When the measure points got closer to the chisel edge of the drilling tool, an increase in feed force was measured. A reason for this rise could constitute in a squeezing effect between the tool and the workpiece based on inappropriate cutting conditions in this area.

Temperature Measurements

The temperature measurements using ring shaped workpieces to detect the temperature evolution at defined points at the main cutting edge are shown in Fig. 10. As expected, a higher cutting speed leads to a higher thermal load at the main cutting edge and consequently in the workpiece. This is caused by a higher number

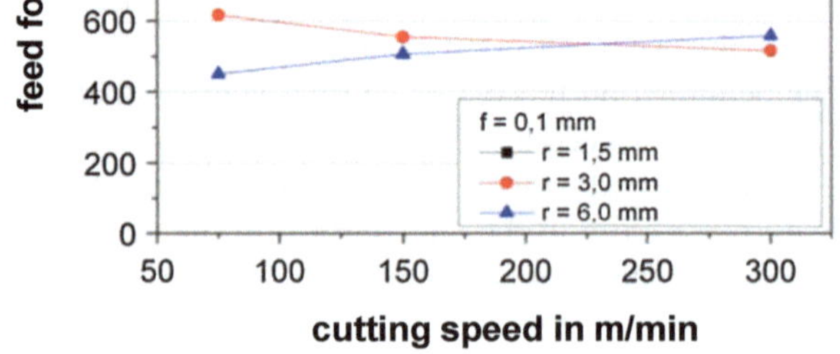

Fig. 9 Feed force measured in different positions along the main cutting edge [9]

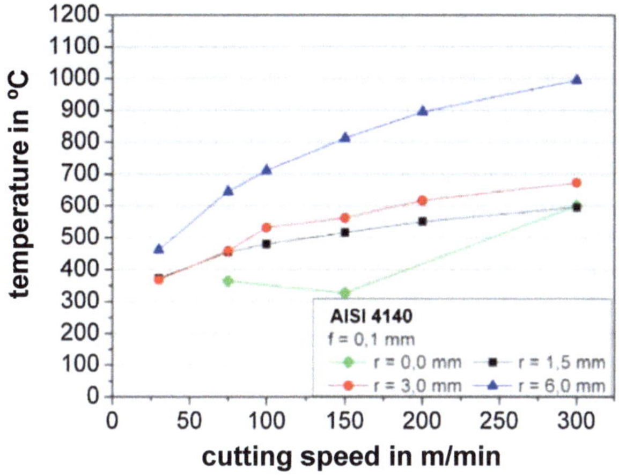

Fig. 10 Temperature in relation to the cutting speed for varying position on the main cutting edge [9]

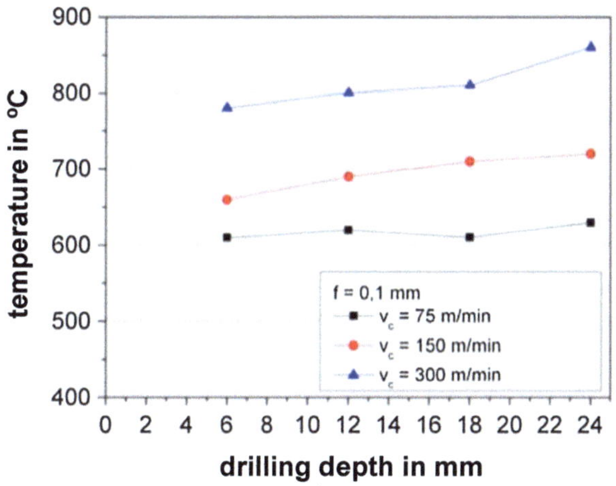

Fig. 11 Temperature depending on the drilling depth [9]

of cuttings according to the higher rotational speed of the tool. The crossing section at r = 6.0 mm between main and minor cutting edge is the most thermally loaded area of the drilling tool. With a smaller radial position of the measuring point the temperature level is decreasing. Figure 10 additionally shows a non-linear rising of the temperature for increasing cutting speeds.

The temperature measurements of Fig. 7 using the pyrometer for different drilling depths are presented in Fig. 11. The temperature consistently increases with higher cutting speed. The temperature rises for each cutting speed and also increases with the drilling depth. The transition temperature was only exceeded for some cutting speeds at defined drilling depths.

In a final step infrared images (Fig. 12) were taken on the skewed side of the workpiece from Fig. 7 to gain information on the spatial and temporal temperature distribution during drilling. Figure 12 (1) depicts the first contact of the tool with

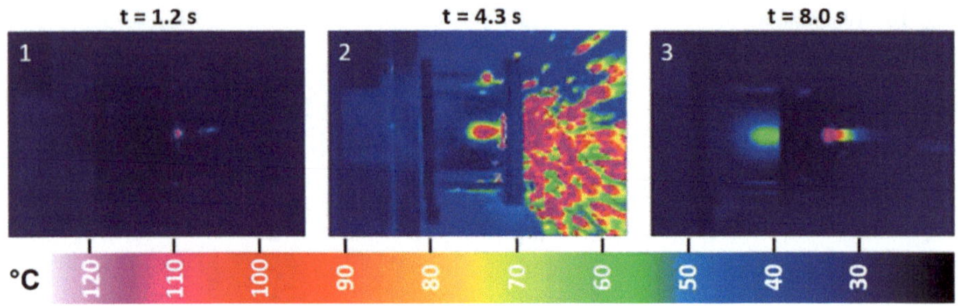

Fig. 12 Infrared images at different times during the drilling process [9]

the workpiece and Fig. 12 (2) demonstrates the reached drilling depth of 30 mm during the drilling process. The cooling of the workpiece is finally represented in Fig. 12 (3) [9].

Metallurgical Investigations—Characteristics of the Surface

Metallurgical investigations were executed on the surface wall of the drilling hole. The results already show the beginning of a white layer formation at a drilling depth of 100 µm for cutting speeds higher than v_c = 300 m/min. In some of the investigated white layers cracks in normal or tangential direction could be detected according to Fig. 18. The thickness of the white layer was measured for cutting speeds lower than v_c = 300 m/min of about 4–6 µm but drastically increased up to 21 µm at a cutting speed of v_c = 300 m/min. As for the turning experiments, an increase in hardness was measured up to three times as high as from the non-transformed microstructure to the white layer transformation [9].

2.3 *Investigation of Friction*

The friction tests in this work focus on developing a friction model for FEM-simulations depending on the temperature and the sliding speed between the tool and the chip. The normal force, the coating, the contact geometry and the cooling were tested as well but were not taken into account for the simulation model in this work.

The experimental investigations were carried out on a laboratory tribometer of the type PLINT TE92HS, see Fig. 13. Basically, this type of laboratory tribometer is designed to perform ball-on-disc tests with changeable materials for the disc and the ball under varying sliding speeds and normal pressures. Ball-on-disc tests are disadvantageous because of the different geometrical shapes, the unrealistic temperature development and the plastic deformation compared to the real tool–workpiece combinations. This fact made it necessary to analyze and compare the measured friction coefficients from ball-on-disc tests with insert-on-disc tests.

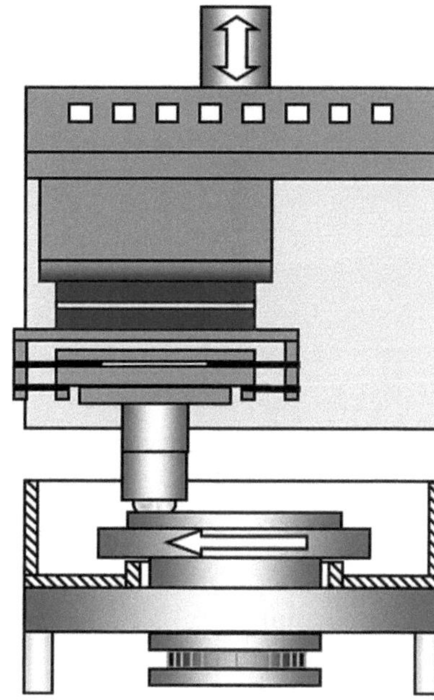

Fig. 13 Laboratory tribometer (schematically)

The friction insert-on-disk tests were performed on a laboratory tribometer of the type CETR UMT3. To overcome an uncertainty of the temperature measurement position two different positions, inside the ball and a direct comparison between ball and insert had been executed.

Each friction test was divided into 100 load cycles including the steps "sliding speed is increased from 0 to "a" m/s (with "a" varying from 1 to 10) within 5 s", "holding for 5 s" and "decreasing again to 0 m/s", see Fig. 14. This amount of repetitions was needed to eliminate roll-in effects. The first 20 load cycles were not taken into account for the analysis of the friction coefficients, to eliminate roll in effects.

Results of the Friction Experiments

Results of friction tests for ball-on-disc and tool-on disc tests are presented. In Fig. 15, the first friction coefficient plot illustrates the differences between a ball and a real cutting tool. The friction in case of the tool remains on a higher level and does not show high influences according to changes in temperature or sliding speed. However, the ball test mechanism has a great influence on the temperature and the sliding speed. Both tests were performed in dry condition with a normal force of 20 N.

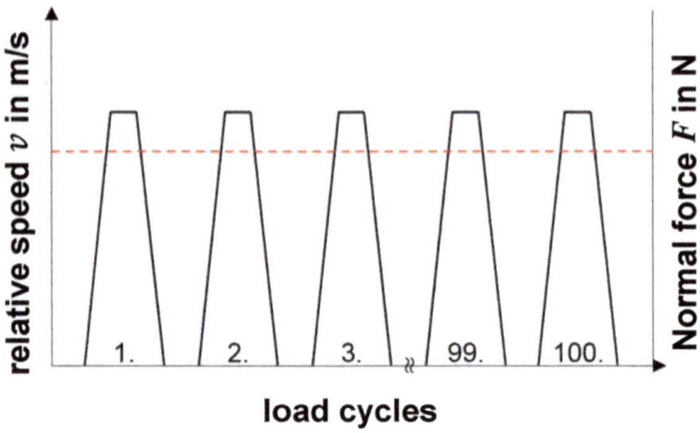

Fig. 14 Load cycles during friction testing

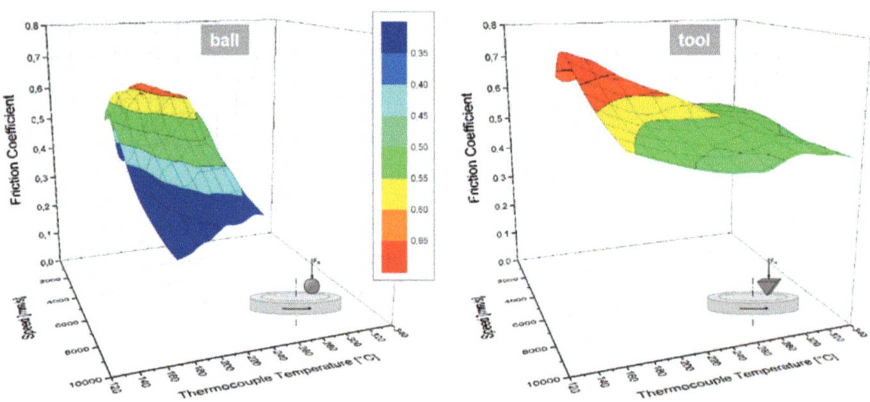

Fig. 15 Comparison of ball-on-disc (*left*) and tool-on disc (*right*) for a normal force of 20 N for AISI4140 QT

In Fig. 16 the influence of the normal force F_N is shown for the tool-on-disc test. The increase in normal force leads to a higher contact pressure resulting in higher temperatures. The lowest measured temperatures ranging from $F_N = 20$ N to $F_N = 40$ N differ in about ΔT = 60 °C. An interesting fact is that the friction plot for $F_N = 20$ N seems to be shifted to roughly ΔT = 60 °C compared to F_N = 40 N. So both normal forces show the same dependence on the temperature and the sliding speed.

Closing the friction investigations, Fig. 17 shows the difference of dry tool-on-disc tests and air cooled tool-on-disc tests. The air cooling leads to higher

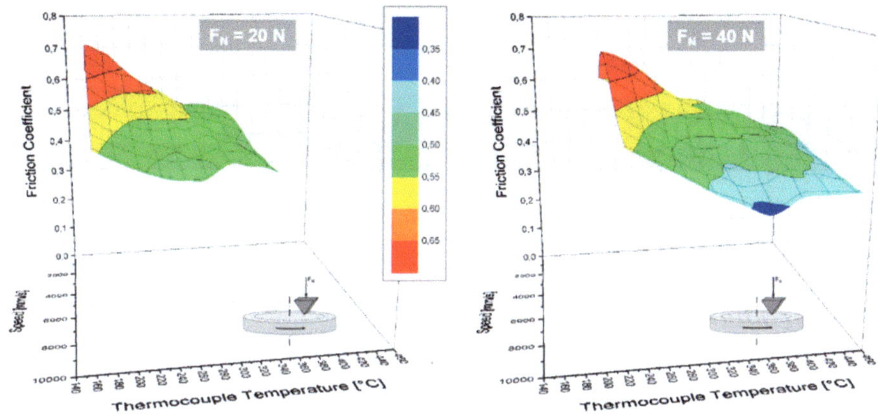

Fig. 16 Friction coefficients for different normal forces for dry testing conditions for AISI4140 QT

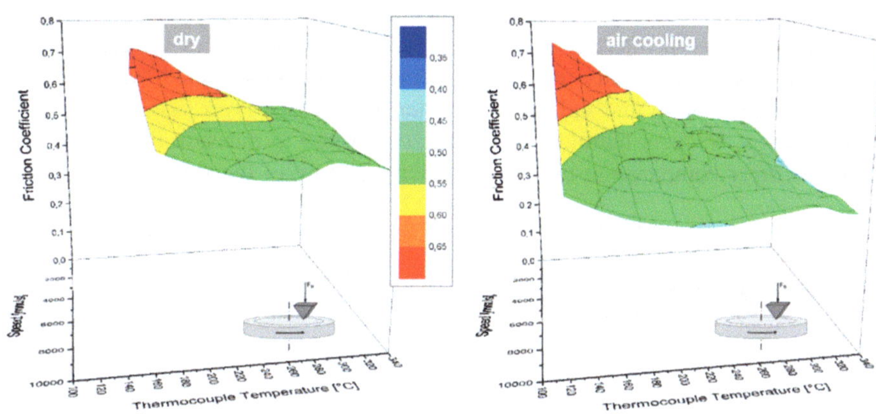

Fig. 17 Friction coefficients for dry and air cooled testing at 20 N contact pressure for AISI4140 QT

friction coefficients at lower temperature but slightly lower friction coefficients at higher temperatures. Also, the friction coefficients for air cooling seem to decrease faster especially for lower temperatures.

Friction measurements for MQL supported insert-on-disc tests are presented in [5].

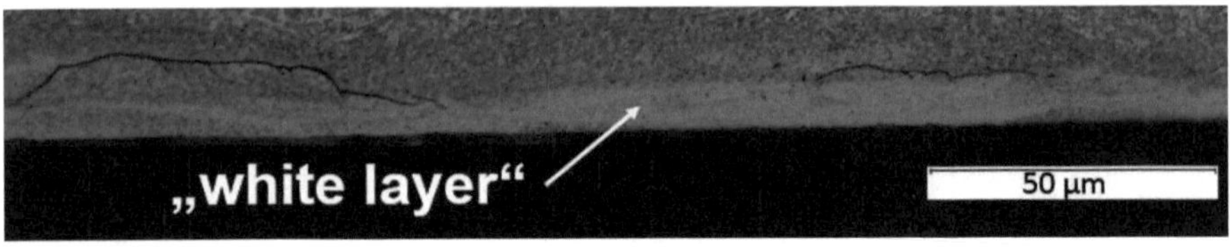

Fig. 18 Exemplary etched cut image with a “white layer” for the drilling parameters v_c = 300 m/min and f = 0.1 mm [9]

2.4 Metallurgical Investigations

Metallurgical investigations were used to characterize the resulting microstructure after the turning and drilling experiments. These investigations were necessary for validating the simulation with real experimental results. As shown in different works [3, 14], the process parameters and their resulting temperature field highly influence the changes in the microstructure. The used characterization methods were light microscopy and hardness testing.

For the light microscopic tests, the turned and drilled test samples were cut into pieces in the areas of interest. After this separation the cut pieces were embedded, etched and analyzed. In case of microstructural changes, so called “white layers” appeared after etching and indicated a martensitic surface layer, see Fig. 18.

In addition, Vickers hardness tests were conducted to measure the hardness of the white layer compared to the present microstructure of the test samples.

3 Simulation Models

In this chapter the different types of simulation models are introduced, based on experimental results, to develop three different methods to compensate shape deviations during drilling of AISI4140 QT. In the following, the material model, the thermo-mechanical load mechanism, the friction model, the phase transformation model and the compensation strategies are presented.

3.1 2D Chip Formation Simulation

The 2D chip formation simulation is used to characterize the cutting process at a micro scale to determine phase transformations, forces and temperatures for the 3D simulation model.

3.1.1 Material Model

The 2D chip formation simulation represents an orthogonal cutting process and was performed using the FEM software ABAQUS/Standard. It was developed for the prediction of phase transformations and residual stresses as well as the deduction of forces and temperature profiles [13].

Machining processes mostly lead to high heating and cooling rates. As a result, the microstructure in the material transforms from one phase to another, each with its own characteristics. The thermally caused strains $\varepsilon(T)$ after the austenitization process had to be implemented in the simulation model with the main aim to predict phase transformations and residual stresses:

$$\varepsilon(T) = f_M \cdot \alpha_{th,M} \cdot T + f_M \cdot \varepsilon_{tr,\gamma \rightarrow M} + f_A \cdot \alpha_{th,A} \cdot T + f_{A,max} \cdot \varepsilon_{th,\alpha \rightarrow \gamma} + f_{F,min} \cdot \alpha_{th,F} \cdot T \quad (1)$$

The variables are the thermal expansion coefficients of martensite and austenite $\alpha_{th,M}$ and $\alpha_{th,A}$, the transformation strains $\varepsilon_{tr,\gamma \rightarrow M}$ and $\varepsilon_{th,\alpha \rightarrow \gamma}$, f_M and f_A the volume of martensite and austenite, $f_{M,max}$ and $f_{A,max}$ the maximum volume of martensite and austenite and finally $f_{F,min}$ the minimum volume of ferrite.

On the other hand, the volume percentage of martensite can be calculated with the knowledge of the strains $\varepsilon(T)$. The relations in Eq. (1) are described in detail in [9, 13].

To describe the elastic-viscoplastic material behavior, an isotropic yield stress model based on the work of Vöhringer and Voce [19, 21] with a flow criterion by Mises was implemented [9]:

$$\sigma_y = \sigma_0^* \left[1 - \left(\frac{T}{T_o}\right)^n\right]^m + \left[\sigma_{G,0} + (\sigma_1 + \theta_1 \bar{\varepsilon}_p)\left(1 - \exp\left(-\frac{\theta_1 \bar{\varepsilon}_p}{\theta_0}\right)\right)\right] \frac{G(T)}{G(0K)} g(T, T_{\ddot{U}}) \quad (2)$$

For more details, refer to [2, 9, 17]. The material parameters for the quenched and tempered steel AISI4140 QT were mostly temperature and phase state dependent [9].

3.1.2 Friction Modeling

The result accuracy of 2D chip formation simulation is based on the material model, the remeshing routine as well as the friction modeled between the tool and the workpiece. According to experimental results for several combinations of tool and workpiece materials, constant friction coefficients can be used in simulation. However, different relative sliding speeds and temperatures exist in the cutting zone. As shown in this work, the friction coefficient varies with different relative

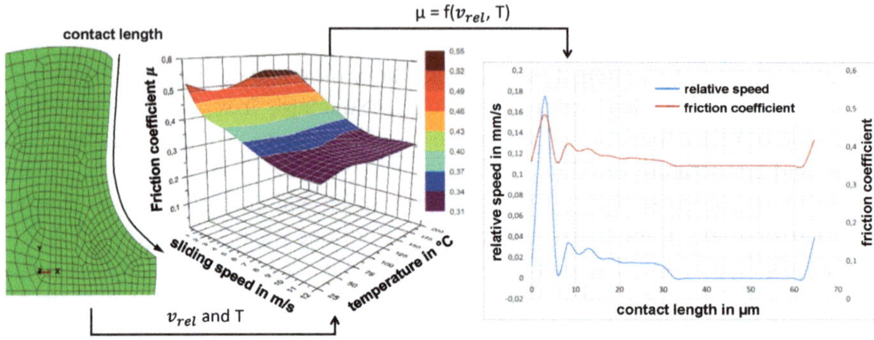

Fig. 19 Modeling of the friction coefficient in FEM-simulation [16]

sliding speeds and temperatures. This makes the assumption of a constant friction coefficient in the cutting zone only an approximation to the reality.

Therefore, a friction model was developed with the aim to take the relative sliding speed and the temperature in the cutting zone into account, see Fig. 19. The dependency of the hydrostatic pressure was neglected, but it is indirectly coupled to the sliding speed.

In a first step the contact length between the tool and the chip was identified for each time step. For every node in the contact tool-workpiece the temperature and the relative sliding speed was extracted and a friction coefficient was assigned to every node with respect to these parameters. The values in Fig. 19 can be implemented as a user subroutine or as a tabular data set. An extrapolation of the friction behavior with a temperature range between 400 and 1200 °C was implemented [9, 16].

3.1.3 Phase Transformations

Many different approaches for the description of phase transformation can be found in literature, e.g. [3, 20]. In this work the validated approach on phase transformation developed by [10] was used to calculate the continuous and isothermal development of austenite applying phenomenological models and was first presented in chip formation simulation by [13] and included a remeshing routine developed by [12] with a starting element size of 1.5 μm.

The transformed amount of austenite f_A after the time t can be calculated based on the isothermal heating equation from Avrami:

$$f_A = 1 - exp[-\{b \cdot t\}^n] \; with \; b(T) = C \cdot exp\left[-\frac{\Delta H}{k \cdot T}\right] \tag{3}$$

In this equation the parameter n is a material parameter depending on time t and $b(T)$ is a temperature dependent term with a velocity constant C, the

Boltzman-constant k and the activation enthalpy ΔH for the transformation process. The required parameters C, n and ΔH can be derived by the amount of austenite [10, 16].

With a modification of the Avrami model the isothermal transformation after a fast heating could also be taken into account. Therefore, the temperature has to be kept for a specific time at its maximum temperature [9, 10]. To simulate the diffusionless transformation to martensite, a modified equation based on the work from Skrotzi was implemented:

$$f_M(T) = f_{M,max}\left(1 - \left(\frac{T - \overline{M_f}}{\overline{M_s} - \overline{M_f}}\right)^{2,5}\right) \tag{4}$$

In Eq. (4), f_M is the temperature dependent amount of martensite, $f_{M,max}$ represents the maximum amount of martensite, $\overline{M_s}$ the martensite start-temperature and $\overline{M_f}$ the martensite finish-temperature for the transformation [17]. The adaptation to experimental results is shown in [10].

3.1.4 Minimum Quantity Lubrication

Heat transfer simulations require a detailed formulation of several material parameters with mostly complex state dependencies. In addition, extensive information about the temperature gradient of the fluid and the solid at the fluid/wall interface are needed. For calculating this heat transfer realistically it is essential to fully resolve the spatial temperature field in the cutting simulation.

The following three-step strategy has been used with the in-house code INSFLA [8] as an approach to master the just mentioned problems and to allow a realistic heat transfer modelling [5]:

1. Fully detailed (direct numerical simulation DNS) and fully resolved numerical simulations of the fluid-wall heat transfer process by using the parameters fluid mass flux, fluid core temperature, wall surface temperature and a contact duration dependent solid-fluid heat transfer for turbulent and laminar fluids. Every phase is considered.
2. Extraction of parameter- and time dependent heat transfer coefficients as a function of the selected parameters, Fig. 20.
3. The FEM-code supplies local wall temperatures, the initial fluid temperature, fluid flow rate, and the contact time between wall and fluid to obtain the heat transfer coefficients from the pre-calculated tabulation.

In the FEM-code a user subroutine was used to implement the extracted heat transfer Eq. (5) as a surface film condition.

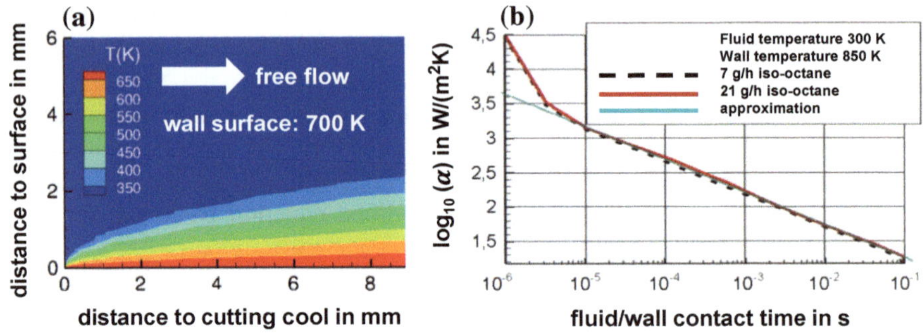

Fig. 20 Temperature field in a boundary layer fluid flow **a** distance to cutting tool and **b** Tabulation of the temporal development of the effective heat transfer rate [5]

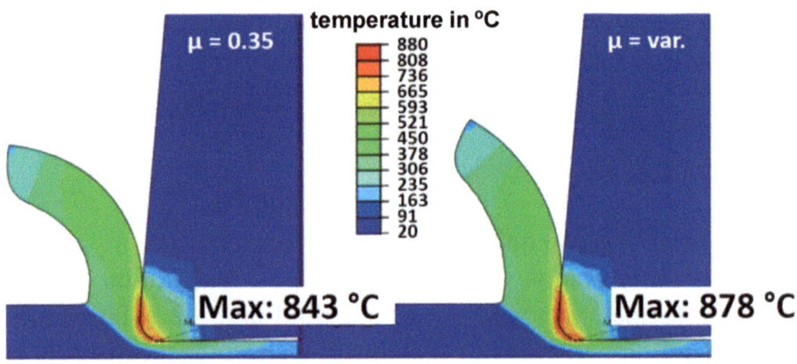

Fig. 21 Effect of different friction modelings on temperature their is just the effect of friction on the temperature shown [9]

$$log_{10}(\alpha) = a - b \cdot log_{10}(t) \quad \text{with } a = 6.2 \text{ and } b = -0.47 \tag{5}$$

The surface film condition is therefore loaded as a node set on the clearance surface of the workpiece. However, it has to be noticed that, Eq. (5) is only valid for the clearance surface and a contact time of the MQL of $t = 10^{-1} - 10^{-5}$ s, Fig. 20 [5]. A deeper look into the MQL modeling is given in [5].

Results of the 2D Chip Formation Simulation

The substantial validation of the 2D simulation is of main importance since it serves as the basis of the developed 3D drilling simulation. Within this validation step, a correct modeling of the friction is shown in Fig. 21. The maximum temperature gets closer to the one measured during turning experiments by applying friction coefficients as a function of the sliding speed and temperature. The related specific cutting forces show best agreement to force measurements during turning, see Fig. 22.

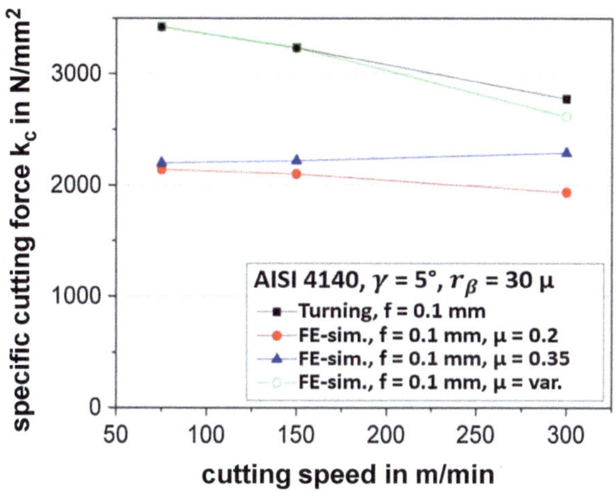

Fig. 22 Specific cutting force [9]

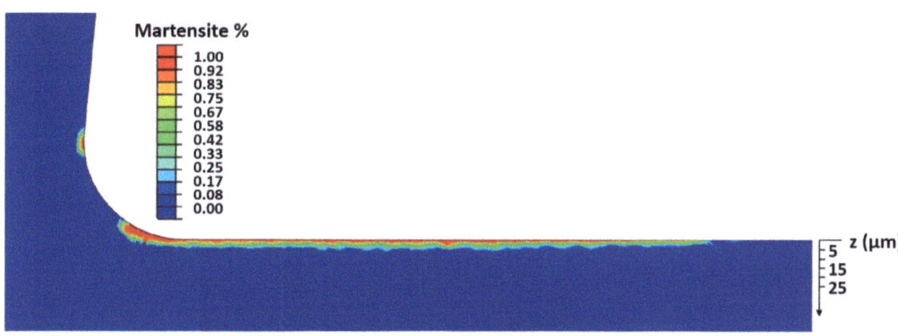

Fig. 23 Calculated martensitic surface [9]

By using the variable friction coefficients the calculated martensitic surface transformation can be calculated, see Fig. 23. As expected the martensitic surface layer starts not before the transition temperature is reached in the simulation model. The thickness of the surface layer increases from 2 to 6 μm until the thermal equilibrium at room temperature is reached.

3.2 3D Drilling Simulation

The following sections explain the drilling simulation model and how the compensation strategies are implemented.

3.2.1 Simplified Drilling Model

A simplified simulation model (Fig. 24) was developed and first introduced by [15]. At the beginning, the thermal and mechanical loads were modeled as surface sets within the FEM software ABAQUS. This approach was not suitable, because after deleting the surface to simulate material removal ABAQUS could not recognize the next element surface as the surface to be thermally and mechanically loaded. Hence, the whole simulation model was divided into many individual parts linked with special types of constraints to overcome the surface recognizing problem. However, this modeling approach was very memory intense and could not be dealt with manually. Additionally simulation results were inaccurate and the simulation time was not economically justified. In the end, the workpiece was modeled as complete part and divided into so called "cells". These cells were capable of modeling the drilling process and its typical process kinematics. The heat transfer coefficient due to the radiation was set to a constant value of 20 W/cm^2 K at 20 °C. Special attention was given to the clamping system. During the drilling experiments the workpiece was clamped at the edge surfaces. Between the workpiece and the clamping system "Pertinax" plates were used to minimize the heat transfer from the workpiece to the clamping system. This made it necessary to set the heat transfer coefficient in these areas to a value of 0.21 W/cm^2*K and at the blackened side for the infrared measurements to 0.98 W/cm^2*K. The implementation was realized by additional element sets. The room temperature was set at 20 °C [9].

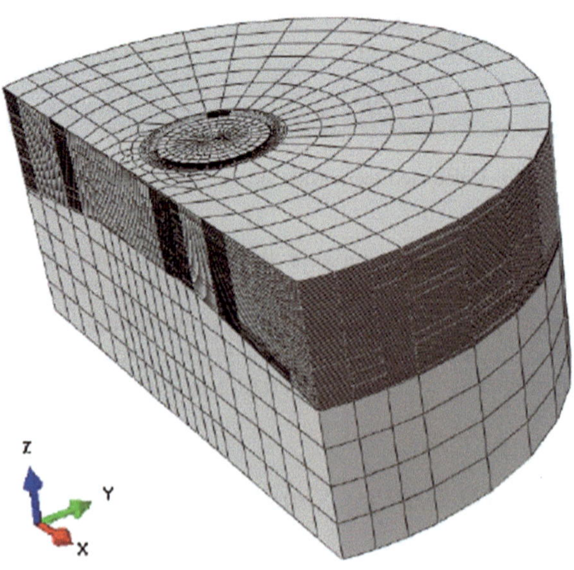

Fig. 24 Simplified workpiece geometry for the 3D FEM drilling simulation [15]

3.2.2 Modeling of Thermal and Mechanical Loads

Modeling the thermal loads during drilling is based on the experimental results and the 2D chip formation simulation. To take the chip removal into account the temperature load was divided into a temperature load along the main cutting edge T_{main} (Fig. 6) and the minor cutting edge T_{minor} (Fig. 7) according to the temperature measurement positions during the drilling experiments. Using regression methods the following functions for the temperature load were developed [9]:

$$T_{main} = 170.4 \cdot f^{0.12} \cdot r_{tool}^{0.23} \cdot v_c^{.023} \cdot t_{drilling}^{0.05} \tag{6}$$

$$T_{minor} = 299 \cdot v_c^{0.13} \cdot t_{drilling}^{0.03} \tag{7}$$

The temperature along the main cutting edge was developed in relation to the feed f, the radius of the drilling tool r_{tool}, the cutting speed v_c and the drilling depth $t_{drilling}$. The temperature profile of Eq. (6) can lead to temperatures below 460 °C. This does not conform with the experimental results. Consequently, it was necessary to limit the minimum temperature to the fixed value of 460 °C at the chisel edge [9].

Analogously to the thermal load modeling, the force modeling was also based on experimental results. The chosen 3D modeling approach required the applied forces to be a bodyforce. The user subroutine DLOAD in ABAQUS offered the possibility to separate the cutting force F_c into its x-(F_{cx}) and y-(F_{cy}) components. In dependence of the position along the main cutting edge the cutting force was calculated using the following equations:

$$F_{cx} = F_c \cdot \frac{y}{\sqrt{(x^2 + y^2)}} \tag{8}$$

$$F_{cy} = -F_c \cdot \frac{x}{\sqrt{(x^2 + y^2)}} \tag{9}$$

Every modulation of the cutting forces F_{cx}, F_{cy} and the feed force F_z could be realized within the subroutine DLOAD. This simplified the modeling of the drilling process in the simulation, because the forces were loaded on defined elements. A Further simplification of the simulation model deactivates the elements after one load cycle, Fig. 25b. The final element-deleting step was done after every five full rotations of the drilling tool in Fig. 25b [9].

Both implemented loads follow the rotation of the drilling tool regarding the cutting speed and while considering the point angle of the drilling tool, Fig. 25a. For this reason, the drilling hole was divided into 32 circular segments with equal loads on opposite segments following the rotation of the drilling tool, Fig. 25b. The force loading is shown as an exemplary element set (Fig. 25c) and the temperature load for the main cutting edge as an exemplary node set, Fig. 25d. The minor cutting edge also representing the chip removal is shown in Fig. 25e [9, 15].

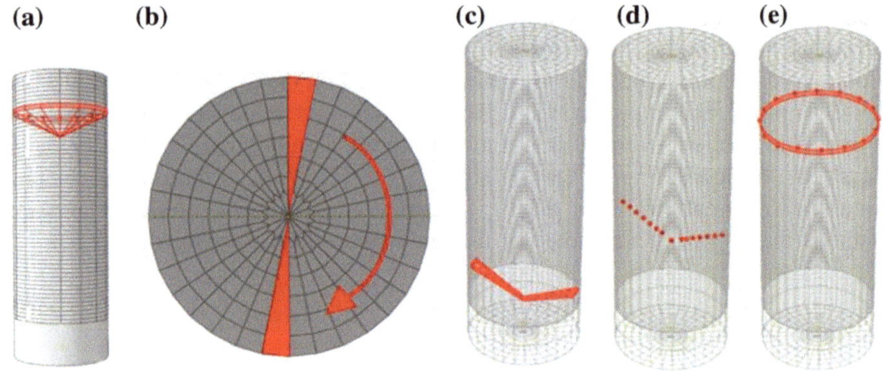

Fig. 25 The loads represented by the point angle of the tool (**a**), circular divided drilling hole (**b**), element set for the force load (**c**), node set for the temperature load (**d**) and node set for the minor cutting edge and chip removal (**e**) [9]

3.2.3 Modeling Phase Transformations

The introduced model to calculate phase transformations can describe phases such as martensite, bainite, austenite or perlite/ferrite for 2D simulations and even 3D simulations. The approach for 2D simulation used plane strain conditions for simulating the elastic- plastic material behavior. This fact allows the calculation of stresses in x-, y-, and z-direction which were needed to predict phase transformations even if one strain direction is blocked [9].

3.2.4 Modeling Minimum Quantity Lubrication

In the 2D chip formation simulation, the calculated heat transfer equation was applied as a surface film condition which could be converted for the 3D drilling simulation. However, this heat transfer equation could not be used directly in the 3D simulation. The introduced 3D model did not consider the tool as a solid body. It takes the temperature and force load of the tool on defined node and element sets into account. This missing solid body contact of the tool and the workpiece neglected the influence of friction. Temperature and force equations for the 3D simulation using multivariate regression functions based on experimental measurements and 2D simulation results with detailed friction and MQL modeling had to be developed to consider friction effects. Consequently, the real friction behavior was expressed in the input parameters for the multivariate regression functions. As a result, all nodes of the 3D drilling model were charged with temperature functions taking friction effects into account [5].

Results of the Simplified Drilling Model

In the following paragraph, the simulation results for the simplified drilling simulation model are discussed. Figure 26 shows the temperature distribution at the drilling hole and clearly indicates the position of the main cutting edge. It also illustrates the rotation of the tool and its thermal load. According to the equations for the main cutting edge the temperature increases from the center of the drilling hole to the drilling hole surface. This temperature gradient changes for the different cutting parameters. The maximum temperature agrees well to the experimental data. A comparison to the results from the infrared camera showed the same tendency in the simulation.

The associated phase transformations are presented in Fig. 27 on the left. The calculated thickness of the martensitic surface layers of 15–20 μm is in the range of the measured thickness between 18 and 21 μm. So the simulation approach is useable for the prediction of phase transformations.

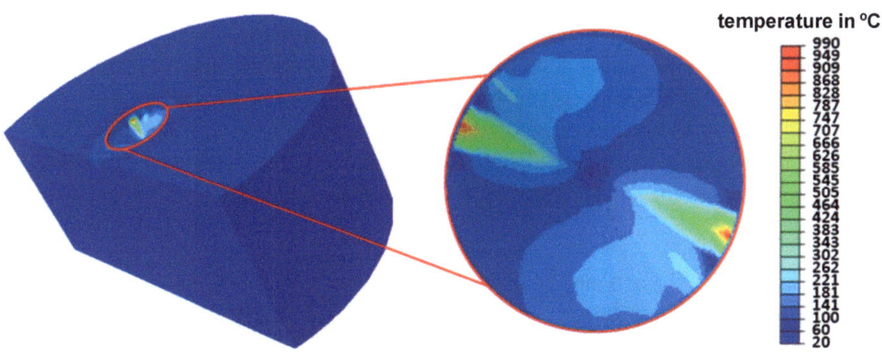

Fig. 26 Temperature distribution within the drilling hole of the simplified drilling model [9]

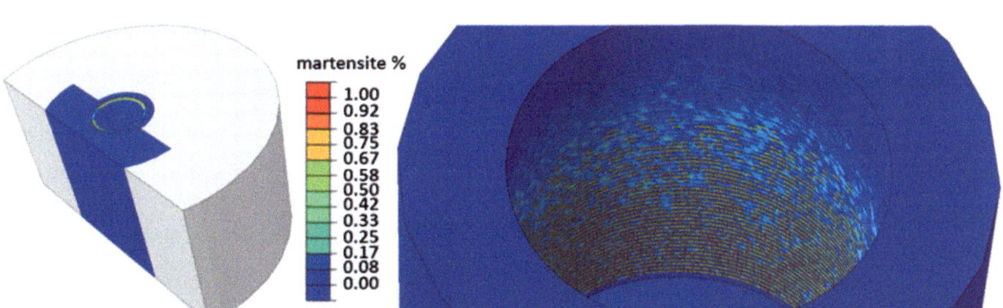

Fig. 27 Martensitic surface layer for the simplified drilling model (*left*, [9]) and advanced drilling model (*right*, [4]) for $v_c = 300$ m/min and $f = 0.3$ mm

3.3 Advanced Drilling Model

The developed advanced drilling model (Fig. 28) aiming at the compensation of thermal effects during machining represents the final step towards the simulation of process results such as distortions or phase transformations on the workpiece scale. The difference between the simplified model and the advanced model is the number of interacting drilling holes. To simulate drilling in the scale of a workpiece, the introduced model needed to be more efficient in its numerical performance and accuracy by improved modeling strategies and increased workpiece complexity. Modeling efficiency was inevitable in order to reduce the long simulation times within the simplified drilling model with only one drilling hole from Sect. 3.2. However, the prediction accuracy conflicted with modeling efficiency and, thus required to be optimized within this work.

3.3.1 Modeling of Thermal and Mechanical Loads

In the simplified drilling model two temperature equations were developed and validated for the temperature load of a drilling tool. However, the total thermal load of a drilling tool is not only given by the main cutting edge and the minor cutting edge but also by the chisel edge of the drilling tool in a 3D simulation. This fact made it necessary to complete the thermal heat input by implementing a temperature equation for the chisel edge of the drilling tool to replace the fixed value of

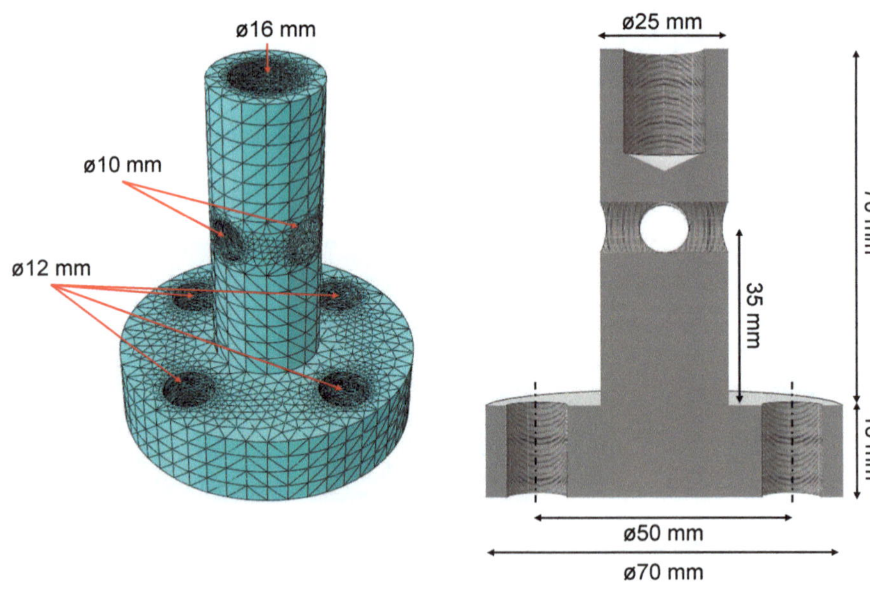

Fig. 28 Advanced drilling model including 7 drilling holes

T = 460 °C from the simplified drilling model. Because shape deviations are in a size range of micrometers, a highly realistic heat input is of importance. Within a 2D chip formation simulation varying cutting speeds, cutting depths and tool geometries were used to develop the following temperature equation for the chisel edge of the drilling tool T_{chisel}:

$$T_{chisel} = 188,82 \cdot v_c^{0,3112} \cdot t_{drilling}^{0,0093} \cdot f^{0,1624} \tag{10}$$

In sum, the main cutting edge, the minor cutting edge and the chisel edge were taken into account to model the thermal load of the drilling tool. The mechanical load was modeled the same way as in case of the simplified model.

3.3.2 Modeling Efficiency

Three different types of heat sources for modeling drilling operations have been developed to find the optimum between the simulation time and the accuracy of the simulation results. These three heat sources differ in their degree of abstraction representing the drilling process.

The first heat source model was based on the simulation approach introduced by [11] and represents the most abstract way of simulating the drilling process. Pabst [11] divided the drilling hole which is already considered by the simulation model into annular ring segments over the complete drilling length. Contrary to Pabst's approach, when applying heat fluxes on selected element sets, a temperature boundary condition was defined. The benefit of this approach constitutes in the fact that the loaded temperature boundary condition equals the measured temperature of experimental results. With the aim of predicting phase transformations, a temperature very close to experimental results is of importance. The heat source temperature boundary condition loads the annular segments on the borehole surface with (Eq. 6) in feed direction. The annular segments move deeper into the drilling hole, see Fig. 29. It must be noted that this model only considers those values of Eq. 6 existing on the surface of the drilling hole and not the complete temperature distribution. Regarding the process kinematics the thickness and the load duration of each ring segment considered the feed rate of the drilling tool [4].

The second heat source model had a higher level of detail due to the implementation of material removal. The underlying approach using conical segments was introduced by [1] to take the point angle into account. As an enhancement of the first heat source model the drilling hole was not already existing in the model and the whole temperature equation T_{main} was applied in the simulation model on the conical segments. As in the first heat source model, the temperature was similarly loaded by a node set. The load duration and the material removal process was divided into a thermal loading step and a deactivation step [4]. Figure 30 shows the material removal process. It can be seen that the removed elements were fitted to the tool geometry.

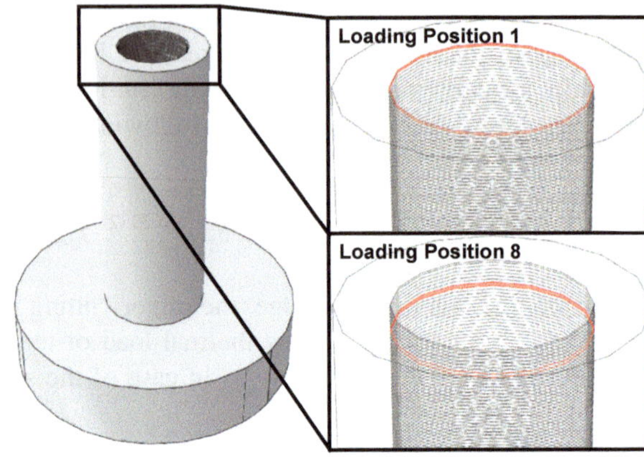

Fig. 29 Loading mechanism of the first heat source model [4]

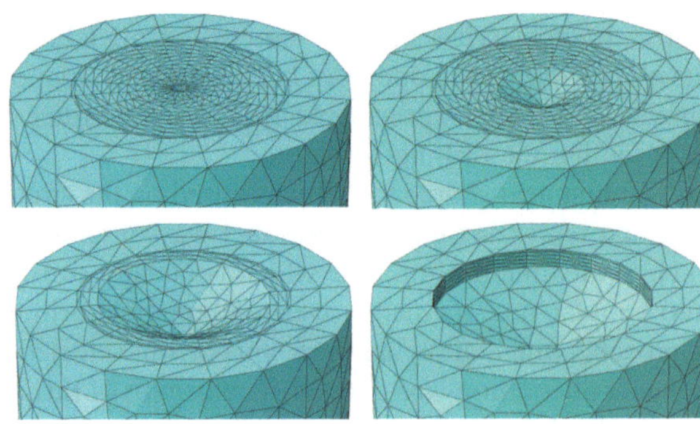

Fig. 30 Material removal process [4]

The third heat source model was the closest to the drilling process considering the whole kinematic of the drilling process by taking the feed rate, material removal and the rotation of the drilling tool into account. In addition to the second heat source model the cone-shaped discs were circular divided into 16 parts (Fig. 31). Compared to the simplified model this amount of circular parts is half of the amount used in the simplified model. The material removal was again separated into a thermal loading step and a deactivation step. However, only one of the circularly divided parts and its direct opposite are loaded and deactivated in a single step. After a 180° rotation the whole segment was deactivated and the segment below was loaded the same way. The rotation speed was adapted to the cutting speed and the feed rate was considered in the segment thickness [4].

Results of the Advanced Drilling Model

In the course of the optimization process the phase transformation routine was implemented into the advanced drilling process, see Fig. 27 on the right. Like

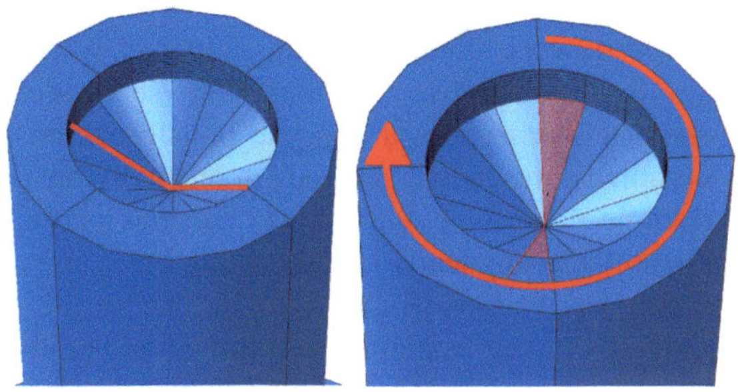

Fig. 31 Temperature node set on the main cutting edge (*left*) and material removal process using element sets (*right*)

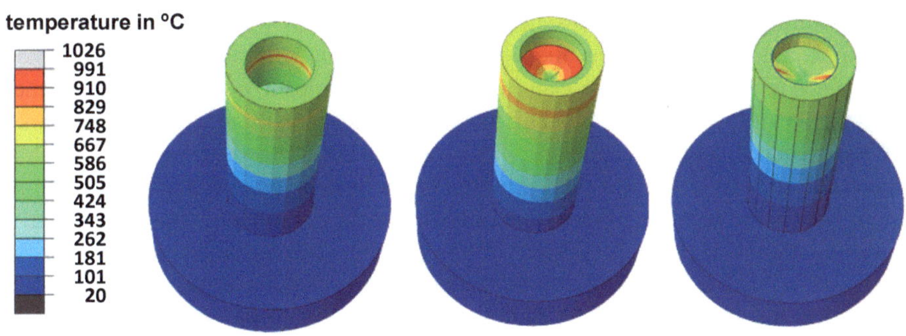

Fig. 32 Temperature distribution for the first heat source model (*left*), second heat source model (*middle*) and third heat source model (*right*) at v_c = 300 m/min and f = 0.3 mm [4]

measured in the metallurgical investigations the transformations do not start until a certain drilling depth is reached. Considering the three optimization models, it has to be mentioned, that the phase transformations are dependent on the temperature boundaries, see Fig. 32. The different temperature distributions and maximum temperatures for the different heat source models are depicted in Fig. 32. The highest temperature level is shown for the second heat source model, followed by the first heat source model and finally the third heat source model. The most realistic temperature distribution was observed for the third heat source model which already was used for the simplified drilling model. The simulation time for the first heat source model is about 10–15 times shorter than for the third heat source model with the kinematics closest to drilling. This fact makes the first heat source model very attractive for simulation studies because the results show similar values to those obtained from the third heat source model. For further findings, see [4].

3.3.3 Compensation Strategies for Thermal Effects

Based on the introduced work three different strategies were tested to derive machining strategies for the purpose of minimizing shape deformations using simulations. The residual stress states and phase transformations had to be controlled in order to optimize dimensional accuracy in one individual drilling hole and for the complete workpiece by achieving a lower thermal loading. This endeavor pursued the aim of manufacturing nearly round drilling holes and correcting curved drilling holes. The benefit of dimensional accuracy due to the compensation of thermal processing influences was implemented through an iterative approximation method, which relies on the result of the three following compensation methods.

Optimization of Machining Strategies

This optimization strategy consisted of the reduction of the resulting overall heat and maximal temperatures in the workpiece. Drilling a multitude of holes into a workpiece one after the other, causes some drilling holes to cool down while others are machined at the same time. This results in contracting and expanding volumes of the material and deformations. The simultaneously proceeding thermal load spectrum strongly promotes the formation of multi-axis, graded stress states, which in turn affect the resulting dimension and form of the component after machining. In addition, the workpiece is subject of intersecting drilling holes where a higher mechanical load is expected due to the possible formation of martensitic surface layers leading to a higher deformation state. To estimate the overall heat balance and resulting material stress, numeric parameter studies for the process parameters feed and cutting speed were performed. The heating and cooling phases were considered separately in order to provide independent results for the following deduction of process strategies. By calculating deformations, residual stresses, maximum temperatures and the temperature fields the strategies were classified.

Results of the Optimization of Machining Strategies

The first target was to reduce the thermal heat input into the workpiece with adapted machining parameters. Figure 33 exemplary shows the results for 3 different cutting speeds in drilling the hole geometry of the flange. The higher maximum temperatures with increasing cutting speeds lead to oval shapes of the presented drilling hole. These oval drilling holes are only observed for the drilling holes of $\varnothing = 10$ mm and $\varnothing = 12$ mm due to position of these drilling holes and the resulting non-symmetrical cooling. The oval shape is also a result of the blocked expansion according to the three-jaw chuck clamping system.

The second target aimed at the variation of the drilling order, with respect to the industrial manufacturing process, see Fig. 34. This means drilling holes of the same

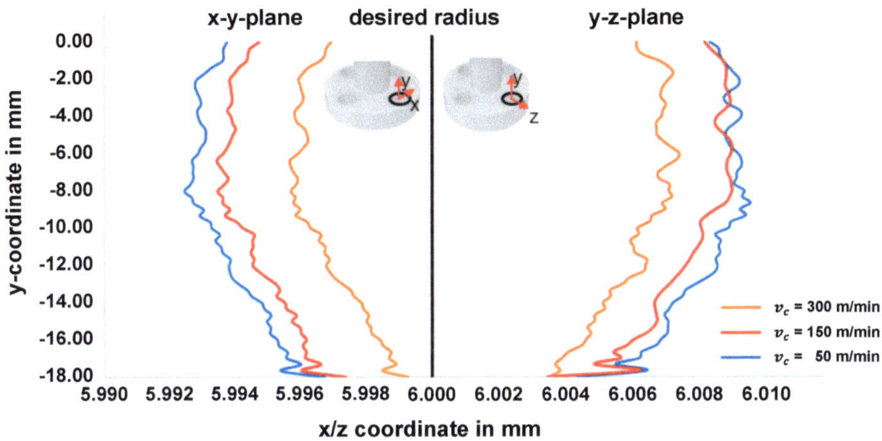

Fig. 33 Simulated shape deviations for different cutting speeds at a feed rate of 0.3 mm for a drilling hole of ∅ = 12 mm

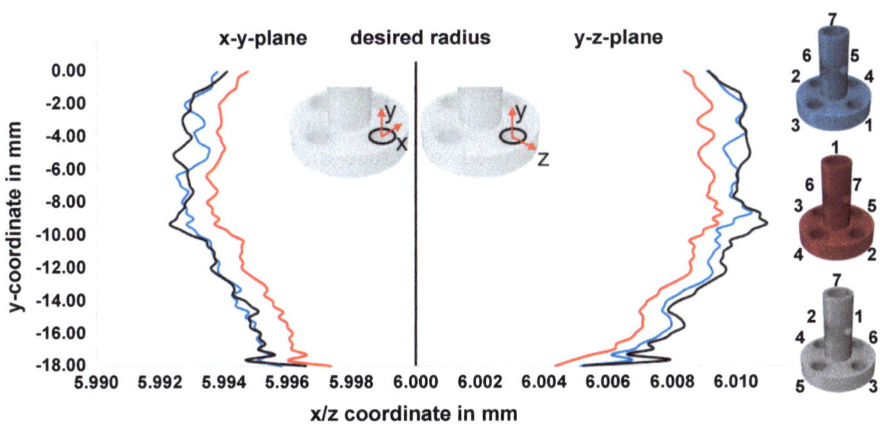

Fig. 34 Simulated shape deviations over the drilling depth for different machining order strategies

diameter are drilled in sequence in order to not change the drilling tool for short manufacturing times. The simulation results showed the smallest shape deviation for strategies where drilling holes with the widest possible distance from each other were chosen. The best strategy started with the four drilling holes on the flange of the workpiece by drilling holes with the maximum distance to each other. Next step in this order, a large drilling hole was drilled on the face side of the cylindrical part. The crossing drilling holes need to be drilled last, beginning with a 90° rotation of the start drilling hole in the flange.

When evaluating these two strategies in terms of their capability to compensate shape deviations, it has to be concluded that only small improvements could be reached for the used demonstrator workpiece. Thus, this approach only reduced the shape deviations of the workpiece, but the aim to compensate shape deviations completely could not be attained.

Thermal Management

The aim of thermal management was to use defined heat sinks like MQL or liquid nitrogen for the controlled setting of distortion states based on the advanced 3D drilling model. The positive influence of cooling with varying heat exchange coefficients and cooling positions on the residual stress and deformation state was already discussed by [18]. They were able to show that controlled cooling of workpieces can reduce the shape deformation of the workpiece. Based on these investigations the heat sinks are deployed to detract heat from the workpiece subsequent or simultaneous to the manufacturing of individual holes and are thereby varied for the parameters position of the heat sink, immersion time, contact area and intensity. For these preliminary studies, the advanced drilling model was again reduced to one drilling hole. The remaining drilling holes were not taken into account. A heat sink represents an element set which was set on the surface of the workpiece or the drilling hole with a constant value or a term depending on time and intensity. Due to the inverse problem of determining the necessary heat sink parameters from an existing residual stress and deformation condition only one parameter was varied at a time in the heat sink. Knowing the present deformation state selected parameters will be used to induce the inverse residual stress and deformation state to compensate the shape deformation of the workpiece.

Results of the Thermal Management

The compensation potential for shape deviations by controlled cooling strategies during drilling is exemplarily shown in Fig. 35. In this graph, different cooling fluids are compared for its potential to reduce shape deviations. It appears, that using higher heat exchange coefficients can reduce shape deviations more effectively than low temperatures. The reason is the better detracting of developed heat during the drilling process compared to low temperatures. This fact can be confirmed by comparing cutting oil at 20 °C ($\alpha \approx 5000$ W/(m^2*K)) and liquid nitrogen at −196 °C ($\alpha \approx 5$ W/(m^2*K)). When high heat exchange coefficients are applied, the remaining plastic energy causing shape deviations is drastically reduced because the heat is effectively detracted and the workpiece does not reach high temperatures. The simulation results show decreasing behavior for the maximum shape deviations, when the used amount of cooling energy (with a high heat exchange coefficient) gets close to the amount of thermal energy produced by the drilling process. The cooling fluids were applied during and subsequent to the drilling process. To sum up, cooling during the drilling process is more effective than cooling subsequent to the drilling process.

Figure 36 shows the compensation results for different cutting fluids over the drilling depth and reflects the high influence of the heat transfer coefficient on the potential correction of the shape of the drilling hole.

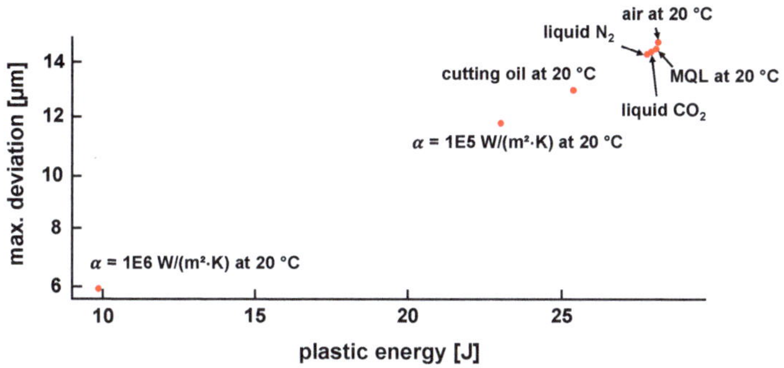

Fig. 35 Different cooling fluids applied during drilling influencing the remaining plastic energy in the workpiece and max. shape deviation in the drilling hole at the face side of the cylindrical part

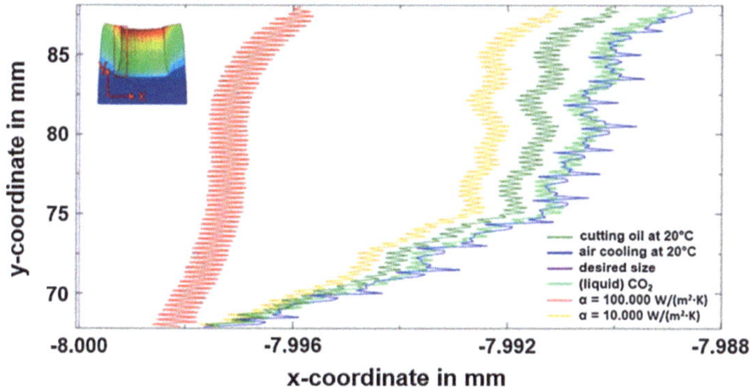

Fig. 36 Shape deviation in the drilling hole after cooling with different fluids

In addition to cooling simulations, heated fluids were tested for their suitability to reduce shape deviations, Fig. 37. The aim using heating fluids (with $\alpha >$ 1000 W/(m^2 K)) with temperatures up to 400 °C is to use their thermal energy to counteract the plastic energy into the workpiece which was introduced to the workpiece by the drilling process. This method was again performed during and subsequent to the drilling process. The simulations showed improvements in shape accuracy for heated fluids, but no benefit compared to the cooling fluids. Consequently, preventing high temperatures seems to be more effective than having a homogenous temperature field in the drilling hole.

Finally, the drilling hole was heated after the drilling process, using a fluid with a high heat exchange coefficient with temperatures up to 600 °C to make sure, that transition temperatures are not exceeded. These fluids were applied for a certain time between 10 and 45 s to ensure, that there was enough time to homogenize the temperature field. After the heating time, the drilling hole was cooled with varying cold fluids of different high heat exchange coefficients. The shape deviations

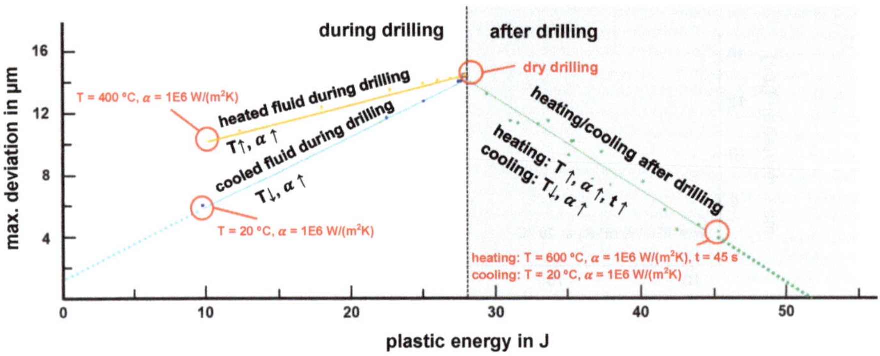

Fig. 37 Comparison of different cooling strategies and their suitability to reduce shape deviations

decreased applying higher heating fluid temperatures and heat transfer coefficients for the fluid. The influence of the heating time showed an asymptotic behavior. Furthermore, it did not significantly improve the results when a heating time of 35 s was exceeded. In addition, longer heating times may also yield to microstructural changes in the material.

The potential of the cooling/heating fluid is depicted in Fig. 37 by linear extrapolation of the results. It shows the possibility for total compensation of shape deviations using heated fluids. The necessary amount of needed plastic deformation energy seems to be a little lower than the energy brought into the workpiece by the drilling process. This could be explained by the option of using some thermal energy of the drilling process for compensation. It seems that a total compensation for the cooling fluids may not be reached, even if the entire plastic deformation energy was reduced. However, this statement may be influenced by the fact that in the trendline only one parameter combination was taken into account for a very high heat transfer coefficient. Another reason could be that the trend line only considered constant and not varying heat exchange parameters over the drilling depth. Therefore, a complete compensation could be reached if a time-dependent heat exchange coefficient or a time-dependent temperature of the fluid were used.

3D Circular Milling

A 3D circular milling model was developed to offer an additional opportunity to compensate thermal effects such as shape deformations. Using the 3D-drilling simulation the manufacturing of drilling holes was selected to be non-circular to achieve residual stresses and deformation states leading to ideal round shapes of the drilling hole after reaching the thermal equilibrium. For this reason, the existing advanced drilling model had to be adapted to the circular milling process to correctly describe the thermal and mechanical load. The constructed simulation model was tested for a reduced number of drilling holes, for example in the efficiency studies to identify residual stress and deformation states after circular milling. Thus various non-circular drilling holes were investigated and iteratively optimized regarding the overall heat balance in the workpiece.

Results of the Circular Milling

The compensation method deployed last was circular milling. It aims at correcting the tool path during the machining process in order to deliberately produce non-circular drilling holes to counteract the elliptical shape of the diameter of the drilling hole. Figure 38 schematically shows the implemented elliptical shape of a circular milling tool path.

For the purpose of correcting the tool path, the shape deviation (ovality from the diameters D_1 and D_2 of the ellipse) was measured and compensated linearly; meaning the tool correction path was used as an ellipse with $D_{1,drilling} = D_{2,circular\,milling}$ and $D_{2,drilling} = D_{1,circular\,milling}$.

The results of this linear compensation strategy is presented in Fig. 39. The black line shows the shape of the drilling hole after drilling and the light blue one indicates the corrected tool path for the circular milling. The final shape after the tool path correction (dark blue) shows an improvement of the roundness of the drilling hole.

Consequently, circular milling represents a good approach for the correction of shape deviations in drilling.

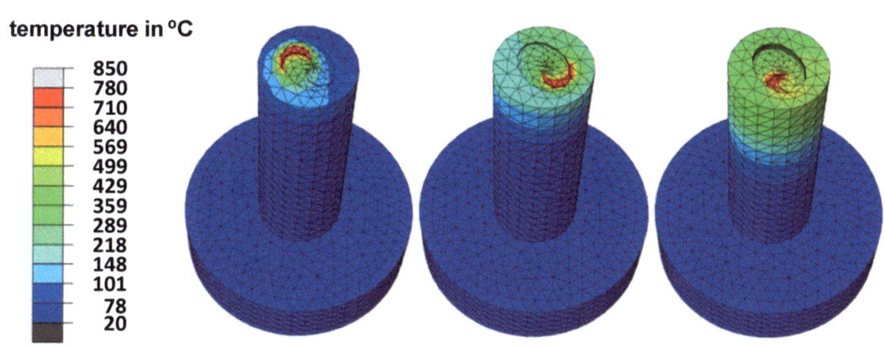

Fig. 38 Circular milling simulation at different time steps (schematically to demonstrate the functionality)

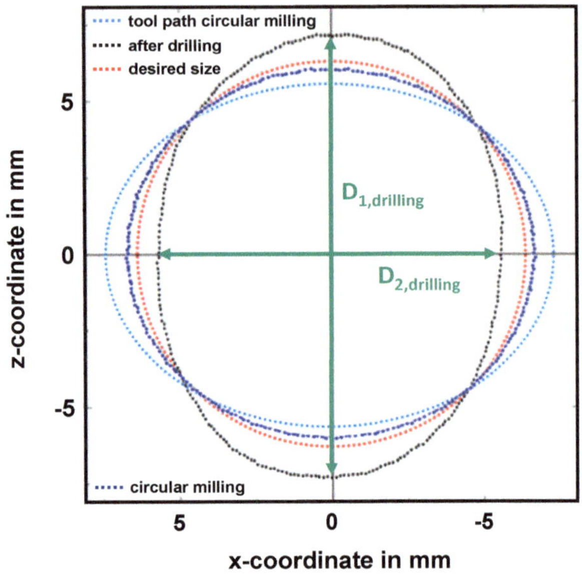

Fig. 39 Comparison of drilled holes and tool path compensation using circular milling (with a deformation scale factor of 300)

Summary and Conclusion

Based on turning experiments, a 2D chip formation simulation model was developed and optimized for the purpose of predicting phase transformations. The thermomechanical load and the calculated phase transformations showed good agreement with the experimental results. Next, the turning investigations and orthogonal cutting simulations were used in combination with drilling experiments to develop a 3D drilling simulation model to reach the process level in simulation. This simulation model includes the feed rate, material removal and rotation of the tool and was used to simulate phase transformations, forces and the temperature distribution during drilling. In a final step, the simulation model was extended to 7 drilling holes to investigate three different kinds of compensation methods and their capability to reduce shape deviations.

The first method used thermal load optimized cutting parameters and adapted drilling sequences of the drilling holes. The results show a decrease in shape deviations for lower cutting parameters and for drilling strategies with the largest distance from one drilling hole to the next. By using this method, minimization of the distortion could be accomplished but no total compensation could be achieved.

The second compensation method used controlled cooling and heating strategies during and subsequent to the drilling process. During the drilling process, the shape deviations can be minimized by adjusting cooling fluids with high heat exchange coefficients to detract heat and consequently deformations. After the drilling process, the hole needs to be heated with a hot fluid of a high heat exchange coefficient. Both approaches show the possibility to compensate shape deviations when adapted cooling or heating parameters are used.

The last and most complex compensation method was to correct tool path using circular milling. The results showed an improvement in roundness of the drilling hole, if the shape of a distorted drilling hole was use to correct the tool path.

In conclusion, adapted cooling or heating fluids and tool path corrections demonstrate the best results for compensating shape deviations.

Acknowledgements The authors gratefully thank the "Deutsche Forschungsgemeinschaft DFG" for the support of the priority program 1480 "Modelling, Simulation and Compensation of Thermal Effects for Complex Machining Processes".

References

1. Al Jourdi, A.: Adaption des Trockenbohrens über thermomechanische Modellierung zur Erhöhung der Fertigungsgenauigkeit. Dissertation, Technische Universität Berlin (2012)
2. Autentieth H.: Numerische Analyse der Mikrozerspanung am Beispiel von normalisiertem C45E. Dissertation, Universität Karlsruhe (TH) (2010)
3. Bartek, T., Menzel, A.: A finite deformation framework for martensitic phase transformations interacting with plasticity based on representative deformation directions. submitted for publication
4. Bollig, P., Köhler, D., Zanger, F., Schulze, V.: Effects of different levels of abstraction simulating heat sources in FEM considering drilling. Procedia CIRP **46**, 115–118 (2016)
5. Bollig, P., Faltin, C., Schießl, R., Schneider, J., Maas, U., Schulze, V.: Considering the influence of minimum quantity lubrication for modelling changes in temperature, forces and phase transformations during machining. Procedia CIRP **31**, 142–147 (2015)
6. Galanis, N.I., Manolakos, D.E., Vaxevanidis, N.M.: Comparison between dry and wet machining of stainless steel. In: Proceedings of the 3rd International Conference on Manufacturing Engineering, pp. 91–98 (2008)
7. Gulpak, M., Sölter, J.: Development and validation of a hybrid model for the prediction of shape deviations in dry machining processes. Procedia CIRP **31**, 346–351 (2015)
8. Maas, U.: Detailed numerical simulation of chemically reacting flows. In: Proceedings of the International Symposium on Computational Fluid Dynamics p. 741 (1991)
9. Michna, J.: Numerische und experimentelle Untersuchung zerspanungsbedingter Gefügeumwandlungen und Modellierung des thermo-mechanischen Lastkollektivs beim Bohren von 42CrMo4. Dissertation, Karlsruher Institut für Technologie (KIT) (2014)
10. Mioković, T.: Analyse des Umwandlungsverhaltens bei ein-und mehrfacher Kurzheithärtung bzw, Läserstrahlhärtung des Stahls 42CrMo4. Dissertation, Universität Karlsruhe (TH), (2005)
11. Pabst, R.: Mathematische Modellierung der Wärmestromdichte zur Simulation des thermischen Bauteilverhaltens bei der Trockenbearbeitung Dissertation, Karlsruher Institut für Technologie (KIT), (2008)
12. Schulze, V., Zanger, F.: Numerical analysis of the influence of Johnson-Cook material parameters on the surface integrity of Ti-6A-4V. Procedia Eng. **19**, 306–311 (2011)
13. Schulze, V., Michna, J., Schneider, J., Gumbsch, P.: Modelling of cutting induced surface phase transformations considering friction effects. Procedia Eng. **19**, 331–336 (2011)
14. Schulze, V., Michna, J., Zanger, F., Pabst, R.: Modeling the process-induced modifications of the microstructure of workpiece surface zones in cutting processes. Adv. Mater. Res. **223**, 371–380 (2011)
15. Schulze, V., Zanger, F., Michna, J., Lang, F.: 3D-FE-Modelling of the drilling process—prediction of phase transformations at the surface layer. Procedia CIRP **8**, 233–238 (2013)

16. Schulze, V., Michna, J., Zanger, F., Faltin, C., Maas, U., Schneider, J.: Influence of cutting parameters, tool coatings and friction on the process heat in cutting processes and phase transformations in workpiece surface layers. J. Heat Treat. Mater. **68**, 22–31 (2013)
17. Schulze, V., Uhlmann, E., Mahnken, R., Menzel, A., Biermann, B., Zabel, A., Bollig, P., Ivanov, I.M., Cheng, C., Holtermann, R., Bartel, T.: Evaluation of different approaches for modeling phase transformations in machining simulation. Prod. Eng. Res. Devel. **9**(4), 437–449 (2015)
18. Schüttenberg, S., Hunkel, M., Fritsching, U., Zoch, H.-W: Controlled quenching of steel parts in fluid jet fields. In: 15th IFHTSE and SMT 20, pp. 25–29 (2006)
19. Tomé, C.N., Canova, G.R., Kocks, U.F., Christodoulou, N., Jonas, J.: The relationship between macroscopic and microscopic strain hardening in F.C.C. polycrystals. Acta Metall. **32**(10), 1637–1653 (1984)
20. Uhlmann, E., Mahnken, R., Ivanov, I.M., Cheng, C.: FEM modeling of hard turning with consideration of viscoplastic asymmetry and phase transformation. J. Mach. Eng. **13**(1), 80–92 (2013)
21. Voce, E.: The relationship between stress and strain for homogeneous deformation. J. Inst. Met **74**, 537–562 (1948)

Modelling, Simulation and Compensation of Thermomechanically Induced Deviations in Deep-Hole Drilling with Minimum Quantity Lubrication

D. Biermann, H. Blum, I. Iovkov, A. Rademacher, K. Rosin and F.-T. Suttmeier

Abstract This chapter summarises interdisciplinary research work on deep-hole drilling of aluminium casting using twist drills and minimum quantity lubrication (MQL). The high thermal conductivity of the workpiece material, the low cooling performance of MQL and the long main time of the process lead to a significant thermal load onto the workpiece. The thermal distortion of the machined component causes a systematic deflection of the long drilling tool and thus a high straightness deviation of the produced bore. Based on extensive technological investigations with a focus on the heat input into the workpiece a reliable simulation-based prediction of the thermomechanical distortion and the resulting bore deviations is presented. Thereby a finite element (FE)-model of the workpiece was coupled with a simple analytic representation of the deep-hole drilling tool. Using the fictitious domain method, adaptive techniques, and massive parallelisation, the FE-simulation is most efficiently realised. In order to compensate the occurring deviations, a novel approach based on radial tool path adjustment was developed, which allows for the drilling direction to be controlled during the process. The sophisticated simulations

D. Biermann (✉) · I. Iovkov
Institute of Machining Technology, TU Dortmund University,
Baroper Straße 303, 44227 Dortmund, Germany
e-mail: biermann@isf.de

I. Iovkov
e-mail: iovkov@isf.de

H. Blum · A. Rademacher · K. Rosin
Chair of Scientific Computing, TU Dortmund University, Vogelpothsweg 87,
44227 Dortmund, Germany
e-mail: heribert.blum@math.tu-dortmund.de

A. Rademacher
e-mail: andreas.rademacher@math.tu-dortmund.de

K. Rosin
e-mail: korinna.rosin@math.tu-dortmund.de

F.-T. Suttmeier
Work Group Scientific Computing, University of Siegen,
Walter-Flex-Str. 3, 57068 Siegen, Germany
e-mail: suttmeier@mathematik.uni-siegen.de

D. Biermann and F. Hollmann (eds.), *Thermal Effects in Complex Machining Processes*, Lecture Notes in Production Engineering,
DOI 10.1007/978-3-319-57120-1_10

enable the determination of the optimal NC-path for the deep-hole drilling tool, which is not possible based on experiments. The validation results show a great potential of the developed methods for the simulation-based minimisation of the thermomechanically induced straightness deviation in deep-hole drilling.

1 Introduction

Machining without metalworking fluid (MWF) allows for significant savings not only due to the direct decrease of costs for their purchase, maintenance and disposal, but also more and more for health and environment protection reasons [19, 26]. In many cases, the energy consumption related to the MWF is the major consumer within the production system and since the energy price is continuously rising, dry cutting and minimum quantity lubrication provide a great potential to enhance the efficiency and reduce the component related energy consumption of the production process [1, 7]. The high adhesion between typical cutting tool materials and hypoeutectic aluminium-silicon casting alloys [16], which are wide spread in automotive powertrain applications, e.g. cylinder head and crankcase, leads to insufficient process reliability in dry drilling [10, 24]. Particularly, deep-hole drilling is very demanding due to the challenging chip evacuation. Hence, these operations are only feasible using minimum quantity lubrication (MQL) as a semi-dry MWF-concept. The compressed air used as a carrier medium for the oil mist in MQL exhibits very low thermal conductivity and density in comparison to typical water-based emulsions, which leads to a low cooling performance of the MQL. The major cooling effect results mainly from the achievable reduction of friction between cutting tool and chip as well as machined surface, based on the high lubrication ability of oil. The heat input into the workpiece is particularly higher in drilling than in other machining operations [21], because the chip formation process takes place within the part, which shifts the thermal balance more towards the workpiece. The relatively long primary machining time in deep-hole drilling forces the heat input as well, so that the high thermal load onto the workpiece increases the risk of thermally induced deviations due to these operations.

In the following section the technological investigations regarding the thermomechanical aspects of the deep-hole drilling process under MQL conditions are presented and the resulting workpiece deflections as well as deviations are discussed depending on the cutting parameters and the bore eccentricity. The second part of the contribution describes the finite element (FE)-based simulation of the thermoelastic workpiece behaviour during the deep-hole drilling process. Using empirical models of the heat input, a highly efficient FE-simulation is realised to predict the workpiece behaviour also for complex geometries and several bores in a reasonable simulation time. To this end, the fictitious domain method is applied to reproduce the material removal without mesh dependencies. Furthermore, adaptive techniques and massive parallelisation enhance the efficiency of the FE-simulation.

Building on this, the prediction of the resultant bore deviations is performed using an analytical approach for representing the drilling tool. Furthermore, a novel compensation method, its development and validation are introduced at the end of this chapter. The procedure allows for the efficient simulation-based prediction and NC-compensation of thermomechanically induced deviations in deep-hole drilling.

2 Technological Investigations

The fundamental investigations on deep-hole drilling of cast aluminium with MQL serve as a basis for the technological evaluation and advancement of the process as well as for the modelling and simulation activities, performed in order to make the prediction and compensation of thermal workpiece deviations possible. The firstly discussed mechanical tool load is one of the most important process variable, as this represents the entire effective power and energy (mechanical work), which is converted into heat during the machining operation. One part of the generated heat flows into the workpiece, where it causes a temperature increase and thermal expansion. The deflection of the part due to these thermal effects can on the one hand impair the precision of subsequent machining steps; on the other hand it may affect the accuracy of the currently drilled bore as well. In addition to these aspects the technological development of a high-performance deep-hole drilling process is introduced, since the feed increase allows to shift the heat balance towards the chip and thus to reduce the thermal load on the workpiece.

2.1 Experimental Conditions

The experiments were carried out on a 4-axis machining centre GROB BZ600 using the widespread aluminium cast alloy AlSi9Cu3 (EN-AC46000). For the workpiece a special shape with a longitudinal slot has been developed, which allows the clamping of the lower section and the drilling experiment to be performed in the upper part of the workpiece (Fig. 1). This reflects typical industrial applications, in which the thermal expansion of the bore area is not suppressed by the clamping system. In this way, it was possible to minimise the heat flow into the clamping system as well.

Solid carbide twist drills with a diameter of $d = 10\,\text{mm}$ were used for the investigations. The tools have four margins and a special point design for the MQL-machining of aluminium with polished flutes for enhancement of the chip evacuation. The length-to-diameter ratio $l/D = 30$ of the produced bores was constant in all experiments and a pilot hole with a depth of $l = 30\,\text{mm}$ was applied. The internal MQL supply of the tool was performed by a special MQL device with an increased air pressure of $p_{\text{mql}} = 15\,\text{bar}$. Thereby the MQL mixing nozzle, which generates the

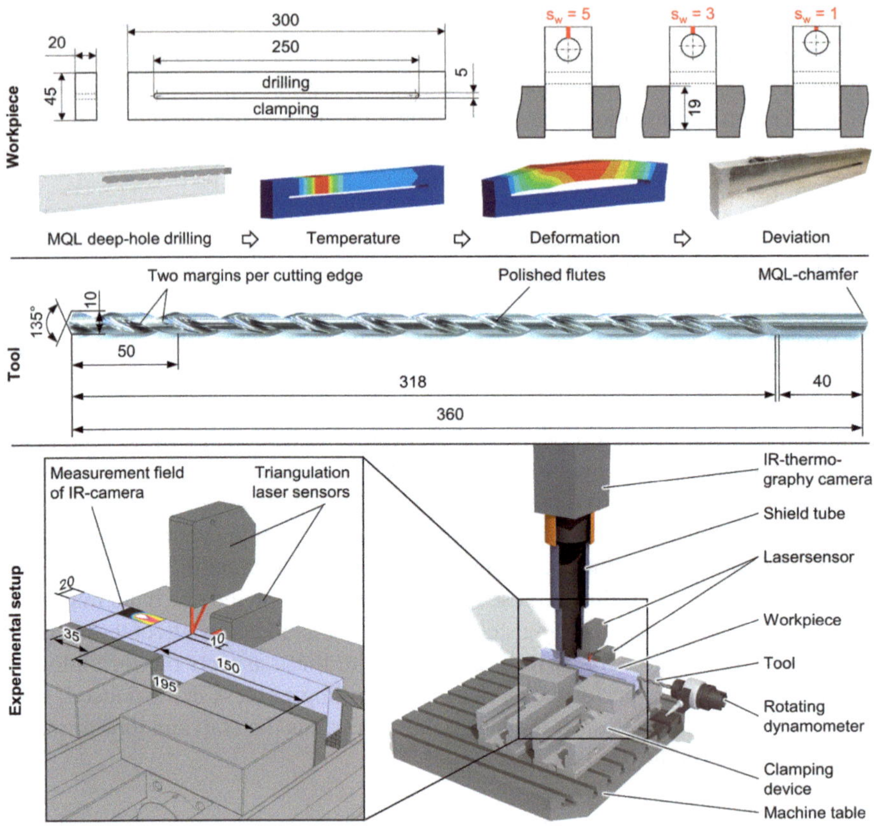

Fig. 1 Workpiece concept (*top*), deep-hole drilling tool (*middle*) and experimental setup including measurement equipment (*bottom*)

oil mist, was placed directly behind the spindle of the machine tool to ensure a high quality of the MQL aerosol and reliable process design.

The cutting forces were determined by a piezoelectric rotating dynamometer KISTLER 9125A. For the workpiece temperature measurement an infrared (IR) thermography camera InfraTec ImageIR 8300 was used with optics allowing a resolution of about $a = 50\,\mu\text{m/pix}$. In order to enhance the poor emissivity of the aluminium surface, the field of measurement was coated in a matt black colour, which increased the emission ratio of the workpiece to $\varepsilon \approx 0.98$ in combination with the shield tube avoiding external radiation. The thermomechanically induced deformations of the workpiece were detected using two laser triangulation sensors, which measure the radial part displacement in the middle of the drilling depth ($l = 150\,\text{mm}$).

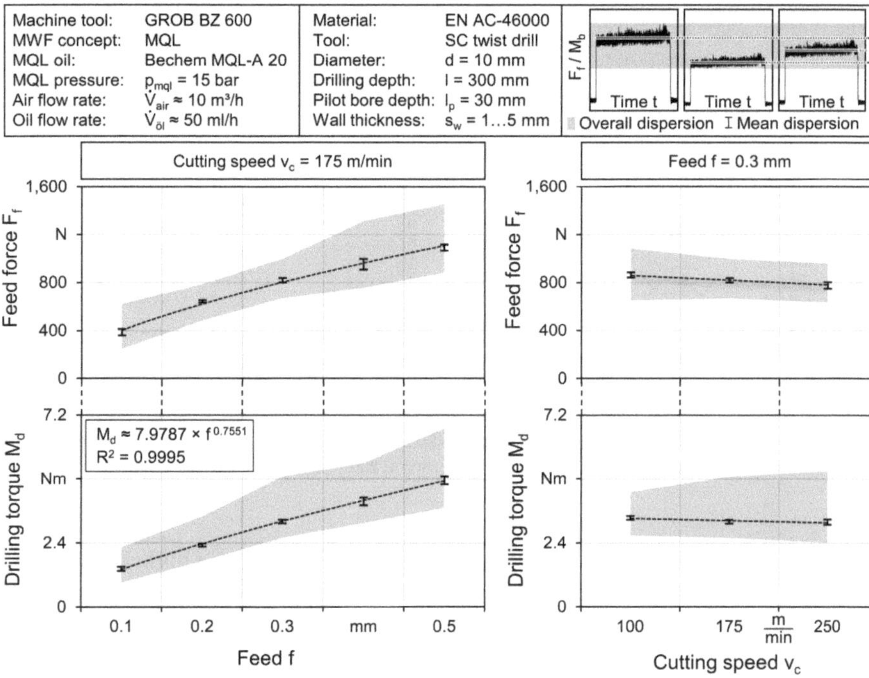

Fig. 2 Mechanical load at different cutting conditions

2.2 *Thermomechanical Workpiece Load*

The mechanical tool load is summarised in Fig. 2 for the investigated range of the cutting speed $v_c = 100 \ldots 250\,\mathrm{m/min}$ and the feed $f = 0.1 \ldots 0.5$ mm. The middle point of the experimental plan at $v_c = 175\,\mathrm{m/min}$ and $f = 0.3$ mm corresponds to the recommendation of the tool manufacturer for the deep-hole drilling of aluminium under MQL conditions. It is obvious that the feed has the major influence on the feed force and the drilling torque, whereas the cutting speed does not affect the mechanical load in a significant way. The increase of the cutting speed from $v_c = 100\,\mathrm{m/min}$ to $v_c = 250\,\mathrm{m/min}$ leads to a minor decrease of the feed force, caused by the thermal softening of the material, but the drilling torque of $M_d \approx 3.2$ Nm remains nearly constant for values $v_c > 175\,\mathrm{m/min}$, which can be explained by the increasing friction with higher cutting speed. Higher feed leads to a higher uncut chip thickness and therefore the drilling torque rises degressively as a power function of the feed $M_d \approx 7.9787 \cdot f^{0.7551}$. This relation correlates very well with the theory of *Kienzle* regarding the specific cutting force. In both cases of increasing the feed or increasing the cutting speed, the resulting effective power $P_e \approx 2 \cdot \pi \cdot n \cdot M_d$, required to machine the bore, rises. In the case of feed increase, this results from the higher torque at constant spindle speed n. Whereas in the case of the cutting speed, the

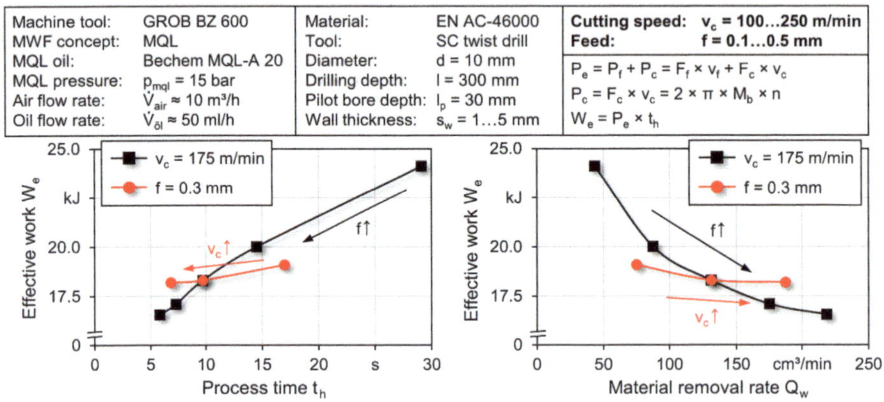

Fig. 3 Influence of the cutting parameters on the effective work per bore as a function of the primary machining time (*left*) and the material removal rate (*right*)

forces are nearly unchanged and only the higher speed leads to a higher power consumption. Considering the particular process time t_h needed to drill a bore, it can be shown that the energy or work per drilling operation can be significantly decreased when drilling with a higher feed rate (Fig. 3). The results for the effective work as a function of the primary machining time and the material removal rate demonstrate that the increase of the cutting speed leads to a balanced decrease of time and rise of the effective power, so the bore-related work remains nearly constant for $v_c > 175\,\mathrm{m/min}$. In contrast, the higher productivity achievable by increasing the feed allows for a considerable energy saving in comparison to identical productivity enhancement (process time decrease or material removal rate increase), generated by the increase of the cutting speed.

The thermal load on the workpiece, induced by the deep-hole drilling process, exhibits an excellent agreement with the presented investigations on the energetic aspects of the machining process (Fig. 4). The experiments using emulsion demonstrate the high cooling performance of the water-based MWF, which provides a reliable prevention of thermal workpiece expansion. According to the expectations, the high thermal load when applying MQL leads to a significant temperature increase of the machined part. Low feeds and thin-walled drilling situations cause local surface temperatures of more than $T = 100\,°\mathrm{C}$. A higher feed rate leads to a lower thermal loading of the workpiece, since more heat is redirected into the chip and the shorter process time reduces the heat exchange between tool, chips, MQL and workpiece. It is obvious that the first step of the cutting speed increase generates a minor reduction of the heat input, whereas the second increase from $v_c = 175\,\mathrm{m/min}$ to $v_c = 250\,\mathrm{m/min}$ exerts only a negligible influence on the resulting workpiece temperature. With regard to the bore position it can be shown that the eccentric drilling situations with resulting wall thicknesses of $s_w = 3\,\mathrm{mm}$ and $s_w = 1\,\mathrm{mm}$ cause a much higher temperature level in the machined part. Due to the reduced mass and heat capacity of these workpiece areas the same heat source, resulting from the

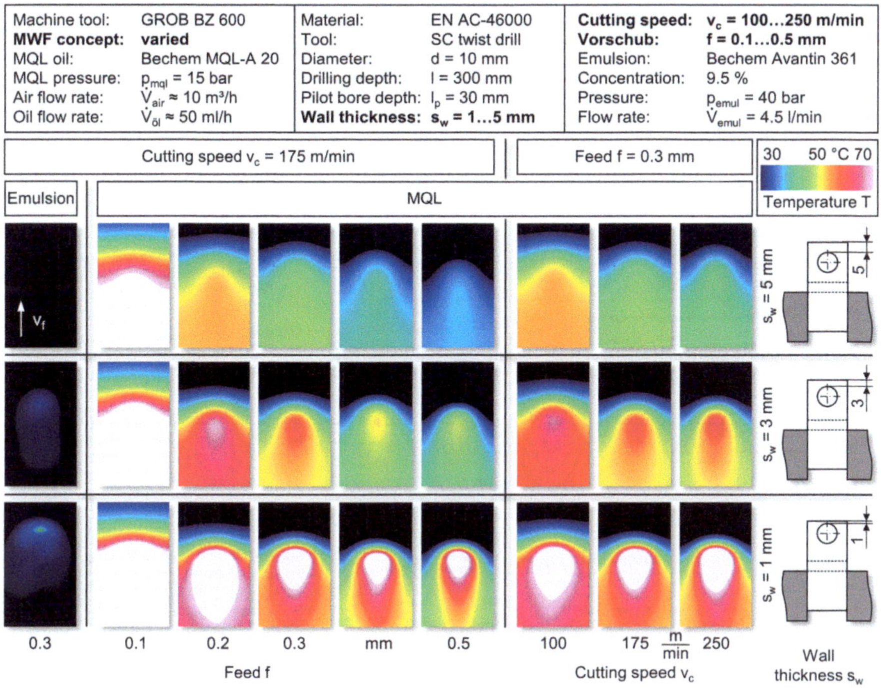

Fig. 4 Measured workpiece temperature distribution when the tool tip passes the measuring field of the IR-camera (see Fig. 1), depending on the cutting parameters and the bore position during deep-hole drilling using emulsion and MQL

working zone, and thus the same amount of heat flowing into the workpiece lead to a higher temperature increase within the thin-walled section. Further investigations and evaluations confirm that the integrated heat input of the process is almost independent from the bore position. The reason for that is the minor convective heat exchange with the environment and the fast homogenisation of the temperature field within the machined part. Therefore, the accumulated heat input depends mainly on the cutting parameters, particularly on the feed. The time dependent analysis of the results shows that apart from the primary heating due to the working zone, there exists a second thermal load on the bore wall, which is responsible for a great part of the entire heat input into the workpiece. Figure 5 summarises the mechanisms of the thermal loading in deep-hole drilling with MQL. The primary heat source corresponds to the working zone, where the shear and friction work is generally converted into heat. Higher feed per tooth leads to an increase of the overall heat generation and thus more heat flows in absolute terms into the workpiece as well. The reason for the lower relative heat partition with higher undeformed chip thickness is on the one hand the greater spatial distance between the primary shear zone and the workpiece surface in statu nascendi. On the other hand the primary heat flow into the

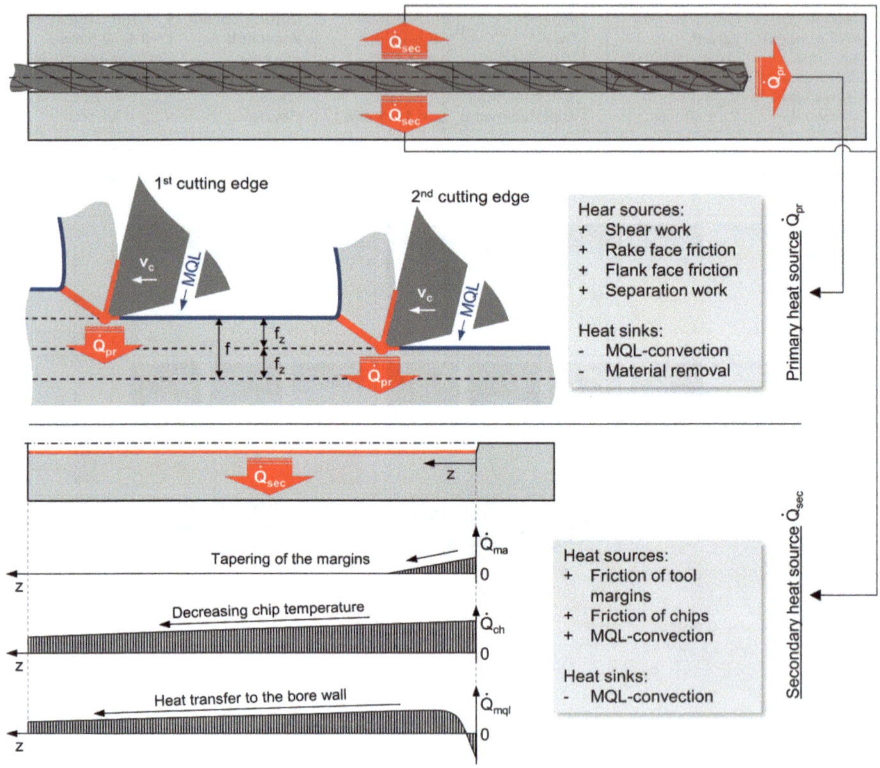

Fig. 5 Mechanisms of heat input into the workpiece in deep-hole drilling with MQL

bore ground is redirected to a great extent into the chip by the next cutting edge engagement, which in drilling immediately follows the prior one.

The secondary heat source results from friction between the tool margins, the chips and the bore wall as well as from an extensive heat exchange with the MQL flow. In contrast to emulsion, the flow rate of the MQL air is approximately one to two hundred times higher, which leads to a turbulent flow and very high convection coefficient. Therefore the MQL flow fulfils a cooling function only in the first section immediately after streaming out of the coolant channels of the tool. Thereby its temperature increases very fast as a result of the low heat capacity of air and due to the temperature gradient the MQL flow becomes a heat source at the bore wall. This secondary heating of the bore can be responsible for more than 50% of the total heat input into the workpiece Q_{wp} (Fig. 6). These values were calculated applying the heat capacity equation to a representative section of the workpiece after achieving a homogeneous temperature distribution within the workpiece.

The results illustrate that the increase of cutting speed above $v_{\mathrm{c}} = 175\,\mathrm{m/min}$ is not an appropriate strategy to reduce the thermal workpiece load. In contrast, the highest tested feed of $f = 0.5\,\mathrm{mm}$ allows not only a considerable decrease of the

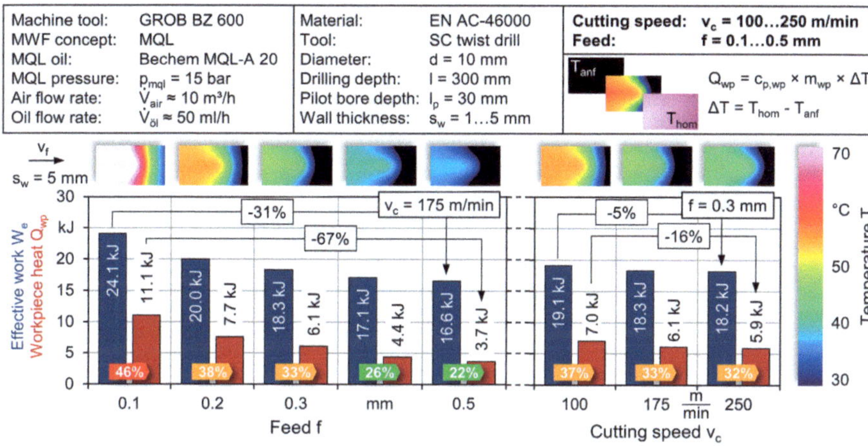

Fig. 6 Total heat input into the workpiece calculated after homogenisation of the temperature field

effective work per bore of $\Delta W_e \approx -31\%$, but also a substantial reduction of the resulting workpiece heat of $\Delta Q_{wp} \approx -67\%$ in comparison to the lowest feed of $f = 0.1$ mm.

2.3 *Workpiece Distortion and Deviations*

The considerations with regard to the heat input into the workpiece are confirmed by the in situ displacement measurement as well (Fig. 7). Due to the special workpiece shape, the main deformation occurs in *y*-direction and thus the *x*-direction will not be further considered. The workpiece distortion in *y*-direction shows a good conformance with the heat input calculations, so the feed increase from $f = 0.1$ mm to $f = 0.5$ mm leads to a decrease of the resulting deformation of approximately $\Delta u_y \approx -62\%$. Long term measurements of the displacement after completion of the drilling process have shown that the deflection of the workpiece returns to near zero values. Hence, it can be concluded that the distortion measured immediately after drilling $u_{y,end}$ is a result of the thermal expansion based on the accumulated heat amount in the workpiece.

The influence of the bore wall thickness and the feed on the thermomechanical workpiece deformation during the drilling process and the resulting straightness deviation of the bore is presented in Figs. 8 and 9, respectively. Whereas the displacement curves at the lowest feed of $f = 0.1$ mm are almost identical, the difference between the centric and eccentric bore positions become greater with higher feed. One possible explanation for this effect is the different speed of heat conduction in relation to the feed velocity. In the case of low feed, the heat front is much faster than the drilling process and thus the entire cross section at the particular drilling

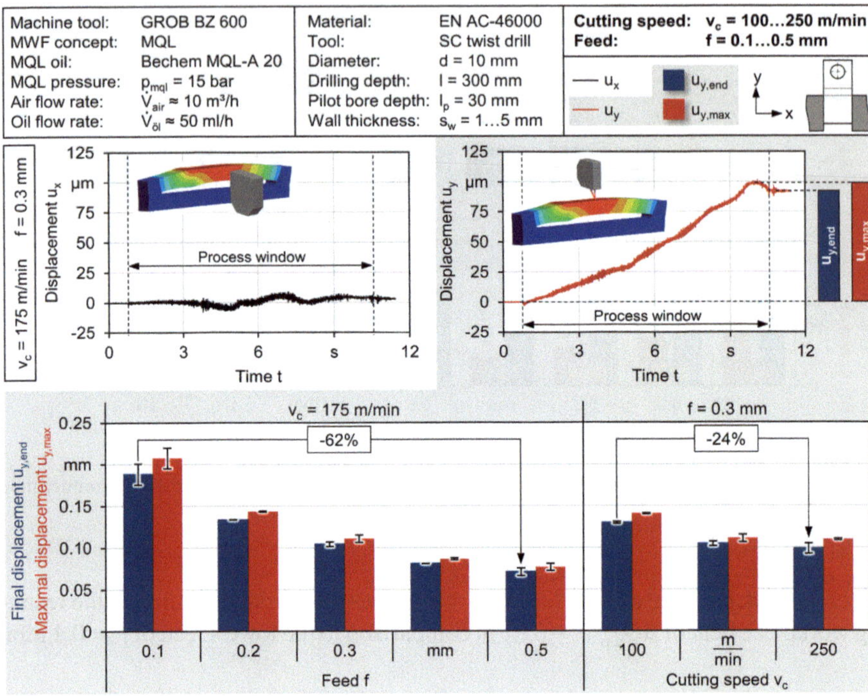

Fig. 7 Thermomechanically induced workpiece distortion during MQL deep-hole drilling

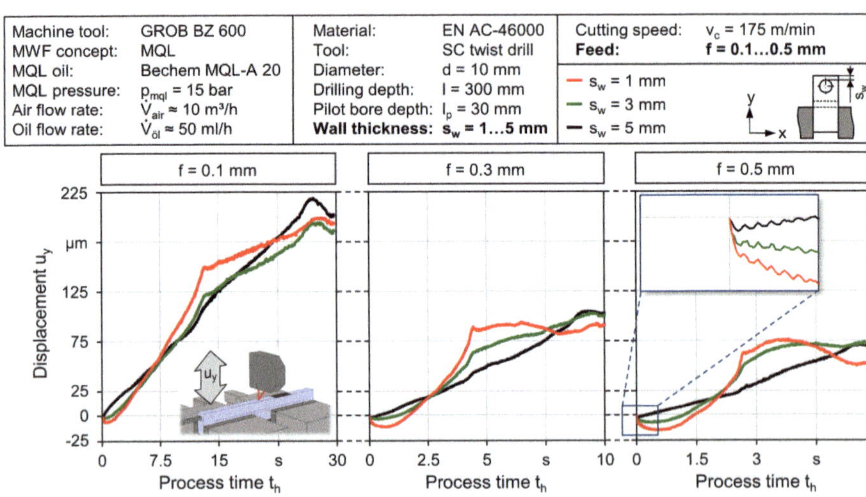

Fig. 8 Characteristic workpiece displacement graphs for different feed values and bore positions

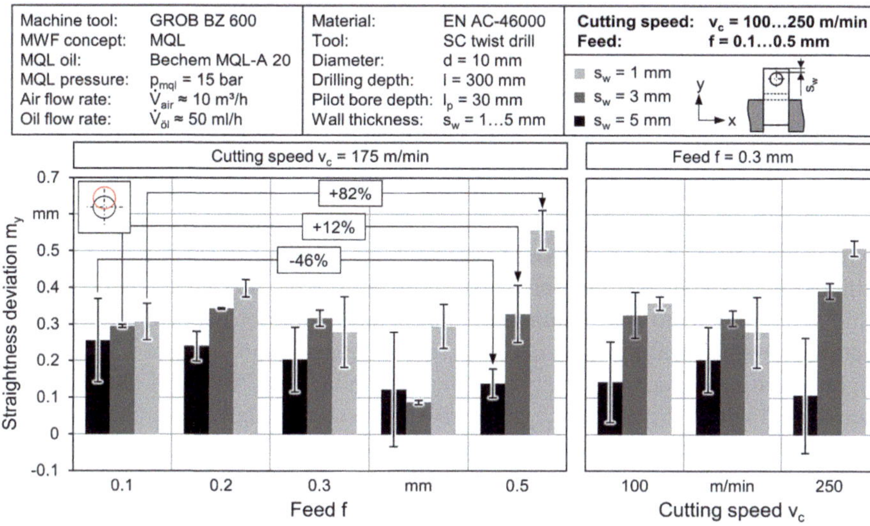

Fig. 9 Straightness deviation of the bores depending on the feed, the cutting speed and the bore position

position exhibits a nearly uniform temperature profile. By contrast, the high feed velocity when drilling with a feed of $f = 0.5$ mm leads to a high temperature gradient within the cross section drilled, so the thin-walled area, where a strong local heat accumulation occurs, also effects a high thermal expansion of this particular workpiece area. Due to these complex interactions the displacement initially decreases and becomes negative when drilling thin-walled bores with a high feed. In this case, the displacement increases rapidly and at the middle drilling depth the distortion reaches values, which are two to three times higher than the corresponding deformation in the centrical bore situation. This rapid rise of the displacement curve is the main reason for the strong increase of the straightness deviations, as illustrated in Fig. 9. Regardless of the cutting parameter values, significant systematic straightness deviations of the bore of several tenth of a millimetre result in the positive y-direction, which obviously corresponds to the thermally induced extension of the workpiece in the same direction. Further investigations using a solid part clamping along the entire drilling depth confirmed that the straightness deviation is not a result of the thermal softening even in thin-wall (thickness $s_w = 1$ mm) drilling situations. In fact, the long drilling tool is deflected by the thermomechanical workpiece deformation due to the guiding by the margins at the accruing bore wall.

2.4 High-Performance Deep-Hole Drilling

The presented results illustrate that the feed is the major influencing factor regarding the heat input into the workpiece. An increase of the cutting speed was identified to be not an appropriate measure for the reduction of the thermal load on the workpiece. Therefore high-performance deep-hole twist drills were developed in close collaboration with a tool manufacturer, in order to provide a much higher feed. To avoid an abrupt increase of the axial tool force, the high-performance drilling process was initiated with a standard feed of $f = 0.3\,\text{mm}$ up to a drilling depth of $l = 80\,\text{mm}$, which corresponds to $l = 50\,\text{mm}$ feed path behind the pilot bore. In the following section between a drilling depth of $l = 80\,\text{mm}$ and $l = 130\,\text{mm}$ the feed was increased slightly using a linear ramp function. The remaining $l = 170\,\text{mm}$ of the bore were then drilled with the final feed f_{end}, which was increased stepwise in order to determine the maximal feasible performance. Using the most sophisticated tool a final feed of $f_{\text{end}} = 4.0\,\text{mm}$ could be achieved (Fig. 10). The mechanical load increases following the current feed and reaches a feed force of $F_f \approx 8\,\text{kN}$ and a torque of $M_d \approx 30\,\text{Nm}$ at the highest feed. Thereby, the feed velocity exceeds the value of $v_f = 22,000\,\text{mm/min}$ and the feed drive has to provide a feed power of $P_f \approx 3\,\text{kW}$. The high feed process demands also for relative high spindle power to deliver the required cutting power of $P_f \approx 17.5\,\text{kW}$. The force curves illustrate also the extreme productivity enhancement. For example, when drilling with a final feed $f > 1\,\text{mm}$ the

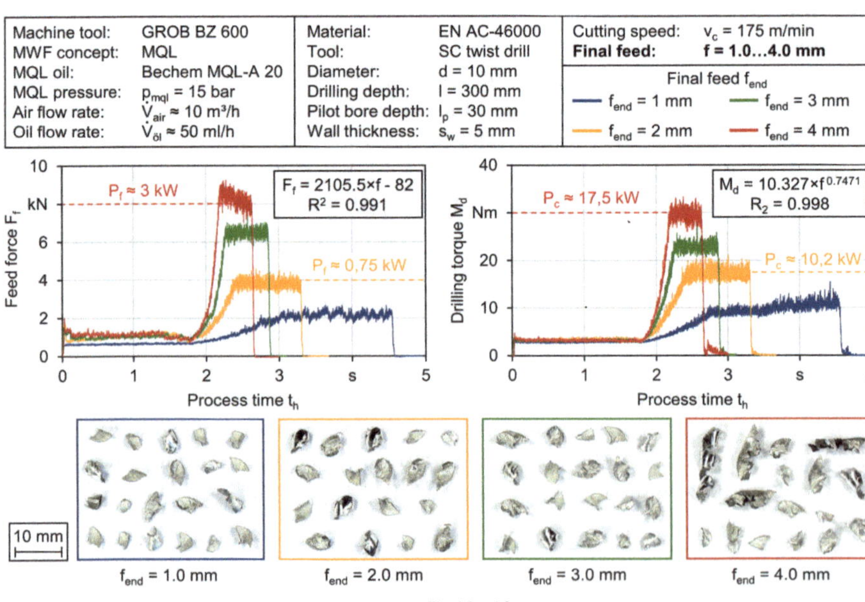

Fig. 10 Typical measuring curves of the mechanical load (unfiltered) and chip formation in high-performance deep-hole drilling

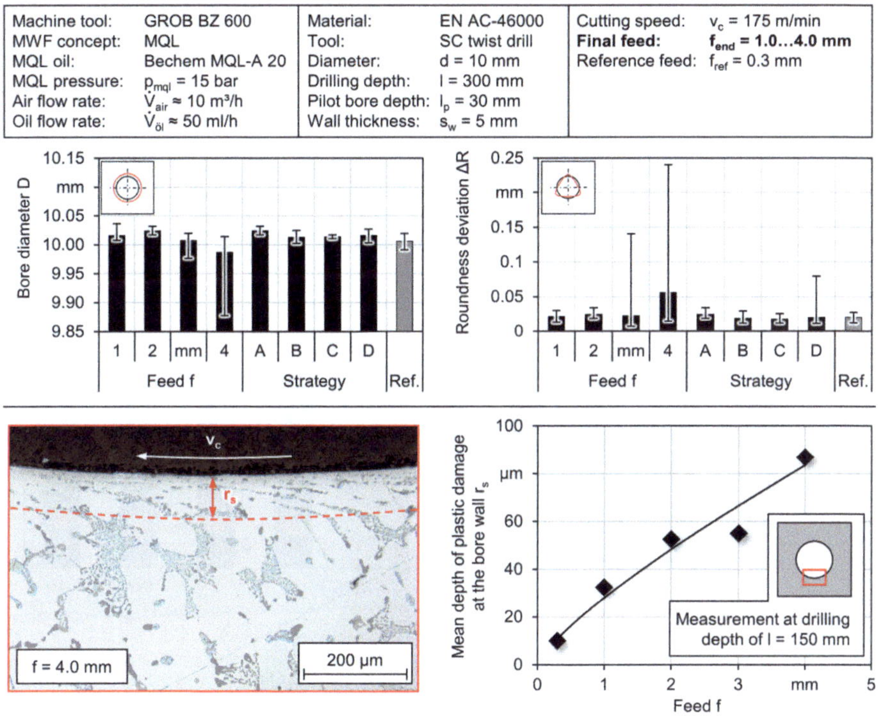

Fig. 11 Bore quality in high-performance MQL deep-hole drilling

initial feed section ($f = 0.3$ mm), which has a length of only $l = 50$ mm takes much longer machining time than the following ramped and high-feed sections together. The analysis of the chip formation shows that up to a final feed of $f_{end} = 3.0$ mm a reliable chip breakage and evacuation is feasible, whereas at a final feed of $f_{end} = 4.0$ mm the chips concatenate in the area of the tool's negative protective chamfer. The evaluation of the bore quality (Fig. 11) shows that even at a feed of $f_{end} = 3.0$ mm the diameter and the roundness deviations strongly increase, which is an indicator for a rough plastic deformation of the bore wall area. The metallographic analysis of the cross section is shown in the bottom of Fig. 11 and it confirms that the high dimensional and shape deviations of the bores results from the extensive plastic forming zone at the bore wall.

The energetic considerations with regard to the development of a high-feed drilling process are summarised in Fig. 12. Due to the slightly different grinded point shape of the tool a minor offset can be seen in the cutting work curve between the standard and the high-feed range. The decreasing tendency exists also in the high-performance domain, but the effective work as a sum of the cutting and the feed work $W_e = W_c + W_f$ remains with values of $W_e \approx 16$ kJ nearly constant in the entire final feed range investigated. There exists also an apparent increase of the feed work, which is no more negligible as usual, but has to be considered within the

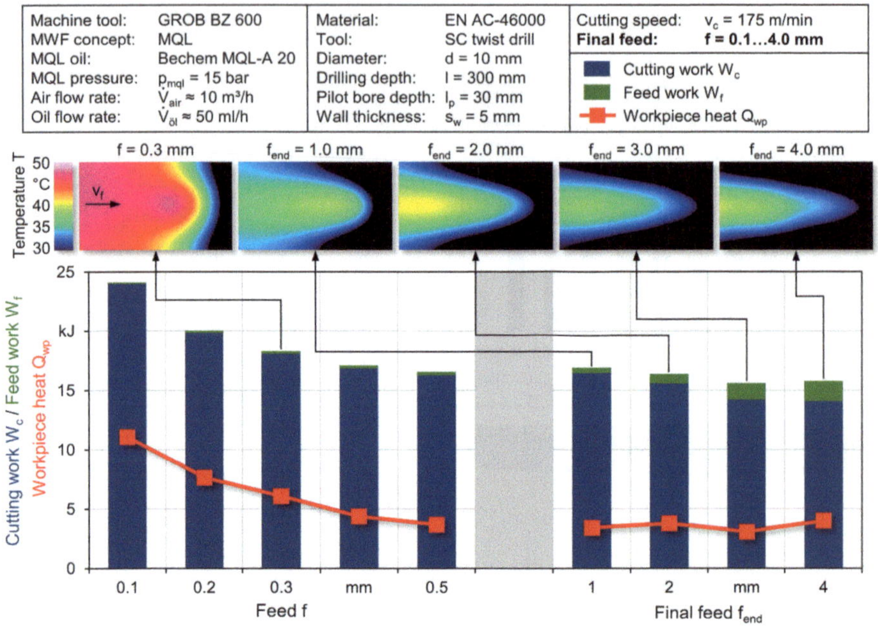

Fig. 12 Energetic considerations in high-feed deep-hole drilling

high-performance considerations as well. The temperature images at the moment when the drilling tool passes through the measurement field of the IR-camera demonstrate that the feed increase within the high-feed range effects a slightly different distribution of the temperature gradients, but the total heat input into the workpiece can not be reduced significantly by the feed increase. The resulting amount of heat, accumulated within the part during the deep-hole drilling process, remains for all final feeds between $f_{end} = 1.0\,\text{mm}$ and $f_{end} = 4.0\,\text{mm}$ at the same level of approximately $Q_{wp} \approx 3.5\,\text{kJ}$. Due to the high-speed plastic deformation in the bore wall area and the bigger chip conglomerations the total heat in the workpiece is even slightly increasing ($Q_{wp} \approx 4\,\text{kJ}$) in the maximum feed test with $f_{end} = 4.0\,\text{mm}$.

The in-process displacement and the resulting straightness deviation of the produced bores are presented in Fig. 13 for the case of centric bore position. The right side of the figure shows the comparison for the investigated high-feed range $f_{end} = 1 \ldots 4\,\text{mm}$. At the left side, different feed increase ramps are summarised, which were developed in order to investigate the influence of the initial low-feed part of the drilling process on the guiding of the tool and the resulting deviations. The feed variation reveals identical deflection remaining after the high-feed processes, relatively independent from the particular final feed. That correlates with the total heat input (cf. Fig. 12), which also could not be reduced by increasing the feed. The higher feed values of $f_{end} = 1 \ldots 2\,\text{mm}$ lead to a reduction of the straightness deviation in comparison to the reference process with a feed of $f = 0.3\,\text{mm}$. The reason

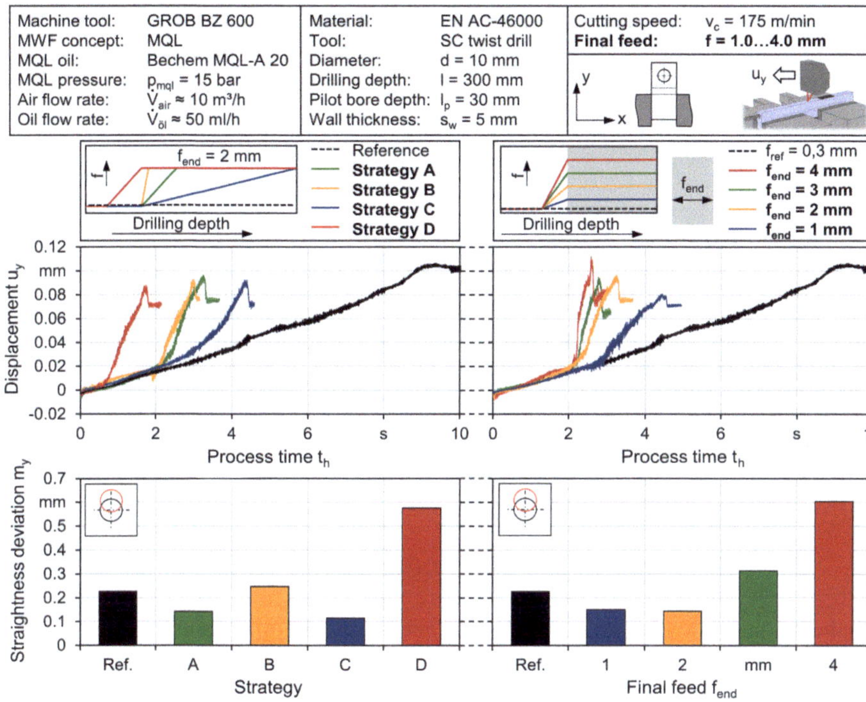

Fig. 13 Workpiece deflection and resulting straightness deviations in deep-hole drilling with high feeds

for this can be found on the one hand in the lower thermal distortion of the workpiece and on the other hand in the better tool centering due to the higher passive force component at the both cutting edges. A further increase of the feed to $f_{\text{end}} = 3 \ldots 4\,\text{mm}$ causes a significant rise of the deviations, which can be explained by the extremely increasing forces and thus also workpiece deformations. The comparison of the different ramp feed strategies (see left side of Fig. 13) shows that the initial low-feed guiding section has a major influence on the accruing bore straightness (cf. strategies A and D) and the most accurate bore can be generated by a smooth increase of the feed along the entire drilling depth (strategy C).

3 Microscopic Modelling and Simulation

Beside technological investigations of the deep hole drilling process, we aim for a highly efficient FE-simulation of the heat dissipation during the chip formation. For such a simulation, we have to consider an appropriate model.

The drilling process is obviously a topic of solid mechanics. On the microscopic level, the drill's cutting edge moves with high speed and high pressure on the cutting zone. Therefore, we have to consider frictional contact as well as plasticity. At last, a model, which allows chip formation, is introduced.

3.1 Thermoplastic Model with Frictional Contact

We begin with a model based on the Johnson-Cook approach for the thermoviscoplastic deformation, cf. [15]. For a clearer notation, we neglect temporarily the effects that are induced by frictional contact.

The quantities of interest are the deformation $\mathbf{u} : \mathbb{R}^d \times \mathrm{I} \to \mathbb{R}^d$ and the temperature $\theta : \mathbb{R}^d \times \mathrm{I} \to \mathbb{R}$ in the d-dimensional space and on a given time interval $I = [0, T]$. First of all we have to consider that the deformation and the temperature are in a state of equilibrium, i. e.

$$-\operatorname{div} P(\mathbf{u}, \theta) = 0. \tag{1}$$

The heat equation with two heat flows ω_e and ω_p is taken into account, too.

$$\omega_e + \omega_p + \rho c_p \dot{\theta} - \kappa \Delta \theta = 0.$$

In the first equation, $P(\mathbf{u}, \theta) = \Sigma(\mathbf{u}) - dK_{\text{bulk}} \alpha(\theta - \theta_0) \, \mathrm{Id}$ denotes the first Piola-Kirchhoff tensor. The term $\Sigma(\mathbf{u})$ describes the mechanical stress or rather the elasto-(visco)plastic material law, which may depend also on inner variables. Another source of stress is the thermal stress $dK_{\text{bulk}} \alpha(\theta - \theta_0)$, where θ_0 is the initial temperature of the workpiece. The constant α denotes the thermal extension coefficient and K_{bulk} the bulk modulus, which itself can be derived from the Young modulus E and the Poisson ratio ν, i. e. $K_{\text{bulk}} = \frac{E}{3(1-2\nu)}$. In the non-static heat equation, the density ρ, the specific heat capacity c_p and the thermal conductivity κ come into play as usual. The origin of the additional heat flows are the elastic deformation, which produces the heat flow

$$\omega_e = \mathscr{H}(\dot{\mathbf{u}}) = dK_{\text{bulk}} \theta_0 \alpha \operatorname{div}(\dot{\mathbf{u}}),$$

and the plastic deformation, which generates the heat flow

$$\omega_p = \beta_q \sqrt{\frac{3}{2}} \| \Sigma^D \| \dot{\varepsilon}_p.$$

Note that the elastic heat flow ω_e depends on the initial temperature θ_0. Of course the plastic one depends on the plastic strain rate $\dot{\varepsilon}_p$ and the stress tensor's deviator

σ^D. The Taylor Quinney coefficient β_q describes the ratio of the plastic work that is turned into heat.

Analysing the microscopic behaviour, we cannot neglect the contact constraints either, which induce a driving force. The cutting edges induce normal and tangential stresses on the contact area. For our modelling, we assume that the deformation of the cutting edge is neglectable. Consequently, the cutting edge is treated as (moving) rigid obstacle. The geometric contact conditions are given by

$$\mathbf{u}_n \leq g, \qquad \Sigma_{nn} \leq 0, \qquad \Sigma_{nn}\left(\mathbf{u}_n - g\right) = 0.$$

Here, $\mathbf{u}_n$ denotes the displacement on the contact boundary in normal direction, g the gap function, and Σ_{nn} the surface stress in normal direction. We refer to [17, Chap. 2] for more details. We are considering the linearised contact conditions, which are appropriate for small time steps using a suitable linearisation. Furthermore, we have to consider the frictional conditions

$$\begin{aligned} &\left\|\Sigma_{nt}\right\| \leq s, \\ &\left\|\Sigma_{nt}\right\| < s \Rightarrow \dot{\mathbf{u}}_t = 0, \\ &\left\|\Sigma_{nt}\right\| = s \Rightarrow \exists \xi > 0 : \dot{\mathbf{u}}_t = \xi \Sigma_{nt}, \end{aligned}$$

see [17, Chaps. 10 and 11], where Σ_{nt} denotes the surface stress in tangential direction, s the frictional resistance, which can depend even non-linear on different quantities like Σ_{nn} or θ, and $\dot{\mathbf{u}}_t$ the tangential velocity. The finite element analysis is based on the weak formulation of the above mentioned problem. To shorten the presentation, we skip the discussion of the weak formulation and instead refer to [17].

3.2 *Chip Formation*

A typical problem in the simulation of machining processes is the description of chip formation. One approach to handle chip formation is based on large deformations without material separation. This procedure leads to largely deformed elements in finite element methods, which require a regular remeshing of the current domain. A second approach is based on mesh separation, which necessitates the doubling of edges and faces in the mesh. Our approach to address this problem is to put the situation inside the framework of discontinuous Galerkin (dG) methods. In particular, this includes the extension of the continuous problem setting by special stabilisation or penalty terms, which at first enforce continuity approximately. We refer to [20] for dG-methods for elastoplastic problems. Mesh separation is now accomplished by not penalizing edges, which are to be opened. The downside of using the described ansatz is definitely the huge amount of degrees of freedom implicated by the dG-method. In order to resolve this issue, local mesh refinement based on a posteriori error estimators is inevitable and helps reducing the size of the non-linear systems to be solved while keeping high accuracy. Also the combination of continuous and discontinuous elements within the same mesh is highly advisable. We use discontinuous elements only in regions of the mesh where mesh separation is expected and continuous elements everywhere else, see [23]. Numerical results are also presented therein.

3.3 Discretising Elasto-Plastic Contact Problems Efficiently

In this part of the project we deepen our studies on the numerical treatment of systems of partial differential equations, where the solution is subjected to inequality constraints. We employ the FE-Galerkin method to obtain approximate solutions of such systems, which for instance typically arise in the field of continuum mechanics. Examples are torsion problems, where certain norms of the derivatives are bounded or contact problems, where the displacement is restricted by a rigid obstacle.

The basis for applying an FE-discretisation is a suitable mathematical setting, which in the topics under consideration takes the form of variational inequalities (VI). Especially, we focus on Lagrange-settings, which can be employed to handle the given constraints. In this way additional auxiliary variables are introduced, which are determined simultaneously to the original primal solution within a so-called mixed system. This approach turns out to offer some advantages: On the one hand, direct access is established to terms hidden in the original setting, e.g. for contact problems additional Lagrange parameters contain information about contact forces. Based on this, concepts for a posteriori error control as well as adaptive mesh refinement are reviewed and refined. A detailed discussion of this part of the project can be found in [6] considering obstacle problems involving a linear differential operator. The extension to non-linear differential operators and numerical results are discussed in [18, Chaps. 4 and 8]. Here, the obstacle problem is used as model problem for complex contact problems. The focus of [23] is on the generalisation of these results to dG discretisations of Signorini's problems involving elasto-plastic material laws. By the adaptive methods, we achieve a given accuracy at approximately minimum number of degrees of freedom.

3.4 Solving Elasto-Plastic Contact Problems Efficiently

Even using adaptive methods, a realistic, three dimensional simulation of the process is still expensive due to a large number of unknowns. To achieve the needed accuracy of the solution in an acceptable time frame, we have to apply very efficient solution algorithms and to utilize modern techniques of parallel computation.

We use the techniques described for instance in [27] for solving the frictional contact problem. They are based on a discretization of the Lagrange multipliers concerning the contact constraints by Mortar finite elements. By exploiting the biorthogonality property, one can construct very efficient solution algorithms based on the active set strategy. The elasto-plastic problem is solved by the techniques discussed e.g. in [11], where a non-smooth projection onto the yield surface is the basis of the applied Newton scheme. Due to the restrictions of the process, we have to choose very small time steps. Thus, we can introduce some explicit approaches. On the one hand, we consider the information of the last time step to evaluate the frictional resistance and have only to consider friction of Tresca type in the solution algorithm, which is much easier and faster to solve. On the other hand, we decouple the thermal part from the mechanical one. We first solve the mechanical part using the temperature distribution of the last time step. Then the thermal part is considered without coupling back the information.

Beside an efficient solution algorithm also a highly efficient implementation is needed. To conquer this challenge, one can utilize the modern techniques of parallel computation, where hundreds or more processor cores keep the computation time sufficiently small. For this task, sufficient tools are provided by the open source DEAL.II finite element library [3, 4], which uses the TRILINOS library [14] for its parallel linear algebra and P4EST [8] for parallel mesh handling. Our approach for elasto-plastic contact problems is described in detail in [12], where especially the strong scalability is investigated based on numerical evaluations. We found that the overall computation time is almost halved, if the number of cores is doubled. Thus an optimal scaling is achieved. However, there is always an overhead due to the necessary communication work between the cores. Since the communication work grows, when we use more cores, the speed up decreases progressively.

4 Macroscopic Modelling and Simulation

In addition to the microscopic modelling and its simulation, we also consider the macroscopic modelling. As we want to compensate the effects of the deep-hole drilling process onto the whole workpiece, a macroscopic view is preferable. For this purpose, we reformulate the material modelling.

We carry the ideas of MQL, which are presented in Sects. 2.1 and 2.2, forward and fit our model to the macroscopic level. The comparatively high input of heat can lead to thermally induced deviations. This is an elastic deformation and influence of the plasticity vanishes in this context. Hence, we use a thermoelastic model and respectively mixed elliptic and parabolic partial differential equations.

For the numerical investigations two different FE-software tools were used, namely the commercial program Ansys and the open source library deal.II, cf. [3]. The first one was used for initial simulations, whereas the latter asserts itself useful, since it provides parallelisation tools as well as the possibility to integrate other aspects presented in subsequent paragraphs.

First results are published in [5]. We give a résumé of the model, which was presented therein.

4.1 Thermoelastic Model with Secondary Heat Source

We consider the domain $\Omega \subset \mathbb{R}^3$ that is given by the workpiece. Most of the components are known from the microscopic model, cf. (1).

$$\begin{aligned} -\operatorname{div}(P(\mathbf{u},\theta)) &= 0 && \text{in } \Omega \times (0,T),\\ \rho c\dot{\theta} - \kappa \Delta\theta &= 0 && \text{in } \Omega \times (0,T),\\ \mathbf{u} &= 0 && \text{on } \Gamma_{D_1} \times (0,T),\\ P(\mathbf{u},\theta)\cdot n &= F_{\mathrm{f}}/|\Gamma_N| && \text{on } \Gamma_N \times (0,T),\\ \theta(\cdot,0) &= \theta_0 && \text{in } \Omega,\\ \partial_n\theta &= P_{\mathrm{W}}/|\Gamma_N| && \text{on } \Gamma_N \times (0,T),\\ \kappa\partial_n\theta + \gamma_{\mathrm{OS}}(\theta - T_{\mathrm{bulkOS}}) &= 0 && \text{on } \Gamma_{D_2} \times (0,T),\\ \kappa\partial_n\theta + \gamma_{\mathrm{W}}(\theta - \theta_s) &= 0 && \text{on } \Gamma_{D_3} \times (0,T). \end{aligned} \tag{2}$$

Since we can neglect the plasticity in the macroscopic model, the heat flow ω_p is dropped from the heat equation in the second line of (2). This would leave the heat flow $\omega_e = \mathscr{H}(\dot{\mathbf{u}}) = dK_{\mathrm{bulk}}\alpha\theta_0 \operatorname{div}(\dot{\mathbf{u}})$ as the only coupling term in the heat equation. However, numerical results shown in [5] suggest that this term is here negligible, too.

Furthermore, the deformations are small due to the macroscopic point of view, such that the usage of the linearised strain tensor $\varepsilon(\mathbf{u}) = \frac{1}{2}\left(\nabla\mathbf{u} + \nabla\mathbf{u}^T\right)$ is justified. Additionally, an isotropic linear elastic material behaviour is assumed, because of the small stresses. This gives the formulation $\sigma = C\varepsilon(\mathbf{u}) = \lambda\,\mathrm{tr}(\varepsilon(\mathbf{u}))\,\mathrm{Id} + 2\mu\varepsilon(\mathbf{u})$ with the Lamé coefficients $\lambda = \frac{E\nu}{(1+\nu)(1-2\nu)}$ and $\mu = \frac{E}{2(1+\nu)}$.

The Dirichlet boundary Γ_{D_1} is given by the clamping area in the experimental setup, cf. Sects. 2.1 and 2.2. The heat transmission on the outer surface Γ_{D_2} is taken into account by the second to last equation in the system (2), which depends on the bulk temperature T_{bulkOS} and the heat transfer coefficient γ_{OS}.

The chisle and cutting edges exert a feed force F_{f} per area on the ground Γ_N of the borehole and act as primary heat source P_{W}, too. This is illustrated in Fig. 15. The chips, which are not explicitly modelled, transport the removed, but heated material in the MQL along side the wall of the drilled hole Γ_{D_3}. This results in a convective heat transfer

$$\kappa\partial_n\theta + \gamma_{\mathrm{W}}(\theta - \theta_s) = 0 \text{ on } \Gamma_{D3} \times (0,T),$$

where the function θ_s is defined by the two linear graphs in Fig. 14.

For a clearer presentation, the boundary conditions are not explicitly notated in this section. The time interval $(0,T)$ is decomposed in N timesteps by $0 = t_0 < t_1 < \ldots < t_N = T$ with timestep size $k_n = t_n - t_{n-1}$. Due to the material removal, we have to mix two discretisations. During the drilling process, we have to apply an implicit Euler method for the time discretization in the heat equation (4), because the load is specified only at the current time. The Crank-Nicolson method can be used otherwise, e.g. in cooling phases. Then the time discretised equations read

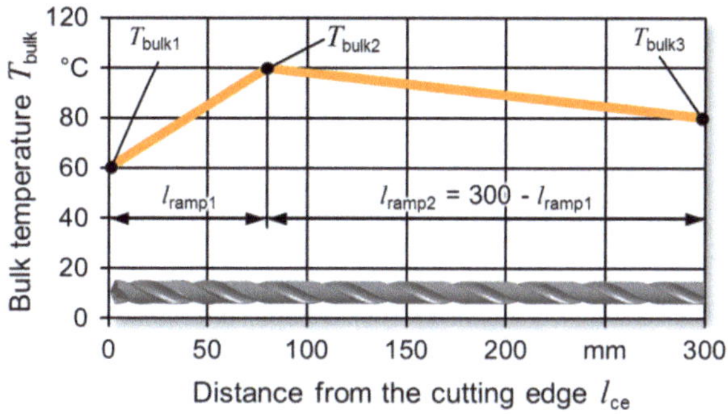

Fig. 14 Boundary condition at the drilled hole wall, applied as a moved (beginning at the current drilling position) double ramped convective heat transfer, in order to summarize the MQL-stream and both the friction effects of the guiding chamfers and the chips along the wall of the drilled hole. This picture is taken from [5]

$$
\begin{aligned}
-\operatorname{div}\left(P(\mathbf{u}^n, \theta^n)\right) &= 0 && \text{in } \Omega, \\
\rho c \frac{\theta^n - \theta^{n-1}}{k_n} - \beta \kappa \Delta \theta^n &= (1-\beta)\kappa \Delta \theta^{n-1} && \text{in } \Omega
\end{aligned}
$$

with the notation $u^n = u^n(x) = u(x, t_n)$ and $\beta \in [0, 1]$, where $\beta = 1$ corresponds to implicit Euler and $\beta = 0.5$ to Crank-Nicolson, see for instance [13]. Since the Crank-Nicolson scheme is of second order in contrast to implicit Euler, which is of first order, larger timesteps can be used.

Multiplying with the timestep size k_n and collecting all terms, which involve the unknowns $\mathbf{u}^n$ and θ^n on the left side, leads to

$$-\operatorname{div}(P(\mathbf{u}^n, \theta^n)) = 0 \qquad \text{in } \Omega, \tag{3}$$

$$\rho c \theta^n - k_n \beta \kappa \Delta \theta^n = \rho c \theta^{n-1} + k_n (1-\beta)\kappa \Delta \theta^{n-1} \qquad \text{in } \Omega. \tag{4}$$

Since the reduced heat equation does not depend on the mechanical variable u^n any longer, one can first solve the heat equation and in a subsequent step the elastic equation.

4.2 *Fictitious Domain Approach*

Since the ambition is to simulate a deep hole drilling process as realistic as possible, the removed material has to be handled in the modelling. Cells in the FE-simulation are deactivated according to the corresponding material removal. In this context deactivation is realized by multiplying the local Lamé coefficients for equation (3) as well as the thermal conductivity for (4) each by 10^{-6}. Due to numerical reasons it is not advisable to set these parameters to zero. Furthermore, this

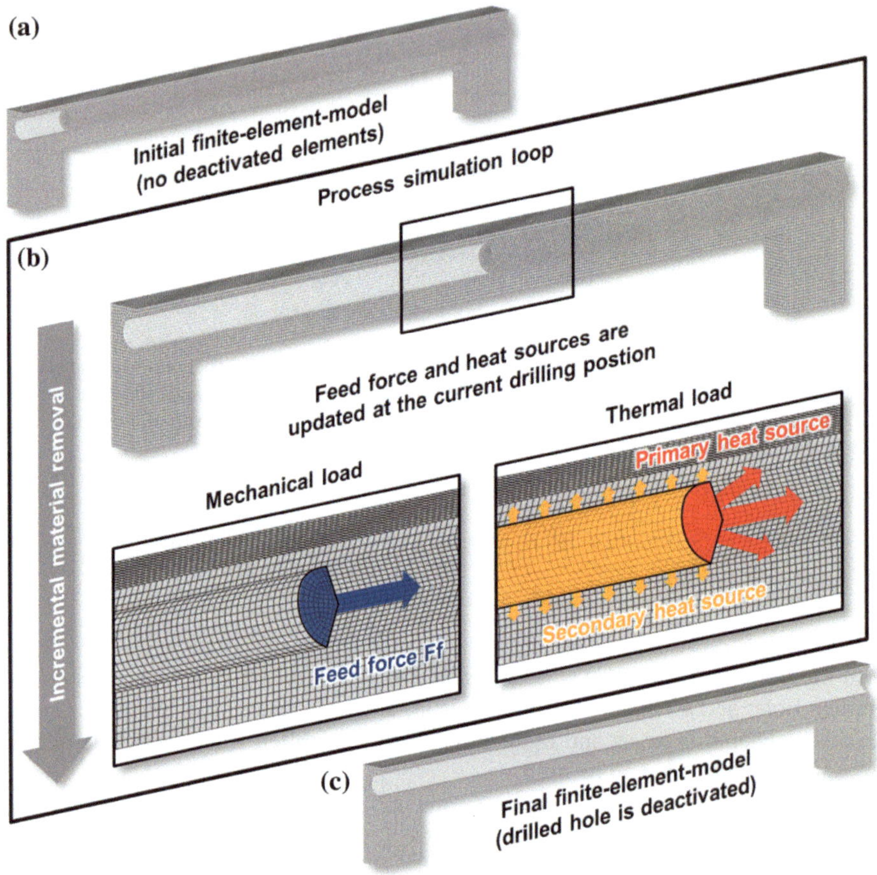

Fig. 15 3D-FE-model in sectional view: **a** initial stage with pilot hole, but no deactivated elements. **b** load application within the process loop, in which elements are incrementally deactivated according to the current drilling time and depth. **c** final FE-model. This picture is taken from [5]

deactivation implies an error in the modelling, which is neglectable in this context. The meshes are created, for now, in such a way that a cell belongs either to the removed material or to the remaining material completely. This implies that from one time step to another a whole layer of cells, at the bottom of the hole, must be deactivated. Furthermore the drill defines the geometry of the mesh, since the mesh has to fit the respective geometry of the drilling tool. Additionally, we have to consider also the motion of the tool, which is given by the velocity and timestep size. Since we perform our simulation on a fictitious domain, this very approach is called fictitious domain method.

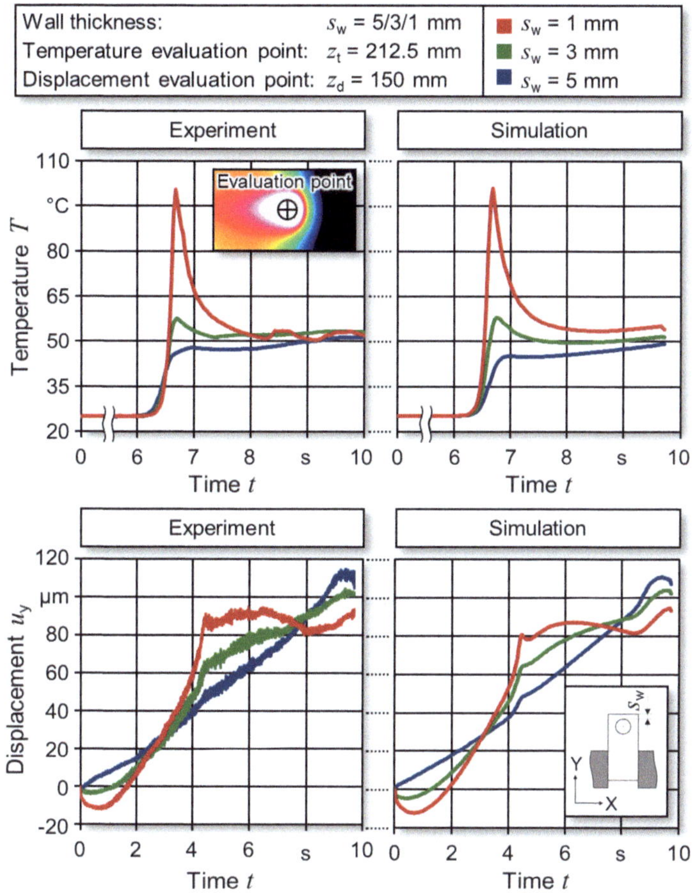

Fig. 16 Comparison between the experimental measured and simulated values of the temperature at point (0, 45.0, 212.5) and the displacement into y-direction at point (0, 45.0, 150.0). This graphic is taken from [5]

In [5] first numerical results are presented. We investigated the deformation as well as the temperature development and observed a good coincidence of the experimental and simulative results, see Fig. 16.

A major disadvantage of the current model, or rather the used discretisation in time and space, is that the discretisation in space has to fit to the one in time as well as the motion of the drilling tool, which is given by the velocity. From one time step to another a whole layer of cells must be removed. Taking a different time step size into account implies that a different mesh has to be used for the simulation. Furthermore, we cannot consider the distortion of the bore during our simulation, since we have no a priori knowledge, where the drill leaves its expected path.

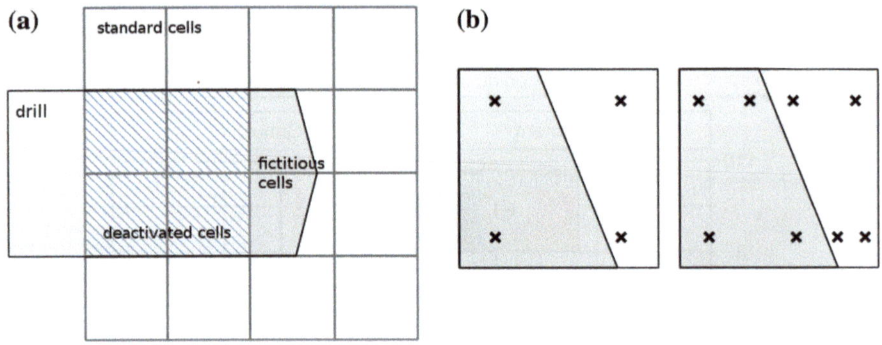

Fig. 17 **a** Removal of material, **b** Adaptation of a quadrature

To overcome these issues, we apply a more sophisticated fictitious domain method, a variant of the finite cell method, see for instance [9] and the references therein. The treatment of different types boundary conditions are discussed e.g. in [22]. The basic idea is rather easy. We use a simple triangulation, which does not match the actual geometry, for the FE-discretisation; but we use a perfectly fitting mesh for the underlying calculations on each cells. This is achieved with a dynamic quadrature formula for the numerical integration. The drill may cut off only parts of a cell, cf. Fig. 17a. For the numerical computations, a suitable quadrature formula is needed. Therefore, we move the quadrature points, which are also called Gauß points, accordingly to the fictitious cells within the true grid, as seen in Fig. 17b.

With the newly independence of time and spatial discretization, we are able to use an error based refinement strategy.

4.3 A Residual Error Estimator

As in the microscopic case discussed in Sect. 3.3, we apply adaptive methods to increase the efficiency of the underlying discretisation. Our approach is based on a residual error estimator developed for a standard heat equation with homogeneous Dirichlet boundary values in [25, Chap. 6] adapted to our problem. Additionally, a second residual error estimator, cf. [25, Chap. 4], provides information about the error in the mechanical equation (3). It should be remarked that due to the fictitious domain approach we obtain only heuristic error indicators, i. e. the reliability and efficiency are not proven.

Here we use the notation $(\theta_h^n, \mathbf{u}_h^n)$ for the solution of the discretised version of the equation system (2) in the n-th timestep. The indicators for the temporal discretization at each time t_n are implied by the heat equation (4) and read as

$$\eta^n_{\theta,k} = \left[\sum_{E\subset\Gamma_{D_2}} k_n\gamma_{\mathrm{OS}} \|\theta^n_h - \theta^{n-1}_h\|^2_{0;E} + \sum_{E\subset\Gamma_{D_3}} k_n\gamma_{\mathrm{W}} \|\theta^n_h - \theta^{n-1}_h\|^2_{0;E} \right.$$
$$\left. + \sum_{K\in\mathcal{T}_{h,n}} k_n \left\| \kappa\,\nabla(\theta^n_h - \theta^{n-1}_h) \right\|^2_{0;K} \right]^{1/2} .$$

There $\mathcal{T}_{h,n}$ denotes the set of all cells K of the triangulation in the n-th timestep and E is an arbitrary face of the triangulation. In contrast to the temporal indicators, the spatial ones consider also the error of the mechanical equation. Hence two different indicators have to be considered on each cell, i. e.

$$\eta^n_h = \left[\sum_{K\in\mathcal{T}_{h,n}} \eta^n_{h;K} \right]^{1/2}, \qquad \eta^n_{h;K} = \eta^n_{u,h;K} + \eta^n_{\theta,h;K}.$$

The first part is implied by the local error on a cell K in the mechanical equation (3) and the second part by the local error in the heat equation (4). They are defined by some residuals, which have a common structure and are given by

$$\eta^n_{u,h;K} = k_n h_K^2 \|R_{u,K}(\mathbf{u}^n_h)\|^2_{0;K} + \frac{k_n}{2} \sum_{E\subset\partial K} h_E \|R_{u,E}(\mathbf{u}^n_h)\|^2_{0;E}$$

and

$$\eta^n_{\theta,h;K} = k_n h_K^2 \left\| R^n_{\theta,K} \right\|^2_{0;K} + \frac{k_n}{2} \sum_{E\subset\partial K} h_E \left\| R^n_{\theta,E} \right\|^2_{0;E}.$$

Let Γ_I denote the set of all inner faces (in 3d) that are not part of the Neumann boundary Γ_N, and n_E the normal vector on a face E. Then the mechanical residuals are given by

$$R_{u,K}(\mathbf{u}^n_h) = dK_{\mathrm{bulk}}\alpha\left(\theta^n_h - \theta_0, \operatorname{div}\mathbf{u}^n_h\right)_K + \beta \operatorname{div}\sigma(\mathbf{u}^n_h) \qquad \text{in } K,$$
$$R_{u,E}(\mathbf{u}^n_h) = \begin{cases} \left[\sigma(\mathbf{u}^n_h)\cdot n_E\right] & \text{on } E\in\Gamma_I, \\ \left[F_f/|\Gamma_N| - \sigma(\mathbf{u}^n_h)\cdot n_E\right] & \text{on } E\subset\Gamma_N, \\ 0 & \text{on } E\subset\partial\Omega\backslash\Gamma_N, \end{cases}$$

and the jump

$$\left[\sigma(\mathbf{u}^n_h)\cdot n_E\right] = (\sigma(\mathbf{u}^n_h)\,|_{K_1} - \sigma(\mathbf{u}^n_h)\,|_{K_2})\cdot n_E \qquad \text{for } E = K_1\cap K_2.$$

For the thermal residuals we define Γ_J by the inner faces that do not belong to neither of Γ_N, Γ_{D_2} nor Γ_{D_2}. It holds

$$
R^h_{\theta,K} = -\rho c \frac{\theta^n_h - \theta^{n-1}_h}{k_n} + \kappa(\beta \Delta\theta^n_h + (1-\beta)\Delta\theta^{n-1}_h)
$$

$$
R^h_{\theta,E} = \begin{cases} \kappa\left[\left(\beta\nabla\theta^n_h + (1-\beta)\nabla\theta^{n-1}_h\right)n_E\right)\right] & \text{on } E \in \Gamma_J, \\ \kappa\left[P_W/|\Gamma_N| - \left(\beta\nabla\theta^n_h + (1-\beta)\nabla\theta^{n-1}_h\right)n_E\right] & \text{on } E \subset \Gamma_N, \\ \left[\gamma_{\mathrm{OS}}\left(\theta_{\mathrm{OS}} - \beta\theta^n_h - (1-\beta)\theta^{n-1}_h\right)\right. & \\ \left. -\kappa\left(\beta\nabla\theta^n_h + (1-\beta)\nabla\theta^{n-1}_h\right)n_E\right] & \text{on } E \subset \Gamma_{D_2}, \\ \left[\gamma_{\mathrm{W}}\left(\theta_{\mathrm{W}} - \beta\theta^n_h - (1-\beta)\theta^{n-1}_h\right)\right. & \\ \left. -\kappa\left(\beta\nabla\theta^n_h + (1-\beta)\nabla\theta^{n-1}_h\right)n_E\right] & \text{on } E \subset \Gamma_{D_3}, \\ 0 & \text{on } E \subset \partial\Omega. \end{cases}
$$

At this point, we have local and global error indicators for the space variable only. The global indicator for the temporal variable is given by the maximum over all timesteps, i. e.

$$
\eta_k = \max_n \eta^n_{\theta,k} \quad \text{and} \quad \eta_h = \max_n \eta^n_h .
$$

Both values are used for a simple error balancing in the algorithm. If η_k is larger than η_h by a factor greater than ten, then only a refinement of the temporal discretisation is performed. Otherwise refine the mesh adaptively and if η_h is larger than η_k by the same factor, do not change any timestep sizes.

We chose to perform the adaptive refinement only after the whole process simulation and start a new process cycle afterwards. Furthermore we do not change the mesh during one simulation cycle. On the one hand, we do not need to transfer the discrete solution between differing meshes. On the other hand, we would sustain a loss in the efficiency of the algorithm. This is because in a reasonable parallelisation the computational load must be balanced fairly uniformly over all processors. Hence the load balancing must be restored repeatedly, leading to an immense overhead in the simulation. This approach is hence based on the time global error indicator on each cell

$$
\eta_{h,K} = \max_n \eta^n_{h,K} .
$$

The error estimator is applied to a rather coarse mesh, which contains 3,488 cells, for a better visualization of the indicators. In the simulation 150 timesteps, 100 for the actual process and 50 for the cooling phase, were performed. The visualization tool ParaView, [2], was used to clip the workpiece along the bore such that the inner of the workpiece can be seen. As might be expected, the numerical experiments show that the error is largest near the contact zone of the drilling tool, cf. Figs. 18 and 19b, since there is the largest heat gradient as well as the biggest stress. Furthermore in the cells close to edges of the workpiece we need to refine our mesh, too.

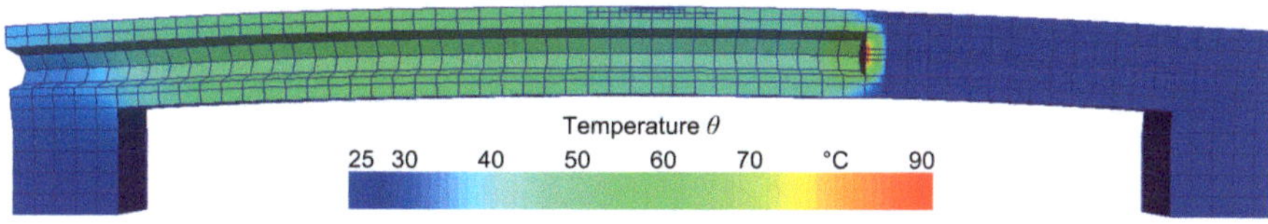

Fig. 18 Temperature distribution and deformation, which is scaled by 100. Completely deactivated cells are not shown

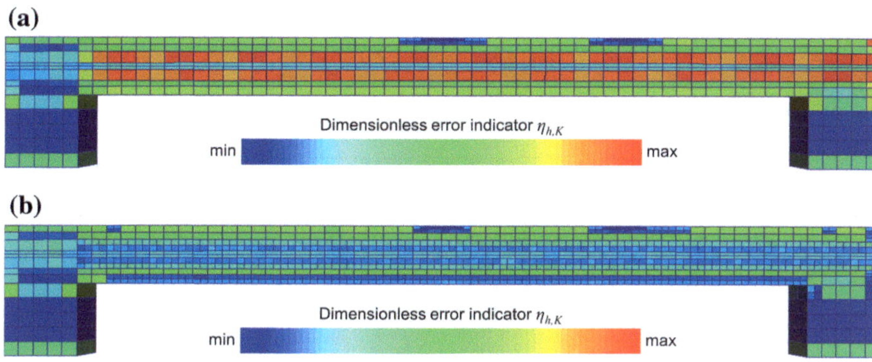

Fig. 19 **a** Indicator distribution before refinement, **b** Indicator distribution after refinement

Fig. 20 Strong scalability of the error estimator. The *lower line* represents optimal strong scaling. The *upper line* shows the scaling of the estimation during the simulative process

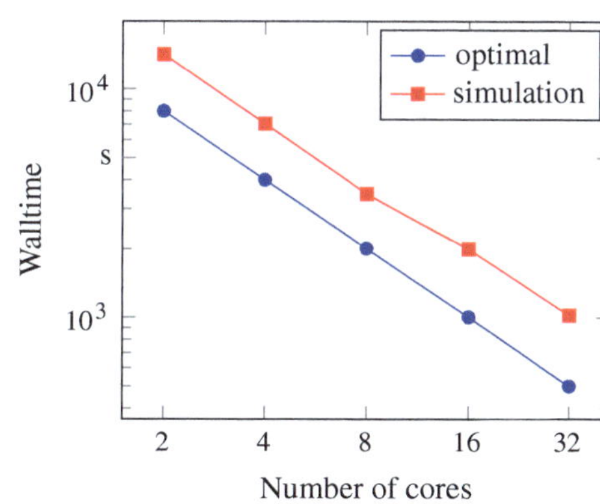

Since we focus on the efficiency of the solution algorithm, the estimator has to be parallelised optimally, too. Figure 20 shows that our implementation of the error estimator is also strongly scalable, i. e. when doubling the number of cores the process time is halved.

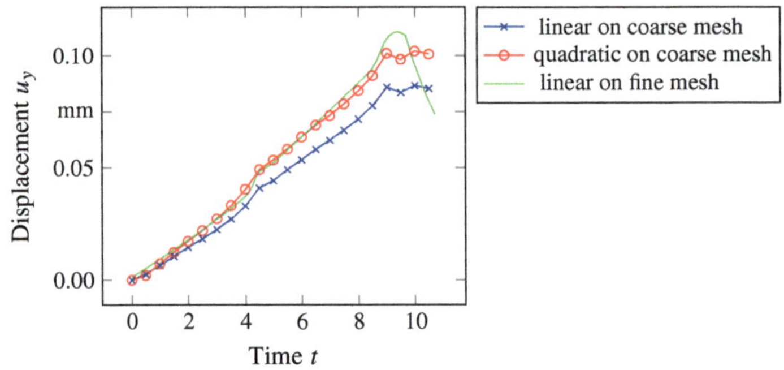

Fig. 21 Deformation at point (0, 45.0, 150.0) over time. The wall thickness is 5 mm

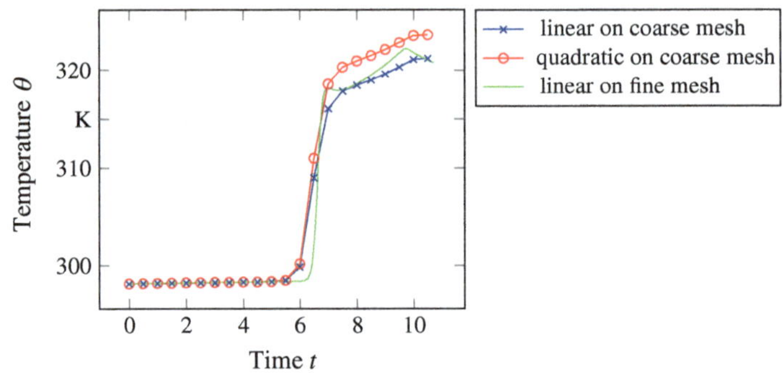

Fig. 22 Temperature at point (0, 45.0, 212.5) over time. The wall thickness is 5 mm

4.4 Further Studies

In further studies we increased the order of the FEs, i. e. instead of trilinear ansatz functions we used triquadratic ones. The advantage is that those functions match the geometry of the drilling tool better. However the drawback is a higher computational cost due to the increase in number of degree of freedoms per cell. Furthermore the quadrature formula needs more points too, since the numerical integration has to be exact for the respective ansatz functions. The numerical results show that as expected the higher order simulation has a larger runtime than the trilinear simulation. The comparison of the evaluation in the displacement and temperature point is presented in Figs. 21 and 22. The reference simulation is the one presented in Sect. 4.1 on a fine mesh with 291 timesteps. Note that we did not use the fictitious domain method in the reference simulation, but here we do.

The higher order elements are much better suited for solving the mechanical equation (3). We match the reference results, and hence the experimental evaluation, rather well, even on this coarse mesh and with a small number of time steps. The heat equation is not reproduced quite as good, neither with the linear nor the quadratic

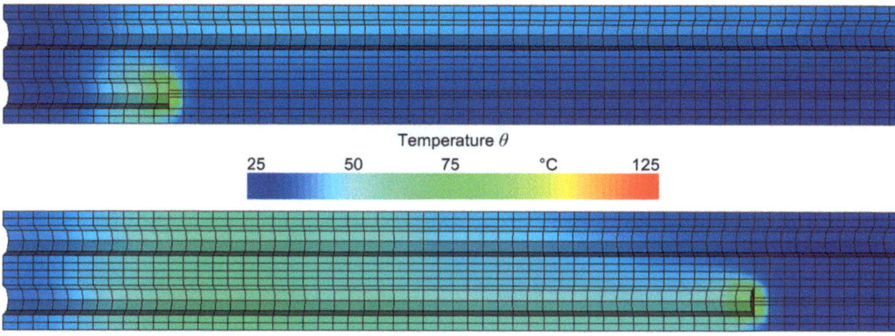

Fig. 23 Temperature distribution of two bores at $t = 1.0$ and $t = 7.0$ in the second bore phase. The workpiece is clipped for $x > 35.0$

ansatz functions, but the trend is reflected even so we used ≈7.6% of the original number of time steps.

A more complex situation is given for instance when more than one bore are performed. The simulation is expanded to simulate at least two bores subsequently. The switch over time between the bore phases is a small cooling phase. Figure 23 shows the temperature distribution on the clipped surface of the workpiece. The upper part of the bores, for $x > 35.0$ was cut off. One can see that there is still an increased temperature from the first bore in the upper right part, which declines with the time passing. But the heating from the second drilling process distributes over the whole workpiece, as one can observe for the later snapshot.

An efficient simulation demands for fast solvers. Therefore, we tested different solvers, which support the parallelisation in deal.II, and determine optimal solving parameters. It was observed that the right choice depends strongly on the temporal and spatial discretization. We use a conjugate gradient method, which is preconditioned for instance with an algebraic multigrid method adapted to the special problem setting.

5 Simulation and Compensation of Thermomechanically Induced Deviations

Building on the FE-simulation of the thermomechanical workpiece behaviour, further development was necessary to enable the simulation-based prediction of the in-process accruing deviations. Therefore the tool as a second indispensable part of the complex deep-hole drilling process was modelled using a simple analytical representation based on the bending beam theory.

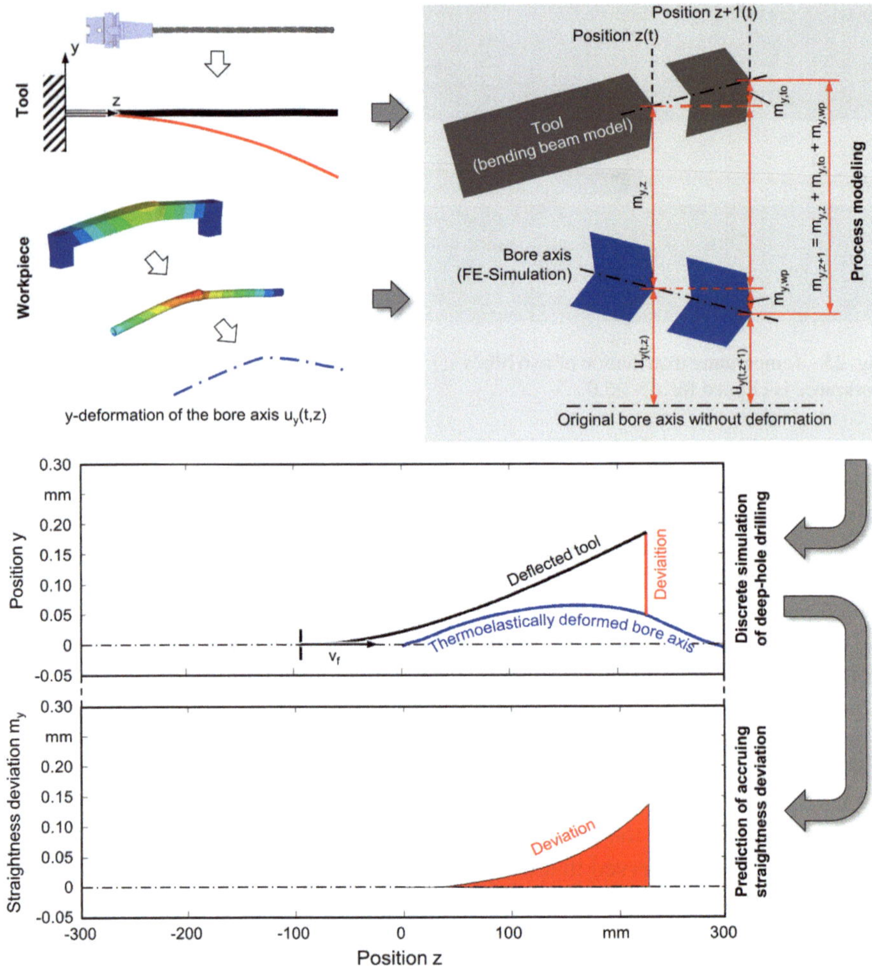

Fig. 24 Coupling of FE-workpiece model with an analytic tool model to a geometric-kinematic process simulation for the prediction of straightness deviations in deep-hole drilling

As shown in Fig. 24 the tool and the workpiece model are coupled by a kinematic approach, which considers the thermomechanical distortion of the workpiece during the process and the deflection of the drilling tool. For each discrete time step the results of the workpiece FE-simulation are reduced to the centre line (deflected bore axis), calculated based on the coordinates of the deformed bore wall. The tool bending model is deflected by the thermal workpiece displacement of the bore wall, where the margins guide the drill point. The second part of the tool deflection results from the straightness deviation itself. When the tool tip position is known, the additionally accruing discrepancy between the bore axis and the drilling direction can be

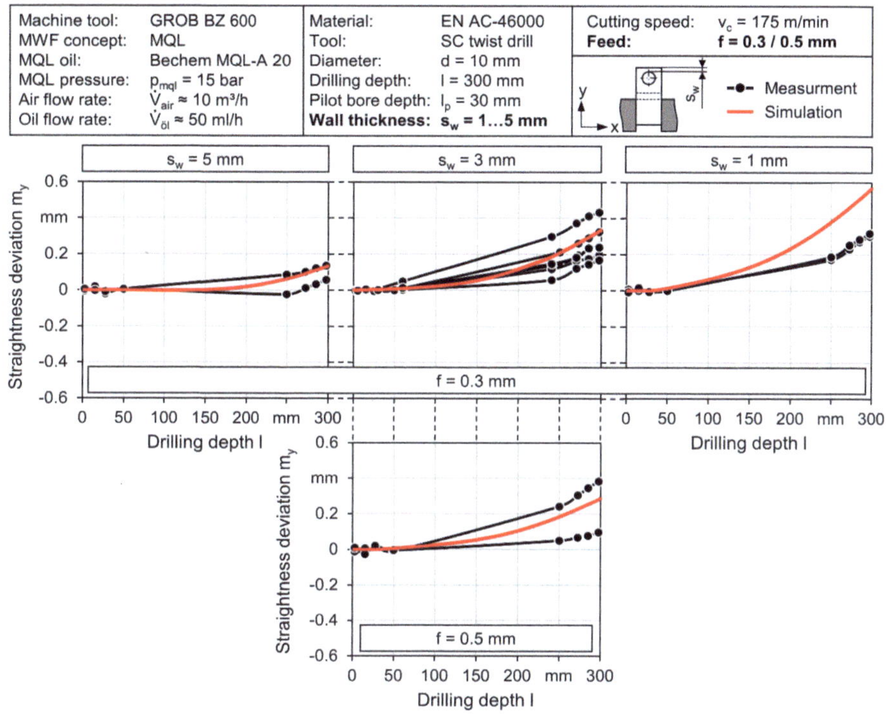

Fig. 25 Comparison between measured and simulated straightness deviation

calculated, so the continuously increasing straightness deviation can be simulated and predicted using this discrete approach. The comparison between the simulated straightness deviation curves and the measurements is presented in Fig. 25 for different feeds of $f = 0.3$ mm and $f = 0.5$ mm as well as for the different bore positions. It is typical that the straightness deviation exhibits a relatively high variation due to material inhomogeneity effects, however it still can be shown that the simulation allows for a good prediction of the straightness error.

5.1 *Compensation Procedure*

In order to actually compensate and not only to simulate the thermomechanically induced deviations of the bore, a novel method for the directional control of the drilling tool and thus the compensation of the systematic part of the straightness error is introduced (Fig. 26). Thereby, a minor radial movement of the spindle and the tool clamping respectively is superimposed to the common axial feed. The guiding of the tool front at the bore wall by means of the margins and the radial displacement of

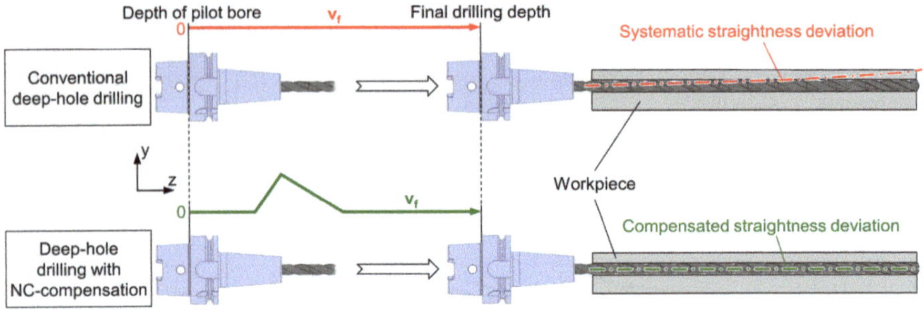

Fig. 26 Method for NC-based compensation of straightness deviation using superimposed radial path of the axial spindle movement

the clamping position lead to a deflection of the drill, which can be used to adjust the drilling direction along the bore length. For example, if the thermomechanical load of the workpiece causes a systematic distortion in positive *y*-direction, the tool is also deflected in the same direction and after the part has cooled down to room temperature again, the straightness error has a particular magnitude in positive *y*-direction as well. To reduce this systematic deviation of the bore axis, the tool can be deflected in such a way that a minor adjustment of the drilling direction in negative *y*-direction occurs. To realise this, the clamping point of the tool has to be moved in the positive *y*-direction as well due to the guiding of the drill front area at the bore wall. To avoid a collision between the tool shank and the bore wall the radial compensation displacement have to be certainly limited to the particular available clearance.

A wide variety of different radial NC-paths strategies was investigated in order to determine a reliable procedure for the compensation of the straightness deviations. In Fig. 27 the systematic error in the positive *y*-direction is shown as a reference process in the left side graphs, where the top chart illustrates the feed movement in the *yz*-plane and the bottom chart represents the bore straightness measured after machining. In the central diagrams the strategy based on trapezoidal NC-paths is presented and the right side shows the triangle based compensation curves. The results indicate that the manipulation of the straightness error using this method is feasible and the uncompensated positive *y*-deviation can be converted in a negative bore axis displacement in the *y*-direction. It is also obvious that all the analysed radial NC-paths effect the drilling direction far too much in the required direction and therefore the resulting straightness error even increases. Nevertheless the main finding is the fact that it is possible to exert a reliable reproducible control over the drilling direction in deep-hole drilling without any restrictions regarding the process stability.

One disadvantage of the experimental determination of the radial compensation NC-path is the extensive effort required for trials and measurements. Therefore the geometric-kinematic simulation was further developed to allows for the calculation of precise NC-paths according to the particular bore deflection (Fig. 28). The most

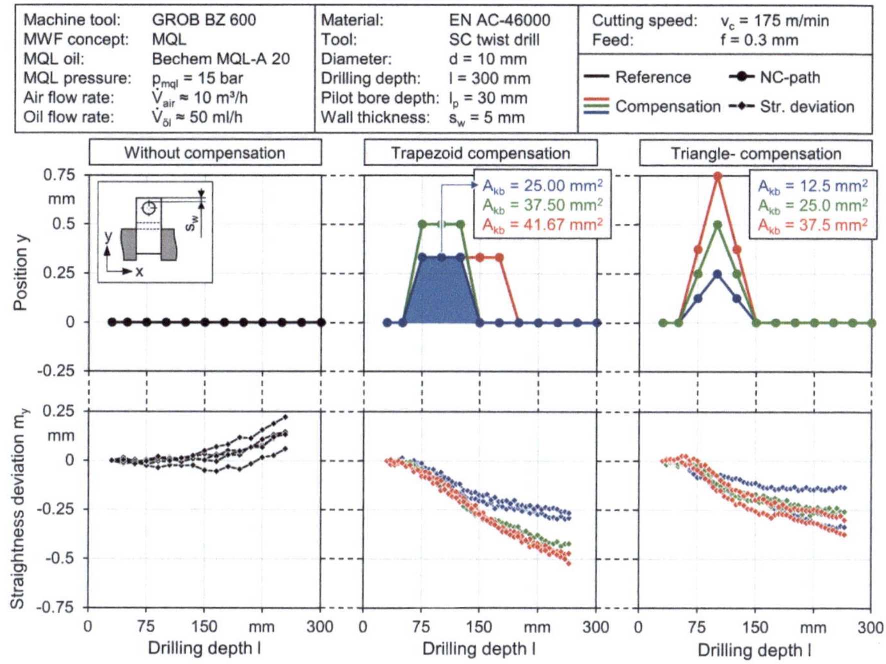

Fig. 27 Result comparison of different NC-compensation strategies

important modification are the new boundary conditions applied on the bending model of the tool. Thereby the radial *y*-position was set to be equal to the current distortion of the bore at each particular drilling position. Furthermore the drilling direction, which corresponds to the deflection angle at the front of the bending beam, was also defined to be equal to the distortion gradient at the corresponding position of the bore. In this way, the bending model was used to calculate the required radial displacement of the tool clamping, which allows for the drilling tool to exhibit the right position and deflection at every discrete point of the drilling process.

After the transfer of the simulation-based calculated compensation NC-paths to the machine tool, the accuracy of the synchronous radial-axial feed movement was analysed to ensure the precision of the NC-compensation. The comparison of the measured, simulated and compensated straightness deviations is presented in Fig. 29. The simulation-based approach provides a reliable minimisation of the occurring bore position error.

The systematic part of the straightness deviations could be compensated to a great extent and the high reliability of the developed methods could be verified for different feed values, bore positions and tool diameters. In the case of the thin-walled drilling position with a thickness of $s_w = 1$ mm, the simulated straightness deviation is higher than the measured errors and therefore also the calculated NC-compensation path leads to a greater compensation effect and to an opposite straightness error. One

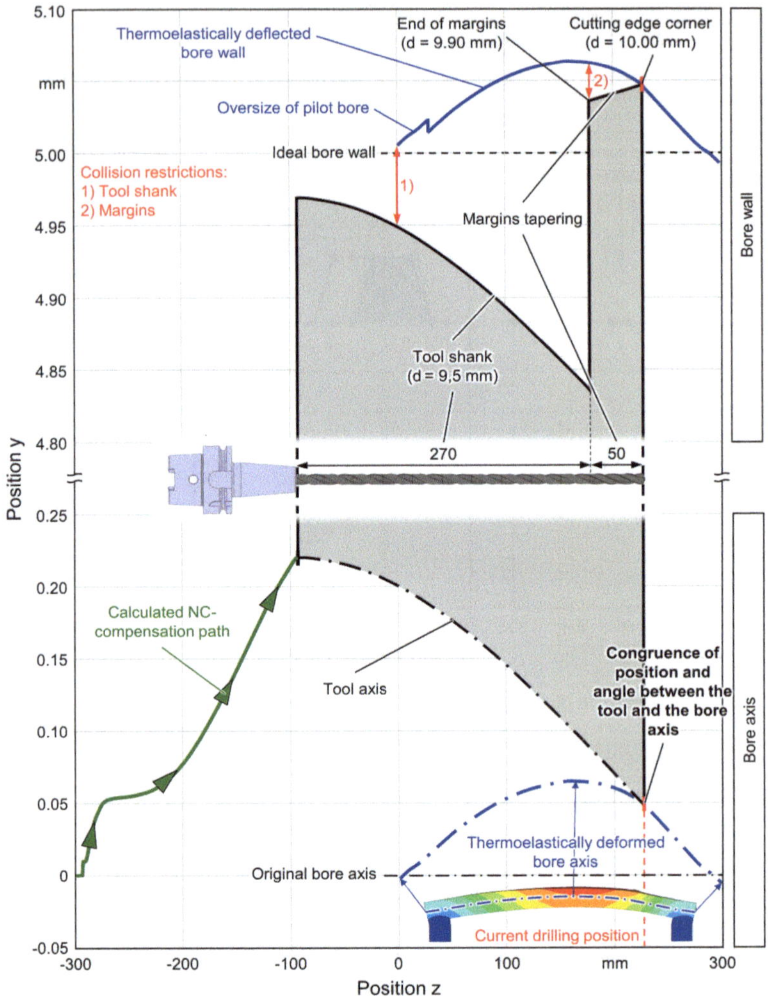

Fig. 28 Simulation-based calculation of optimal radial compensation NC-path

possible reason for this inaccuracy is the limited local resolution and thus precision of the FE-simulation in the case of very low thickness values. On the other hand, this attests that the developed compensation methodology is fundamentally capable to reproducibly control and minimise the thermomechanically induced deviations of the workpiece, occurring in deep-hole drilling with MQL.

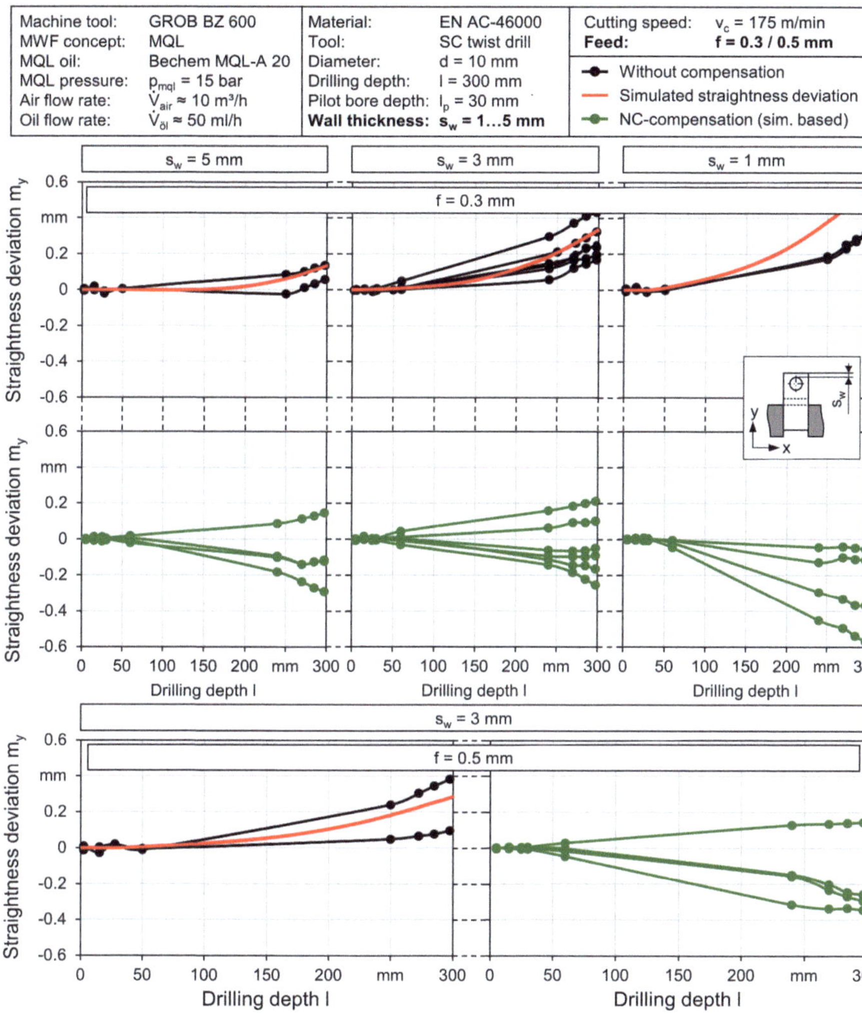

Fig. 29 Comparison of measured, simulated and compensated straightness deviations in MQL deep-hole drilling of aluminium casting

6 Conclusions

In this chapter interdisciplinary collaboration work focused on the thermal workpiece distortion in deep-hole drilling of aluminium cast alloy is presented. In comparison to classic emulsion MWF-concepts MQL leads to a significant heat input into the part machined, which can cause thermal distortion during the process and deviations of the drilled bore. Furthermore, subsequent operations can be negatively affected by the accumulated heat within the workpiece. Therefore extensive deep-hole drilling experiments were carried out, in order to determine the thermal loads on the workpiece and the occurring deflections and deviations. The investigations have shown that the secondary heat source at the bore wall, resulting from the chip friction and the heated MQL-flow, is not negligible, but plays an important role for the total heat loading of the workpiece. The feed was identified to be the major influencing factor and thus a high-performance drilling tools were developed with the aim of increasing the feed and hence reducing the heat input. Thereby an extreme productivity enhancement could be achieved, which allows for deep-hole drilling with feed velocity values of $v_f \approx 10 \ldots 20\,\mathrm{m/min}$. The heat input into the workpiece could be decreased by the developed high-feed strategies in a significant manner as well.

The experimental investigations were accompanied by simulations on the microscopic and the macroscopic level. On the microscopic level, a complex frictional contact problem involving thermo-visco-elasto-plastic material laws had to be considered. To speed up the simulation, adaptive FE-methods were developed, which were efficiently implemented in parallel. On the macroscopic level, a linear thermoelastic problem setting is an appropriate model. Using element deactivation to represent the material removal and empiric models for the primary and secondary heat sources, it was simulated in a commercial FE-software. The restrictions posed by the element deactivation were circumvented by fictitious domain methods, which were implemented in an adaptive highly parallel FE-code. Thus, a very efficient simulation on the macroscopic level were created, too. It is especially able to reproduce the straightness deviations.

Based on the coupling of a FE-workpiece analysis with an analytic model of the tool using a geometric-kinematic approach, for the first time the prediction of the development of thermomechanically induced straightness deviations along the bore was achieved. Furthermore, the developed modelling technique allows for the determination of optimal radial NC-compensation paths, which cannot be identified by experimental tests and enable the control of the drilling direction to avoid straightness errors.

Acknowledgements This contribution is based on investigations in the research projects "Experimental and FE-based analysis of the thermal load on the tool and the workpiece in deep hole drilling using twist drills" (BI 498/24) and "Numerical analysis and efficient implementation of complex FE-models for the simulation of deep-hole drilling processes" (BL 256/11) within the Priority Programme 1480 "Modelling, Simulation and Compensation of Thermal Effects for Complex Machining Processes", which are kindly funded by the German Research Foundation (DFG).

References

1. Abele, E., Kuhrke, B., Rothenbücher, S.: Energieeffizienz spanender werkzeugmaschinen. Forschen—Wissenschaftsmagazin der TU Darmstadt, Frühjahr, 64–67 (2011)
2. Ayachit, U.: The ParaView Guide: A Parallel Visualization Application. Kitware Inc, USA (2015)

3. Bangerth, W., Davydov, D., Heister, T., Heltai, L., Kanschat, G., Kronbichler, M., Maier, M., Turcksin, B., Wells, D.: The deal.II Library, Version 8.4. J. Numer. Math. (2016). Appeared online
4. Bangerth, W., Hartmann, R., Kanschat, G.: deal.II—a general purpose object oriented finite element library. ACM Trans. Math. Softw. 33(4):24/1–24/27 (2007)
5. Biermann, D., Blum, H., Frohne, J., Iovkov, I., Rademacher, A., Rosin, K.: Simulation of MQL deep hole drilling for predicting thermally induced workpiece deformations. In: 15th CIRP Conference on Modelling of Machining Operations in Procedia CIRP (2015)
6. Biermann, D., Blum, H., Iovkov, I., Klein, N., Rademacher, A., Suttmeier, F.-T.: Stabilization techniques and a posteriori error estimates for the obstacle problem. Appl. Math. Sci. **7**, 6329–6346 (2013)
7. Brecher, C., Herfs, W., Heyers, C., Klein, W., Triebs, J.B., Beck, E., Dorn, T.: Ressourceneffizienz von werkzeugmaschinen im fokus der forschung: effizienzsteigerung durch optimierung der technologien zum komponentenbetrieb. wt Werkstattstechnik online **100**(7/8), 559–564 (2010)
8. Burstedde, C., Wilcox, L.C., Ghattas, O.: `p4est`: scalable algorithms for parallel adaptive mesh refinement on forests of octrees. SIAM J. Sci. Comput. **33**(3), 1103–1133 (2011)
9. Dauge, M., Düster, A., Rank, E.: Theoretical and numerical investigation of the finite cell method. J. Sci. Comput. **65**(3), 1039–1064 (2015)
10. Eisenblätter, G.: Trockenbohren mit Vollhartmetallwerkzeugen. Dissertation, RWTH Aachen (2000)
11. Frohne, J.: FEM-Simulation der Umformtechnik metallischer Oberflchen im Mikrokosmos. Ph.D. thesis, Universität Siegen (2011)
12. Frohne, J., Heister, T., Bangerth, W.: Efficient numerical methods for the large-scale, parallel solution of elastoplastic contact problems. Int. J. Numer. Methods Eng. **105**, 416–439 (2016)
13. Grossmann, C., Roos, H.-G.: Numerical treatment of partial differential equations. Universitext. Springer, Berlin (2007). Translated and revised from the 3rd (2005) German edition by Martin Stynes
14. Heroux, M.A., Bartlett, R.A., Howle, V.E., Hoekstra, R.J., Hu, J.J., Kolda, T.G., Lehoucq, R.B., Long, K.R., Pawlowski, R.P., Phipps, E.T., Salinger, A.G., Thornquist, H.K., Tuminaro, R.S., Willenbring, J.M., Williams, A., Stanley, K.S.: An overview of the trilinos project. ACM Trans. Math. Softw. **31**(3), 397–423 (2005)
15. Johnson, G.R., Cook, W.H.: A constitutive model and data for metals subjected to large strains, high strain rates and high temperatures. In: Proceedings of the 7th International Symposium on Ballistics, pp. 541–547. The Hague, The Netherlands (1983)
16. Kamiya, M., Yakou, T., Sasaki, T., Nagatsuma, Y.: Effect of si content on turning machinability of al-si binary alloy castings. Mater. Trans. **49**(3), 587–592 (2008)
17. Kikuchi, N., Oden, J.: Contact problems in elasticity: a study of variational inequalities and finite element methods. SIAM Studies in Applied Mathematics. SIAM, Society for Industrial and Applied Mathematics, Philadelphia (1988)
18. Klein, N.: Consistent FE-analysis of elliptic variational inequalities. Ph.D. thesis, Universität Siegen (2013)
19. Klocke, F., Eisenblätter, G.: Dry cutting. CIRP Ann.—Manuf. Technol. **46**(2), 519–526 (1997)
20. Liu, R., Wheeler, M., Dawson, C., Dean, R.H.: A fast convergent rate preserving discontinuous galerkin framework for rate-independent plasticity problems. Comput. Methods Appl. Mech. Eng. 32133226 (2010)
21. Pabst, R.: Mathematische Modellierung der Wärmestromdichte zur Simulation des thermischen Bauteilverhaltens bei der Trockenbearbeitung. Dissertation, Karlsruher Institut für Technologie (2008)
22. Ramière, I., Angot, P., Belliard, M.: A fictitious domain approach with spread interface for elliptic problems with general boundary conditions. Comput. Methods Appl. Mech. Eng. **196**(4–6), 766–781 (2007)
23. Taebi, K.: Adaptive unstetige Finite Elemente Methoden für elastoplastische Kontaktprobleme. Ph.D. thesis, TU Dortmund (in preparation) (2016)

24. Thamke, D.: Technologische und ökonomische Aspekte der Trocken- und Minimalmengenbearbeitung am Beispiel des Einlippentiefbohrens. Dissertation, Universität Dortmund (1999)
25. Verfürth, R.: A posteriori error estimation techniques for finite element methods. Numerical Mathematics and Scientific Computation. Oxford University Press, Oxford (2013)
26. Weinert, K., Inasaki, I., Sutherland, J.W., Wakabayashi, T.: Dry machining and minimum quantity lubrication. CIRP Ann.—Manuf. Technol. **53**(2), 511–537 (2004)
27. Wohlmuth, B.: Variationally consistent discretization schemes and numerical algorithms for contact problems. Acta Numerica **20**, 569–734 (2011)

Thermomechanical Deformation of Complex Workpieces in Milling and Drilling Processes

B. Denkena, P. Maaß, A. Schmidt, D. Niederwestberg, J. Vehmeyer, C. Niebuhr and P. Gralla

Abstract In this chapter the work of an interdisciplinary collaboration for modeling thermomechanical deformations in dry milling and drilling processes is presented. The simulation based approach allows the prediction of structural workpiece deformation due to thermal effects occurring during the machining process. Geometric changes of the workpiece volume and the current engagement of tool and workpiece are included in the developed model. The model is applied for the optimization of NC paths with respect to workpiece deformations. The combined model is assembled by individual sub-models, which are coupled to account for interactions with each other. A dexel model is applied for contact zone analysis of tool and workpiece and also allows efficient geometric modeling of the workpiece surface. Geometric process variables of the contact zone are passed to the process model which provides thermal and mechanical loads for the thermomechanics. The thermomechanical behaviour is numerically approximated using finite elements on the time depending co-domain of the dexel model together with thermal and mechanical loads provided by the process model. In all, a closed loop between Boolean material removal, process forces, heat

B. Denkena (✉) · D. Niederwestberg
Institute of Production Engineering and Machine Tools, Leibniz Universität Hannover, 30823 Garbsen, Germany
e-mail: denkena@ifw.uni-hannover.de

D. Niederwestberg
e-mail: niederwestberg@ifw.uni-hannover.de

P. Maaß · A. Schmidt · J. Vehmeyer · C. Niebuhr · P. Gralla
Center for Industrial Mathematics, University of Bremen, 28334 Bremen, Germany
e-mail: maass@math.uni-bremen.de

A. Schmidt
e-mail: schmidt@math.uni-bremen.de

J. Vehmeyer
e-mail: vehmeyer@math.uni-bremen.de

C. Niebuhr
e-mail: niebuhr@math.uni-bremen.de

P. Gralla
e-mail: gralla@math.uni-bremen.de

D. Biermann and F. Hollmann (eds.), *Thermal Effects in Complex Machining Processes*, Lecture Notes in Production Engineering,
DOI 10.1007/978-3-319-57120-1_11

flux and thermo-elastic deformation is established and allows an accurate prediction of workpiece deformation and shape deviations. Furthermore, the simulation operates as a forward model for an NC path optimization. A significant improvement of form deviation is achieved with the approach.

1 Introduction

Shape deviations caused by thermal and mechanical loads of the workpiece are a major challenge in dry machining. Manufacturing errors are based on thermal expansion of the workpiece during the process. Therefore, conventional simulation approaches have to be extended to consider these effects in the simulation of machining processes.

Several investigations discuss the thermal and mechanical effects in machining operations. Simulation-based prediction of cutting forces is a basic approach in modeling of milling and drilling processes. In most investigations the cutting forces are calculated time dependently to analyze the process. The consideration of the workpiece shape deviation is important in machining of thin walled workpieces. Therefore, in [1] a linked simulation of vector based cutting forces and time-dependent elastic deformation of the workpiece was introduced.

For the analysis of thermal effects in machining operations the Finite Element Method (FEM) is a common approach. The thermal load on the FEM mesh is predicted by special descriptions of the process kinematics. Including the Boolean material removal process by deleting simplizes [2–4], the resulting model simulates the thermomechanical shape deviation during drilling and turning processes. These approaches are similar in the description of the process kinematics. Several investigations exist for modeling of the process by defined thermal load profiles for more complicated process kinematics like milling of free-form workpieces. Therefore, a mesh of the finished workpiece is constructed. The load is projected onto the finished workpiece. Changes of the workpiece size during the processes are not included [5–7]. A similar approach was shown in [8]. A FEM mesh of a thin walled workpiece was loaded by forces and heat flux to predict shape deviations. For more flexibility, in [9], a CSG-based modeling of the workpiece load linked with a voxel model for arbitrary machining operations was introduced.

In this paper, a method aiming to predict and to minimize the manufacturing errors based on thermal expansion is presented. Experimental investigations are carried out to analyze the mechanical and thermal workpiece load. Further, the basic simulation approach is presented. Deduced from the state-of-the-art, the NC-Simulation is extended by several properties to emulate thermomechanical effects. These extensions can be divided analytically into two components, the process model and the workpiece model (Fig. 1).

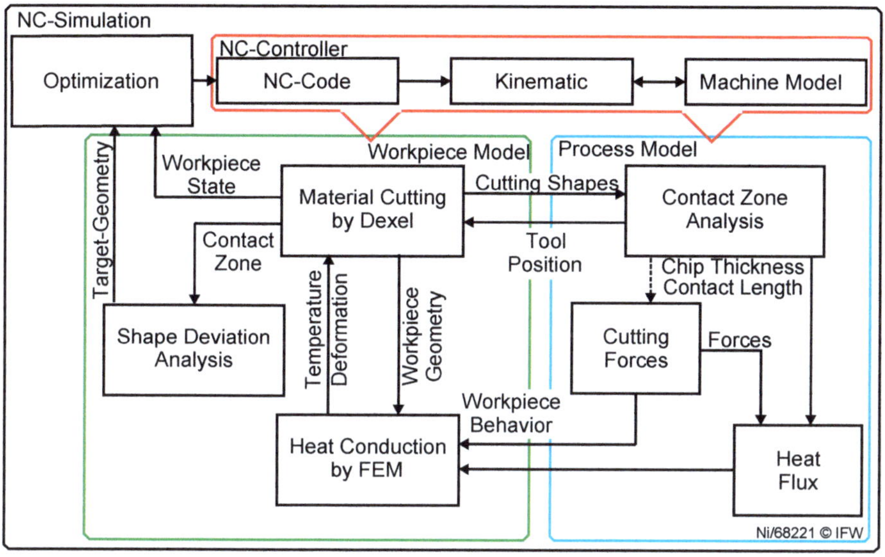

Fig. 1 Detailed information flow

The workpiece model describes the current state of the workpiece in geometrical and thermomechanical aspects (Fig. 1 left). Additional Boolean removal algorithms allow changing the geometrical representation and the boundary conditions of the workpiece, while thermal conduction algorithms allocate the thermal condition of the workpiece.

The process model has several calculation steps (Fig. 1 right). The cutting conditions are calculated from the intersection of tool and workpiece. With this data, the cutting force and heat flux are predicted and the mechanical and thermal loads are calculated. This changes the boundary conditions of the workpiece model. The shape deviations generated by the material cutting process are predicted based on the calculated temperature and deformation of the workpiece. Finally, algorithms to minimize the shape deviation will be presented.

2 Experimental Investigation

The process modeling is separated in two major aspects, the mechanical and thermal load of the workpiece. The models are empirical and base on experimental investigations. In a first step, the experimental setups are presented. In a second step, results and deduced models are discussed.

2.1 Mechanical Load

In order to determine the mechanical workpiece load cutting force measurements were performed with a dynamometer (milling: Kistler 9257B, drilling: Kistler 9272). A KENNAMETAL 90° shell-type milling cutter (Type 40A04RS90ED14D) with inserts (EDPT140404PDERHD/KCPK30) for milling of 1.1191 (C45EN) steel was used. To avoid the effects of interaction of different teeth and tool-runout, the number of inserts was reduced to one. Slot milling has been chosen as a reference process and forces were measured on the workpiece side. Using the rotational angle φ of the tool the forces can be transformed from measured feed force F_f, feed normal force F_{fN} and passive force F_p to rotating cutting force F_c, normal force F_{cN} and passive force F_p. A KENNAMETAL modular drill (KSEM200R3WN25M) with inserts (KSEM2000HPGM/KC7315) was used for drilling. The 4-component dynamometer enables the direct measurement of drilling torque and feed force. Cutting power P_c and cutting work W_c can be determined in good approximation for both processes by the cutting speed v_c or spindle speed n and feed v_f by the following equations:

$$\begin{aligned} \text{Milling:} \quad P_c &= F_c \cdot v_c, \\ \text{Drilling:} \quad P_c &= F_v \cdot v_f + M_z \cdot 2\pi \cdot n, \end{aligned} \tag{1}$$

$$W_c = \int_{t=0}^{t_{end}} P_c \, dt. \tag{2}$$

2.2 Thermal Load

For the detection of the thermal workpiece load, the heat flux into the workpiece has to be determined. The assumption of the equalization temperature is used on that account. This assumption is based on measuring heat by the temperature before and after the process. The temperature of the workpiece is measured continuously at different positions of the workpiece by thermocouples. Two thermocouples (TC4-TC5, Type K, NiCr-Ni, Reckmann Mess + Regeltechnik, diameter = 0.5 mm) are positioned at the opposite corners of the workpiece farthest from the heat source and three thermocouples (TC1-TC3) are placed along the feed direction in a distance to the processes of 2 mm (drilling) or 2–12 mm (milling). Heat flux paste was used to fix the thermocouples in the workpiece and improve the contact at the measuring point of the thermocouples. For measuring the temperature difference with minimal heat losses, dimensioning and insulation of the workpiece is an important task. Therefore, pre-experiments with different workpiece dimensions were conducted and a size of $50 \times 50 \times 50$ mm^3 (drilling) and $25 \times 50 \times 50$ mm^3 (milling) was selected. This size leads to a reliable measurable temperature difference in the workpiece due to low heat losses. For the insulation of the workpiece to the clamping, bakelite with a thickness of 3 mm was used.

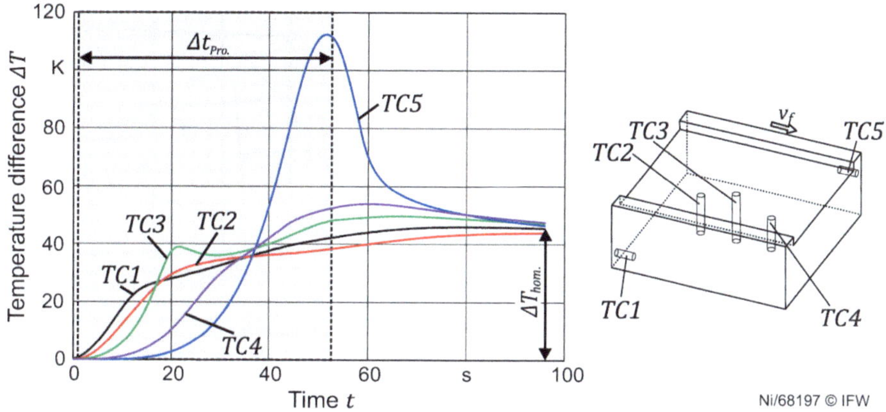

Fig. 2 Temperatures during heat measuring in milling

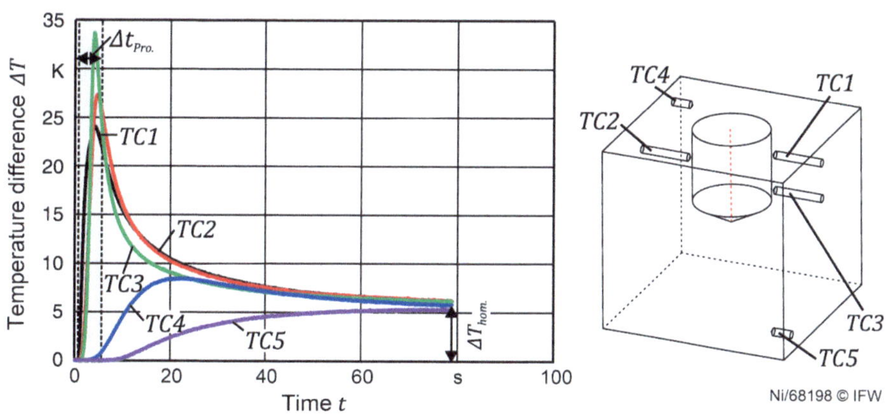

Fig. 3 Temperatures during heat measuring in drilling

In Figs. 2 and 3 exemplary measurements of an experiment are plotted. It is shown that for the milling and drilling process, the signal of the thermocouples near to the process one after another react to the moving heat source with distinctive peaks when the tool passes them. After the process heat conduction leads to an equalization of the temperatures in the workpiece. Nearly 40 s (milling) or 75 s (drilling) after the process the steady-state temperature of the workpiece is reached and the induced energy can be calculated with the additional input parameters of mass and specific heat capacity

$$Q = \Delta T_{hom} \cdot c_p \cdot \rho \cdot V_{wp}. \tag{3}$$

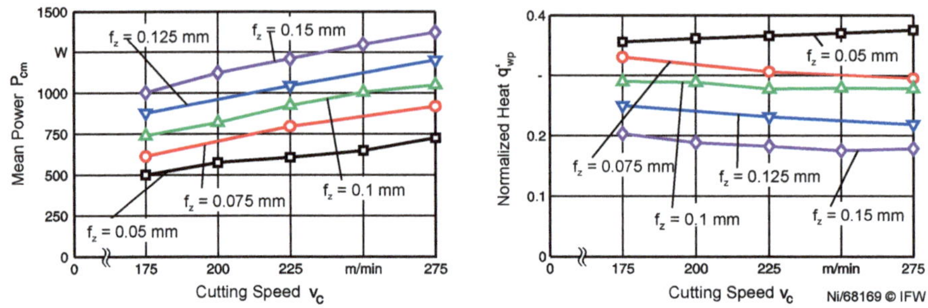

Fig. 4 Measured cutting power and heat partition for milling operation

2.3 *Thermal Balancing*

For thermally balancing the processes, it is necessary to know which percentage of the total generated heat flows into the workpiece. For this purpose the measured heat flow into workpiece has to be divided by the cutting power calculated from the cutting forces. This is called the normalized heat

$$q'_{wp} = \frac{\dot{Q}_{wp}}{P_c} \tag{4}$$

All experiments were done with various combinations of cutting speed and feed for a clearer understanding of the effects on the thermal balance.

The results showed that in milling (Fig. 4) and drilling (Fig. 5) the cutting power rises nearly linearly with cutting speed and feed. The normalized heat in the workpiece decreases with cutting speed and feed. In milling operations the feed is the major influence. The milling process is characterized by varying chip thickness during the process and almost constant cutting speed, because of the trajectory of the cutting edge of a single experiment.

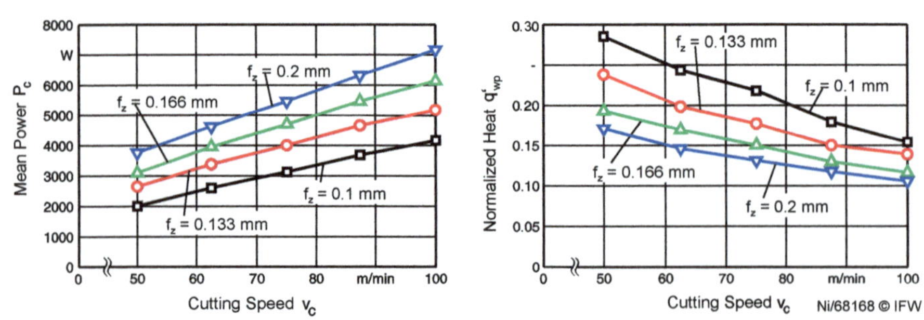

Fig. 5 Measured cutting power and heat partition for drilling operation

In Drilling operations the effect of the cutting speed is more distinct, because of the lower absolute cutting speeds and higher relative cutting speed range of the experiment. Further, in drilling operations the cutting speed changes along the cutting edge, while the undeformed chip thickness is constant for a single experiment.

These results are influenced by several effects that have to be divided. These effects are local in the shear plane and global in the tool leading zone. There is an interaction between localized heating of the workpiece and removing the heated material by the process.

On the one hand the rising cutting speed and feed lead to more heat generation during the chip formation. On the other hand, with rising cutting speed, less of the primary shear zone generated heat flows into the workpiece. The material is removed with the chip before the heat can flow into an area that will not be removed. Equal to that effect with rising feed the mean distance of the heat generated in the primary shear zone increases. In that case less energy can flow into the mentioned area. Another effect is the influence of the cutting speed on the friction in the separative zone. With rising cutting speed the generated heat in this area also rises. In addition to these local effects, the global effects of cutting speed and feed have to be named. In most cutting processes, the surface to be machined was created by another cut only a short time before. The heat could not flow into an area that will not be machined. This time decreases with rising cutting speed and the depth of the cut will be changed by the feed. Especially in drilling this seems to be the major influence, because for large tools the part of the energy that can flow further than the diameter of the tool decreases.

Due to the continuous thermal load on the workpiece, the temperature of the workpiece rises. To investigate the influence of workpiece temperature on the heat partition, further experiments with preheated workpieces were carried out. The experimental setup is similar to Sects. 2.1 and 2.2. It was found that in the range of 100 K the cutting power decreases about 5%. In this range the material properties change significantly. The heat partition, however, remains constant. According to this, the workpiece temperature has to be considered in the further modeling.

3 Process Model

The process model provides mechanical and thermal loads for the heat conduction problem. The data-driven approach is closely related to the experimental investigations of the previous section. As depicted in the flow chart, see Fig. 1, the process model is positioned between the dexel based material removal model and the thermomechanical FE simulation. In the process model the geometric process data is mapped to energetic process output. The most essential input is the chip thickness $h = h(t, s)$, but also tool geometry and tool kinematics are influencing factors. In a first step, the input variables are mapped to mechanical loads by the force model. In a second step, the forces are mapped to thermal loads by an empirical heat partition function.

In the following sections, the main tool characteristics are determined. In further steps, the force model is formulated and a method to identify the specific cutting forces from measurement data is presented. Finally, the empirical partition function for the induced heat is obtained by regression.

3.1 Geometric Tool Modeling

To perform a machining simulation by consideration of the shape of the cutting edge, the common simulation approach of using a body of revolution has to be extended. Therefore, a full three-dimensional model of the tool is required.

The modeling of the tool starts with the discretization of the cutting edge. The cutting edges of the CAD-model are sorted in a continuous sequence. Iterating this sequence, every edge is discretized by equidistant linear sections represented by discretization points $P_{ED}(i)$ (Fig. 6). During the further procedure of the method every section of the discretized edge represents a cutting edge element with individual characteristics. For every discretization point $P_{ED}(i)$, interpolation points $P_{REV,i}(j)$ are created by rotating the discretization point in a defined angle ω around the tool rotational axis to model the body of revolution. By triangulation of the discretization and interpolation points of two following sections a consistent surface is generated. For further analysis of the contact zone every triangle is allocated to a cutting edge segment. The consistent body of revolution is modelled repeating this triangulation for all cutting edge segments.

Based on data determined by the tool preparation, the tool reference plane $P_r(i)$ for the i-th cutting edge segment is calculated. The rotation of the tool reference plane $P_r(i)$ from the basic reference plane $P_{rb}(i)$ of the Tool Coordinate System (*TCS*) is identical with the rotation of the force application point $P_{ED}(i)$ and is called lag angle $\psi(i)$. The lag angle can also be used to characterize the tool geometry and is calculated in relation to a defined start angle point P_{ref} in the coordinate plane.

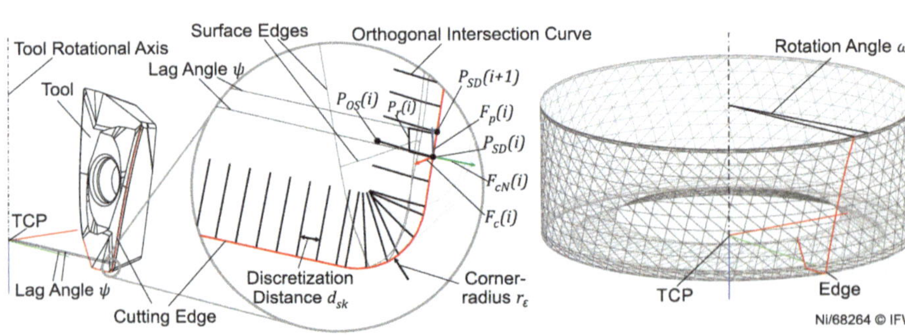

Fig. 6 *Left side* Cutting insert (milling); *Detailed view* Cutting edge discretization; *Right side* Body of revolution for machining simulation

The tool cutting edge angle $\kappa_r(i)$ is the cutting edge's orthogonal angular deviation in the reference plane $P_r(i)$, which can be obtained by the vectors connecting the cutting edge discretization points $P_{ED}(i)$. The individually calculated angles (κ_r, ψ) enable the consideration of the local cutting edge of the tool in processing of the local cutting conditions. For the calculation of the resulting cutting forces, individual cutting force coordinate systems for every cutting edge segment are implemented. The cutting force $F_c(i)$ is oriented in the direction of the cutting speed v_c, which is equal to the normal of the tool reference plane n_P. The normal cutting force $F_{cN}(i)$ is equal to the normal of the surface of the modelled tool n_P. The passive force $F_p(i)$ can be modelled orthogonal to $F_c(i)$ and $F_{cN}(i)$.

As a result of this section a technological extended tool model for the thermomechanical machining simulation is developed. The method is identical for drilling and milling processes.

3.2 Contact-Zone Analysis

Each functional part of the tool surface is investigated separately to perform an analysis of the contact zone. Each face is cut with the multidexel model and at each intersection point of a dexel, called Dexel Cut Point (*DCP*), the triangulated surface is assigned to the corresponding cutting edge element (Fig. 7). For the *i*-th cutting edge element all *DCPs* are analyzed. The first step in the contact analysis is to calculate the feed direction of the tool. By iterating the movement of all axes the movement of the Tool Center Point (*TCP*) can be calculated in a global coordinate system. The

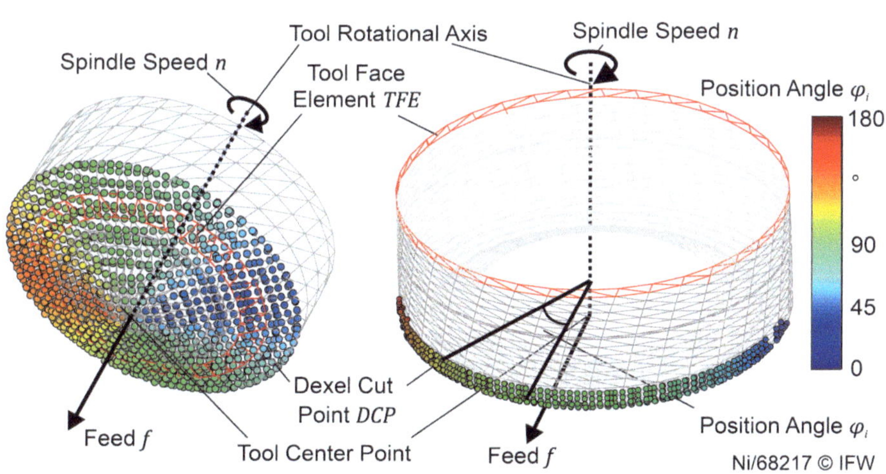

Fig. 7 Contact analysis of *DCPs* on the tool of revolution for drilling (*left*) and milling (*right*)

feed direction of the tool corresponds to the difference of the *TCP* from the last to the current simulation step.

For the milling process in the next step, the feed direction of the tool is projected on a plane parallel to the X-Y plane of the tool and defined by the *DCP*. The angular deviation of the *DCP* from the projected feed direction characterizes the contact condition of the *DCP*. By the following relation, the corresponding position angle can be calculated:

$$\varphi_{DCP} = \sphericalangle(v_{f,i}, DCP). \tag{5}$$

In drilling operations the position angle is calculated independent on feed direction but dependent on the predefined direction of the cutting edge.

For further analysis the *DCPs* allocated to a tool face element are sorted from entry φ_e to exit φ_a angle by a quicksort algorithm. The workpiece load has to be modelled to simulate and to compensate workpiece related thermal and thermo-mechanical effects. Using the calculated feed the ideal undeformed chip thickness $h_{ideal,DCP}$ can be calculated for every *DCP*:

$$\begin{aligned} &\text{Milling:} \quad h_{ideal,DCP} = f_z \cdot sin(\varphi_{DCP}) \\ &\text{Drilling:} \quad h_{ideal,DCP} = f_z. \end{aligned} \tag{6}$$

In order to process the undeformed chip thickness h_{DCP}, the individual cutting edge characteristics, i.e. cutting edge angle of the tool are considered. Further in milling operations, the ramping angle α_i, which is the deviation of the feed direction from spindle axis orientation (*SAO*) of the tool, has to be calculated. The ramping angle is calculated by the following equation:

$$\alpha_i = \sphericalangle(v_{f,i}, SAO) - 90°. \tag{7}$$

Furthermore, the undeformed chip thickness is calculated by

$$\begin{aligned} &\text{Milling:} \quad h_{DCP} = h_{ideal,DCP} \cdot cos(\kappa_i - \alpha_i) \\ &\text{Drilling:} \quad h_{DCP} = h_{ideal,DCP} \cdot cos(\kappa_i). \end{aligned} \tag{8}$$

The contact analysis can be applied to any process for drilling and milling operations. In every simulation step the dexel based workpiece representation is intersected by the modelled tool. For the discretization of an exemplary workpiece ($48 \times 48 \times 195$ mm^3) $64 \times 64 \times 256$ dexels are chosen. Figure 8 shows the contact zone of a single simulation step. The detected *DCP*s are allocated with respect to the position on the cutting edge and the position angle. Responding to the movement of the tool the corresponding undeformed chip thickness is calculated for each *DCP*. The entry and exit angle of the simulation step is described as a polygon, connecting the minimum and maximum position angle.

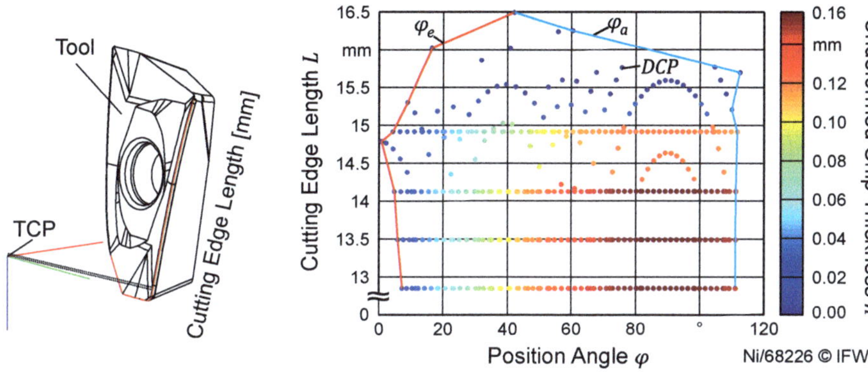

Fig. 8 Contact analysis of a single simulation step

During the theromechanically linked simulation the contact-zone is influenced by the deformation of the workpiece. The consideration of the deformation is shown in Sect. 4.4. Further, a detailed discussion of tool modeling and contact-zone analysis is given in [10].

3.3 The Force Model

The force model serves as basis for quantifying the effects of thermal and mechanical loads. Regression has been a main aspect of statistics and time series analysis for many years and is also the basic tool in the next sections for the prediction of cutting forces.

Often, the shear friction model [13] is applied, which assumes a constant specific cutting force k_c and a friction constant k_e. The main emphasis of the presented work is the prediction of specific cutting forces under varying cutting conditions. The simulation based approach allows the consideration of the exact tool geometry, compare Sect. 3.1, and kinematics as well as the current uncut chip thickness. Furthermore, a time synchronization is performed for an easy practical application. The model is based on linear regression with higher order terms, so that nonlinearities, i.e. a curvature of the function of specific cutting force, can be modeled correctly. The investigation has been published in [11] and is discussed extensively there. In the following, the outline and the main results of the work are presented.

Under the assumption of the existence of a specific cutting force k_c, the task is a function approximation. The force acting on an idealized cutting edge can generally be written as

$$F_C = f(k_c, b, h), \tag{9}$$

where b is the width of the undeformed chip and h is the undeformed chip thickness.

In the following, infinitesimal forces are approximated by a polynomial of uncut chip thickness and the resulting total cutting force is obtained by integration along each cutting edge. Furthermore, the local forces are transformed into stationary coordinates to allow a comparision with measured forces.

The infinitesimal force contributions are

$$dF = \sum_{i=0}^{p} k_i h^i, \quad p > 0, \tag{10}$$

where k_i is a coefficient vector of the three force directions according to Sect. 3.1. The force contributions are transformed according to the local force directions and process kinematics into stationary coordinates by $R(t, s) \in SO(3)$ and integrated along each cutting edge:

$$F(t) = \int_{\gamma} RdF = \int_{0}^{L} R(t,s) \sum_{i=0}^{p} \begin{pmatrix} k_{i,cN} \\ k_{i,c} \\ k_{i,p} \end{pmatrix} h^i(t,s)ds. \tag{11}$$

The identification of specific cutting force coefficients $k_i \in \mathbb{R}^3$ for $i = 0, \ldots, p$ leads to a linear model

$$||Ak - F_{\text{meas}}||_2^2 \rightarrow \min_k, \tag{12}$$

where A is the discrete integral operator resulting from (11) and F_{meas} consists of measured forces. The vector k represents the specific cutting force coefficients and generally results from the least squares solution

$$k = (A^T A)^{-1} A^T F_{\text{meas}}. \tag{13}$$

For the numerical treatment of the identification problem and time synchronization of the measurement, the reader is referred to [11]. The method is simple and efficient for both prediction of cutting forces and identification of specific forces. It can be applied for a large range of process parameters and various cutting conditions as well as tool geometries. In Fig. 9 the time courses of measured and simulated forces for linear tool retraction within the machining of the reference workpiece are shown. A four tooth cutter ($Z = 4$) was used. In Fig. 10 the equivalent tangential and radial specific cutting forces are shown. Furthermore, a second retraction process is shown and compared with simple slot milling ($Z = 1$ and $Z = 4$) as a reference. The slight nonlinear behaviour is modeled by a polynomial degree $p = 2$. The good agreement of the four specific cutting forces demonstrates the possibilities of the identification method and the contact-zone analysis.

In Fig. 11 measured and simulated forces for a drilling process are shown. The specific cutting force is shown together with the tangential forces of the milling process in Fig. 10. During the spot drilling the torque rises until the tool is fully

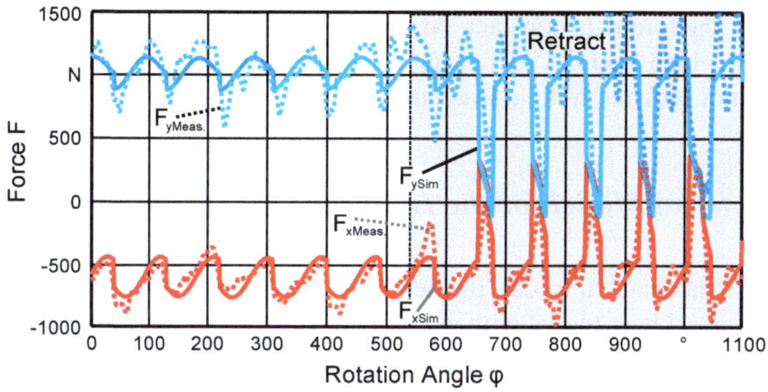

Fig. 9 Forces F_x and F_y: Simulation and time synchronized measurement for linear retraction in the reference process

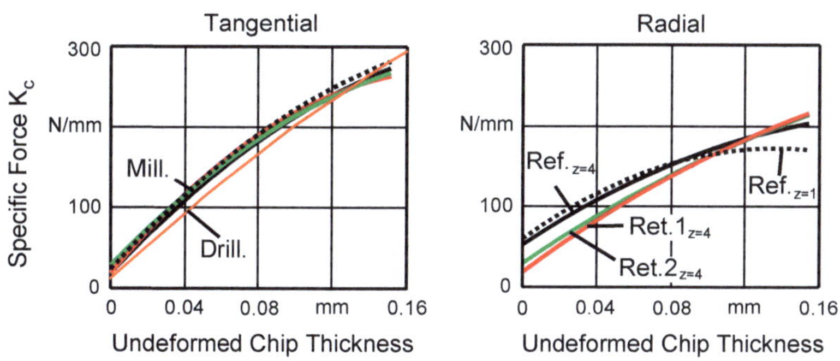

Fig. 10 Identified specific cutting forces ($p = 2$) for two different retraction processes in comparison with simple slot milling for a single tooth cutter ($Z = 1$) and a four tooth cutter ($Z = 4$). Tangential (*left*) and radial (*right*) forces

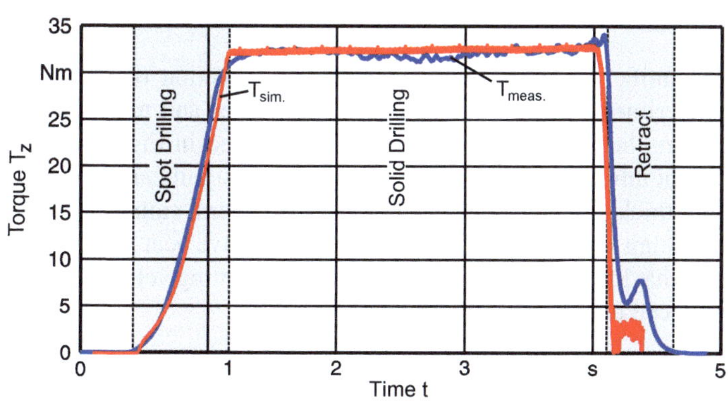

Fig. 11 Tangential forces of drilling acting on the axis of rotation as torque T_Z. Simulation and time-syncronized measurement

engaged and the torque is constant. At the end of the solid drilling process the feed slows down, which leads to decreasing torque. When the tool is retracted from the bore contact of tool and hole, it leads to a further torque plateau.

3.4 Empirical Model for the Heat Induced by the Process

In contrast to mechanical forces, the heat flux cannot be determined directly. Therefore, an empirical heat model with an indirect calibration campaign was developed. Generally, the induced heat flows partly into the workpiece and into the tool while a part is removed from the process with the chips. The total heat is given from the mechanical work resulting from cutting forces. The splitting of total heat into the portion which flows into the workpiece is of interest for the thermomechanical behavior and is presented in the following.

In Sect. 2 measurements of the induced heat Q were determined according to (3) for different combinations of process parameters. The primary influencing factors are the cutting velocity v_c and the nominal uncut chip thickness h. Due to the sinusoidal course of the function $h(t)$, individual means of $\bar{h}$ were generated by changing the parameters of feed f_z and radial engagement a_e.

The total work produced by a single turn is calculated from the process forces:

$$W_{\text{turn}} = \int_0^{1/N} \left(\sum_i k_i h^i \right) v_c \, dt = \frac{1}{N} \sum_i k_i \bar{h}^i v_c. \tag{14}$$

The assumption is that the work W_{turn} is completely converted into heat. The next step is to equate simulated and measured heat to calculate the quantifier w:

$$Q = w \, W_{\text{turn}}. \tag{15}$$

The final quantifier $w = w(h, v_c)$ is a regression function of 61 (milling) or 40 (drilling) experiments with different cutting velocities and mean chip thicknesses $\bar{h}$. The resulting heat partition functions are presented in Fig. 12. For the milling process a slight influence of cutting velocity v_c and a significant nonlinear influence of uncut chip thickness h can be stated. From the general course of the graph it can be stated that high values for h and v_c reduce the relative heat input. For the drilling process the behavior is similar, but the influence of cutting velocity is nonlinear and more pronounced due to the large range of v_c.

3.5 Workpiece Load

In order to use the process models in the machining simulation, the mechanical forces and thermal loads for the determined contact zone of the Boolean intersection are

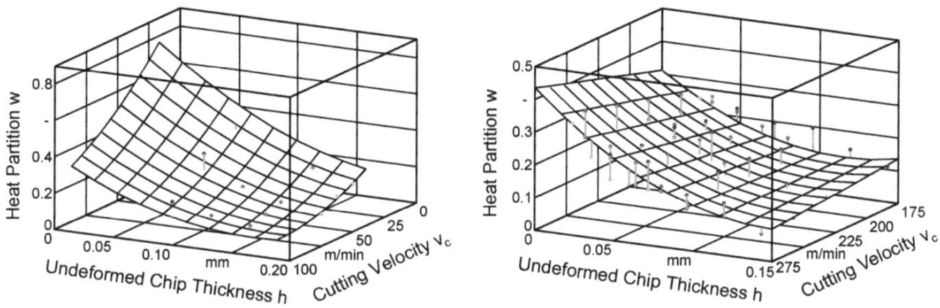

Fig. 12 Proportion of the total induced heat flowing into the workpiece for drilling (*left*) and milling (*right*) processes

calculated. For this, they are assigned to each Dexel Cut Point (DCP) in place and time on the respective active component of the total load. Each DCP is assigned to the time t_{DCP} and to the cutting edge position s_{DCP}.

The mechanical load on an individual part of the contact zone is given for all *DCP*s by the normalized vector:

$$F_{DCP} = \frac{1}{|A_{DCP}|} \int_{t_{DCP-1}}^{t_{DCP}} \int_{s_{DCP-1}}^{s_{DCP}} dF. \tag{16}$$

The area of action A_{DCP} for a *DCP* is limited by the operation area of the cutting edge segment b_i and the distance d_{DCP} to the following *DCP* of the segment:

$$|A_{DCP}| = b_i(\varphi_{DCP-1} - \varphi_{DCP})d_{DCP}. \tag{17}$$

Starting point for the thermal model is the force model of Sect. 3.3. Its main aspects are described in the following. The quantifier $w(h, v_c) \in [0\%, 100\%]$ gives the proportion of the total heat (mechanical work) flowing into the workpiece. With the cutting force F_c the resulting power is $P_c = F_c v_c$. The local induced heat flux for each *DCP* is calculated as follows:

$$\dot{Q}_{DCP} = w(h_{DCP}, v_{c,DCP})\, F_{DCP}\, v_{c,DCP}. \tag{18}$$

Further, the thermal workpiece load is modelled as heat input. Therefore, the local dissipated heat flux magnitude has to be converted into the heat with respect to the time of action:

$$Q_{DCP} = \dot{Q}_{DCP}\, \Delta t, \quad \text{with } \Delta t = (\varphi_{DCP-1} - \varphi_{DCP})\frac{1}{n}. \tag{19}$$

Overall, this method allows the local discretized simulation of the thermal and mechanical workpiece load.

4 Workpiece Model

The workpiece model combines the material removal simulation by a dexel model with the simulation of thermomechanics by the FEM. As depicted in the data flow diagram (Fig. 1), the workpiece model is disposed on the left side and includes the dexel model and the FE simulation. Especially the interchange of data between the two models is essential. Exchanged data are workpiece geometry, thermal and mechanical load plus temperature and deformation behavior, to identify thermal effects, like shape deviations. The dexel model and the linking of the workpiece behavior is described, followed by the formulation of the thermomechanical model. The linking of the two models into the workpiece model, especially in view of the finite element formulation, is presented in the main part. Finally, the identification of the thermomechanical effects is given.

4.1 Thermomechanical Extended Dexel-Model

The component of the workpiece model used for the material removal simulation is discretized by a dexel model. To consider the thermomechanical effects in the material removal simulation and the combined process model, the geometric representation of the workpiece by the dexel model has to be extended by thermomechanical information. Both temperature and deformation of the workpiece are important data which are used as input for the models. The current temperature of the workpiece has a major influence on the specific characteristics of the material like shear yield stress and therefore on the resulting cutting forces. The deformation of the workpiece changes the local (cross-section of undeformed chip A) and global (depth a_p and width a_e of cut) geometric contact conditions between tool and workpiece.

The data structure of the dexel model enables a geometric representation of the workpiece surface by discrete start- and endpoints of the dexel (Fig. 13). In cutting

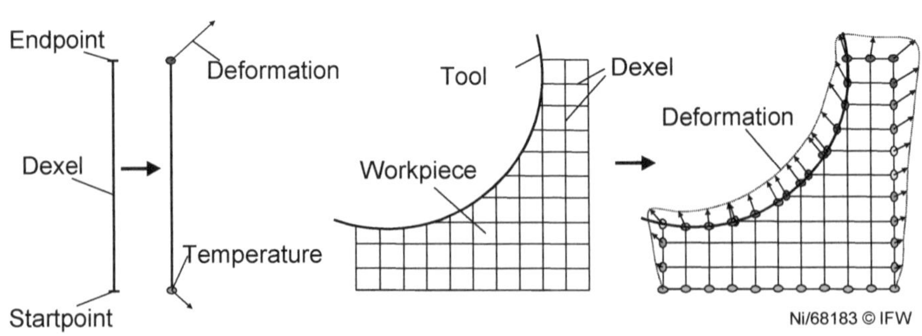

Fig. 13 Thermomechanically extended dexel-structure

operations only near-surface areas are machined. Thus, only information in these areas of the workpiece is relevant for a material removal simulation and is sufficient to extend the existing structure in the start- and endpoints of the dexel. To consider temperature and deformation, a dexel point was extended by a temperature value and a deformation vector. Using a deformation vector instead of alternative methods like shortening and lengthening of the dexel or shifting of dexel points, gives essential advances. These are:

1. Dexel fields are parallel to the coordinate axes. This accelerates intersection computation.
2. Easy linking to the FE model structure, which is based on the formulation in reference coordinates which coincides with the original dexel representation.

4.2 Thermomechanical Finite Element Model

The second part of the workpiece model is the simulation of thermomechanics to get information about the effects of thermal and mechanical loads during machining operations. This is based on FEM and implemented by using the toolbox ALBERTA [14].

In this context a new implementation of moving boundaries on a time-dependent domain and the consideration of large domain modification during the process is given. The used toolbox allows here an automatical adjustment of local grid refinement on the time-dependent domain and its moving boundary.

Classical formulations of heat equation and quasi-stationary elastic deformation equation on a time-dependent domain formed the basis of the thermoelastic model. To realize this time-dependent domain, a fixed domain $\Omega = \Omega_t \cup \Omega_m \cup \Gamma_t \subset \mathbb{R}^3$ is divided into different time-dependent domains. Ω_t denotes the time varying workpiece, Ω_m the removed material of the workpiece and Γ_t the moving boundary between these domains. The subdomain boundary $\partial\Omega_t = \Gamma_t \cup \partial\Omega \backslash \partial\Omega_m$ is subdivided in parts with different boundary conditions. For the heat equation $\partial\Omega_t = \Gamma_N(t) \cup \Gamma_R(t)$ is divided in parts with Neumann or Robin conditions and for deformation equation $\partial\Omega_t = \Xi_D(t) \cup \Xi_N(t)$ is divided in parts with Dirichlet or Neumann conditions.

The data at the boundaries are given by the process model, see Sect. 3. Heat fluxes and process forces have to be given on $\partial\Omega_t$, which is at least partly composed of interior sides of elements of the original triangulation of Ω.

The complete system of partial differential equations (PDE) for temperature θ and deformation vector v is given as follows:

$$
\begin{aligned}
\rho(\theta)c_e(\theta)\frac{d\theta}{dt} - \operatorname{div}(\kappa(\theta)\nabla\theta) &= 0 && \text{in } [0,T]\times\Omega_t\\
-\kappa(\theta)\nabla\theta\cdot n &= q_N && \text{on } \Gamma_N(t)\\
-\kappa(\theta)\nabla\theta\cdot n &= \delta(\theta-\theta_{ext}) && \text{on } \Gamma_R(t)\\
\theta(\cdot,0) &= \theta_0 && \text{in } \Omega = \Omega_t|_0\\
-\mu(\theta)\Delta v - (\mu(\theta)+\lambda(\theta))\nabla\operatorname{div}(v) &= \rho(\theta)f && \text{in } \Omega_t\\
v &= 0 && \text{on } \Xi_D(t)\\
(\lambda(\theta)\,(tr(\epsilon(v)))\,I + 2\mu(\theta)\epsilon(v))\cdot n &= g_N && \text{on } \Xi_N(t)\\
\epsilon(v) &= \frac{1}{2}\left(\nabla v + (\nabla v)^T\right)
\end{aligned}
\tag{20}
$$

with the temperature dependent material parameters density ρ, specific heat c_e, thermal conductivity κ and Lamé coefficients μ and λ, which result from the elastic modulus and the Poisson ratio. Ignoring acceleration effects, which operate on a much faster time scale than the thermomechanics, the deformation is modeled quasi-stationary. q_N and g_N denote the fluxes and forces provided by the process model (which depend nonlinear on temperature, deformation, and machining tool parameters). The volume force f could include gravity forces or the like if necessary.

The result is a system of parabolic/elliptic initial and boundary value problems, which model the relevant effects during machining processes.

4.3 Linking of Dexel and FE Models

The dexel and thermomechanic model use totally different spatial discretizations and time step sizes. Typically the FEM time steps are larger than the time steps of material removal simulation. The linking of both models is based on the interchange of parameters and values: The surface of workpiece geometry and its normal vectors as well as heat flux and process forces produced by the cutting process are transfered to the FE model, which supports the dexel model with resulting temperature and deformation in return.

Both models and their data structures had to be extended in order to account for this linking. A flexible interface was built using a general data interchange format (XML), while an efficient exchange of information could be realized by a restriction to (locally) relevant data.

Local operations, based on both geometrical mesh coherencies, instead of global search operations, are used for fast localization and identification of dexel points in the FEM mesh and vice versa. Local interpolation of data from both models can then easily be used for data exchange.

Computations for a FEM time step involve the solution of large linear systems of equations, thus needing much more computation time than a material removal time

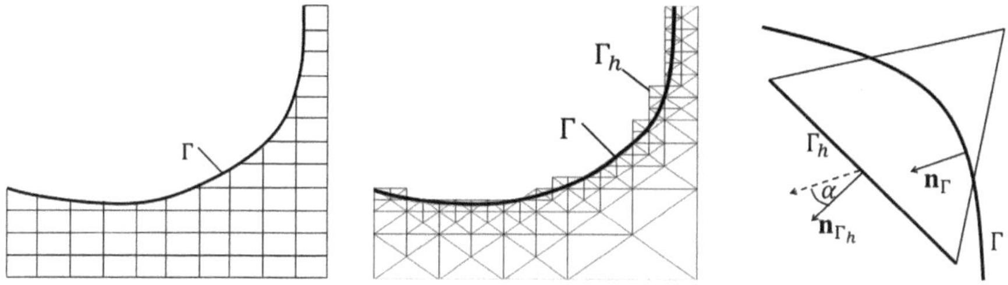

(a) Exact boundary in 2D dexel field (b) Exact and discrete boundary overlay (c) Normal projection on simplex

Fig. 14 Sketches of boundary projection in FEM

step. As thermomechanical effects on the macroscopic scale considered here are relatively slow, it makes sense to use much larger time steps for the FE simulation than in the material removal simulation by the dexel model. In order to account for these different time step sizes, a temporal accumulation of fluxes and forces is necessary. This was implemented in the dexel model, thus additionally reducing the amount of data to exchange, see Sect. 3.5.

As mentioned in Sect. 4.1, the dexel model was extended by thermal and mechanical properties in order to provide the material removal simulation and the process model with corresponding data.

The thermomechanical model was extented by time-dependent modifications of the workpiece geometry, which is given by the dexel points in different directions (x-, y-, z-direction), see Fig. 14a for a 2D sketch. For each dexel field, the subgraph of the bilinear interpolation of dexel endpoint coordinates is defined. The intersection of all three subgraphs defines the current workpiece geometry Ω_t with cutting surface Γ_t. The discrete domain $\Omega_{s,h}(t)$ for the finite element method is then given by all mesh elements S of the triangulation $\mathscr{S}_h$ with a non-empty intersection with Ω_t: $\bar{\Omega}_{t,h} = \bigcup\{S \in \mathscr{S}_h \, : \, S \cap \Omega_t \neq \emptyset\}$. For a good approximation of the current domain, the mesh is adaptively locally refined near the cutting boundary Γ_t. Nevertheless, the discretized bound $\Gamma_{t,h} \subset \partial\Omega_{t,h}$ is typically a larger surface than the relatively smooth surface given by the geometrical information from the dexel model, see Fig. 14b. Using the flux density given by the process model on this larger surface, too much heat would be transfered into the workpiece. Thus, an adjustment of the fluxes and forces is necessary. A correction by the scalar product of the cosine of the enclosed angle α between the two normal vectors is proposed (Fig. 14c). Normal vectors n_Γ^k at dexel endpoints are provided by the dexel model together with heat fluxes Q_{DCP}^k.

Through the different discretizations in FEM and dexel model in space and time, only information from few dexel cut points (DCP) could be assigned to the boundary of one simplex S. By bilinear interpolation over each dexel field k and each quadrature point x, there are four supporting points $\tilde{q}_i^k, i = 1, \dots, 4$ for heat flux data

$$\tilde{q}_i^k = Q_{DCP}^k \cdot \cos\alpha(n_{\Gamma_{N,h}}, n_{DCP}^k) \tag{21}$$

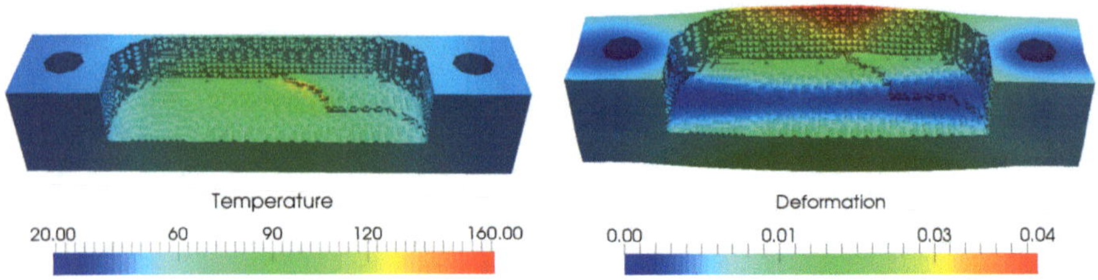

Fig. 15 Temperature field (*left*) and deformation behaviour with scale 200 (*right*) of the reference workpiece (Sect. 5.5) during milling visualzed by FEM

with thermal load Q^k_{DCP} (see (19)) given on each *DCP* and normal vector n^k_{DCP} of the continuous surface Γ_t. The discrete Neumann boundary condition is finally

$$\int_{\Gamma_{N,h}(t)} -\kappa_h \nabla \Theta(x,t) \cdot n \, \psi(x) \, \mathrm{d}x = \int_{\Gamma_{N,h}(t)} q_h(x,t)\psi(x) \, \mathrm{d}x \qquad (22)$$

with temperature Θ FE test functions ψ and approximated Neumann flux $q_h(x,t)$ at quadrature point x,

$$q_h(x,t) = \frac{1}{\Delta t} \sum_{k=1}^{d} I_h(x, \tilde{q}^k_1, \tilde{q}^k_2, \tilde{q}^k_3, \tilde{q}^k_4) \qquad (23)$$

with the interpolation operator I_h, the number of given dexel fields d and time step Δt. The division by the time step size is necessary, as the fluxes given by the process model are accumulated values over time. As mentioned above this is necessary due to the provision for different time step sizes in material removal and FEM computations. The Neumann boundary condition for the deformation behavior is analogous.

Using these modifications for the boundary data from the dexel model, this gives good approximations (see [17]) of the thermomechanical deformation and workpiece temperature in the FEM, see Fig. 15. A detailed description of the linking is given in [15–17].

4.4 Shape Deviation Prediction

The thermomechanical data in the dexel structure enables the consideration of thermomechanical effects in the NC-Simulation. These effects are caused by the local temperature and the local deformation of the workpiece in the contact zone. The temperature leads to a significant reduction of cutting power and thus to the resulting cutting power as shown in Sect. 2.3. The deformation influences the dimension of the contact zone and the geometric representation of the workpiece. The resulting shape deviation can be predicted by additional shortening of the cut dexel by the length of the local deformation. Therefore, the tool is translated inversely to the

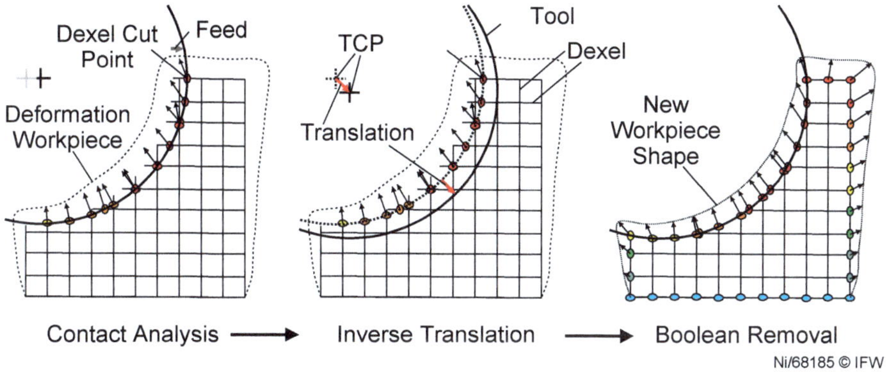

Fig. 16 Tool translation for consideration of the workpiece deformation

analyzed deformation in the contact zone (Fig. 16). In addition, for every cut dexel the current vector of deformation is stored in the dexel field for a detailed analysis of the resulting shape.

As a result of this chapter, there is a closed simulation circle, which enables the prediction of the temperature related workpiece deformation and the resulting shape errors. In this case, the thermal and thermomechanical interactions are taken into account by the linking of the individual modules. In the following, this basic approach is used to examine an optimization of the process and thus the machining result. The simulation approach was validated in [16–18].

5 Optimization

In this section, the developed simulation approach is extended by an optimization method. The aim is the thermomechanical adaption of conventional NC paths to ensure the desired shape of the workpiece. The shape of thermal distortions depends on different influencing factors, e.g. the tool, cutting parameters and the process strategy.

All of these factors are considered in the simulation and are therefore potential control parameters. The moving tool operates as heat source, controlled by the cutting parameters a_e, a_p, v_c, v_f and furthermore, material is removed which changes the shape of the current workpiece Ω_t. The interaction of tool and workpiece is characterized by the cutting parameters in two ways. The volume occupied by the tool is configured by the process parameters and resulting parameters of the contact-zone are mapped by the process model to thermal and mechanical loads, which effect the thermal expansion of the workpiece, especially in the contact zone.

The co-simulation with the three sub-systems is used as forward operator A_S which maps the control u to the state y

$$y = A_S u. \tag{24}$$

The state y consists mainly of the deformation v but also values like temperature distribution or process forces might be considered. In addition, economic factors like process time or path length must be considered.

Due to the high computational costs of the operator and the high dimension of the optimization problem, a greedy approach is formulated. The basic idea is to divide the high dimension problem into a large number of low dimension problems, which can be solved with little effort and thus lead to a low total calculation effort.

The time interval is divided into n stages. The variables of the optimal control problem are the discrete control

$$u = (u_0, \dots, u_{n-1})^T \text{ with constant controls } u_i \in \mathbb{R}^l, \quad i = 0, \dots, n-1$$

and the state

$$y = (y_1, \dots, y_n)^T, \text{ where } y_i \in \mathbb{R}^{m_i}, \quad i = 1, \dots, n.$$

The problem is to find the optimal control u^* and the optimal state y^* with subject to

$$\min_{(y,u)} J(y, u) = \sum_{i=0}^{n-1} J_i(y_i, u_i) + J_n(y_n) \tag{25}$$

and constraints

$$y_{i+1} = A_{i+1}(y_i, u_i), \quad i = 0, \dots, n-1 \tag{26}$$

with given initial state $y_0 \in \mathbb{R}^{m_0}$. Furthermore, the target functions

$$\begin{aligned} J_i &: \mathbb{R}^{m_i} \times \mathbb{R}^l \to \mathbb{R}, \quad i = 0, \dots, n-1, \\ J_n &: \mathbb{R}^{m_n} \to \mathbb{R} \end{aligned} \tag{27}$$

and the forward problems

$$A_i : \mathbb{R}^{m_i} \times \mathbb{R}^l \to \mathbb{R}^{m_i}, \quad i = 1, \dots, n \tag{28}$$

must be sufficiently smooth, which is investigated due to the black-box manner of the co-simulation by a sensitivity analysis. The main idea behind greedy methods is to find an optimal control for each state separately, whereas the previous states are fixed. Thus, the original minimization problem is now solved separately for each state, i.e.

$$\min_{(y_k,u_k)} J_k(y_k, u_k), \quad k = 1, \dots, n-1. \tag{29}$$

By fixing previous states, the dimension of the minimization problem is reduced for each state. To find an overall solution each state will be minimized in succession of the previous one. This is solved numerically by using a Newton-like method. With this method, a control u_k is determined such that $\nabla J_k(y_k, u_k) = 0$ holds and a local minimum is found.

In the following, minimization problems with respect to different target functions J are solved. In the next section a control for parallel rough milling is investigated with the aim to minimize the overall induced heat and distortion during roughing. This control allows a thermally sensible process design for roughing and gives a good precondition for the final finishing step. In Sect. 5.2, the NC path optimization is generally investigated and in Sect. 5.3, the reference process is subject of the optimization.

5.1 Optimal Roughing

In this section an optimal control for simple raster milling is investigated. The aim is to find the constant control $u = (a_e, a_p, v_c)$ to prevent thermal deformation of the workpiece. Therfore, deformation vectors are investigated and the largest magnitude of deformation is minimized. The problem is represented by a single stage in context of the optimization scheme, i.e. the minimization of

$$J(y, u) = ||v||_\infty + \alpha T \ \text{ s.t. } \ y = A_S u \tag{30}$$

with respect to the machining restriction, formulated by the constraint $Ru \leq b$, must be calculated. Furthermore, a penalty is used in (30) to enforce short process durations.

The sensitivity of the target function is examined by a local sensitivity analysis, thus the changes of the target function are investigated, caused by the change of one input parameter, while the other input parameters are fixed. Two developments over time of the largest magnitude of deformation vectors in the workpiece domain as well as in the contact-zone are visualized in Fig. 17. Generally, the deformation $||v||_\infty$ becomes larger with progress of the process and thus increases with increasing workpiece temperature in both cases. This is particularly clear to see in the case of the global deformation of the workpiece. The optimization result of rough milling for a sliced reference workpiece is presented in Fig. 18. The warping of the workpiece during the process and the remaining shape error of the cooled workpiece are significantly reduced by choosing the determined roughing parameters.

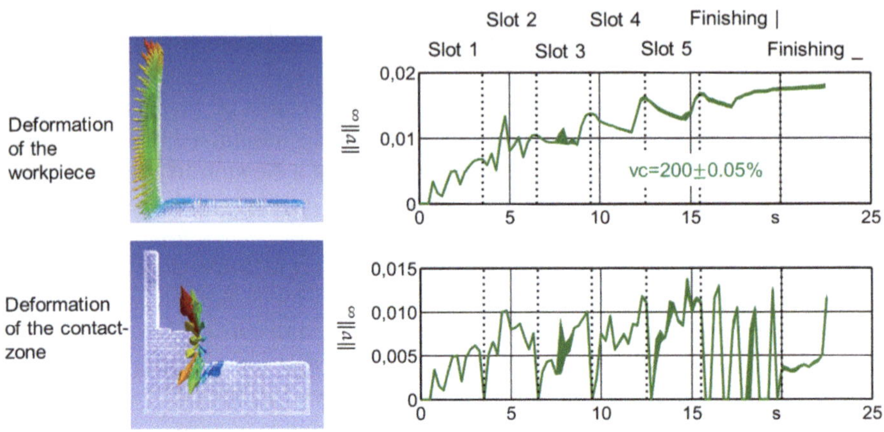

Fig. 17 Visual method for the quantification of the input $v_c = 200$ m/min $\pm 0.05\%$ to the resulting deformation vector; $a_e = 20$ mm, $a_p = 10$ mm. Deformation of the workpiece and of the contact-zone

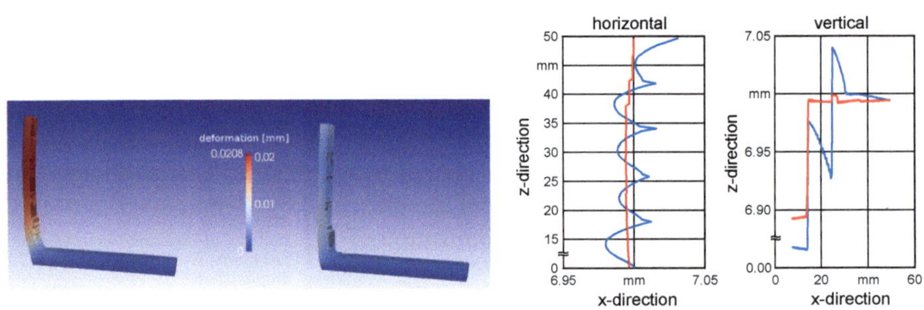

Fig. 18 Sliced reference workpiece. *Left* Deformed workpiece (FE-model) at the end of cutting for $a_p = 8$ mm, $a_e = 8$ mm, $f_z = 0.2$ mm, $v_c = 200$ m/min. *Centre* Deformed workpiece after cutting with optimal parameters ($a_p = 4.3$ mm, $a_e = 20$ mm, $f_z = 0.12$ mm, $v_c = 200$ m/min). *Right* Form deviation of cooled workpiece for both parameter sets

5.2 NC Path Optimization

The aim of a NC path optimization is a manipulation of given control points and cutting velocities v_c in such a manner that the state of the final workpiece becomes as close as possible to the target state. With the greedy approach for each stage a comparison of the current and the desired state $y_{i+1}^{(0)}$ is taken into account:

$$J_i(y_i, u_i) = ||A_i u_i - y_{i+1}^{(0)}||_x^2 + \phi(u_0, \dots, u_i), \quad x = 2, \infty, \; i = 0, \dots n-1. \tag{31}$$

In order to ensure the target shape, the states must be formulated in a way that a comparison of two volumes is possible. For example a ray casting with dexel lines of the volumes can be performed. The comparison of two dexel representations is described in Sect. 5.4. Furthermore, a penalty function ϕ of previous controls is applied. In order to consider all possible effects, it is necessary to continue the simulation until the workpiece has cooled down. If roughing is also subject of the path optimization, the comparison cannot be calculated with help of the target geometry (CAD model) and a ghost simulation under ideal conditions would be necessary.

The problem is simplified considerably if only thermo-elastic effects are taken into account. The idea is to compensate elastic deformations of the contact zone by a movement of the tool, so that undercuts and overcuts cancel each other out.

In the following, the problem is addressed from a geometrical point of view. A parameterized rigid transformation of the swept tool volume is subject of the optimization of each stage. This problem includes solving the interaction between the workpiece and the swept volume. The procedure has to iterate the loop of calculating workpiece load, temperature distribution and deformation and the impact on the contact-zone.

The transformation has six degrees of freedom and has influence on a_p and a_e. The cutting speed is treated separately and the cutting velocity v_f is fixed in the approach. The problem is to minimize a function in order to eliminate overcuts and undercuts:

$$\min_{\omega_i \in \mathbb{R}^6} d(\Omega_i, S_i, \omega_i) + \phi(\omega_0, \dots, \omega_i), \quad i = 0, \dots, n-1, \tag{32}$$

where d is a distance measure, $\omega_i \in \mathbb{R}^6$ the transformation, Ω_i the workpiece and S_i and swept volume at state i. Furthermore, ϕ is the penalty function to ensure small rotations and to suppress corrections in the direction of feed speed. The distance d can be efficiently implemented with distance fields. The rotary volume of the tool is represented with help of the function

$$\tau(x) = \min\left(-\sqrt{x_1^2 + x_3^2} + (d/2)^2\ ,\ x_2\right), \quad x = (x_1, x_2, x_3) \in \mathbb{R}^3. \tag{33}$$

This function has, for each $x \in \mathbb{R}^3$ and especially for $x \in \{DCP\}$, the property

$$\begin{aligned} \tau(x) < 0 &: x \text{ is outside of the tool volume (undercut)},\\ \tau(x) > 0 &: x \text{ is inside the tool volume (overcut)},\\ \tau(x) = 0 &: x \text{ lies on the hull of the tool volume (correct cut)}. \end{aligned}$$

In the next section the compensation is described in detail and is applied to the reference process.

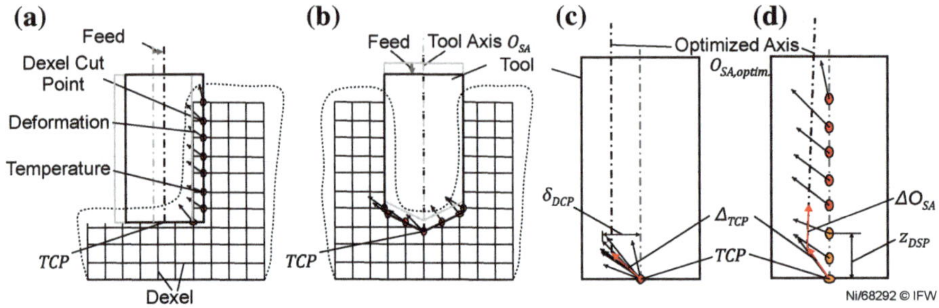

Fig. 19 Deformation extended contact zone analysis of milling (**a**) and drilling (**b**), tool axis orientation optimization for 3-axis (**c**) and multiaxis (**d**) machining of a single simulation step

5.3 Tool Path Optimization by Tool Axis Adaption

The optimization of the machining parameters leads to a considerable reduction of the thermal load on the workpiece. However, this cannot be completely prevented. Therefore, an adjustment of the tool paths is necessary to compensate the remaining shape error. The main challenge are varying shape deviations in the contact zone of a single simulation step. A minimal resulting shape deviation in the contact zone is the final goal of the tool path adaption. As there are 3-axis and multiaxis machine tools the optimization is subdivided.

In 3-axis machining operations (Fig. 19c) only the tool center point *TCP* can be adjusted to minimize the resulting shape deviation, while in multiaxis machining the orientation of the tool can also be adjusted. In the case of 3-axis machining the resulting error of every dexel cut point *DCP* in the contact zone can be described by the difference of the translation of the tool center points Δ_{TCP} to the vector of deformation of the *DCP*, called δ_{DCP}. The total error of contact zone can be described by the least square sum of all *DCP*s. Caused by varying distances of the *DCP*s they are weighted by the representing contact area A_{DCP} in the contact zone. This finally leads to the description of the resulting error δ_{tot} for a single simulation step:

$$\delta_{tot} = \min \sum_{DCP=0}^{DCP} \frac{(\Delta_{TCP} - \delta_{DCP})^2 A_{DCP}}{A_{tot}} \tag{34}$$

In addition for milling processes, the adaption of the orientation of tool axis (multiaxis) (Fig. 19d) can improve the results of the tool path adaption. Therefore, the *DCP*s have to be projected orthogonal onto the tool axis. Based on this transformation the total resulting error δ_{tot} is described along the tool axis by the axial distance of the *DCP* to the *TCP* called z_{DCP}. δ_{tot} is related to the transformation of the tool center point Δ_{TCP} and the change of the orientation of the tool axis ΔO_{SA} as follows:

$$\delta_{tot} = \min \sum_{DCP=0}^{DCP} \frac{\left(\left(\Delta_{TCP} + \Delta O_{SA} z_{DCP}\right) - \left(\delta_{DCP} + z_{DCP}\right)\right)^2 A_{DCP}}{A_{tot}} \tag{35}$$

The minimization of the total error in the contact zone requires a fast and reliable optimization algorithm. In this work, a downhill simplex algorithm is used [19]. It converges very fast for the shown minimization problem. As start point for the algorithm the last optimization result is taken, what additionally accelerates the optimization. Overall, the computational time of a single optimization step is less than 0.1 s.

5.4 Target Workpiece Comparison

Based on the method shown, a minimization of the resulting shape deviation in the whole contact zone is possible. In machining only small sections of the contact zone generate the final shape and thereby the shape deviations of the workpiece. To improve the minimization of the shape deviation on the final part surface, the relevant sections of the contact zone have to be detected. To reach this goal the workpiece model was extended by a target workpiece (Fig. 20), representing the final shape deviation free part.

This target workpiece can be imported from CAD as STEP, IGES or similar file formats. Ahead of the simulation the CAD model is converted to an equivalent target dexel model. In every simulation step all *DCP*s are compared to the target model. Only the *DCP*s which are in the tolerance of the target workpiece are relevant for the final shape of the part. For the further minimization only these *DCP*s are taken into account. The optimization is discussed in detail in [18].

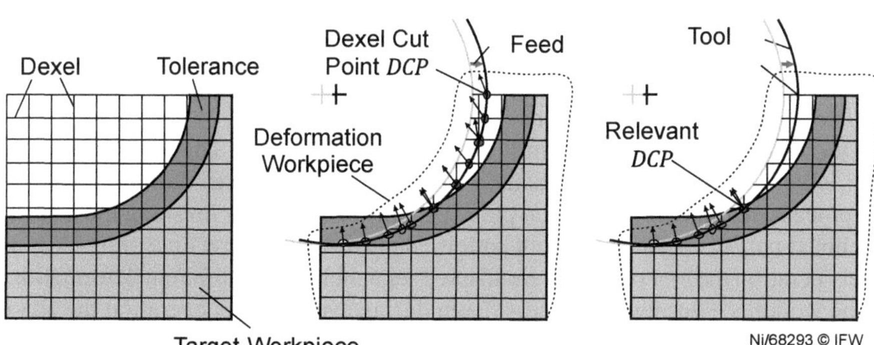

Fig. 20 Target comparison and relevance analysis during contact zone analysis of a single simulation step

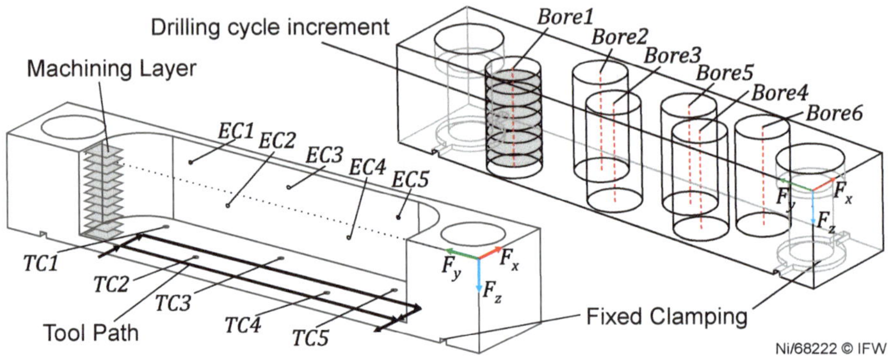

Fig. 21 Reference workpiece for milling and drilling processes

5.5 Reference Processes

The milling of a thin walled part and drilling of a bore sequence are chosen as reference processes (Fig. 21). The blank part are rectangular workpieces. The parts are clamped on two sides, with no degree of freedom of torsion and translation. The sides are fixed on a dynamometer, which enables the measurement of the fixture forces.

The milling strategy is z-level constant. The roughing process is divided in two steps for every level. The finishing of the thin wall is performed in one cut. This process results in a percentage of machined material of about 60%. The drilling reference workpiece has six bores positioned in variant distance to each other and the workpiece sides. All bores are drilled in sequences of 5 mm depth.

5.6 Analysis and Axis Adaption of the Reference Processes

Knowledge about the tool path adaption can be gained by applying the minimization method for the reference processes (Fig. 22).

During the roughing of the milling workpiece almost no adaption of the tool path is necessary. The deformation of the workpiece increases during the roughing process, but mostly no relevant parts are machined. During finishing most of the machined part is relevant. The tool path is adapted in three directions. For every z-level operation the tool path must be adapted separately due to different deformations of the workpiece during the process. Even during every z-level operation the adaption varies along the tool path caused by locally varying deflection and the thermal imbalance in the part. The tilt of the tool axis is small during all operations. Only during finishing of the bottom, where the deformation is different along the width of the contact zone, a significant tilt of the tool axis is necessary.

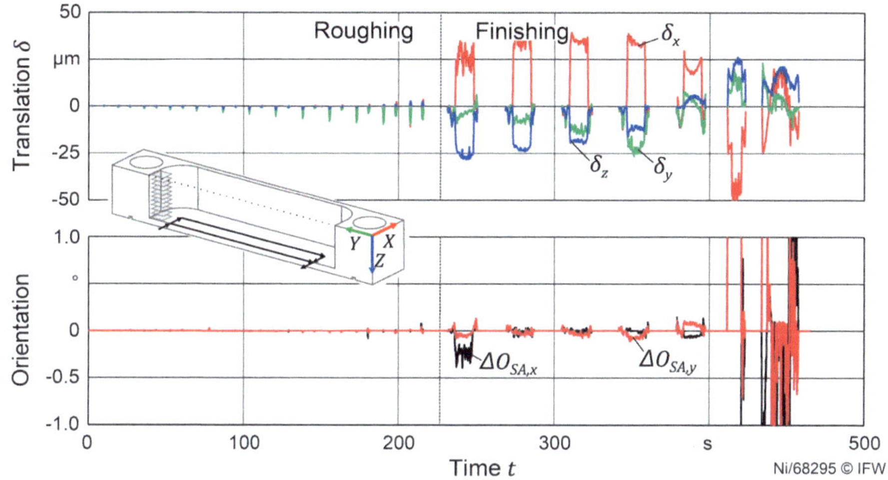

Fig. 22 Multi axis adaption of the tool path for the whole reference process based on the relevant contact zone analysis

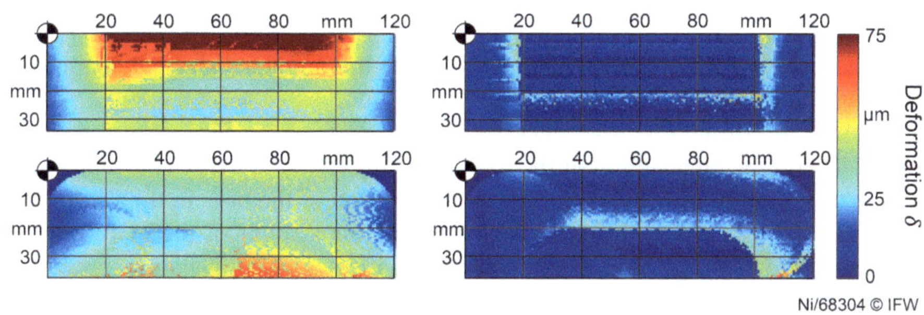

Fig. 23 Simulated shape deviations of workpiece wall (*top*) and bottom (*down*) with (*right*) and without (*left*) tool path adaption

Figure 23 shows the simulated shape deviations after the finishing operation. The shape deviations of the back wall (*top left*) are decreasing from the top to the bottom of the workpiece and are caused by the higher stiffness at the area near the bottom. On the bottom (*bottom left*) the shape deviations in the middle of the part are at a minimum, while they are increasing to both sides.

The shape deviations after the tool path adaption are shown on the right hand side of the figure. As shown, the deviations are reduced significantly at the wall and bottom of the part, but they could not be eliminated entirely. Especially in the corner where the whole tool is in contact an elimination cannot be reached. Further, during retraction of the tool, the differences of the shape deviations in the contact zone increase, so that the maximum shape deviation error results from this final cut. A

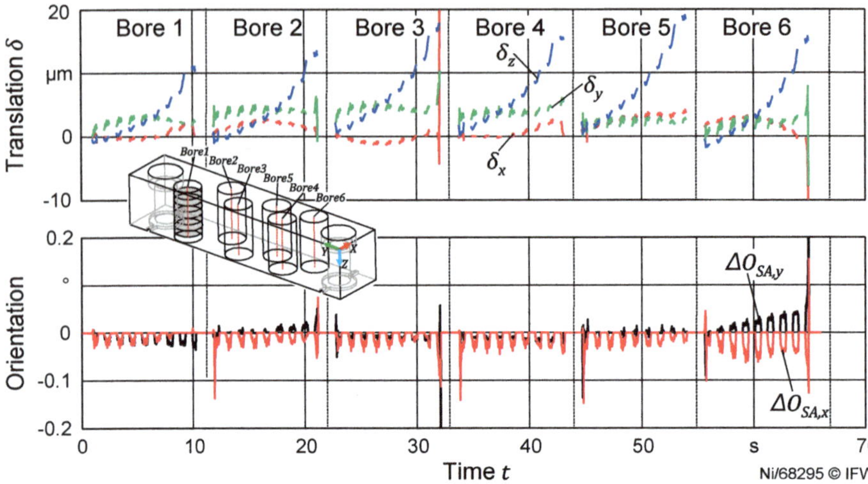

Fig. 24 Multi axis adaption of the tool path for the whole reference process based on the relevant contact zone analysis

mean accuracy of <5% was detected, comparing the predicted shape deviations to measured data.

During the drilling process, generally relevant parts of the final surface are generated. Therefore, an adaption of the toolpath is necessary in every contacting simulation step of all bores (Fig. 24). Due to the position of the bore in the workpiece, especially the distance to the workpiece side, the translations of the TCP in x and y direction are varying. Also along the depth of the hole the translation is varying. The translation is reaching a maximum in the middle and at the exit of the hole. The tilt of the tool axis is small with a value that is always less than 0.2°.

Figure 25 shows the simulated shape deviations of all bores without (*top*) and with (*bottom*) tool path adaption. The shape deviations consist of two major aspects: Straightness and diameter deviation along the depth of the bore. Without tool path adaption the straightness of the bore varies along the depth with a maximum straightness deviation of a few micrometers. This small straightness deviation can be reduced by the tool path adaption. Especially the arched form is compensated. The diameter deviations of the bores are much more distinct and very similar in shape for all bores. At the entrance of the bore the deviation is less than 10 μm. Due to the generated heat the deviation increases until a steady state is reached. At the exit of the hole the deviation increases further, which is caused by the heat accumulation. The diameter deviations cannot be reduced by tool path adaption.

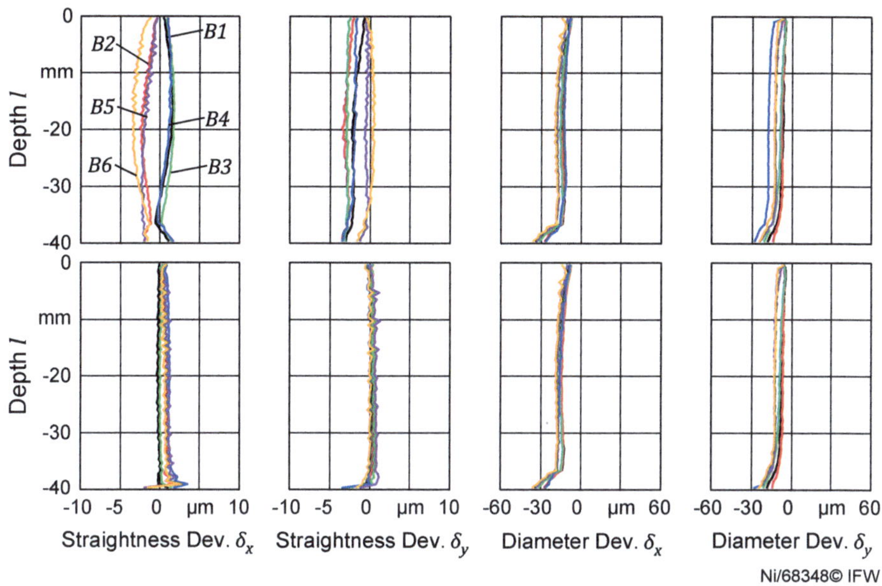

Fig. 25 Simulated straightness and diameter deviation without (*top*) and with (*bottom*) multi axis adaption of the tool path

6 Conclusion and Discussion

In the present work a universal simulation approach has been proposed for the prediction and compensation of temperature-related shape deviations for machining processes. A material removal simulation was linked to process models and a FEM simulation. The removal simulation determines the tool engagement while taking into account the current thermomechanical workpiece state. The process models describe the thermomechanical load of the workpiece. The FEM simulation calculates the temperature and the deformation distribution of the workpiece, based on the current workpiece shape and the load. This thermomechanical workpiece state is used in the material removal simulation. The linking of the modules leads to a closed circle that allows the consideration of all relevant effects and interactions of the process. The approach described above was used in order to carry out targeted process improvements. By optimizing the adjustment parameters, the thermal load of the work could be significantly reduced. The thermal shape deviations could be reduced by a compensation approach. Here, the tool follows the deformation of the workpiece, so that, after cooling, only a small dimensional error remains.

In practice, the developed approach can be used along the entire chain for process planning. First, for a specific material optimum cutting parameters and tool configurations can be set. Subsequently, various production strategies for the preparation of the component can be tested virtually. Finally, the remaining shape deviations can be compensated. In summary, this represents a significant improvement of the production results.

Acknowledgements The presented results have been obtained within the research project "Thermomechanical Deformation of Complex Workpieces in Drilling and Milling Processes" (DE 447/90-3, MA 1657/21-3) within the DFG Priority Program 1480 "Modeling, Simulation and Compensation of Thermal Effects for Complex Machining Processes". The authors would like to thank the DFG for its financial and organizational support of the project.

References

1. Kersting, P., Biermann, D.: Modeling techniques for the prediction of workpiece deflections in NC milling. Proc. CIRP **2**, 83–86 (2012)
2. Bono, M.: Jun Ni: the effects of thermal distortions on the diameter and cylindricity of dry drilled holes. Int. J. Mach. Tools Manuf. **41**(15), 2261–2270 (2001)
3. Isbilir, O., Elaheh, G.: Finite element analysis of drilling of titanium alloy. Proc. Eng. **10**, 1877–1882 (2011)
4. Zimmermann, M., et al.: Analysis of thermo-mechanical effects on the machining accuracy when turning aluminium alloys (2012)
5. Fleischer, J., Pabst, R., Kelemen, S.: Heat flow simulation for dry machining of power train castings. CIRP Annals-Manuf. Technol. **56**(1), 117–122 (2007)
6. Neugebauer, R., et al.: Thermal interactions between the process and workpiece. Proc. CIRP **4**, 63–66 (2012)
7. Schmidt, C.: Einflussgrössensensitive Simulation und Überwachung von Fräsprozessen., Dr.-Ing. Dissertation, Leibniz Universitt Hannover, PZH Produktionstechnisches Zentrum (2011)
8. Denkena, B., Schmidt, C., Krüger, M.: Experimental investigation and modeling of thermal and mechanical influences on shape deviations in machining structural parts. Int. J. Mach. Tools Manuf. **50**(11), 1015–1021 (2010)
9. Surmann, T., et al.: Simulation of the temperature distribution in NC-milled workpieces. Adv. Mater. Res. **223**, 222–230 (2011)
10. Niederwestberg, D., Denkena, B.: Simulation of thermal and mechanical workpiece load. CIRP J. Manuf. Sci. Technol. **7**(4), 315–323 (2014)
11. Denkena, B., Vehmeyer, J., Niederwestberg, D., Maaß, P.: Identification of the specific cutting force for geometrically defined cutting edges and varying cutting conditions. Int. J. Mach. Tools Manuf., 42–49 (2014)
12. van Luttervelt, A., et al.: Present situation and future trends in modelling of machining operations progress R. CIRP Annals—Manuf. Technol., **47**, 587–626 (1998)
13. Altintas, Y.: Manufacturing Automation: Metal Cutting Mechanics, Machine Tool Vibrations, and CNC Design. Cambridge university press (2012)
14. Schmidt, A., Siebert, K.G.: Design of Adaptive Finite Element Software: The Finite Element Toolbox ALBERTA. Springer, Heidelberg (2005)
15. Niebuhr, C., Niederwestberg, D., Schmidt, A.: Finite element simulation of macroscopic machining processes—implementation of time dependent Domain and Boundary Conditions, University of Bremen, ZeTeM-Report 14-01 (2014)
16. Denkena, B., Schmidt, A., Henjes, J., Niederwestberg, D., Niebuhr, C.: Modeling a thermomechanical NC-simulation. Proc. CIRP **8**, 69–74 (2013)
17. Denkena, B., Schmidt, A., Maaß, P., Niederwestberg, D., Niebuhr, C., Vehmeyer, J.: Prediction of temperature induced shape deviations in dry milling. Proc. CIRP **31**, 340–345 (2015)
18. Denkena, B., Niederwestberg, D.: Tool path adaption for compensation of thermal workpiece deflections based on virtual machining. In: 4th international conference on virtual machining process technology (2015)
19. Lagarias, J.C., Reeds, J.A., Wright, M.H., Wright, P.E.: Convergence properties of the Nelder-Mead simplex method in low dimensions. SIAM J. Optim. **9**(1), 112–147 (1998)

Compensation Strategies for Thermal Effects in Dry Milling

M. Gulpak, H. Wernsing, J. Sölter and C. Büskens

Abstract In dry machining of high precision parts shape deviations mainly arise due to thermo-elastic expansion during cutting and machining induced residual stresses. In order to meet higher quality standards when applying the ecologically and economically favourable concept of dry machining, predictive compensation strategies have to be developed which take into account the aforementioned causes for shape deviations. This work presents a newly developed hybrid model integrated in an optimisation algorithm that allows to systematically identify such compensation strategies. Both mechanisms contributing to the shape deviations of machined workpieces are implemented as sub-models applying the Finite Element Method. Aiming at a minimisation of total shape deviations the Simultaneous Analysis and Design approach is utilised. The overall optimisation procedure shows to be time-efficient and is able to find milling strategies leading to a significant reduction of shape deviations. When considered individually, the hybrid model is able to predict each of the two mechanisms, deformation due to machining induced residual stresses and uneven material removal due to thermo-elastic expansion, very well. Experiments based on the predicted milling strategies lead to a substantial reduction of shape deviations and show that additional mechanisms should be implemented for a further process optimisation.

M. Gulpak (✉) · J. Sölter
Department of Manufacturing Technologies, Foundation Institute of Materials Science, University of Bremen, Badgasteiner Str. 3, 28359 Bremen, Germany
e-mail: gulpak@iwt-bremen.de

J. Sölter
e-mail: soelter@iwt.uni-bremen.de

H. Wernsing · C. Büskens
Optimization and Optimal Control, Center for Industrial Mathematics, University of Bremen, Bibliothekstraße 5, 28359 Bremen, Germany
e-mail: heinrich@math.uni-bremen.de

C. Büskens
e-mail: bueskens@math.uni-bremen.de

D. Biermann and F. Hollmann (eds.), *Thermal Effects in Complex Machining Processes*, Lecture Notes in Production Engineering,
DOI 10.1007/978-3-319-57120-1_12

Latin

$\mathbf{A}$	System matrix
a	Width of a premachined and normalised bar
a_e	Width of cut (mm)
$\mathbf{a_e}$	Vector of widths of cut
$a_{e,i}$	Width of cut of the i-th milling path (mm)
a_i	Coefficient of the i-th regression model
a_p	Depth of cut (mm)
b	Height of a premachined and normalised bar
$\mathbf{b}$	Right-hand side vector
b_i	Coefficient of the i-th regression model
c	Cost-term factor
c_i	Coefficient of the i-th regression model
c_p	Specific heat capacity (J/(kg K))
d_i	Coefficient of the i-th regression model
E	Modulus of elasticity (N/m^2)
e_i	Coefficient of the i-th regression model
F	Objective function (K^2 or µMicrosoft^2)
$\mathbf{G}$	System of equations
f_i	Coefficient of the i-th regression model
f_z	Feed per tooth (mm)
g_i	Coefficient of the i-th regression model
h	Lever (mm)
h_i	Coefficient of the i-th regression model
k	Thermal conductivity (W/(m K))
K	Plate stiffness (Nmm)
l	Length of a premachined and normalised bar
M	Number of measured temperatures
M_x	Bending moment along x-axis
M_{xy}, M_{yx}	Torsion moment in xy-plane
M_y	Bending moment along y-axis
n	Rotational speed or number of milling paths
$\mathbf{n}$	Normal vector
N	Size of $\mathbf{T}$, $\mathbf{u}_{source}$ or $\mathbf{u}_{tot,z}$
$\dot{q}_{wp}$	Heat flux into the workpiece (W/mm^2)
$\mathbf{S}$	Vector of model states
T	Temperature (K) or plate thickness (mm)
$\mathbf{T}$	Vector of simulated temperatures (K)
$\bar{\mathbf{T}}$	Vector of measured temperatures (K)
$\bar{T}_i$	i-th measured temperature (K)
$\tilde{T}_i$	i-th simulated temperature corresponding to $\bar{T}_i$ (K)
u	Nodal displacement
u_{source}	Source stresses induced displacement vector field (µm)
$\mathbf{u}_{source}$	Vector of discrete displacements induced by source stresses (µm)
$u_{source,i}$	i-th entry of $\mathbf{u}_{source}$ (µm)

$u_{source,\hat{\imath}}$	Source stresses induced displacement in z-direction in the unsupported corner (μm)
$u_{source,x}$	Source stresses induced displacement in x-direction (μm)
$u_{source,y}$	Source stresses induced displacement in y-direction (μm)
$u_{source,z}$	Source stresses induced displacement in z-direction (μm)
u_{source}^{max}	$\max_{i=1,\dots,N} \lvert u_{source,i} \rvert$
u_{therm}	Thermo-elastic displacement (μm)
u_{tot}	Total displacement (μm)
$\mathbf{u}_{tot,z}$	Vector of discrete total displacements in z-direction (μm)
$u_{tot,z,i}$	i-th entry of $\mathbf{u}_{tot,z}$ (μm)
u_{tot}^{max}	$\max_{i=1,\dots,N} \lvert u_{tot,z,i} \rvert$
v_c	Cutting velocity (100 m/min or mm/min)
$\mathbf{v_c}$	Vector of cutting velocities
$v_{c,i}$	Cutting velocity of the i-th milling path (m/mm)
v_f	Feed velocity (m/min or mm/min)
$\mathbf{v_f}$	Vector of feed velocities
$v_{f,i}$	Feed velocity of the i-th milling path (mm/min)
w	Displacement in z-direction
$\mathbf{X}$	Vector of SAND optimisation variables
$\mathbf{X}^*$	Vector of final SAND optimisation variables
z_0	Layer thickness of source stresses
$\mathbf{Z}$	Vector of NAND optimisation variables

Greek

α	Rotation angle of the milling paths (rad) or coefficient of thermal expenasion (1/K)
Γ	Workpiece surface
λ	Lamé's first parameter (N/m^2)
μ	Lamé's second parameter (N/m^2)
ν	Poisson' ratio (–)
ρ	Density (kg/m^3)
$\sigma_{source.m}$	Source stress tensor field (N/mm^2)
$\boldsymbol{\sigma}_{source.m}$	Vector of discrete source stresses (N/mm^2)
$\hat{\boldsymbol{\sigma}}_{source.m}$	Fixed source stress tensor (N/mm^2)
$\sigma_{source.m,x}$	Normal source stress in x-direction (N/mm^2)
$\sigma_{source.m,y}$	Normal source stress in y-direction (N/mm^2)
$\sigma_{source.wp,i}$	Workpiece inherent source stress tensor in layer i (N/mm^2)
$\sigma_{source.wp,i,x}$	Workpiece inherent source stress in layer i in x-direction (N/mm^2)
$\sigma_{source.wp,i,y}$	Workpiece inherent source stress in layer i in x-direction (N/mm^2)
$\tau_{source.m,xy}$	Shear source stress in x-y-plane (N/mm^2)
$\tau_{source.wp,i,xy}$	Workpiece inherent source stress in layer i in x-y-plane (N/mm^2)
Ω	Workpiece domain

1 Introduction

In nearly all metal cutting processes workpieces are exposed to a significant thermal load resulting from shearing and friction between the cutting tool and the workpiece [1]. Depending on the machining procedure, a varying percentage of the generated heat is dissipated into the workpiece and the cutting tool leading to a temperature induced deformation and therefore a diminished geometric accuracy of the machined workpiece [2]. This is especially true when dry machining is applied [3]. Also, the generation of machining induced residual stresses is dependent on the thermo-plastic behaviour of the work material and therefore on the heat dissipated to the workpiece.

The modelling approaches for the simulation of thermal effects in machining can generally be subdivided into microscopic modelling, e.g. chip formation, and macroscopic modelling. The latter is more suitable for the simulation of the machining of a whole workpiece. In this approach transient temperature fields are calculated in a finite element (FE) model of a moving surface heat source. When the minimisation of shape deviations is pursued, it is further necessary to calculate the mechanical response of the workpiece and couple it with a kinematic simulation of the tool movement [4]. In these simulations the knowledge of the intensity and distribution of the moving surface heat flux is of special importance, since the simulation results strongly depend on them [5].

One common procedure for the determination of heat flux densities in machining is the temperature matching technique. It incorporates a model of the physical and geometrical workpiece properties enabling a calculation of the internal heat flow and resulting temperature fields. The heat flow into the workpiece is found from the simulation delivering the best fit with the measured temperature fields.

Besides thermal effects, unwanted shape deviations due to a change in the residual stress state is introduced in every metal cutting process. The change in the residual stress state is caused by cutting induced near surface plastic deformations and the removal of material layers containing residual stresses at the same time [6]. Depending on the rigidity of the workpiece, residual stresses can be dominant in regard to distortion. Although models for the prediction of residual stresses exist and calculation of resulting shape deviations is possible this method is in most cases limited to two dimensions. As an alternative method the effects of machining induced residual stresses can be described by effective source stresses [7]. They can be calculated directly from the deformation of machined specimens and represent the distortion potential of the machining process for a given set of parameters. This is a comparatively efficient approach. However, the actual residual stress state is reflected only to a certain extent and the workpiece inherent stresses are not yet considered.

Because high precision parts gain increasing relevance in industrial manufacturing and the trend of dry machining continues, predictive compensation strategies for the efficient production of workpiece shapes that are sensitive to thermal effects and distortion due to residual stresses will be required in the future. Therefore, a hybrid model incorporating both described effects is developed to be used in an optimisation algorithm to find optimal machining strategies with regard to accuracy and productivity [8].

2 Objectives, Procedure and Working Programme

2.1 Objectives

The main objective is a predictive minimisation of geometry deviations in dry machining of distortion sensitive high precision components. For these components the effects of geometry deviations due to thermo-elastic strains during the machining process and due to residual stresses (thermo-plastic strains) are assumed to be dominant. Therefore, for the accurate prediction of geometry deviations an advanced hybrid model, which incorporates both effects, has been developed. Subsequently, the hybrid model was implemented into an optimisation algorithm minimising the geometry deviations described by an objective function. Solving the resulting nonlinear optimisation problem yielded optimal machining strategies minimising the workpiece distortion.

A plate made of 42CrMo4 with a pattern of boreholes served as an abstract model for a linear rail guide (Fig. 1).

Due to its comparatively low rigidity this workpiece is easily distorted when machining induced plastic deformations alter the residual stress state of the workpiece. In addition, the lean geometry benefits the accumulation of heat during machining and is therefore more sensitive to thermal expansion and the related accuracy. The machining objective is to face mill a thin plate with a suitable machining strategy avoiding torsion, minimising bending and at the same time compensating the uneven material removal due to thermal expansion by predictively adapting the depth of cut. Subsequently, the boreholes need to be drilled in an order leading to a minimal deviation from their nominal position caused by thermal expansion. The influence on the residual stress state can be neglected in drilling due to the rotational symmetry of the affected region.

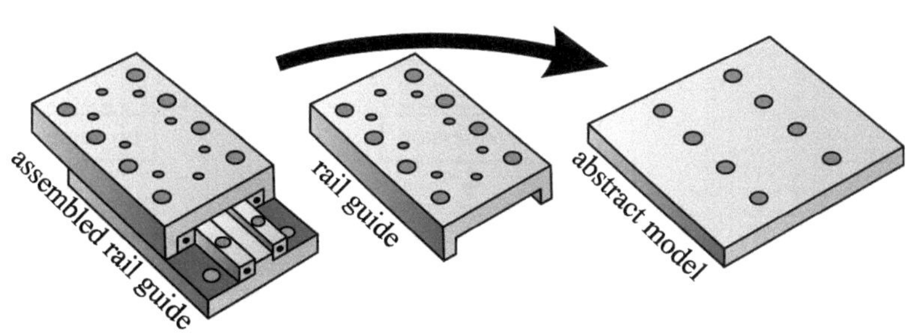

Fig. 1 Model workpiece

This work focuses on shape deviations induced by the milling process. A detailed analysis of the heat input in drilling can be found in [9] while the optimisation of drilling induced shape deviations is under development.

2.2 Procedure

2.2.1 Hybrid Model and Optimisation Procedure

The overall procedure can be divided into the development of the hybrid model and the implementation and utilisation of the hybrid model within an optimisation algorithm for the predictive minimisation of shape deviations (Fig. 2).

The hybrid model's structure for milling is made up of two initially separated sub–models in which shape deviations are calculated. Sub–model 1 is concerned with thermo-plastic effects in machining, where due to the associated near surface plastic deformation, residual stresses are introduced leading to workpiece deformations.

The residual stresses are modelled as effective source stresses evenly distributed within the machining influenced top layer of the modelled plate. The variables describing the stress state in this layer are $\sigma_{source.m,x}$, $\sigma_{source.m,y}$ and $\tau_{source.m,xy}$. The bending source stresses $\sigma_{source.m,x}$ and $\sigma_{source.m,y}$ act in x- and y-direction respectively. $\tau_{source.m,xy}$ is a torsion source stress responsible for a twist of the workpiece.

Sub–model 2 is associated with thermo-elastic effects and calculates shape deviations due to an uneven material removal caused by temperature induced strains in the workpiece. They are calculated by a model of a moving heat source in conjunction with a kinematic model (KM). The magnitudes of the strains depend on the calculated temperature distributions and therefore on the heat flux densities $\dot{q}_{wp,i}$, $i = 1, \ldots, 4$ of the moving heat source.

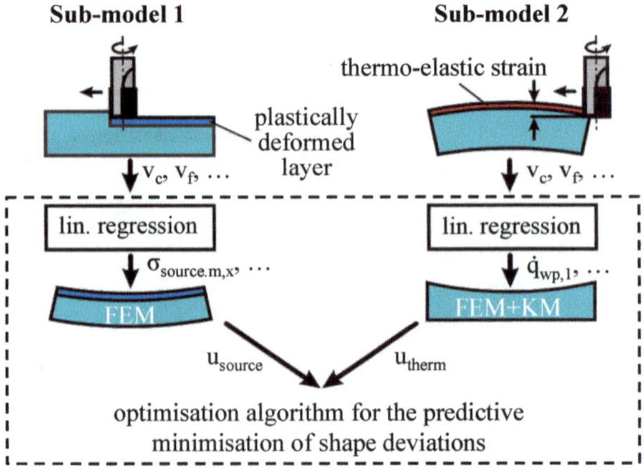

Fig. 2 Composition of the hybrid model

Both sub–models utilise the Finite Element Method (FEM). Each FE model requires a representation of the workpiece, a set of boundary conditions and process specific loads, e.g., stress fields or heat flux densities in order to calculate the resulting workpiece shape. The relation between the calculated node displacements u and the loads applied to the models can be summarised as follows:

$$\text{Sub}-\text{model 1}: u_{source} = f_1(\sigma_{source.m,x}, \sigma_{source.m,y}, \tau_{source.m,xy}) \tag{1}$$

$$\text{Sub}-\text{model 2}: u_{therm} = f_2(\dot{q}_{wp.1}, \ldots, \dot{q}_{wp.4}) \tag{2}$$

The loads vary with the machining parameters and their distribution and magnitude are essential for the optimisation procedure. Therefore, they have to be determined and modelled as accurate as possible, e.g., in form of a linear regression model [8]. The coefficients for each regression model are determined experimentally. For each of the needed relations a multilinear regression model was chosen to map the effects of the machining parameters on the applied loads. In the experiments two levels (+ and −) were used for the cutting velocity v_c and the feed velocity v_f. The width of cut a_e was varied on four levels. The required regression models for the hybrid model can be summarised as follows:

$$\text{Sub}-\text{model 1}: \ \sigma_{source.m,x} = f_3(v_c, v_f, a_e) \tag{3}$$

$$\sigma_{source.m,y} = f_4(v_c, v_f, a_e) \tag{4}$$

$$\tau_{source.m,xy} = f_5(v_c, v_f, a_e) \tag{5}$$

$$\text{Sub}-\text{model 1}: \ \dot{q}_{wp,i} = f_{6,i}(v_c, v_f), i = 1, \ldots, 4 \tag{6}$$

The calculated shape deviations from each sub–model are validated separately and used in a subsequent optimisation procedure.

The optimisation of shape deviations is separated in two steps (see Fig. 3). Step 1 utilises Sub–model 1 to minimise deformations u_{source}, which are caused by source stresses by addressing all parameters describing the milling strategy and all machining parameters except from the depth of cut. Step 2 utilises Sub–model 2 which considers the thermo-elastic effects of machining causing local expansions. Thermo-elastic deformations u_{therm} and resulting deformations in step 1 are compensated in step 2 by adjusting the depth of cut and minimising the total deformations u_{tot}.

Previous investigations show that the deformations induced by source stresses are mainly affected by the position as well as the width of the milling paths [10] and are mostly independent from the depth of cut within the considered range (a few micrometers) [11]. Hence, it is reasonable to optimise the depth of cut separately. The essential benefit of this separation is its computational efficiency. Since step 1

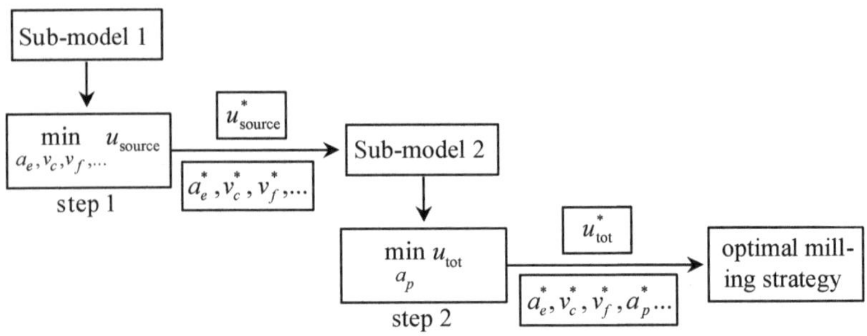

Fig. 3 Optimisation procedure; optimisation results are marked with an asterisk

is considered to be independent of the depth of cut, thermal expansions do not have to be calculated for each set of machining parameters during optimisation. This reduces the duration of the optimisation process from several weeks to less than a day.

2.2.2 Simultaneous Analysis and Design

The optimisation problems occurring in this report are in general nonlinear optimisation problems (NLP), e.g. the minimisation of deformations (see Fig. 3). The models appearing here are mostly given as partial differential equations (PDE) and are needed for calculating states like temperatures in heat transfer analyses or displacements in stress analyses. There are basically two approaches to include a PDE based model within the optimisation problem [12].

The first approach is called Nested Analysis and Design[1] (NAND), Fig. 4. It is the conventional approach in optimisation and uses an external PDE solver to calculate the model states which are needed by the optimisation algorithm to minimise the objective function. The derivatives of the PDE model which are needed by gradient based optimisation algorithms can rarely be given analytically and are usually computed by finite differences. Additional constraints concerning the model states, e.g. temperature restrictions, are also realised by model evaluations. As a consequence in NAND the PDE has to be solved several times.

In the second approach, Simultaneous Analysis and Design[2] (SAND), the discretised states are considered as optimisation variables in addition to the common NAND optimisation variables. The discretised PDE is added as additional constraints in the optimisation problem (see Fig. 4). Thus, the PDE is solved

[1]Also called black-box optimisation.

[2]Also known as PDE-constrained optimisation.

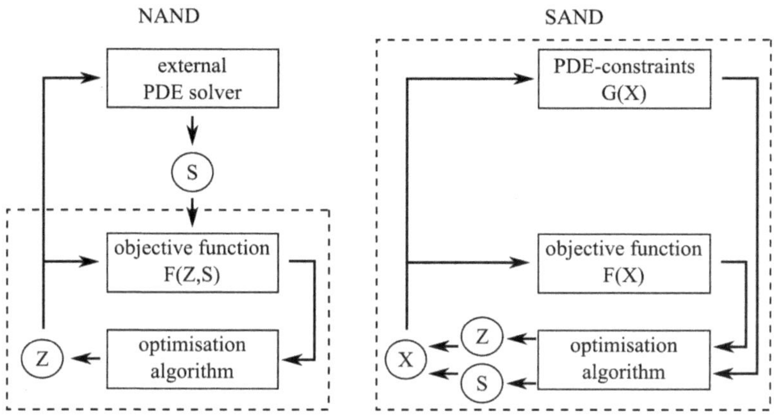

Fig. 4 Differences of NAND and SAND in handling the PDE model and the optimisation variables, where S is a vector of model states, Z is a vector of the common NAND optimisation variables and X is the vector of the SAND optimisation variables comprising S and Z

simultaneously during the minimisation of the objective function and must only be solved exactly in the final iteration of the optimisation.

Apart from an increased effort to implement the discretised PDE as constraints, SAND has several benefits. The derivatives of the PDE model can be calculated analytically, which can improve the convergence of the optimisation algorithm. Constraints regarding the model states are easily considered by so-called box-constraints, since the states are part of the optimisation variables. If the model is based on a nonlinear PDE a reduction in computational time can be observed. Therefore, the SAND approach is used for the optimisation problems considered in this report. Suitable methods for solving a SAND formulation of an NLP are numerical methods like Sequential Quadratic Programming or the Interior Point method. Advanced implementations of these algorithms are provided by *WORHP* [13] and *IPOPT* [14].

2.3 Working Programme and Tool Data

A full factorial design of experiments was chosen for both sub–models. The varied parameters were the cutting velocity v_c, the feed velocity v_f and the width of cut a_e. The cutting velocity and feed velocity were varied on two levels and the width of cut on four levels (Table 1).

The used milling tool was an indexable square shoulder face mill with inserts made of tungsten carbide (Table 2).

Table 1 Design of experiments, milling

Level	v_c/m/min	v_f/mm/min	a_e/mm
–	255	2000	0.95
+	373	2800	3.66
++	–	–	7.72
+++	–	–	12.5

Table 2 Milling tool and cutting inserts, manufacturer´s data

Milling tool		Cutting inserts	
Tool manufacturer	Sandvik	Tool manufacturer	Sandvik
Model	R390-025B25-11H	Model	R390-11 T3 08E-PL 1010
Cutting diameter	25 mm	Substrate	Tungsten carbide
Number of cutting edges	4	Coating	PVD
Tool cutting edge angle	90°	Corner radius	0.8 mm

3 Modelling Workpiece Shape Deviations Caused by Milling Induced Source Stresses

3.1 Analytical Model for the Calculation of Source Stresses

From plate theory it is known that bending moments acting on a plate correspond to a particular curvature response (Fig. 5) [15].

If one assumes that constant effective source stresses act on the machined surface up to a constant penetration depth z_0 on an otherwise residual stress free workpiece, the effective source stresses $\sigma_{source.m,x}$, $\sigma_{source.m,y}$ and $\tau_{source.m,xy}$ can be calculated from the respective bending curvature $\frac{\partial^2 w}{\partial x^2}$, $\frac{\partial^2 w}{\partial y^2}$ and torsion $\frac{\partial^2 w}{\partial x \partial y}$ [16].

$\sigma_{source.m,x}$ and $\sigma_{source.m,y}$ are machining induced source stresses acting along the x-axis and y-axis respectively and $\tau_{source.m,xy}$ is a torsion source stress responsible for a twist of the plate.

3.2 Multilayer Model for Source Stresses

If a residual stress free workpiece cannot be assumed it is more accurate to consider the workpiece inherent residual stresses approximated by a workpiece model consisting of several layers each containing a different set of constant source

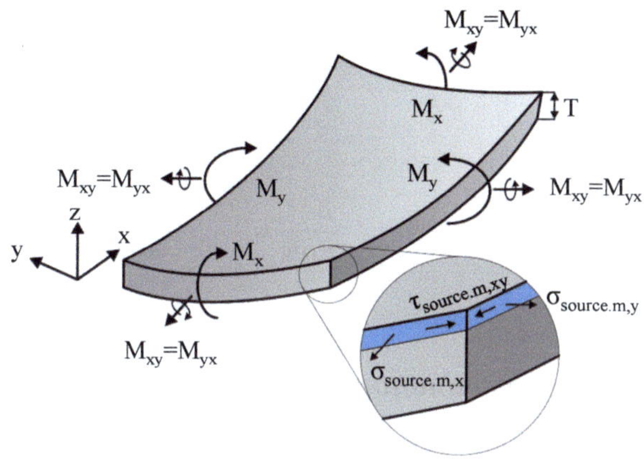

Fig. 5 Bending moments acting on a plate

stresses. In this case every layer contributes to the overall bending and torsion moments deforming the plate.

For this purpose it is necessary to solve the source stress equations in [16] for the curvature terms $\frac{\partial^2 w}{\partial x^2}$, $\frac{\partial^2 w}{\partial y^2}$ and $\frac{\partial^2 w}{\partial x \partial y}$ respectively:

$$\frac{\partial^2 w}{\partial x^2} = \frac{\left(\sigma_{source.m,x} \cdot z_0 - \nu \cdot \sigma_{source.m,y} \cdot z_0\right) \cdot \left(\frac{T}{2} - \frac{z_0}{2}\right)}{K \cdot (1 - \nu^2)} \tag{7}$$

$$\frac{\partial^2 w}{\partial y^2} = \frac{\left(\sigma_{source.m,y} \cdot z_0 - \nu \cdot \sigma_{source.m,x} \cdot z_0\right) \cdot \left(\frac{T}{2} - \frac{z_0}{2}\right)}{K \cdot (1 - \nu^2)} \tag{8}$$

$$\frac{\partial^2 w}{\partial x \partial y} = \frac{\tau_{source.m,xy} \cdot z_0 \cdot \left(\frac{T}{2} - \frac{z_0}{2}\right)}{K \cdot (1 - \nu)} \tag{9}$$

In Eqs. (7)–(9) the curvature and torsion terms are expressed as functions of the forces per unit width, e.g. $\sigma_{source.m,x} \cdot z_0$, the lever arm $h = \left(\frac{T}{2} - \frac{z_0}{2}\right)$ and the terms $K \cdot (1 - \nu^2)$ and $K \cdot (1 - \nu)$, where K denotes the plate stiffness $K = \frac{E \cdot T^3}{12 \cdot (1 - \nu^2)}$.

If now a plate is considered that consists of n layers of thickness $z_{0,i}$, each containing (workpiece inherent) source stresses $\sigma_{source.wp,x}$, $\sigma_{source.wp,y}$ and $\tau_{source.wp,xy}$, the curvature terms and the torsion can be expressed as:

$$\frac{\partial^2 w}{\partial x^2} = \frac{\sum_{i=1}^{n} \left(\sigma_{source.wp,i,x} \cdot z_{0,i} - \nu \cdot \sigma_{source.wp,i,y} \cdot z_{0,i}\right) \cdot h_i}{K \cdot (1 - \nu^2)} \tag{10}$$

$$\frac{\partial^2 w}{\partial y^2} = \frac{\sum_{i=1}^{n} \left(\sigma_{source.wp,i,y} \cdot z_{0,i} - \nu \cdot \sigma_{source.wp,i,x} \cdot z_{0,i}\right) \cdot h_i}{K \cdot (1 - \nu^2)} \tag{11}$$

$$\frac{\partial^2 w}{\partial x \partial y} = \frac{\sum_{i=1}^{n} \tau_{source.wp,i,xy} \cdot z_{0,i} \cdot h_i}{K \cdot (1 - \nu)} \tag{12}$$

3.3 Determination of Source Stresses

With the milling process, newly generated residual stresses modelled as effective source stresses are induced into the layer under the machined surface causing a specific bending and torsion moment. At the same time workpiece deformations are expected due to the removal of material layers containing residual stresses.

For the determination of machining induced source stresses the multi-layer source stress model (Sect. 3.2) was applied to workpieces with a thickness of $T = 12$ mm modelling four symmetric layers of 1 mm thickness on the upper side as well as on the lower side and a core layer of 4 mm in the centre of the workpiece (Fig. 6).

Each workpiece was machined four times, during which one layer of $a_p = 1$ mm was removed from the top surface. For each step the corresponding layer was removed from the model and the layer beneath the machined surface was superimposed with the machining induced effective source stresses $\sigma_{source.m,x}$, $\sigma_{source.m,y}$

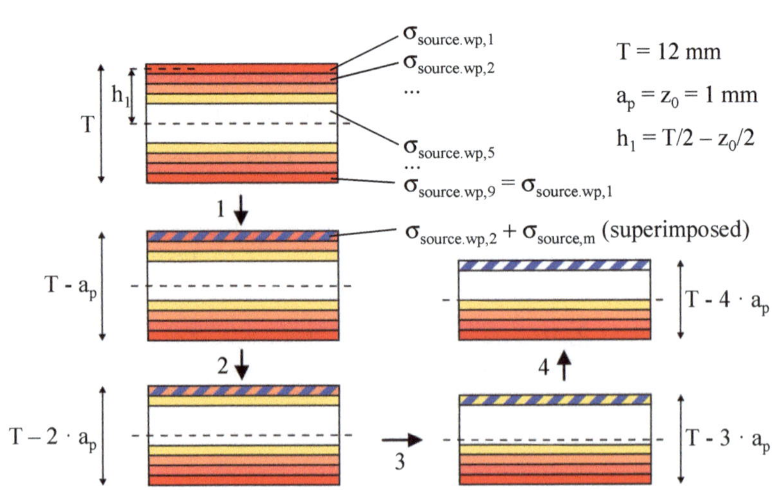

Fig. 6 Machining and modelling steps for the determination of source stresses

and $\tau_{source.m,xy}$ resulting from the milling operation. The workpiece shapes were measured on the lower not machined side before and after each machining step. The overall procedure is summarised in Fig. 6.

With the multi-layer model this procedure leads to a system of linear equations which can be solved with three additional equations, ensuring mechanical equilibrium prior to machining:

$$0 = \sum_{i=1}^{9} \sigma_{source.wp,i,x} \cdot z_{0,i}, \tag{13}$$

$$0 = \sum_{i=1}^{9} \sigma_{source.wp,i,y} \cdot z_{0,i}, \tag{14}$$

$$0 = \sum_{i=1}^{9} \tau_{source.wp,i,xy} \cdot z_{0,i} \tag{15}$$

and the following assumptions:

- In each step the milling process is carried out with similar machining parameters leading to the same induced source stresses $\sigma_{source.m,x}$, $\sigma_{source.m,y}$ and $\tau_{source.m,xy}$ evenly distributed.
- The workpiece inherent source stresses in the first layer removed, $\sigma_{source.wp,1,x}$, $\sigma_{source.wp,1,y}$ and $\tau_{source.wp,1,xy}$ are known, e.g., from residual stress measurements.
- The depth of cut a_p of each cut is considerably higher than the plastically deformed surface layer of the previous cut.

3.3.1 Experimental Programme

The workpieces were cut from normalised and premachined bars of 42CrMo4 with a width $a = 145.2$ mm, a height $b = 145.2$ mm and a length $l = 500$ mm. The workpieces were sawn from the central region of the bars ($\Delta l = T = 12$ mm) ensuring a symmetrical cutting operation from both sides and a neglectable variation of material characteristics over the specimen thickness.

The milling experiments were conducted in down-milling mode with feed direction along the width of the workpiece. For each cut the cutter was set back in feed direction and shifted lengthwise according to the width of cut a_e. This procedure was repeated until the whole upper workpiece surface was machined. During the experiments, the cutting inserts were replaced after machining five workpieces.

Initially a series of 20 experiments was conducted at a constant cutting velocity of $v_c = 373$ m/min, a constant feed velocity of $v_f = 2800$ mm/min and a constant depth of cut of $a_p = 1$ mm varying the width of cut a_e on four levels ($a_e \in \{0.95$ mm; 3.66 mm; 7.72 mm; 12.5 mm$\}$). Each parameter set was repeated five times.

The source stresses in the first layer $\sigma_{source.wp,1,x}$, $\sigma_{source.wp,1,y}$ and $\tau_{source.wp,1,xy}$ were obtained by the removal of the first 1 mm thick layer by electrolytic etching and measuring the shape change of one specimen.

From this set of experiments the workpiece inherent source stresses for every workpiece layer could be obtained as well as machining induced source stresses for varying widths of cut.

In order to complete the linear regression model for the machining induced source stresses, the results of the described procedure were combined with another set of experiments where the cutting velocity v_c and the feed velocity v_f were also varied on two levels [10]. In these less complex experiments the premachining of the workpieces aimed at a residual stress free state by normalisation of the workpiece material which could not be achieved to a satisfactory extent. This could be overcome by linking both experimental results. The link was a shared variation of the width of cut a_e at a matching cutting velocity $v_c = 373$ m/min and feed velocity $v_f = 2800$ mm/min assuming that these conditions resulted in the same machining induced source stresses in both experimental sets.

3.3.2 Source Stress Regression Model

Three regression models for the functional relationship between machining parameters and source stresses are developed assuming a multilinear dependency:

$$
\begin{aligned}
\sigma_{source.m,x} &= a_1 + b_1 v_c + c_1 v_f + d_1 a_e + e_1 v_c v_f + f_1 v_c a_e + g_1 v_f a_e + h_1 v_c v_f a_e \\
\sigma_{source.m,y} &= a_2 + b_2 v_c + c_2 v_f + d_2 a_e + e_2 v_c v_f + f_2 v_c a_e + g_2 v_f a_e + h_2 v_c v_f a_e \\
\tau_{source.m,y} &= a_3 + b_3 v_c + c_3 v_f + d_3 a_e + e_3 v_c v_f + f_3 v_c a_e + g_3 v_f a_e + h_3 v_c v_f a_e.
\end{aligned}
\tag{16}
$$

All parameters are scaled to a similar magnitude before calculating the regression coefficients by applying the method of least squares. Therefore, the input units for the models are $[v_c] = 100$ m/min, $[v_f] = 1$ m/min and $[a_e] = 1$mm. The resulting 24 coefficients are shown in Table 3.

Table 3 Coefficients of the source stress regression model

i	a_i	b_i	c_i	d_i	e_i	f_i	g_i	h_i
1	−14.71	2.57	5.55	1.88	−1.17	−0.41	−0.34	0.09
2	−27.04	4.06	10.86	3.41	−2.47	−0.66	−1.05	0.27
3	2.86	−1.36	−2.52	−0.76	0.60	0.35	0.39	−0.12

3.4 *Finite Element Formulation of the Source Stress Model*

3.4.1 Derivation of the Mathematical Model

In order to calculate shape deviations a model describing the mathematical relations between source stresses and displacements is needed. The stationary Lamé-Navier-equations[3] give a relation between the vector field $u_{source}\colon \Omega \to \mathbb{R}^3$ of displacements induced by source stresses and the body force field $f\colon \Omega \to \mathbb{R}^3$ on the domain Ω [17–19]. The body force can be defined as $f = -\operatorname{div}\sigma_{source.m}$ using the equilibrium equations, where $\sigma_{source.m}\colon \Omega \to \mathbb{R}^{3\times 3}$ is a source stress tensor field. Due to the fact that in the machining experiments the workpiece is fully supported from below and thus expected gravity induced displacements are small and can be neglected. Altogether, this yields the desired relation:

$$\mu\,\Delta u_{source} + (\lambda + \mu)\nabla(\operatorname{div} u_{source}) = \operatorname{div}\sigma_{source.m}, \tag{17}$$

where λ and μ are the Lamé parameters, which can be written in terms of the Poisson's ratio ν and the modulus of elasticity E (see Table 6 in Sect. 4.2), when an isotropic body is assumed. This holds for the interior of the domain $\Omega = [0,\ 145.2] \times [0,\ 145.2] \times [0,\ 11]\ \text{mm}^3$, which represents the entire workpiece. Equation 17 is a stationary partial differential equation (PDE) for which boundary conditions are additionally specified. Assuming no external loads on the outside, Neumann-boundary conditions on the workpiece surface $\Gamma\colon = \partial\Omega$ can be formulated with its outer normal vector $\mathbf{n}$:

$$\sigma_{source.m} \cdot \mathbf{n} = 0. \tag{18}$$

Furthermore, to avoid rigid body translation and rotation the workpiece has to be constrained in six degrees of freedom in at least three points, which must not lie on the same line. This can be stated as Dirichlet-boundary conditions:

$$\begin{aligned} u_{source} &= 0 \quad \text{on } (0|145.2|0)^{\mathrm{T}} \\ u_{source,x} = 0,\quad u_{source,y} &= 0 \quad \text{on } (0|0|0)^{\mathrm{T}} \\ u_{source,z} &= 0 \quad \text{on } (145.2|0|0)^{\mathrm{T}}. \end{aligned} \tag{19}$$

For a given source stress tensor field in Ω the resulting displacement can be calculated by solving the Eqs. (17)–(19). A numerical solution by the FEM is favoured. A setup with hexahedral cells and quadratic elements is utilised to generate a mesh with equidistant nodes with the software library *deal.II* [20, 21], which is an open source FEM-library. This numerical approach yields a system of equations:

[3]Also known as Navier-Cauchy-equations or elastostatic equations.

$$\mathbf{G}(\mathbf{u}_{source}) = \mathbf{b}(\boldsymbol{\sigma}_{source.m}) = \mathbf{b}(\boldsymbol{\sigma}_{source.m}(\mathbf{a_e}, \mathbf{v_c}, \mathbf{v_f})), \tag{20}$$

where $\mathbf{G}$ is the system of equations and $\mathbf{b}$ an appropriate right-hand side vector of dimension N denoting the approximate correlation between the displacement vector $\mathbf{u}_{source}$ and a source stress vector $\boldsymbol{\sigma}_{source.m}$, which hold the respective approximated values of u_{source} and σ at the nodes of the mesh.

Together with the regression model of source stresses (see Sect. 3.3.2) this correlation can be expanded by the machining parameter vectors $\mathbf{a_e}$, $\mathbf{v_c}$ and $\mathbf{v_f}$ of the milling operations.

3.4.2 Model Verification

Before the model is used in the optimisation it is verified using the analytical model of [16]. First, a fixed source stress tensor

$$\hat{\boldsymbol{\sigma}}_{source.m} := \begin{pmatrix} \sigma_{source.m,x} & \tau_{source.m,xy} & \tau_{source.m,xz} \\ \tau_{source.m,xy} & \sigma_{source.m,y} & \tau_{source.m,yz} \\ \tau_{source.m,xz} & \tau_{source.m,yz} & \sigma_{source.m,z} \end{pmatrix} = \begin{pmatrix} 3 & 1 & 0 \\ 1 & 2 & 0 \\ 0 & 0 & 0 \end{pmatrix} /\mathrm{N/mm^2} \tag{21}$$

is prescribed to the upper layer of the workpiece and the resulting displacements are calculated. Next, by curve fitting of the deformed surface the curvature parameters are obtained which lead to the source stresses by using [16]. For verification the total number of mesh nodes is varied from 3,381 up to 193,821.

Initially, the best approximation of the fixed source stress tensor can be observed whenever the number of partitions of the z-coordinate is a multiple of 11. This coincides with the thickness of the stress induced layer which is an 11th ($z_0 = 1$ mm) of the whole workpiece thickness. After fixing this discretisation value, only the discretisation level of the other two coordinates is increased. The convergence is observed to be very fast (see Fig. 7). Once the number of partitions of

Fig. 7 Convergence of the stress values by refining the FE-mesh

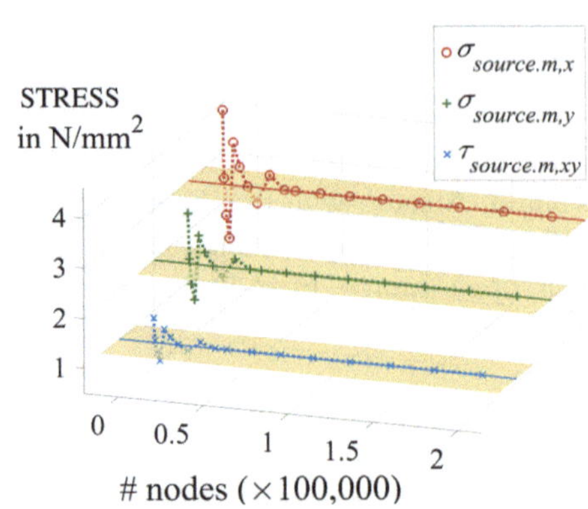

each coordinate is greater than 11, the error of all stress values remains smaller than 0.01 N/mm^2. To additionally take into account that the upper *x*-*y*-plane of the surface is divided in milling paths, the *x*- and *y*-coordinates are discretised into 16 partitions. Together with the 11 partitions of the *z*-coordinate this leads to a total number of 75,141 nodes.

4 Modelling Workpiece Shape Deviations Caused by Thermo-Elastic Strains

4.1 Determination of Heat Flux Densities

The following approach is an extension of the method presented in [22]. In this work transient temperature fields are calculated in an FE model of a moving surface heat source. The heat flux into the workpiece $\dot{q}_{wp,i}$ is determined in an inverse optimisation procedure in which the heat flux densities are iteratively changed until simulated temperature fields agree with temperature fields from dry machining experiments [23]. The cutting arc is virtually discretised into four domains assuming that in each domain the related heat flux density $\dot{q}_{wp,i}$ is approximately constant.

Temperature fields have been measured in milling experiments in which the width of cut a_e was varied between four levels, each covering an engagement angle of 22.5° (Fig. 8).

The iterative procedure starts with the identification of the heat flux density $\dot{q}_{wp,1}$ corresponding to the smallest width of cut of $a_{e,1} = 0.95$ mm. The finite element model is modified to resemble the cutting experiment using $a_{e,1} = 0.95$ mm with only the domain of the moving heat source corresponding to $\dot{q}_{wp,1}$ present. The heat flux density $\dot{q}_{wp,1}$ is changed by an optimisation algorithm until the comparison of the simulated and measured temperature field provides the best possible match. The

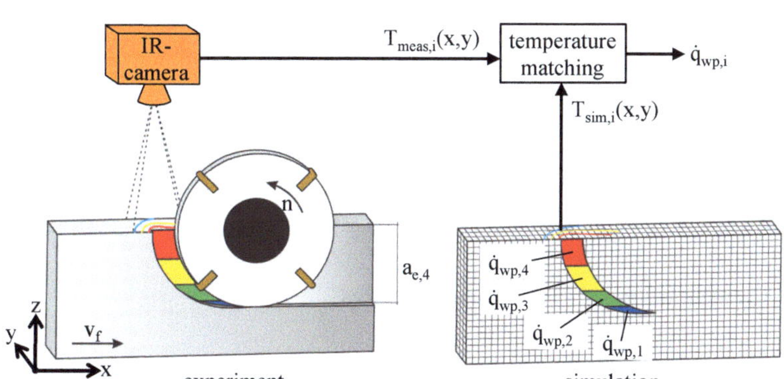

Fig. 8 Procedure for the determination of heat flux densities into the workpiece

value of $\dot{q}_{wp,1}$, which yields the best result, is considered in the following simulations.

After $\dot{q}_{wp,1}$ has been identified it is used as a constant value in the simulations for the determination of $\dot{q}_{wp,2}$. This procedure is repeated until the whole cutting arc of $a_{e,1}$ to $a_{e,4}$ with the appropriate heat flux densities $\dot{q}_{wp,1}$ to $\dot{q}_{wp,4}$ is determined.

4.1.1 Experimental Programme

Workpieces were made of normalized AISI 4140 (42CrMo4) and were premachined to a size of 150 mm · 19 mm · 20 mm. The workpiece face perpendicular to the machining face was painted black to ensure a constant emissivity for temperature measurements.

Dry down milling experiments were conducted on an NC machining centre varying the cutting velocity v_c and the feed velocity v_f on two levels and the width of cut a_e on four levels in a full factorial design of experiments. Each of the 16 combinations of machining parameters was realised five times (Sect. 2.3).

4.1.2 Finite Element Model

The heat transfer due to thermal impact is described mathematically by the three-dimensional heat equation

$$\rho c_p \frac{\partial T}{\partial t} = \operatorname{div}(k \nabla T), \tag{22}$$

which is a PDE.

It can be solved with the Finite Element Method. The FEM is applied with linear elements using the software *deal.II* [20, 21]:

$$\mathbf{A}\mathbf{T} = \mathbf{b}(\dot{q}_{wp,i}), \tag{23}$$

where $\mathbf{T} = (T_1, \ldots, T_N)$ is the vector of temperatures to be simulated with a suitable sorting of the discretisations of the three spatial and one time dimensions. $\mathbf{A}$ is the appropriate $N \times N$ system matrix and $\mathbf{b}$ is the appropriate right-hand side vector of dimension N. A more detailed description as well as a validation of the model can be found in [24].

4.1.3 Identification of Heat Flux Densities

To find the best match of measured and simulated temperature fields the quadratic difference between simulated and measured temperatures is defined as the objective function:

$$F(\mathbf{T}) := \frac{1}{M}\sum_{i=1}^{M}(\tilde{T}_i(\mathbf{T}) - \bar{T}_i)^2, \qquad (24)$$

where M is the number of measurement points to be compared, $\bar{T}_i$ represents the i i-th measured temperature and $\tilde{T}_i(\mathbf{T})$ describes a suitable approximation of the temperature at the same spatial position and time position as $\bar{T}_i$.

The SAND formulation of the optimisation problem is:

$$\begin{aligned} \min_{\mathbf{T}, \dot{q}_{wp,i}} \quad & F(\mathbf{T}) \\ \text{s.t.} \quad & \mathbf{A}\mathbf{T} - \mathbf{b}(\dot{q}_{wp,i}) = 0, \end{aligned} \qquad (25)$$

which expresses a minimisation of the objective function (24) by varying the optimisation variables $\mathbf{T}$ and $\dot{q}_{wp,i}$ subject to (s.t.) the constraints (23). Equation 25 is solved for each experiment (see 4.1.1) to identify the heat flux density. The results are given in detail in [24].

4.1.4 Thermo-Elastic Regression Model

Four regression models for the functional relationship between machining parameters and the heat flux densities are developed assuming a multilinear dependency:

$$\dot{q}_{wp,i} = a_i + b_i v_c + c_i v_f + d_i v_c v_f, i = 1, \ldots, 4. \qquad (26)$$

All parameters are scaled to a similar magnitude before identifying the regression coefficients by applying the method of least squares. Therefore, the input units for the models are $[v_c] = 100\,\text{m/min}$ and $[v_f] = 1\,\text{m/min}$. The 16 resulting coefficients are shown in Table 4.

4.2 *Modelling Workpiece Shape Deviations Caused by Thermo-Elastic Strains*

The simulation of shape deviations in Sub–model 2 consists of three steps. In a first step the three dimensional heat transfer model is discretised with 4–node linear tetrahedron elements. Because the simulation of machining a complete workpiece is

Table 4 Resulting coefficients of the thermal regression model for milling

	Model coefficients (× 100)			
i	a_i	b_i	c_i	d_i
1	−13.45	6.32	5.15	−2.19
2	−9.33	5.22	3.63	−1.77
3	−3.53	2.32	1.34	−0.75
4	−3.55	2.10	1.35	−0.69

Table 5 Material data for 42CrMo4 at 20 °C [25]

Description	Symbol	Value	Unit
Thermal conductivity	k	45.1	W/m/K
Heat capacity	c_p	461	J/kg/K
Density	ρ	7830	kg/m^3

Table 6 Material data for 42CrMo4 at 20 °C [25]

Description	Symbol	Value	Unit
Modulus of elasticity	E	212×10^9	N/m^2
Poisson`s ratio	ν	0.3	–
Coefficient of thermal expansion	α	11.5×10^{-6}	K^{-1}

time-consuming, constant material data (Table 5) and adiabatic boundary conditions are used.

For the simulation of the material removal, the geometrical representation of the workpiece in the premachined state is initially divided into the final workpiece shape and a varying number of volume elements that are removed during the machining simulation. With this procedure a constant mesh region can be ensured at the newly generated surface of the workpiece. When material removal is triggered during the simulation the current temperature field is mapped to the new geometry. In the constant mesh region the nodes of the new mesh are congruent with the nodes from the previous geometry and the temperature values can be assigned directly. In the other mesh region the temperature values are mapped by linear interpolation.

The heat source moves on the surface generated by the material removal computation until it reaches the end of the current removal increment and initiates the next material removal.

In a second step the same mesh and the previously calculated temperature fields in each time increment are used in a stress analysis to calculate the mechanical response of the workpiece to the heat input of the milling process with the material data given in Table 6.

The nodal displacements of the lower surface of the model are set to zero along with the rotational degrees of freedom in order to reproduce the clamping on a magnetic chuck.

Shape deviations are calculated in a third step which comprises a kinematic simulation of the tool movement in conjunction with the calculated node displacements of step two. The analysis of the used milling tool and the cutting inserts shows that cutting can only occur if the position of the currently evaluated node is between 10.2 and 12.5 mm away from the centre axis of the cutting tool (Fig. 9).

In this region, which moves with the tool for every time increment, the maximal vertical displacement of every affected node is saved. After the kinematic simulation is completed, the stored values for maximal vertical displacement of each node represent the inverted topology of the machined surface.

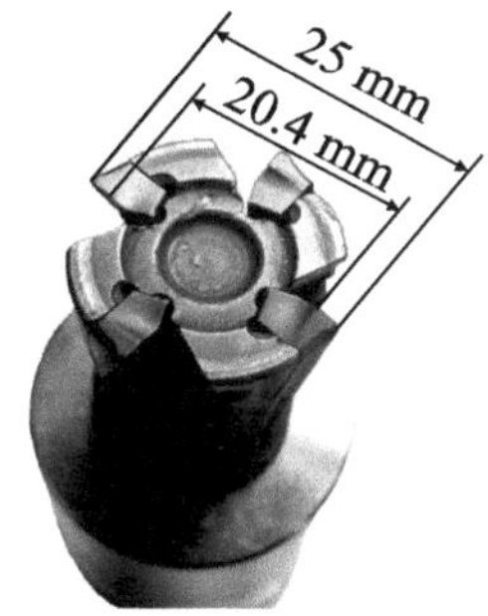

Fig. 9 Indexable milling tool with inserts [4]

4.3 Model Reductions for the Thermo-Elastic Simulations

The three-dimensional simulation of the moving heat source showed to be relatively time consuming, because for each incremental shift of the heat source the underlying equation system in the FE model has to be solved completely. This way the coverage of a complete workpiece requires several hundreds of time increments when the moving heat source is only displaced by the distance of feed per tooth f_z.

The simulation time depends further on different variables and the general simulation approach. The fastest simulations could be achieved with a separate simulation of heat transfer and mechanic response. Also the discretised model equations should be linear. This is assured by using temperature independent material properties and adiabatic boundary conditions.

In order to simulate the machining of a whole workpiece, the modelling of the material removal and the moving heat source had to be further abstracted. For this purpose benchmark simulations were carried out comparing the simulated temperatures and displacements in relation to simulations of the finest possible order. These benchmarks lead to a compromise between simulation time and simulation accuracy (material removal increment of $f_z \cdot 32$ and heat source width of $f_z \cdot 8$) which was used for the subsequent simulations.

5 Optimisation of Milling Strategies

5.1 Distinction Between Discrete and Continuous Variables

For the following optimisation the variables describing the machining process are utilised as optimisation variables. These are, for instance, the machining parameters or the number and order of the milling paths. While e.g. the feed velocity can be changed continuously, the increase or decrease of the number of milling paths is a discrete change. In this research discrete variables are investigated by comparing different milling strategies, which are varied by changing the number, orientation

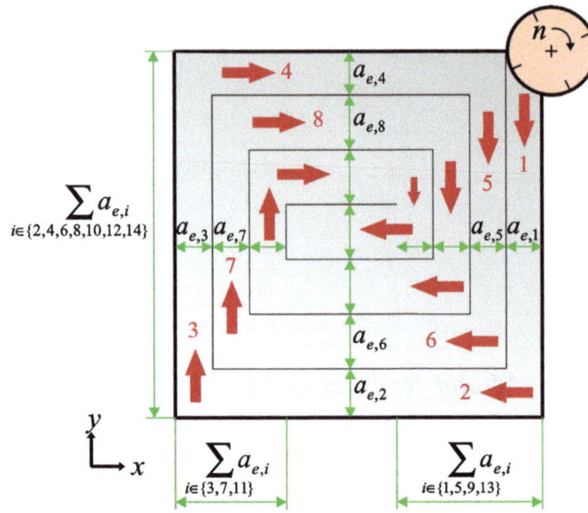

Fig. 10 Example for a milling strategy with fixed quantities (*red*) and variable quantities (*green*)

and order of the single operations, while continuous variables are optimised by nonlinear optimisation techniques.

Figure 10 shows a milling strategy with four subareas that differ from each other in orientation and position. Discrete quantities (red) are the order and orientation of the paths and are fixed during the continuous optimisation process. The sizes of the subareas are variable (green) and can be changed by adjusting the widths of cut. Feed velocity, cutting velocity and depth of cut are also continuous parameters which are optimised by the algorithm.

5.2 *Minimisation of Shape Deviations Caused by Source Stresses—Optimisation Step 1*

5.2.1 Investigation of Discrete Variables—Milling Strategies

The discrete variables are investigated by comparing different machining strategies, which differ from each other in the number, orientation and order of the single milling paths. Figure 11 shows the considered machining strategies.

- A Type **A** provides a trivial strategy with the milling paths rotated by an angle α relative to the workpiece coordinate system. It allows the direction of the effective principle stresses to coincide with the coordinate axes and thus to facilitate the minimisation of shape deformations.
- B Type **B** is a composite of four machining areas. The milling paths are parallel to the workpiece edges. This strategy is designed to offer the optimisation algorithm a large degree of freedom for varying the fraction in which the workpiece is machined in a certain direction.

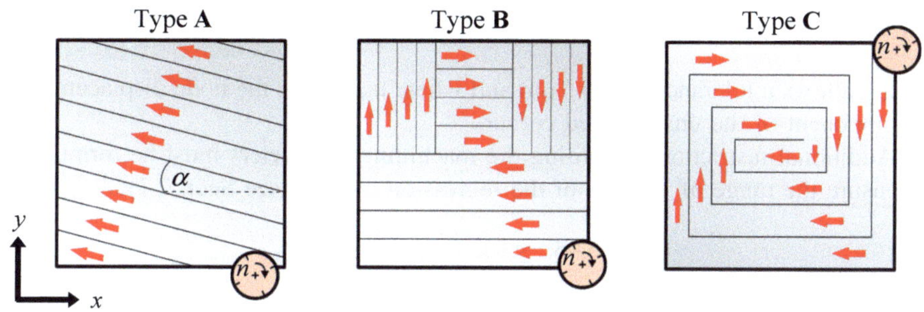

Fig. 11 Considered machining strategies

C Type **C** represents an economic strategy with the shortest total path length of the tool. It is assumed that adjusting the machining parameters decreases shape deviations as well.

Additionally, the total number of milling paths are varied for the three strategies. Each strategy has three different numbers of paths in order to vary the mean width of cut, which yields a total number of nine strategies that are compared.

5.2.2 Mathematical Formulation of the Optimisation of Continuous Variables

The continuous optimisation is approached by nonlinear programming, which optimises the machining parameters a_e, v_c and v_f as well as α. To minimise the workpiece shape deviations the following PDE-constrained (compare Eq. 20) optimisation problem is formulated:

$$\begin{array}{llll} \min\limits_{\mathbf{a_e},\mathbf{v_c},\mathbf{v_f},\mathbf{u}_{source}} & F(\mathbf{u}_{source}) & := & \frac{1}{N}\sum\limits_{i=1}^{N} u_{source,i}^2 \\ \text{s.t.} & 0 & = & \mathbf{G}(\mathbf{u}_{source}) - \mathbf{b}(\boldsymbol{\sigma}_{source.m}(\mathbf{a_e},\mathbf{v_c},\mathbf{v_f})) \end{array}, \quad (27)$$

where $\mathbf{a_e}$, $\mathbf{v_c}$ and $\mathbf{v_f}$ are vectors containing the machining parameters for each milling path and $\mathbf{u}_{source} \in \mathbb{R}^N$ denotes the displacements in each coordinate direction for each node.

The optimisation problem in step 2 (see Sect. 2.2.1) is simplified if the workpiece torsion is eliminated in step 1. Since there is no deviation in z-direction (normal to the machined surface) at the supported corners, the node displacement in z-direction at the unsupported corner quantifies the torsion of the workpiece. The torsion is constrained by:

$$-0.5 \leq u_{source,\hat{\imath}} \leq 0.5 \text{ in } \mu\text{m}, \tag{28}$$

which allows a tolerance of ±0.5 μm and where $u_{source,\hat{\imath}}$ is the node displacement's z-component of the unsupported corner.

Additional restrictions regarding the machining parameters must be formulated to ensure the range of validity of the regression models (see Sect. 2.3):

$$\begin{array}{rcccl} 0.95 & \leq & a_{e,i} & \leq & 12.5 \;, \\ 255 & \leq & v_{c,i} & \leq & 373 \;, \\ 2000 & \leq & v_{f,i} & \leq & 2800 \;, \end{array} \tag{29}$$

where $a_{e,i}$ is the width of cut of the i-th path, $v_{c,i}$ and $v_{f,i}$ are the respective velocities.

Furthermore, the paths must cover the whole top surface. Therefore, the widths of cut have to satisfy additional constraints (see Fig. 11). For the three strategies these are:

Strategy	Constraints	
A	$\sum_{i=1}^{n} a_{e,i} = 145.2\,(\sin\alpha + \cos\alpha)$ $0 \leq \alpha \leq \pi/4$	
B	$\begin{array}{rcl} & \sum_{i=1}^{1/4n} a_{e,i} + \sum_{i=3/4n+1}^{n} a_{e,i} & = 145.2 \\ 0 \leq & \sum_{i=1/4n+1}^{3/4n} a_{e,i} & \leq 145.2 \end{array}$	(30)
C	$\begin{array}{rcl} 0 \leq & \sum_{i=1}^{n/2} a_{e,2i-1} & \leq 145.2 \\ & \sum_{i=1}^{n/2} a_{e,2i} & = 145.2 \end{array}$	

The number n of milling paths as well as the order of the milling paths influences the strategy's profitability, which is taken into account when the strategies are assessed.

The whole optimisation problem is defined by Eqs. 27–30.

5.2.3 Numerical Optimisation Results and Discussion

For reasons of simplicity the following variable is introduced:

$$u_{source}^{\max} := \max_{i=1,\ldots,N} |u_{source,i}|. \tag{31}$$

For the optimisation of milling processes a source stress free workpiece with a height of 12 mm is assumed. The surface area with a size of 145.2×145.2 mm^2 is

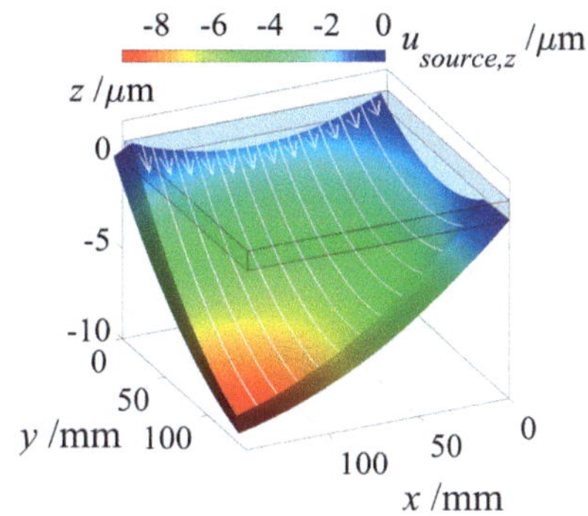

Fig. 12 Simulated shape deviations of the trivial milling strategy (*solid*); non-deformed workpiece (*transparent*)

machined by down-milling to a workpiece height of 11 mm. It is assumed that the milling only induces source stresses into the upper layer of 1 mm thickness.

Benchmark

For benchmarking constant machining parameters are used:

$$a_{e,i} = \frac{145.2}{n}\ \text{mm},\ v_{c,i} = 300\ \text{m/min},\ v_{f,i} = 2400\ \text{mm/min}\ \forall i = 1, \ldots, n. \tag{32}$$

Figure 12 shows the shape deformation induced by the source stresses of the trivial machining strategy which is of type **A** with $\alpha = 0$ and 12 milling paths.

The deviation in this case is relatively large. The objective value[4] is $F(\mathbf{X}) \approx 7.8\ \mu\text{m}^2$ and the maximal deviation is $u^{\max}_{source} \approx 9.2\ \mu\text{m}$. Also the workpiece is twisted considerably.

Type A, 16 Paths

For 12 paths it turns out, that constraint (28) cannot be satisfied, hence the number of paths is increased to $n = 16$ and the initial width of cut is decreased to $a_{e,i} = 9.075$ mm. All other parameters are left unchanged. The optimisation algorithm eliminates the torsion by alternating narrow and wide paths (see Fig. 13 a), which neutralise each other's opposite shear stress values $\tau_{source.m,xy}$ (see Fig. 13 b).

The normal stresses $\sigma_{source.m,x}$ and $\sigma_{source.m,y}$ cannot be compensated completely. This results in a slight bending of the workpiece. The bending along the y-axis is bigger than the bending along the x-axis, since $\sigma_{source.m,y}$ is on average higher than $\sigma_{source.m,x}$ (see Fig. 13b). All other machining parameters $v_{c,i}$ and $v_{f,i}$ converge to the limits of their feasible interval, which is due to the linearity of the regression model of source stresses.

The rotation angle α is found to be zero, since a small angle $\alpha \neq 0$ would result in an increase of the average width of cut and regarding the regression model to a higher bending (see Fig. 14), which is avoided by the optimisation.

[4]The objective value corresponds to the arithmetic mean of the squared deviations [μm²].

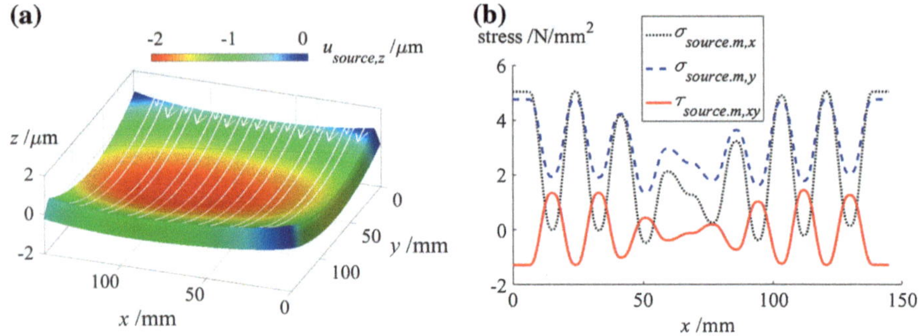

Fig. 13 **a** Minimised shape deviations and optimised paths (strategy **A**, 16 paths), **b** Source stress distribution within the workpiece along the *x*-axis (stress distribution is constant along the *y*-axis)

Fig. 14 Summed up path widths with rotation and without rotation

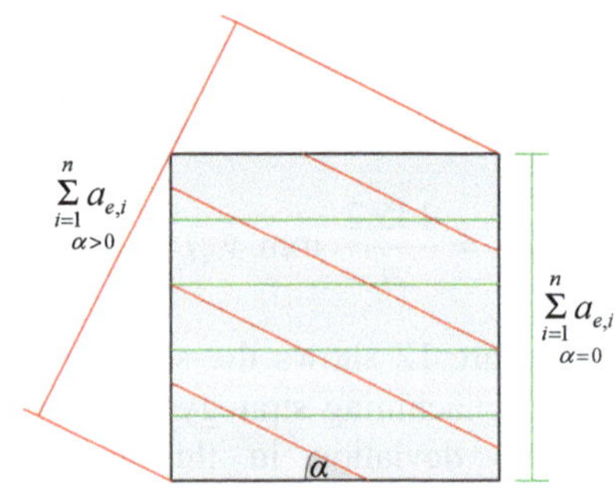

With an objective value of $F(\mathbf{X}^*) \approx 0.7\ \mu\text{m}^2$ the trivial strategy is improved significantly. The maximal deviation is reduced by 78 % to a value of $u^{max}_{source} \approx 2.1\ \mu\text{m}$. However, from an economic point of view this is not an advanced strategy, since the total number of paths is increased and the milling tool has to interrupt the milling process between subsequent paths.

Type B, 40 Paths

In strategy **B** with $n = 40$ paths all machining parameters for the optimisation start are chosen again by reference to (32). The optimised deformations and the optimised paths are shown in Fig. 15a. Due to the large number and the varying orientations of the paths the final objective value almost vanishes ($F(\mathbf{X}^*) \approx 0.0005\ \mu\text{m}^2$) and the maximal deviation decreases by 99 % to $u^{max}_{source} \approx 0.1\ \mu\text{m}$.

Moreover, the mean width of cut is reduced, which leads to a reduction of normal stresses as well. Figure 15b shows the normal source stress distribution of $\sigma_{source.m,y}$ within the workpiece. In this milling strategy negative normal stresses also occur. The balanced distribution of negative and positive values prevents bending almost completely. Since conventional machining centres allow just a

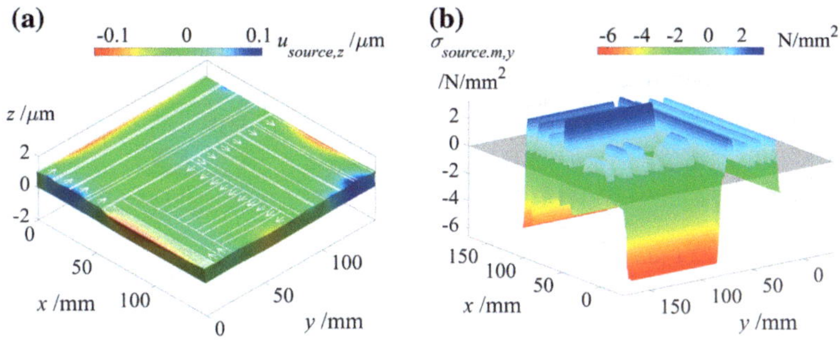

Fig. 15 **a** Minimised shape deviations and optimised paths (strategy **B**, 40 paths), **b** Source stress distribution of $\sigma_{source.m,y}$ within the workpiece; the grey plane marks the zero-level

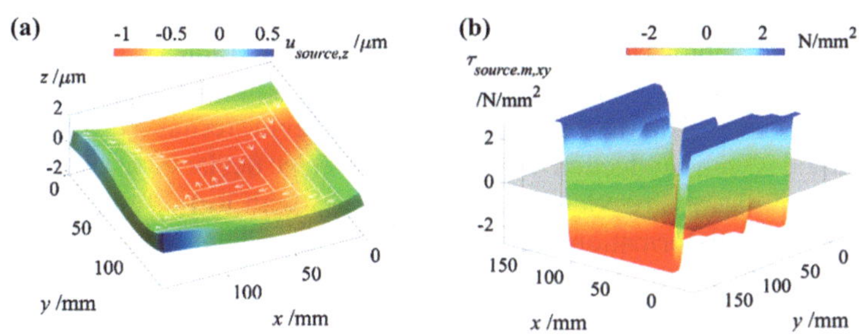

Fig. 16 **a** Minimised shape deviations and optimal paths (strategy **C**, 24 paths), **b** Shear stress distribution $\tau_{source.m,xy}$ within the workpiece; the *grey* plane marks the zero-level

control accuracy of 1 μm, this optimised strategy provides less deformation than it is demanded.

With 40 paths and the interruptions for the tool movements between the single milling processes, this is the most time-consuming of all considered strategies.

Type C, 24 Paths

For strategy **C** with $n = 24$ paths the deformations of the optimised strategy can be found in Fig. 16a. There is no need to alternate the widths of cut, since the rotation of the paths of 90° already inverts the torsion stress values $\tau_{source.m,xy}$ (see Fig. 16b).

The quantities describing deformations are $F(\mathbf{X}^*) \approx 0.2\ \mu\text{m}^2$ and $u^{max}_{source} \approx 1.1\ \mu\text{m}$, which is a reduction by 88 %, and fulfil the required precision (machining centre control accuracy of 1 μm). Particularly noteworthy is that this strategy, which is common in industrial milling processes, has the shortest milling path length and no interruptions.

Table 7 Optimisation results of the considered nine strategies.n denotes the number of paths and $F(\mathbf{X}^*)$ the final objective function value, where $\mathbf{X}^*$ is the final optimisation variable vector. u_{source}^{max} denotes the maximal shape deviation of the machined workpiece

Type A			Type B			Type C		
n	$F(\mathbf{X}^*)$	u_{source}^{max}	n	$F(\mathbf{X}^*)$	u_{source}^{max}	n	$F(\mathbf{X}^*)$	u_{source}^{max}
12	Infeasible		24	0.276 μm^2	1.9 μm (−79%)	24	0.198 μm^2	1.1 μm (−88%)
16	0.73 μm^2	2.1 μm (−78%)	32	0.001 μm^2	0.2 μm (−98%)	32	0.002 μm^2	0.2 μm (−98%)
20	0.015 μm^2	0.5 μm (−95%)	40	0.0005 μm^2	0.1 μm (−99%)	40	0.0004 μm^2	0.1 μm (−99%)

Summary of the Results

The results of all optimised strategies are summarised in Table 7. In most cases the maximal shape deviation is less than the common control accuracy in machining. It turns out that the more paths a strategy contains the better the results are. Strategy **C** with 40 paths provides with a reduction of 99 % the least shape deviation as well as the most efficient way of machining. Nevertheless, strategies with just a few paths might be preferred, since the deformations are sufficiently small and the economic efficiency is better.

5.3 Minimisation of Total Shape Deviations—Optimisation Step 2

5.3.1 Mathematical Formulation

Optimisation step 2 utilises the depth of cut a_p to minimise total shape deviations in z-direction. The deformations induced by source stresses can be taken from the preceding optimisation results, since these are calculated with a model which is independent of a_p. The thermo-elastic deformations are calculated in a heat transfer and stress analysis (see Sect. 4.2). From both deformations, the presented kinematic model of Sect. 4.2 calculated the total displacements $\mathbf{u}_{tot,z}$.

For reasons of simplicity the depth of the paths is represented as parabolas, which additionally ensures the practicability of the optimisation results on a CNC machining centre. Each parabola can be expressed by three variables describing the height of the parabola in z-direction at the start-, mid- and end-point, which leads to $3n$ optimisation variables, summarised in $\mathbf{X}$. The objective function contains a sum over the squared total shape deviations in z-direction $u_{tot,z,i}$:

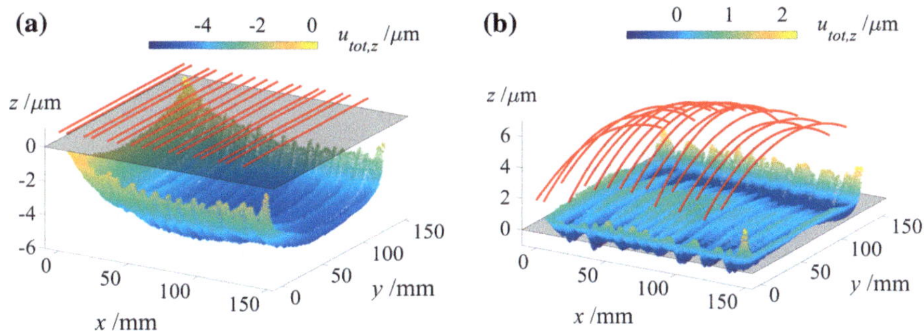

Fig. 17 Milling paths (*red lines*) and total shape deviations of the *upper* surface (strategy **A**, 16 paths): **a** Constant depths of cut, **b** Optimised depths of cut; the *grey* plane marks the ideal surface

$$\min_{\mathbf{X}} \quad F(\mathbf{X}) := \sum_{i=1}^{N} u_{tot,z,i}(\mathbf{X})^2 + c \sum_{i=1}^{3n} X_i^2, \tag{33}$$

where N is the number of discrete displacement values in the z-direction. The second sum in (33) with small weight factor c is introduced for reasons of stability and to ensure that the paths are close to the workpiece surface.

5.3.2 Numerical Optimisation Results and Discussion

For reasons of simplicity the following variable is introduced:

$$u_{tot}^{\max} := \max_{i=1,\ldots,N} |u_{tot,z,i}|. \tag{34}$$

For this investigation two efficient and practically relevant strategies are chosen. The first strategy considered is type **A** with 16 paths. Figure 17 a shows the total shape deviations of the upper surface of the workpiece if a constant depth of cut of $a_p = 0.999$ mm is used.

Two effects occur, which complement each other: Source stresses, which lead to shape deviations at the centre of the workpiece (see Fig. 13a) and thermal expansion of the machined surface layer in the positive z-direction. The expansion increases towards the centre and increases with a proceeding milling process towards the positive x-direction due to heat accumulation. Without compensation the tool removes too much material, which results in shape deviations at the centre as well. This leads to a maximal deviation of $u_{tot}^{\max} \approx 6.0$ μm, where about one third of the shape deviations are caused by source stresses (see Fig. 13a) and two thirds result from an uneven material removal due to thermo-elastic effects. The objective value is $F(\mathbf{X}) \approx 1.7$ μm^2.

The optimised cutting strategy and the resulting shape deviations are shown in Fig. 17b. The convex-shaped tool paths compensate to a large extent the shape

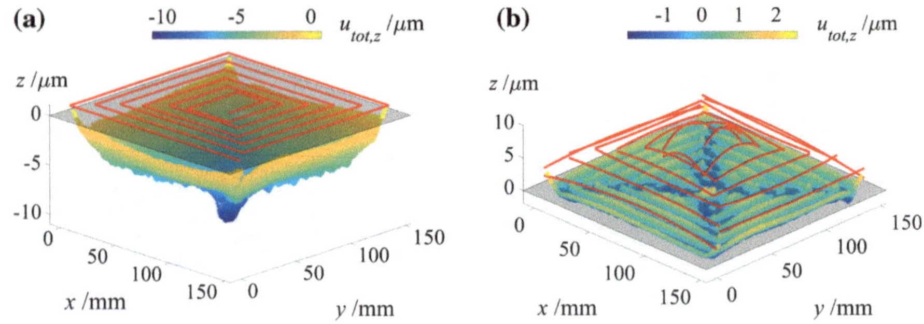

Fig. 18 Milling paths (*red lines*) and total shape deviations of the *upper* surface (strategy **C**, 24 paths): **a** Constant depths of cut, **b** Optimised depths of cut; the *grey* plane marks the ideal surface

deviations of the machined surface. The largest shape deviations occur at the edges of the workpiece and between two tool paths, since the large gradients of the originally deformed surface cannot be compensated completely by a parabolic milling path. Nevertheless, the maximal deviation can be reduced to $u_{tot}^{\max} \approx 2.3$ μm and the final objective value is $F(\mathbf{X}^*) \approx 0.04$ μm^2. This is a reduction of the maximal deviation of approximately 62 % compared to the constant milling paths.

Next we investigate a type **C** strategy with 24 paths. With constant depths of cut the spiral tool path starting at the outside and proceeding inwards generates a heat accumulation at the centre of the part. As a consequence a shape deviation of more than 10 μm occurs (see Fig. 18a).

In this case the deformations induced by source stresses have only a minor influence (see Fig. 16a).

Optimising the depth of cut of $a_p = 0.999$ mm yields a decrease of the maximal deviation of about 76 %, from $u_{tot}^{\max} \approx 10.7$ μm to $u_{tot}^{\max} \approx 2.6$ μm, and the objective function decreases from $F(\mathbf{X}) \approx 2.0$ μm^2 to $F(\mathbf{X}^*) \approx 0.1$ μm^2. Referring to Fig. 18b shape deviations are minimised by milling paths with mostly open downwards parabolas and decreasing depth of cut towards the centre.

Both strategies show a significant improvement with respect to the targeted goal. Since type **C** is economically more efficient, it should be preferred.

6 Validation of the Optimised Milling Strategies

6.1 Procedure

After applying the optimisation procedure described in Sect. 5 the simulated shape deviations were tested by machining two specimens of each machining strategy (Type **A**–**C**). The milling experiments were carried out using specimens from the same batch of workpieces used to determine machining induced source stresses as described in Sect. 3. Half of the specimens were stress relief annealed in order to

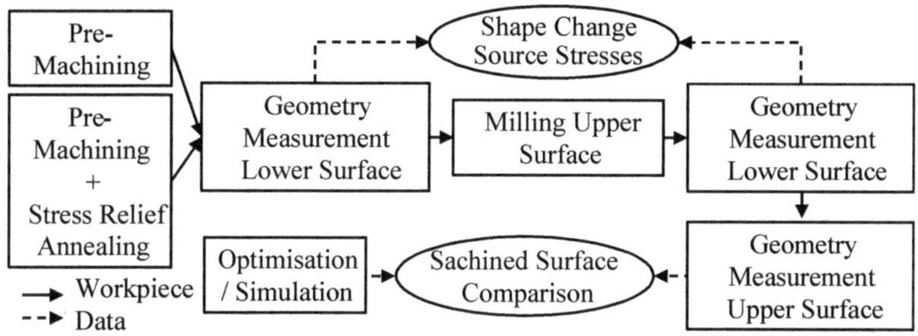

Fig. 19 Procedure for the validation of numerically optimised machining strategies

comply with the optimisation procedure of Sect. 5 in which a stress free workpiece was assumed. The other half were not heat treated and remained in their residual stress state as determined in Sect. 3. In order to test the simulation accuracy of the hybrid model, the optimised machining strategies were used for simulations with a model containing initial residual stresses. The results were compared to the specimens without a stress relief annealing.

The tool paths generated by the optimisation algorithm were transcribed into an appropriate CNC-code which is executable on the utilised Steinel FFZ 200 milling centre. The parabolic milling paths were approximated by linear increments of at least 1 μm in each direction in the CNC code.

The workpieces were measured before and after the experiments on the lower surface which was not machined in order to determine shape deviations caused by the source stresses (workpiece inherent and machining induced) only. After the milling experiments the workpieces were also measured on the machined upper surface to compare the measured workpiece shape with the overall optimisation results (Fig. 19).

6.2 Results

6.2.1 Machining of Stress Relief Annealed Workpieces

The exemplary comparison of the overall shape deviations on the machined side of the Type **C** machining strategy shows a difference of 30 μm between the predictive optimisation and the measurement (Fig. 20).

When plotted in the same range the optimisation results appear nearly flat because the optimisation procedure was able to find machining strategies which lead to an almost perfect shape in the simulations.

A detailed analysis of the measurements of the lower workpiece surface shows that the prediction of the source stress model agrees reasonably well with the

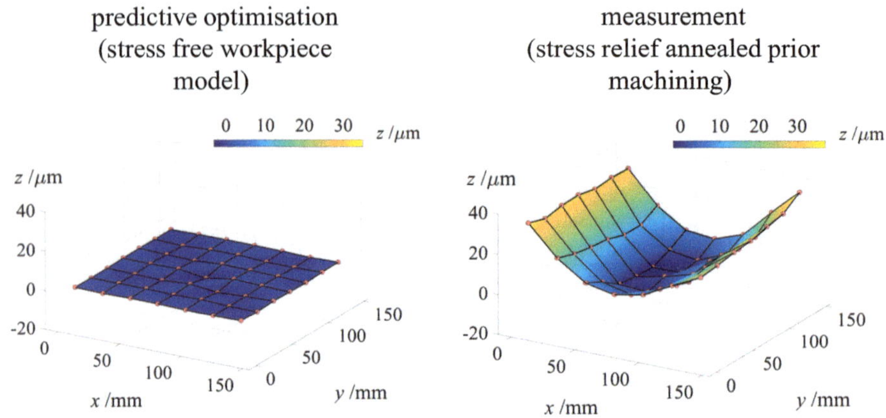

Fig. 20 Comparison of simulation and measurement for the Type C machining strategy, machined surface

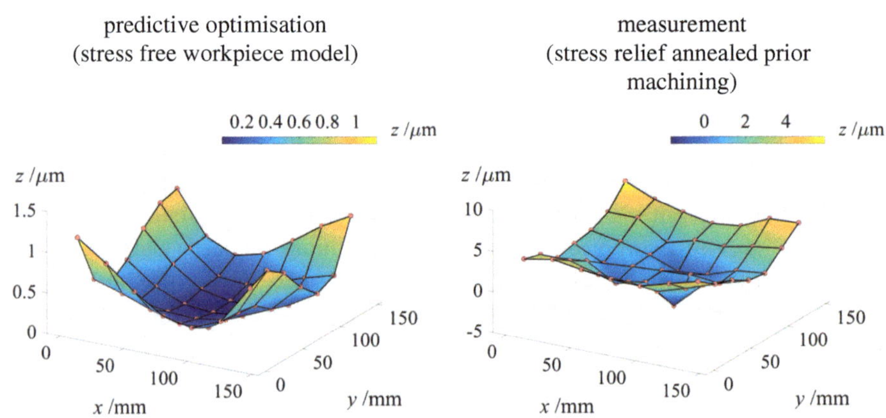

Fig. 21 Comparison of simulated and measured deformation induced by source stresses for the Type C machining strategy, measured shape change on *lower* surface

measurement (Fig. 21). Both sides show a similar deformation with bending along the *x* and *y*-axis. The simulated deformation for the Type C machining strategy lies in the region of 1 μm where the measurement shows a deformation of about 4 μm, indicating a slight remainder of the initial residual stress state. Nevertheless, the stress relief annealing lead to a satisfactory reduction of the workpiece inherent residual stresses and a torsion free workpiece shape could be achieved in the optimisation procedure and in the measurement.

The reason for the difference between the predictively optimised and the experimentally achieved shape deviations cannot be attributed to thermal effects alone, because the simulations and measurements in previous trials indicate thermal effects in the range of 10 μm [4]. Further conclusions may be drawn from the form

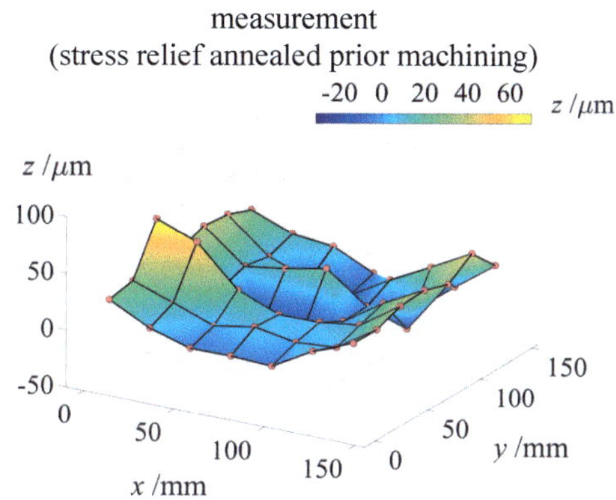

Fig. 22 Measured workpiece shape on the *lower* workpiece side before machining with the Type C machining strategy

of the lower workpiece side before machining (Fig. 22). The measured workpiece shape shows the same characteristic bending in x–direction of around 30 μm found on the machined surface. Because a similar effect can be found on all examined workpieces it is most likely that the magnetic chuck used during machining had an additional influence on the resulting shape of the machined surface.

When the workpieces were put on the magnetic chuck the workpiece shape was measured before and after activation of the magnetic chuck. This lead to a small flattening of the workpieces which lies in the range of 5 μm. After machining the workpieces are released and the flattening is reversed. As a result the workpieces should have an uneven surface caused by the spring back effect. But this does not account for the large shape deviation found in the measurements. From Fig. 22 it is evident that a bending of 5 μm would not completely flatten the workpiece onto the magnetic chuck but only to a certain extent, depending on the magnetic force provided. A certain convex shape of the lower surface remained and allowed the workpiece to roll when machining forces acted perpendicular to the machined surface subsequently leading to an uneven material removal. By this mechanism the shape of the lower workpiece side, to a certain extent, would be transferred to the machined surface.

6.2.2 Machining of Workpieces Containing Residual Stress

The evaluation of the shape of the measured surface indicates that the simulated effects are superimposed with the additional effects caused by the magnetic chuck (Fig. 23).

However, the shape change of the lower surface is matched reasonably well (Fig. 24).

Both, the measurement and simulation show a bending along the x-axis in the same range of 10 μm and a similar torsion. This leads to the conclusion that the multi-layer source stress model describes the influence of the initial residual stress state of the workpiece very well.

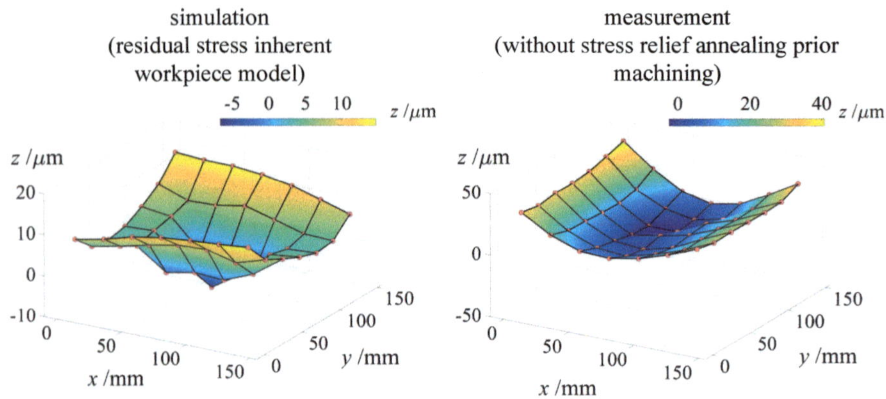

Fig. 23 Comparison of simulation and measurement for the Type C machining strategy, machined surface

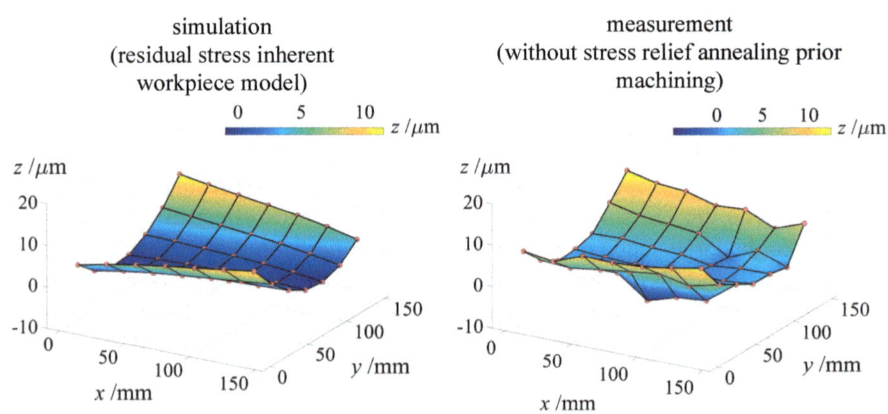

Fig. 24 Comparison of predicted and measured deformation induced by source stresses for the Type C machining strategy, measured shape change on *lower* surface

The detected differences between simulation and measurement on the machined surface suggest that additional mechanisms responsible for shape deviations need to be included into the current model in order to enhance the simulation accuracy. Especially the clamping of the workpiece, machining forces and a possible displacement of the cutting tool due to a temperature rise in the cutting tool and the machine tool should be addressed.

7 Summary and Conclusions

7.1 *Hybrid Model*

In order to simulate shape deviations due to the machining induced change in the residual stress state of the workpiece, a source stress model was employed. This approach was extended to a multi-layer model which can account for workpiece inherent source stresses throughout the whole part. The results indicate that, depending on the initial state of the specimens, workpiece inherent residual stresses can be dominant regarding shape deviations. If workpiece inherent residual stresses are reduced, e.g. by stress relief annealing, the influence of machining induced source stresses increases. In this case a variable milling path layout can be used effectively in an optimisation procedure.

Shape deviations due to thermo-elastic strains in machining mainly depend on the intensity and distribution of the heat input resulting from the interaction between tool and workpiece. The heat flux into the workpiece was determined in a detailed analysis with the temperature matching technique utilising two dimensional thermographic images. For the simulation of the associated shape deviations in milling three simulation steps were necessary to calculate the temperature fields, the mechanical response of the workpiece and the resulting material removal. The calculated workpiece shape matches the experimental results reasonably well if a workpiece shape is chosen which is not sensitive to distortion by residual stresses, clamping or machining forces.

7.2 *Optimisation of Milling Strategies*

The minimisation of shape deviations was sequenced in two steps, which separates the optimisation of source stress induced shape deviations from the optimisation of shape deviations due to thermo-elastic effects.

The presented nonlinear optimisation approach with finite element based models as constraints is a suitable approach for the predictive minimisation of shape deviations in milling. The optimisation of deformations induced by source stresses can reach accuracies which exceed the control accuracy of conventional machining centres. This allows a special focus on the economic efficiency. Spiral paths with few windings allow a precise machining and are economically most efficient.

Optimising the depth of cut mainly prevented the increased material removal due to thermal expansions. Bending left over in the first optimisation step could be compensated as well. Both effects caused a bowl profile of the workpiece which was equalised by parabola paths with high apices. In the presented work a reduction of total shape deviations of up to 76% could be achieved by utilising the newly developed hybrid model combined with the SAND optimisation.

Separating thermo-elastic and source stress induced effects decreases the computation time of the optimisation to a reasonable value. Nevertheless, it would be interesting to combine both optimisation steps allowing even more potential for the minimisation of shape deviations. The necessary reduction of the computation time could be achieved by model simplifications like broadening the moving heat source in addition to a lower spatial and time resolution of the FE discretisation. Further computation time savings could be reached by using a cluster of high-performance computers.

Compared to common NAND implementations the used SAND optimisation approach does not need to solve the PDE model exactly in each iteration step of the optimisation. This becomes particularly important when a nonlinear PDE model is considered. Therefore, using SAND allows a consideration of temperature dependent material parameters without a significant impairment of computation time and with a simultaneous enhancement of model accuracies.

7.3 Validation

The current implementation of the hybrid model is able to predict each of the two considered effects, deformation due to source stresses and uneven material removal due to thermal expansion, very well when considered separately. The measured deformation on the lower, not machined, workpiece side could be predicted precisely for a plate, especially when the initial residual stress state was considered.

The effects of uneven material removal were tested on a workpiece shape where the effects of residual stresses and deformation due to clamping are diminished by the workpiece thickness. These effects lie in the range of 10 μm and were predicted by the thermo-elastic model very accurately [4]. When large workpieces were machined it was found that the temperature induced displacements of the tool began to take effect. In further research this should be incorporated within the hybrid model. Nevertheless, this would require the introduction of an additional longer time scale because cutting tool and the machine are not reset to an initial temperature state when the next workpiece is mounted during production.

Further research should also address shape deviations due to clamping effects. When the machined surface is evaluated, the effects of clamping may override the two effects incorporated in the hybrid model. Although shape deviations due to clamping were initially found to be small by measuring the deformation of the clamped workpiece, this topic needs further attention as additional effects of the workpiece clamping are most likely responsible for the observed differences in measurement and simulation. The dynamic interaction of machining forces, elastic deformations of the workpiece and possible displacements of the workpiece on the magnetic chuck currently lie beyond the scope of the hybrid model but are considered feasible to be taken into account.

In total, the complexity of any model that could incorporate the mentioned effects would be a further step beyond the current state of the art but seems to be within reach from the present perspective.

Acknowledgements The results were obtained within the DFG priority programme 1480 "Modelling, Simulation and Compensation of Thermal Effects for Complex Machining Processes". The authors kindly thank the Deutsche Forschungsgemeinschaft (DFG) for the financial support of the projects BR825/65–1, SO1236/1–2, SO1236/1–3, BU1289/6–1, BU1289/6–2 and BU1289/6–3.

References

1. Komanduri, R., Hou, Z.B.: Thermal modelling of the metal cutting process Part I—Temperature rise distribution due to shear plane heat source. Int. J. Mech. Sci. **42**, 1715–1752 (2000)
2. Kronenberg, M.: Theory and practice for operation and development of machining processes. Machining Science and Application. Pergamon Press, Oxford, New York (1966)
3. Weinert, K., Inasaki, I., Sutherland, J.W., Wakabayashi, T.: Dry machining and minimum quantity lubrication. CIRP Ann.—Manuf. Technol. **53**(2), 511–537 (2004)
4. Gulpak, M., Sölter, J.: Development and validation of a hybrid model for the prediction of shape deviations in dry machining processes. Procedia CIRP **31**(1), 346–351 (2015)
5. Dyck, M.: Beitrag zur Analyse thermisch bedingter Werkstückdeformationen in Trockenbearbeitungsprozessen. Dr. Ing. Dissertation wbk Institute of Production Science Karlsruhe, Shaker Verlag, Aachen (2007)
6. Brinksmeier, E., Sölter, J.: Prediction of shape deviations in machining. CIRP Ann.—Manuf. Technol. **58**(1), 507–510 (2009)
7. Sölter, J.: Ursachen und Wirkmechanismen der Entstehung von Verzug infolge spandender Bearbeitung. Dr.–Ing. Dissertation University of Bremen, Shaker Verlag, Aachen (2010)
8. Sölter, J., Gulpak, M., Brinksmeier, E.: Modellentwicklung zur Minimierung von Geometrieabweichungen in der Trockenbearbeitung. ZWF **107**(4), 224–228 (2012)
9. Gulpak, M., Sölter, J.: Thermal modelling of drilling steel. Adv. Mater. Res. **1140**, 205–212 (2016)
10. Gulpak, M., Sölter, J.: Development of a hybrid model for the prediction of shape deviations in milling. Materialwiss. Werkstofftech. **47**(8), 718–725 (2016)
11. Tönshoff, H.K.: Eigenspannungen und plastische Verformungen im Werkstück durch spanende Bearbeitung, Dr.-Ing. Dissertation TH Hannover (1966)
12. Arora, J.S., Wang, Q.: Review of formulations for structural and mechanical system optimization. Struct. Multidisciplin. Optim. **30**(4), 251–272 (2005)
13. Büskens, C., Wassel, D.: The ESA NLP solver WORHP. In: Fasano, G., Pintér, J.D. (eds.) Modeling and Optimization in Space Engineering. Optimization and Its Applications, vol. 73, pp. 85–110. Springer, Berlin (2013)
14. Wächter, A., Biegler, L.T.: On the implementation of a primal-dual interior point filter line search algorithm for large-scale nonlinear programming. Math. Program. **106**(1), 25–57 (2006)
15. Timoshenko, S.P., Woinkowsky-Krieger, S.: Theory of Plates and Shells. McGraw-Hill Book Company (1959)
16. Gulpak, M., Sölter, J., Brinksmeier, E.: Prediction of shape deviations in face milling of steel. Procedia CIRP **8**, 15–20 (2013)
17. Sadd, M.H.: Elasticity: Theory, Applications, and Numerics. Elsevier Science, Burlington (2009)

18. Kienzler, R., Schröder, R.: Einführung in die Höhere Festigkeitslehre. Springer, Heidelberg (2009)
19. Braess, D.: Finite Elemente—Theorie, schnelle Löser und Anwendungen in der Elastizitätstheorie. Springer, Berlin, Heidelberg (2007)
20. Bangerth, W., Davydov, D., Heister, T., Heltai, L., Kanschat, G., Kronbichler, M., Maier, M., Turcksin, B., Wells, D.: The deal.II library, version 8.4. J. Numer. Math. **24** (2016). doi:10.1515/jnma-2016-1045
21. Bangerth, W., Hartmann, R., Kanschat, G.: deal.II—a general purpose object oriented finite element library. ACM Trans. Math. Softw. **33**(4), 24/1–/27 (2007)
22. Wernsing, H., Gulpak, M., Büskens, C., Sölter, J., Brinksmeier, E.: Enhanced method for the evaluation of the thermal impact of dry machining processes. Prod. Eng. Res. Devel. **8**(3), 291–300 (2014)
23. Sölter, J., Gulpak, M.: Heat partitioning in dry milling of steel. CIRP Ann.—Manuf. Technol. **61**(1), 87–90 (2012)
24. Wernsing, H., Büskens, C.: Parameter identification for finite element based models in dry machining applications. Procedia CIRP **31**, 328–333 (2015)
25. N.N.: Taschenbuch der StahlEisen-Werkstoffblätter. Verein Deutscher Eisenhüttenleute (Ed.), Verlag Stahleisen, Düsseldorf (1997)

Modeling, Simulation and Compensation of Thermomechanically Induced Material Deformation in Dry NC Milling Processes

T. Siebrecht, P. Wiederkehr, A. Zabel, M. Schweinoch, A. Byfut and A. Schröder

Abstract During machining processes, a significant amount of energy is converted into heat at the point of contact of the machining tool and the workpiece. Throughout the process, the workpiece is thermally loaded, resulting in a complex and transient temperature distribution. The latter induces material deformations, which, when not considered in the planning of the machining process, may result in rejects of the manufactured part due to shape deviations or tolerance violations. Compensating thermally induced deformations in machining processes is a challenge for several reasons: The workpiece geometry itself is constantly changing due to the material removal process, which influences the heat flow and dispersion within the semi-finished workpiece. Due to the relative tool-workpiece movement and varying engagement conditions, the heat source changes with respect to location and magnitude, while the change of the surface area affects the cooling of the workpiece. The induced complex and inhomogeneous temperature field within the workpiece leads to an equally complex material response, which may lead to an erroneous material removal when finishing the workpiece in a thermally loaded state. As a result of the transient nature of the thermal loading and the induced deformations, an accurate compensation approach must take into account the specific time-dependent state of the workpiece for a particular engagement condition. For any non-trivial geometry, this requires the use of a simulation system. This paper presents a hybrid simulation system comprised of a geometric physically-based (GP) process simulation and a variation of the fictitious domain method in combination with a (higher-order) finite element (FE) method. The resulting FE simulation is used to predict the thermomechanically induced material response, while the GP simulation provides the appropriate boundary conditions. The hybrid simulation system is able to provide a

T. Siebrecht (✉) · P. Wiederkehr · A. Zabel · M. Schweinoch
Institute of Machining Technology, TU Dortmund University,
Baroper Str. 303, 44227 Dortmund, Germany
e-mail: siebrecht@isf.de

A. Byfut · A. Schröder
Department of Mathematics, Paris Lodron University of Salzburg,
Hellbrunner Str. 34, 5020 Salzburg, Austria

D. Biermann and F. Hollmann (eds.), *Thermal Effects in Complex Machining Processes*, Lecture Notes in Production Engineering,
DOI 10.1007/978-3-319-57120-1_13

detailed analysis of the transient in-process state of the workpiece, which forms the basis for avoiding or compensating an erroneous material removal. The simulation system and the compensation procedure are validated with machining experiments.

1 Introduction

As a manufacturing process, NC milling plays an important role in, e.g., the automotive and aerospace industry. Like many other production fields, the aforementioned ones have seen a steady increase in quality and tolerance requirements. The attained level of sophistication has reached a point where the compliance with these requirements is more strongly affected by the physical phenomena exhibited by the workpiece during the milling process—such as material stress or thermomechanically induced deformations—than by the accuracy of the utilized machine tools and manufacturing equipment. This is particularly true for dry NC milling processes, where the absence of a cooling lubricant results in an overall higher temperature of the workpiece during manufacturing, which in turn results in a more pronounced state of thermomechanically induced deformation.

While the thermomechanically induced deformations recede once the workpiece cools down after manufacturing, the in-process deformations may result in an erroneous material removal, causing a reduction of the workpiece quality and possibly a violation of shape tolerance requirements. In order to adhere to the requirements and avoid rejects, the manufacturing process must be modified to take those thermomechanical deformations of the workpiece into account. Assuming the magnitude of the thermomechanical deformations was given, steps can be taken to avoid the erroneous material removal during the process, such as changing the order of operations or even modifying the NC milling path.

Due to the transient nature of the temperature related phenomena in machining processes, a static analysis of the finished prototype typically cannot yield sufficient data to attempt a compensation. Rather it is necessary to consider the state of the thermal loading and the induced deformation of the semi-finished workpiece at a particular point in time and for the specific tool-workpiece interaction. From a practical perspective, an attempt to analyze in-process phenomena by prototype manufacturing is not feasible: The quality of the in-process measurement technology is limited, and more importantly, the configuration and setup of the measuring equipment may not be possible to the degree that is required for an encompassing analysis. An alternative approach to analyze the milling process with respect to transient phenomena is the use of a simulation system.

The most prominent and proven numerical method in engineering applications is the finite element method (FEM). However, using the (standard) FEM for the analysis of milling processes is problematic: Firstly, the shape of the workpiece changes frequently and severely during milling processes. Further, the frequency of the tool-workpiece engagement is comparatively high, while the simulated material—in metal cutting—is comparatively stiff. When applying the (standard) FEM to simulate a milling process on a chip formation level, these aspects imply a high time and space resolution in order to allow for a stable and accurate simulation. The required computational resources for such a (standard) FE simulation typically limit the application of the method to either comparatively short process time predictions, or require large simplifications, such as low resolution models, which in turn are detrimental for the accuracy. With respect to the geometric representation of milling processes not considering thermomechanical deformations, a number of successful simulation systems have been developed which make use of geometrical methods and empirical models [2]. These GP approaches are considerably less demanding with regard to computational resources, typically allowing for an efficient computation of certain fundamental variables and in-process phenomena, such as cutting forces and heat input prediction, or dynamic tool or workpiece response. Due to the comparatively low computational resources, GP simulations can be used to compute complete, non-trivial milling processes, while retaining a sufficient level of accuracy.

This paper presents a hybrid simulation system for the prediction and compensation of thermomechanically induced erroneous material removal in dry NC milling processes. The simulation system is comprised of a fictitious domain method in combination with a (higher-order) FEM and a GP process simulation. The former allows for an efficient computation of the heat distribution and the corresponding thermomechanical deformations via the (partially decoupled) equations of linear thermoelasticity for a given NC milling process. The efficiency is basically achieved by decoupling for the mesh representation of the time-dependent domain for the (milled) workpiece and the definition of the Ansatz spaces for the (higher-order) FEM via the fictitious domain method. The GP process simulation is used to simulate the material removal process using empirical models to determine the boundary conditions for the FEM. The systems are coupled such that the output of the proposed FE simulation can be used by the GP simulation to compensate the NC path to avoid the erroneous material removal. The hybrid simulation system and the compensated NC paths are validated by experimental results.

Section 2 provides an overview of prior and related work of relevance to the work described herein. A detailed description of the developed simulation system—with a particular focus on the prediction of the erroneous material removal—is given in Sect. 3. The validation of the simulation system is performed by a comparison with experimental results, described and discussed in Sect. 4. In Sect. 5, a compensation approach is presented, along with the validation experiments and measurements. Finally, Sect. 6 concludes with a summary and an outlook on future developments.

2 Related Work

The heat generation as well as the heat input into the workpiece and its effects still remain important problems for all kinds of cutting processes. The prediction of the thermal workpiece behaviour as well as following optimization steps require powerful process modeling approaches which allow for valid and efficient simulations [48].

A lot of work with respect to thermal aspects of cutting processes has been carried out within the Priority Program 1480 of the DFG (German Research Foundation). These interdisciplinary research activities are documented in [36]. A general overview of existing simulation methods for cutting processes and their underlying models is given in [3], which focuses on the chip formation, as an inherently thermomechanical process. Therefore, appropriate models and their parameters with respect to temperature and heat are discussed and as temperature fields and heat transport, both for the tools and the workpieces are major results of such simulations, these effects are also covered.

On the one hand temperature is important with respect to tool life [44] but on the other hand it also determines the behaviour of the remaining material—the final workpiece. In this context the existing simulation approaches can be subdivided in those which are focusing on material properties [6, 23, 37] and those which concentrate on geometric issues [24, 39]. Moving towards the material properties influenced by heat originating from machining processes Bollig et al. present an approach which covers different levels of abstraction for the simulation of heat sources in drilling processes [6]. For non-metallic workpiece materials such as polymers the process-induced heat is of great importance with respect to the material, due to the low heat resistance of plastics. Therefore, a 2D-model and its experimental validation is presented in [34].

One example for simulations aiming at the geometric distortions generated by thermal influences is the multi-scale modeling of thermal workpiece deformations in dry turning [32], where FE models are applied for the computation of the generated heat during chip formation and for the determination of the macroscopic workpiece deformation. Furthermore, a correction methodology for thermally and mechanically induced workpiece deformations is presented in [49], while in [17] a compensation approach for slot milling processes based on the modeled material removal and on the application of a moving heat source model is described. Here, relations to the work presented in this paper can be found.

The heat input in milling, which is generated by the rotating cutting edges, is investigated experimentally and model-based by the authors of [26]. Also in their work, the prediction and compensation of induced form errors is focused. The authors of [13] analyze the influence of machining parameters on the induced heat for milling processes, focusing on avoiding a local overheating of thin-walled workpieces.

Also the accuracy of machine tools is influenced by the thermal process properties, especially with respect to different cooling strategies [31]. In addition, the process heat influences the accuracy of the used tools, which is of special importance for high-precision applications. To improve the situation, the authors of [33] have implemented and calibrated a method for computing and compensating such effects.

Many of the previously mentioned approaches are based, at least partly, on FE models, although there are promising analytical approaches [1]. The presented simulation system consists of a geometric-physical model coupled with a fictitious domain method in combination with a (higher-order) finite element method. In this paper, the latter is employed for the simulation of heat diffusion and thermoelastic deformations in NC milling processes for geometrically complex shapes of a workpiece resulting from the material removal process during NC milling processes. Using this approach, possible violations of critical tolerances can be predicted for a given NC milling process. For this purpose, a variant of the fictitious domain method [4, 14, 18, 30, 42, 47] is developed. In these methods, the actual domain for the problem under consideration is embedded in some covering domain that is represented by some simple, e.g. Cartesian, grid. The latter is used for the definition of piecewise polynomial Ansatz functions as in standard finite element methods. To include the actual domain of interest into the discretization, these Ansatz functions are multiplied with a characteristic function for that domain. Fictitious domain methods seem particularly well suited for the simulation of NC milling processes, since not only the changing domain has to be considered due to the actual material removal, but also the influence of thermomechanical deformations on material removal. The latter dependence yields a major difficulty for the development of this approach.

The modeling of heat diffusion and thermoelastic deformations that occur in NC milling processes is done by the partially decoupled equations of linear thermoelasticity, see also [5, 20, 46]. Well-established generalized trapezoidal finite difference schemes are applied to discretize the parabolic contributions of the variational equations in time, see for instance [22]. For the spatial discretization of the resulting semi-discrete system of equations, a fictitious domain method is developed that utilizes removal-dependent refinements of a hexahedron mesh covering the workpiece under consideration for any given time. The mesh refinements rely on the isotropic bisection of the hexahedrons intersecting the surface of the milling tool at its various positions over time. The resulting fine scale hexahedrons at the milling path interface are further refined into pyramids and tetrahedrons to approximate that interface more accurately. In order to achieve this, refinement patterns corresponding to the interface meshes obtained via the marching cubes algorithm are employed, see [9]. These fine scale elements allow for an elementwise constant definition of the characteristic function for the fictitious domain method. The Ansatz functions of the finite element spaces for the fictitious domain method are defined on some coarse hexahedrons, whereas integration for the corresponding bilinear and linear forms is pursued via the fine scale hexahedrons, pyramids and tetrahedrons, permitting an elementwise evaluation of the characteristic function. Note that due to the definition of the shape functions on (coarse) hexahedrons, the involved application of higher-order shape functions on tetrahedrons and in particular pyramids is avoided. In fact, this definition also enables the exclusive use of tensor product shape functions, which is advantageous for an efficient evaluation of the functions themselves as well as their derivatives. To increase the quality of the approximation, locally refined (coarse) meshes as well as varying polynomial degrees may be employed for the Ansatz spaces. Locally refined meshes may contain hanging nodes resulting from the refinement of hexahedrons without a corresponding refinement of neighboring hexahedrons. Using a conforming discretization approach, the finite element functions have to be continuous across edges and faces. This is realized via the concept of constrained approximation [11, 12, 27, 43] as proposed in [7].

3 Modeling of Workpiece Deformations in Milling Processes

In order to allow for a compensation of thermomechanically induced deformations in dry NC milling processes, it is necessary to be able to simulate these deformations and to be able to consider these deformations for the actual simulation of the NC milling processes. For this purpose, various process parameters have to be determined.

3.1 Finite Element Method

The thermomechanically induced deformations are approximated using a fictitious domain method in combination with a higher-order finite element method applied to the partially decoupled equations of linear thermoelasticity. The following presentation of the modeling and discretization of NC milling process follows the approach proposed in [8].

3.1.1 Modeling

The modeling of the NC milling process basically consists of two parts: The modeling of the changing shape of the milled workpiece and the modeling of the coupling of the heat diffusion in the workpiece with the implied thermomechanical deformation.

Let $\hat{\Omega} := (a_0, b_0) \times (a_1, b_1) \times (a_2, b_2) \subset \mathbb{R}^3$ be some paraxial cuboid that contains the milled workpiece in its reference configuration, i.e., in the absence of any forces, for any time $t \in I := [0, T] \subset \mathbb{R}$ during the milling simulation. Furthermore, let $\Lambda_t \subset \mathbb{R}^3$ denote the closed subset of the domain covered by a milling tool and let $\phi_t : \hat{\Omega} \to \mathbb{R}^3$ describe the deformation of the (milled) workpiece for any time $t \in I$. As usual in linear elasticity, the deformations ϕ_t are assumed to be injective. Given some open subset $\Omega_0 \subset \hat{\Omega}$ for the initial domain in its reference configuration, the deformation-dependently milled workpiece for a time $t \in (0, T]$, before the removal of the material for Λ_t, is formally given by

$$\phi_{t-\delta}(\Omega_0) \cap \bigcap_{s\in(0,t-\delta]} \phi_{t-\delta}\Big(\phi_s^{-1}\big(\phi_s(\hat{\Omega}) \setminus \Lambda_s\big)\Big). \tag{1}$$

Here, ‘before’ refers to some small $\delta > 0$. This definition is motivated by the idea of the following causal chain: For a given workpiece at some initial temperature, some material is removed through the milling. This causes heat input and thereby a thermomechanical deformation that is relevant for the next material removal and so on. For the following modeling of the heat diffusion and the implied thermomechanical deformation, it is necessary to consider the deformation-dependently milled workpiece in its reference configuration. Using the injectivity of ϕ_t, the domain for the latter is defined as

$$\begin{aligned} \Omega_t := \Omega_0 \cap \bigcap_{s\in(0,t-\delta]} \phi_s^{-1}\left(\phi_s(\hat{\Omega}) \setminus \Lambda_s\right) = \Omega_0 \cap \bigcap_{s\in(0,t-\delta]} \hat{\Omega} \setminus \phi_s^{-1}(\Lambda_s) \\ = \Omega_0 \setminus \bigcup_{s\in(0,t-\delta]} \phi_s^{-1}(\Lambda_s) \end{aligned} \tag{2}$$

for all $t \in (0, T]$. Note that this modeling implies $\Omega_s \supseteq \Omega_t$ for $s, t \in I$ with $s \leq t$. For the ease of notation, let $\Gamma_t := \partial\Omega_t$ for all $t \in I$. In NC milling processes, the workpiece is physically clamped at some closed subset of the boundary part $\Gamma_D \subset \Gamma_0$. The clamped boundary part has a positive measure and there holds

$$\Gamma_{D,t} := \Gamma_D \cap \Gamma_t \neq \emptyset \tag{3}$$

for all $t \in I$. The latter condition implies that the workpiece remains clamped at least at some subset of Γ_D throughout the milling process. This is a natural choice for the NC milling processes under consideration.

For the modeling of the coupling of the heat diffusion in the workpiece with the implied thermomechanical deformation for the considered NC milling processes, the linear equations of thermoelasticity [16, 20, 35, 46] are employed. Let $L^2(\Omega)$ and $H^1(\Omega)$ denote the usual Sobolev spaces and $H^1_D(\Omega) := \left\{v \in H^1(\Omega) \mid v = 0 \text{ on } \Gamma_D\right\}$, where the boundary condition is to be understood in the sense of traces. For a displacement field $u \in H^1(\Omega; \mathbb{R}^3)$, the deformation is specified as $\phi := I + u$, the linearized strain tensor as $\varepsilon(u) := \frac{1}{2}\left(\nabla u + (\nabla u)^\top\right)$ and the stress tensor for Hooke's material law as $\sigma(u) := \lambda \operatorname{tr} \varepsilon(u)\mathbb{I} + 2\mu\, \varepsilon(u)$ with the identity tensor $\mathbb{I}$ and Lame's constants λ and μ. The heat flow is given by $\kappa\nabla\theta$ for a heat distribution $\theta \in H^1(\Omega)$ with some heat conductivity $\kappa > 0$. For the following problem of linear thermoelasticity, a displacement field u and a heat distribution θ with $u(0) = \dot{u}(0) = 0, \theta(0) = \theta_0$ in Ω_0 and $u(t) = 0$ on $\Gamma_{D,t}$ have to be found, such that

$$\rho\, \ddot{u}(t) = \operatorname{div}\left(\sigma(u(t)) + m\,(\theta(t) - \theta_0)\mathbb{I}\right) \quad \text{in } \Omega_t, \tag{4}$$

$$c\rho\, \dot{\theta}(t) = \operatorname{div}(\kappa\nabla\theta(t)) + m\theta_0 \operatorname{tr} \varepsilon(\dot{u}(t)) \quad \text{in } \Omega_t, \tag{5}$$

$$\left(\sigma(u(t)) + m\,(\theta(t) - \theta_0)\mathbb{I}\right) \cdot n_t = 0 \quad \text{on } \Gamma_t \setminus \Gamma_{D,t}, \tag{6}$$

$$\kappa\nabla\theta(t) \cdot n_t = g(t, \theta(t)) \quad \text{on } \Gamma_t \tag{7}$$

for almost all $t \in I$. Here, $\theta_0 \in \mathbb{R}$ is an initial constant heat distribution, $m := -\alpha\,(3\lambda + 2\mu) \in \mathbb{R}$ is the stress-temperature modulus, $\alpha \in \mathbb{R}$ is the coefficient of thermal expansion, $c \in \mathbb{R}$ is the specific heat and $\rho \in \mathbb{R}$ is the density of the workpiece. Furthermore, $n_t : \Gamma_t \to \mathbb{R}^3$ denotes the vector-valued function describing the outer unit normal vector with respect to the boundary Γ_t. The function $g(t, \theta(t)) \in L^2(\Gamma_t)$ models the heat transfer into and out of the workpiece on Γ_t for $t \in I$ as specified in the following. Note that no interior heat sources are assumed in Eqs. (5) and (7). Furthermore, no additional volume or surface forces are included in Eqs. (4) and (6), so that deformations may only result from thermal stresses. The parameters α, c, ρ, λ and μ are assumed to be positive constants.

For the modeling of NC milling processes, the coupling of the heat distribution to the strains is neglectable. Hence, the coupling term $m\theta_0 \operatorname{tr} \varepsilon(\dot{u}(t))$ in Eq. (5) may be omitted and there holds the parabolic heat equation

$$c\rho\,\dot{\theta}(t) = \operatorname{div}\big(\kappa \nabla \theta(t)\big) \qquad \text{in } \Omega_t. \tag{8}$$

Considering acceleration effects to be irrelevant, the quasi-static equilibrium equation

$$0 = \operatorname{div}\big(\sigma(u(t)) + m(\theta(t) - \theta_0)\mathbb{I}\big) \qquad \text{in } \Omega_t \tag{9}$$

is obtained as a substitute for the hyperbolic equation (4). These model simplifications lead to a decoupling of the heat distribution from the deformation field and facilitate the discretization of the problem, since only the parabolic heat equation has to be discretized in time. For more detailed explanations, in particular, with respect to further justifications of the model reductions, we refer to the monograph [5].

For the considered NC milling processes, the only source of heat is given by the material removal at the milled surface for a given time. It is modeled by the boundary data $g\big(t, \theta(t)\big)$, which also models the balancing of the workpiece's heat with the ambient temperature. A certain fraction $\mu > 0$ of the actual mechanical cutting work W is assumed to be induced into the workpiece on the contact area with the milling tool in direction of the milling tool movement, before the material for the considered time $t \in I$ is removed. Again, 'before' refers to $t - \delta$ for the same $\delta > 0$ as proposed in (1). For a precise definition of the boundary data, it is necessary to formally define the contact area of the milling tool and the workpiece in direction of the milling tool movement, as depicted in Fig. 1. Given some consistently chosen points $M_{\Lambda_s} \in \Lambda_s$, $s \in I$, e.g. the geometric center of Λ_s, the (normalized) direction of the milling tool movement is approximated by

$$D(s, \varepsilon) := (M_{\Lambda_{s-\varepsilon}} - M_{\Lambda_s}) / ||M_{\Lambda_{s-\varepsilon}} - M_{\Lambda_s}|| \tag{10}$$

for $s \in [\varepsilon, T]$ and some small $\varepsilon > 0$. Hence, the contact area in direction of the milling tool movement is formally defined as

$$\Gamma_{N,t} := \begin{cases} \left\{x \in \Gamma_t \cap \phi_{t-\delta}^{-1}\big(\Lambda_{t-\delta}\big) \mid \langle D(t-\delta, \varepsilon), n_t(x)\rangle \geq \tau \right\}, & t > \delta, \\ \emptyset, & \text{otherwise} \end{cases} \tag{11}$$

for some $\varepsilon > 0$ satisfying $\varepsilon < \delta$ and $\tau \in (0, 1]$. On the remaining part of the workpiece surface, the heat is balanced with the ambient temperature. Using $\Gamma_{N,t}$, this heat transfer is modeled by

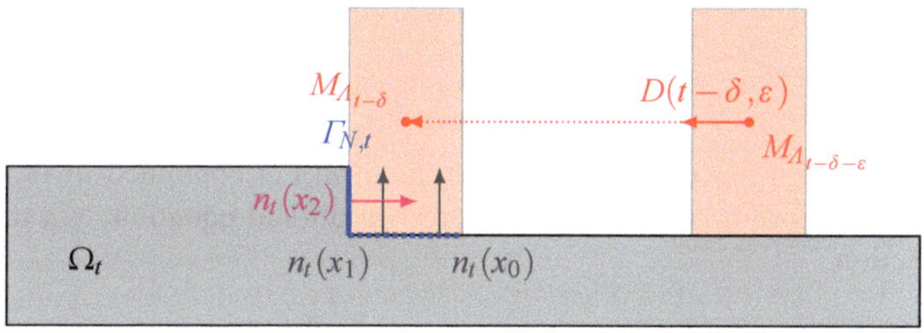

Fig. 1 Illustration of the contact area of a milling tool $A_{t-\delta}$, $\delta > 0$ and a workpiece Ω_t in direction of the milling tool movement as defined in (11) for some $t \in I$. The blue part of the boundary $\partial\Omega_t$ depicts the contact area of the milling tool and the workpiece. The (normalized) normal vectors $n_t(x_0)$ and $n_t(x_1)$ cannot satisfy the angle condition with the (normalized) milling tool movement direction $D(t-\delta,\varepsilon)$, $\varepsilon > 0$ in (11), whereas that condition is satisfied for any $\tau > 0$ for $n_t(x_2)$. Hence, the contact area in direction of the milling tool movement $\Gamma_{N,t}$ is given by the continuous part of the blue boundary

$$g\big(t,\theta(t)\big) := \begin{cases} \mu(t)W(t), & \text{on } \Gamma_{N,t}, \\ a(t)\Big(\theta(t) - \theta_a\big(t,\theta(t)\big)\Big), & \text{on } \Gamma_{R,t} := \Gamma_t \setminus \Gamma_{N,t}, \end{cases} \tag{12}$$

where the definition of the ambient temperature

$$\theta_a\big(t,\theta(t)\big) := f(t)\left(\theta(t) - \bar{\theta}_a\right) + \bar{\theta}_a \tag{13}$$

follows the modeling given in [40] to account for the heat emitted to the environment in the direct vicinity of the workpiece and the corresponding heat up of the former. Here, $\bar{\theta}_a \in \mathbb{R}$ yields the constant ambient temperature away from the workpiece. The parameters

$$a(t) := \begin{cases} a_{\text{air}} & \text{on } \Gamma_{R,t} \setminus \Gamma_D, \\ a_{\text{clamp}} & \text{on } \Gamma_{R,t} \cap \Gamma_D, \end{cases} \qquad f(t) := \begin{cases} f_{\text{air}} & \text{on } \Gamma_{R,t} \setminus \Gamma_D, \\ f_{\text{clamp}} & \text{on } \Gamma_{R,t} \cap \Gamma_D \end{cases} \tag{14}$$

define the thermal conductivity and the heat up ratios via constants $a_{\text{air}}, a_{\text{clamp}} \in \mathbb{R}$ and $f_{\text{air}}, f_{\text{clamp}} \in [0, 1]$ for the workpiece surface in contact with air and the clamp, respectively. For the determination of these paramaters as well as the parameters for the heat input μ and W, see Sect. 3.2. Note that the boundary condition (12) modeling the heat input at the contact area of the milling tool yields a Neumann boundary condition and the condition modeling the balance with the ambient temperature results in a Robin boundary condition.

To apply discretization schemes based upon finite elements, it is necessary to reformulate (8) and (9) along with the boundary conditions (6) and (7) in terms of variational equations. For this purpose, let $(\cdot,\cdot)_{0,\omega}$ denote the usual L^2-scalar product on $L^2(\omega)$, $L^2(\omega;\mathbb{R}^3)$ and $L^2(\omega;\mathbb{R}^{3\times 3})$ for $\omega \subset \hat{\Omega}$, respectively, and define

$$V := W^{2,\infty}\big(I; L^2(\hat{\Omega}; \mathbb{R}^3)\big) \cap L^\infty\big(I; H^1_D(\hat{\Omega}; \mathbb{R}^3)\big),$$
$$W := W^{1,\infty}\big(I; L^2(\hat{\Omega})\big) \cap L^\infty\big(I; H^1(\hat{\Omega})\big).$$

The time-dependent domains Ω_t, $t \in I$ as given in (2) contained in the covering domain $\hat{\Omega}$ are incorporated in the following variational equations via the characteristic function

$$\chi_{\Omega_t} : \hat{\Omega} \to \{0, 1\}, \qquad x \mapsto \begin{cases} 1, & x \in \Omega_t, \\ 0, & x \in \hat{\Omega} \setminus \Omega_t. \end{cases} \tag{15}$$

Thus, the resulting equations yield a fictitious domain approach. In contrast to many of the various fictitious domain approaches reported in the literature, e.g. [4, 14, 18, 30, 42, 47], we define the range of this characteristic function as $\{0, 1\}$ instead of $\{\varepsilon, 1\}$ for some small $\varepsilon > 0$. The latter is well known to introduce a modeling error into variational formulations, whereas the former may cause badly conditioned systems of equations. To account for the conditioning issues, special care is given to the spatial discretization, confirm [8]. Multiplication of (8) and (9) with test functions and using integration by parts leads to the variational formulation to find $(u, \theta) \in V \times W$ with $\theta(0) = \theta_0$, such that

$$\begin{aligned} &\big(\sigma(u(t)), \varepsilon(\chi_{\Omega_t}\varepsilon(v)\big)_{0,\hat{\Omega}} = m\left(\operatorname{div}\big((\theta(t) - \theta_0)\mathbb{I}\big), \chi_{\Omega_t} v\right)_{0,\hat{\Omega}}, \\ &c\rho\,(\dot{\theta}(t), \chi_{\Omega_t}\varphi)_{0,\hat{\Omega}} + \kappa\left(\nabla\theta(t), \nabla(\chi_{\Omega_t}\nabla\varphi)\right)_{0,\hat{\Omega}} - \big(a(t)\big(1 - f(t)\big)\theta(t), \varphi\big)_{0,\Gamma_{R,t}} \\ &\quad = \big(\mu(t)W(t), \varphi\big)_{0,\Gamma_{N,t}} - \big(a(t)\big(1 - f(t)\big)\bar{\theta}_a, \varphi\big)_{0,\Gamma_{R,t}} \end{aligned} \tag{16}$$

for all $(v, \varphi) \in H^1_D(\hat{\Omega}; \mathbb{R}^3) \times H^1(\hat{\Omega})$ and almost all $t \in I$. Note that the differential operators $\varepsilon(\cdot)$ and $\nabla(\cdot)$ are defined with respect to weak derivatives as usual. The existence of a solution for (16) is not clear, cf. [15]. This is particularly true due to the continuously changing domain $\Omega_t \subset \hat{\Omega}$ that also dependents upon the deformations ϕ_s, $s \in (0, t - \delta]$ as defined in (2). It is noted, that some existence and uniqueness results for the solution restricted to Ω_t, $t \in I$ of the linear equations of thermoelasticity are available for certain rather restrictive assumptions, including a not changing domain, see [19]. Recently, the existence and uniqueness for the solution of the heat equation has been proven in [10] for time-dependent domains with Lipschitz continuous boundary. Note however that the latter may not necessarily be satisfied for a given NC milling process, so that the results proven in [10] may not be applicable.

3.1.2 Discretization

The variational formulation (16) is first discretized in time and then in space. For the temporal discretization, also the time-dependently defined domain (2) has to be discretized in time. For this purpose, the time interval I is split into N

sub-intervals $I_n := (t_{n-1}, t_n]$, $n \in \{1, \ldots, N\}$ of length $k_n := t_n - t_{n-1}$ with $0 =: t_0 < t_1 < \ldots < t_{N-1} < t_N := T$. For implementation purposes, a given NC milling process is assumed to be accurately modeled by finitely many milling tool positions $\{\Lambda_{s_{n,0}}, \ldots, \Lambda_{s_{n,M_n}}\}$ for all $s_{n,j} \in I_n$ with $n \in \{1, \ldots, N\}, j \in \{0, \ldots, M_n\}$ and $M_n \in \mathbb{N}$, $M_n < \infty$. Thus, the temporally discretized domains for the (milled) workpieces in their reference configuration are defined in analogy to (2) with respect to the time intervals I_n as

$$\Omega^n := \hat{\Omega} \setminus \hat{\Lambda}^n \tag{17}$$

for $n \in \{0, \ldots, N\}$, where

$$\begin{aligned} \hat{\Lambda}^n &:= \hat{\Lambda}^0 \cup \bigcup_{m\in\{1,\ldots,n-1\}} \bigcup_{j\in\{0,\ldots,M_m\}} \phi_{t_{m-1}}^{-1}(\Lambda_{s_{m,j}}), \\ \hat{\Lambda}^1 &:= \hat{\Lambda}^0 \cup \phi_0^{-1}(\Lambda_0), \\ \hat{\Lambda}^0 &:= \hat{\Omega} \setminus \Omega_0 \end{aligned} \tag{18}$$

yields the entire domain covered by the inversely deformed milling tool for any time until t_{n-1} for $n \in \{2, \ldots, N\}$. Note that $\hat{\Lambda}^0$ is employed to obtain the initial domain Ω_0 in (17). In the following, $\hat{\Lambda}^n$ is refered to as the (inversely deformed) milling tool envelop until $n \in \{0, \ldots, N\}$. Given $\Gamma^n := \partial\Omega^n$, the boundary parts in correspondance to (11) and (12) are defined as

$$\begin{aligned} \Gamma_R^n &:= \Gamma^n \setminus \Gamma_N^n, \\ \Gamma_N^n &:= \begin{cases} \left\{x \in \Gamma^n \cap \left(\hat{\Lambda}^n \setminus \hat{\Lambda}^{n-1}\right) \mid \langle D(t_{n-1}, k_{n-1}), n_t(x)\rangle \geq \tau\right\}, & n \geq 2 \\ \Gamma^n \cap \phi_0^{-1}(\Lambda_0), & n = 1, \\ \emptyset, & n = 0 \end{cases} \end{aligned} \tag{19}$$

for all $n \in \{0, \ldots, N\}$, using the direction of the milling tool movement (10).

For the temporal discretization of (16), the generalized trapezoidal method is employed. For this purpose, let θ^n and v^n denote approximations of $\theta(t_n)$ and $v(t_n) := \dot{\theta}(t_n)$, respectively. These approximations have to satisfy

$$\begin{aligned} &\left(\sigma(u^n), \varepsilon(\chi_{\Omega^n}\varepsilon(v))\right)_{0,\hat{\Omega}} = m\left(\operatorname{div}\left((\theta^n - \theta_0)\mathbb{I}\right), \chi_{\Omega^n} v\right)_{0,\hat{\Omega}}, \\ &c\rho\left(v^n, \chi_{\Omega^n}\varphi\right)_{0,\hat{\Omega}} + \kappa\left(\nabla\theta^n, \nabla(\chi_{\Omega^n}\nabla\varphi)\right)_{0,\hat{\Omega}} - \left(a(t^n)\left(1 - f(t^n)\right)\theta^n, \varphi\right)_{0,\Gamma_R^n} \\ &\quad = \left(\mu(t^n)W(t^n), \varphi\right)_{0,\Gamma_N^n} - \left(a(t^n)\left(1 - f(t^n)\right)\bar{\theta}_a, \varphi\right)_{0,\Gamma_R^n}, \\ &v^n = \frac{1}{\psi k_n}(\theta^n - \theta^{n-1}) - \frac{1-\psi}{\psi} v^{n-1} \end{aligned} \tag{20}$$

for all $(v, \varphi) \in H_D^1(\hat{\Omega}; \mathbb{R}^3) \times H^1(\hat{\Omega})$ and $n \in \{0, \ldots, N\}$, where $\psi \in [0, 1]$. The resulting system of equations implies an unidirectional coupling in the sense that u^n

depends on θ^n, but θ^n does not depend on u^n. Note that the definition of Ω^n, $n \in \{0, \dots, N\}$ in (17) was chosen in analogy to (2) to only yield a dependence on $u^m, m \in \{0, \dots, n-1\}$, i.e., only on displacements 'before' time step $t_n \in I$. The initial data v^0 is obtained from the second equation in (20) for $n = 0$. In general, v^{n-1} has to be computed and stored explicitly since $\left(v^{n-1}, \chi_{\Omega^n}\varphi\right)_{0,\hat{\Omega}} = \left(v^{n-1}, \chi_{\Omega^{n-1}}\varphi\right)_{0,\hat{\Omega}}$ is satisfied only for $\Omega^{n-1} = \Omega^n, n \in \{1, \dots, N\}$.

Given the temporal discretization of the domain (17) as well as of the variational equations (20), an additional spatial discretization is necessary to compute approximations to the heat distributions and the displacements. For the spatial discretization of the domain Ω^n, an octree algorithm is employed to generate hexahedron-based meshes that are refined towards the surface of the inversely deformed milling tool envelop $\hat{\Lambda}^n$. This yields a coarse as well as a fine-scale hexahedron mesh. For a piecewise linear approximation of that surface, the fine-scale hexahedrons intersecting $\partial\hat{\Lambda}^n$ are further refined using the marching volume polytopes algorithm as proposed in [9]. In that algorithm, intersections of $\hat{\Lambda}^n$ with the edges of the fine-scale octree meshes are computed. These intersections are employed to generate a piecewise linear approximation of the surface along with corresponding volume meshes. The resulting fine-scale meshes are employed for the integration of the L^2-scalar products in (20), i.e., for an element-wise evaluation of the characteristic function for the time-dependent domains Ω^n.

For the actual spatial discretization of (20), a standard (higher-order) finite element method is used. The corresponding Ansatz spaces are defined on the coarse-scale hexahedrons of the octree that define the finite element meshes. This allows for the exclusive use of tensor product shape functions, which can be evaluated very efficiently. For a greater flexibility in the definition of the Ansatz spaces, to increase the spatial approximability, these finite element meshes may contain hanging nodes. The latter occur whenever one hexahedron of a given mesh is refined but at least one adjacent hexahedron is not. In addition to the finite element meshes, a polynomial degree distribution is required for the Ansatz spaces. Here, a uniform polynomial degree distribution with polynomial degree one is assumed. A comprehensive definition and explicit construction of the Ansatz spaces for meshes with hanging nodes and given polynomial degree distributions is presented in [7]. Figure 2 depicts the proposed mesh refinements for a cross section of a given domain for a given time step $n \in \{0, \dots, N\}$. For a detailed presentation on the definition of the meshes and the corresponding Ansatz spaces in the context of fictitious domain methods and in particular for the modeling of NC milling processes, see [8].

3.2 GP Process Simulation

The GP simulation presented in this paper is based on the system described in [28]. By combination of different modeling approaches and techniques operating on multiple scales, the system can be used to predict a variety of in-process variables and

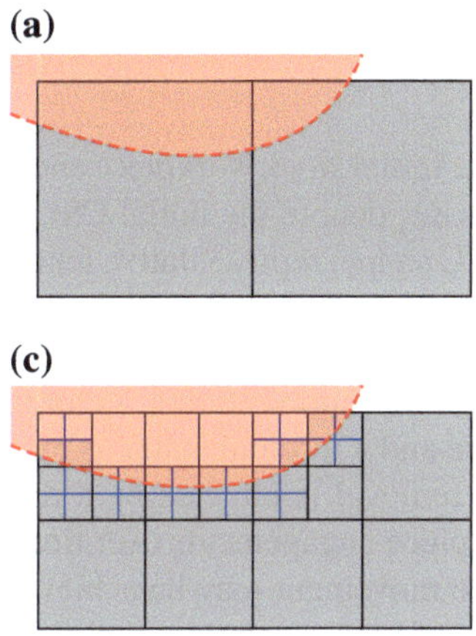

Fig. 2 Visualization of the milling dependent mesh refinement. **a** depicts the cross section, an initial octree, and a milling tool envelop (*red*). **b** shows a refinement of the octree towards the surface of the milling tool envelop that may be employed for the definition of appropriate Ansatz spaces. In **c**, the octree is further refined for a better approximation of the surface of the milling tool envelop. **d** depicts a cross section of the final meshes as obtained via the marching volume polytopes algorithm

effects [45]. Of fundamental importance is the prediction of the cutting forces with respect to the actual engagement condition. Using an empirical model, the simulation system is able to calculate the cutting forces to a high degree of accuracy.

Once the cutting forces have been determined, they can be used as input into other models incorporated in the simulation system. For example, by attaching an oscillator-based model which is excited with the calculated cutting forces, process dynamics can be simulated, allowing for the prediction of surface location errors.

For the purposes of the research described herein, the simulation system was extended with a heat input model [41]. The fundamental idea of the heat input model is based on the notion that a fraction of the cutting energy is inserted into the workpiece as heat. This model developed is an empirical approach which takes the chip thickness into account to determine the heat input fraction for the workpiece. A validation of the heat input model was achieved by extending the simulation system with a three dimensional finite difference model [25, 38]. Using the simulation system, various experiments could be reproduced to a high level of accuracy with respect to the thermal loading of the workpiece. The following subsections outline the principle ideas used in the GP simulation.

3.2.1 Cutting Force Calculation

The modeling technique used in the cutting force calculation is based on Constructive Solid Geometry (CSG), see also [21]. Using CSG, arbitrarily complex geometries can be constructed by combining primitives—spheres, cuboids or cylinders—using

Boolean operations, such as *union* or *intersection*. Being a set-theoretic approach, CSG is very well suited for intersection calculations, which is beneficial with respect to the utilized force calculation model.

In order to achieve the force calculation, the initial stock workpiece and the cutting tool are represented by CSG composites. Let Ω_0 denote the initial CSG workpiece representation. Ω_0 can itself be a complex CSG shape, representative for the physical stock workpiece prior to machining. By not representing the flutes in the CSG shape, the tool model is simplified. Instead, the tool shape is defined to be the envelope of the physical cutting tool. Let Λ_0 denote the CSG tool representation. Figure 3 presents the CSG models for a stock cuboid workpiece and a toroidal cutter as examples.

During the milling process, the material removal is effected by a relative tool-workpiece movement, resulting in tool-workpiece engagements over time. Depending on the specific machine tool, this relative movement may be achieved by moving the workpiece, the tool, or a combination of both. In the simulation system, the workpiece movement is transformed to an appropriate tool movement, resulting in

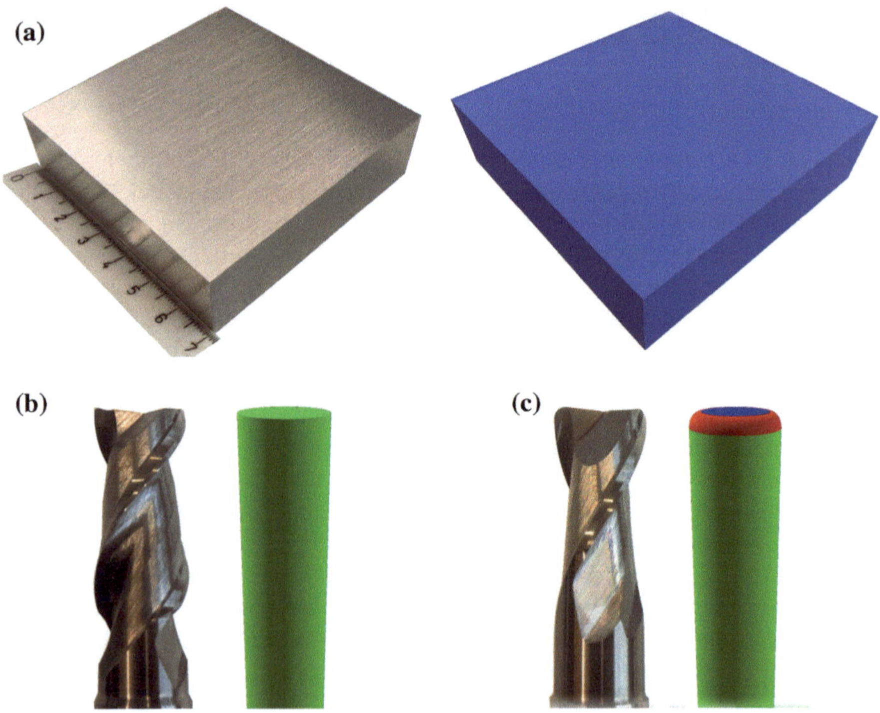

Fig. 3 **a** Example of a simple workpiece of dimensions 70 mm × 70 mm × 20 mm and the corresponding CSG model W_0, which is a cuboid of equal dimensions. **b** Cylindrical cutter with a diameter of $d = 12$ mm and corresponding CSG model, which is simply a cylinder of equal diameter. **c** Toroidal cutter with a diameter of $d = 12$ mm and corresponding CSG model, which is a union of an outer cylinder for the shaft, a torus to capture the corner radius, and an inner cylinder to fill the hole of the torus

the same relative tool-workpiece constellation. Thus, in the described simulation system, the workpiece is considered to be stationary, while the tool is appropriately moved and rotated per simulation step. The simulation is discretized in time according to the tooth feed f_z, thus Λ_i represents the tool shape, displaced by f_z with respect to Λ_{i-1}.

Given the initial workpiece shape Ω_0 as well as the NC program, it is then possible to determine the workpiece geometry W_i and the shape of the uncut chip C_i in the i-th simulation step using (21) and (22), respectively, so that [28]:

$$W_i = W_{i-1} \setminus \Lambda_i = \Omega_0 \setminus \bigcup_{j \in \{0,\ldots,i-1\}} \Lambda_j \tag{21}$$

and

$$C_i = W_i \cap \Lambda_i. \tag{22}$$

Note that the definition of W_i corresponds to (17) not considering deformations of the workpiece. While the deformation of the workpiece influences the undeformed chip thickness in general, the influence is rather small for the presented use case and, thus, this influence is neglected. The cutting force prediction is based on an empirical approach which considers the thickness of the uncut chip [29]. Given C_i, the thickness in a particular direction and at a particular point can be determined by intersecting C_i with an appropriate sampling ray, and calculating the distance between the entry and exit point. Within the simulation, rays are generated perpendicular to the 2D section outline of the tool envelope, as shown by an example for a toroidal cutter in Fig. 4. The number of rays generated in the sectional plane is a user defined parameter and affects the spatial discretization of the simulation, where a higher number of rays corresponds to an increased spatial resolution (Fig. 4a and b). In order to achieve an adequate sampling of C_i, the sampling rays are rotated about the tool axis in steps defined by the user, where a higher number of rotations corresponds to an increased time resolution (Fig. 4c and d).

For every sampling ray which intersects C_i, a cutting force sample can be calculated. The summation of all cutting force samples then yields the total cutting force for the i-th simulation step. This is shown in Fig. 5 for a workpiece machined with a cylindrical cutter.

3.2.2 Heat Input Modeling

As described in Sect. 3.2.1, the cutting force in each simulation step is determined by sampling the uncut chip shape with different sampling rays generated in order to get a discretization of the cutting edge of the tool. This results in cutting force samples associated with those cutting rays. The scheme, by which these sampling rays are generated, effectively corresponds to a subsampling in time (radial subsampling) and space (axial subsampling). For those rays that intersect the uncut chip shape, two

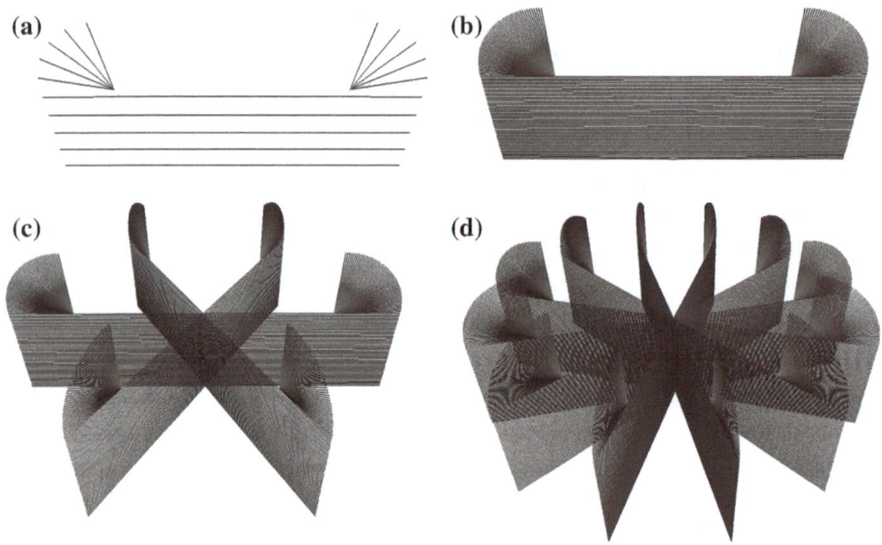

Fig. 4 Examples of scan rays generated for calculating the undeformed chip thicknesses using the CSG model. **a** 2D sectional sampling for a toroidal cutter using 10 rays. **b** 2D sectional sampling for a toroidal cuter using 100 rays, corresponding to an increased spatial resolution. **c** Time subsampling in each simulation step is achieved by a rotation of the generated rays about the axis of the tool, here using 6 subsample planes. **d** By increasing the subsample planes, a higher resolution in time is achieved

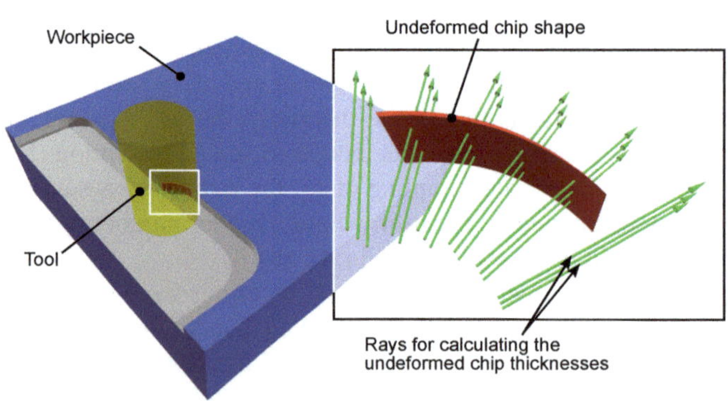

Fig. 5 The ideal workpiece shape W_i is generated by subtracting A_i from W_{i-1}. The uncut chip C_i is sampled at multiple locations, according to the specified spatial and time resolution, as shown in the enlargement. Resulting force samples are summed to yield the total cutting force for the simulation step

points of intersection always exist, i.e. an entry and an exit point with respect to the direction of the ray. From the exit point, associated force sample and displacement with respect to the previous sample in time, it is then possible to calculate the cutting energy, as described in [41].

In each simulation step, a fraction of the total cutting work is inserted into the workpiece as heat. The heat input locations correspond to the sample ray exit points. In order to determine the heat input fraction μW in (12), an empirical model is developed in [25], which considers the thickness of the uncut chip

$$p(h) = k_1 + k_2 \cdot e^{k_3 \cdot h}. \tag{23}$$

Here, h denotes the thickness of the uncut chip, which varies with process time and spatially along the cutting edge. The coefficients k_1, k_2 and k_3 (constant, linear and exponential energy fraction) are specific to a particular tool and workpiece material, and their determination requires several calibration experiments. The fraction of the mechanical cutting work μW is determined by evaluating $p(h)$ for each ray of the CSG model and treating the resulting value a heat source for the local part of the contact area between tool and workpiece represented by the ray.

A finite difference (FD) method was used to simulate the temperature field in a voxel-based approximation of the workpiece for a given simple milling process designed for the parameter fitting and to calibrate the heat input coefficients. The FD model consists of a fixed-resolution 3D voxel grid, with each voxel representing a volume of the workpiece material. Throughout the milling process, voxels are deactivated when the milling tool envelope has covered all eight corner vertices. Active voxels are thermally loaded if they contain any heat input points. After inserting the cutting energy as heat into the FD model, the thermal load of each voxel is updated via its six bounding faces. If a face is shared between two active voxels, then the temperature flow is dependent upon the thermal conductivity of the material and the thermal load of the two voxels. If a face represents the material boundary, then the temperature flows from that voxel to the environment, i.e., to the clamping device or the air surrounding the material.

Using this setup, the heat input model was successfully calibrated and validated by comparing measured and simulated process temperatures for multiple experiments, cf. [25].

3.2.3 Boundary Condition Modeling

In the modeling of the heat diffusion as stated in (16), particular care has to be given to the modeling of the boundary conditions. Clamping, thermally conductive material contact as well as the movement of air surrounding the workpiece affect the temperature distribution in the workpiece. Thus, the modeling has to provide means to address convection and radiation.

Convection can be modeled by the simulation of the heat transfer between the workpiece material and the environment. This is achieved by choosing appropriate heat transfer coefficients $a(t)$ in (14) for the individual parts of the workpiece boundary, i.e., for the heat transfer from the workpiece to the surrounding air and to the clamping via a_{air} and a_{clamp}, respectively. In order to model the convection appropriately, the ambient temperature θ_a in (12) cannot be assumed to be constant throughout the simulation. Even for perfect ventilation, the ambient temperature for the air or the clamping in the immediate vicinity of the workpiece will increase in correspondence to the temperature of the workpiece at least to a certain fraction. Thus, a constant ambient temperature would effectuate an excessive cooling of the workpiece over time, see also (12). To model an increased ambient temperature in the immediate vicinity of the workpiece, a boundary zone model is proposed in [40]. For this model, the ambient temperature θ_a is assumed to be linearly dependent upon the workpiece temperature and the assumingly constant ambient temperature further away from the workpiece, see also (13). The boundary zone model introduces two additional parameters f_{air} and f_{clamp} for the ambient temperature of the air and the clamping, respectively. Using this approach along with appropriately fitted parameters, this modeling has proven to be suitable for the prediction of the thermal load of workpieces for several longer running experiments with high accuracy in [40].

Radiation cannot be directly included into the modeling equations as proposed in (4)–(7). This would require a major extension of the model under consideration. In particular, this would require the modeling of the environment of the workpiece at least in the vicinity of the latter, which would also drastically increase the computational complexity. For the determination of the parameters f_{air} and f_{clamp} of the boundary zone model, a parameter fitting is pursued for a given milling process and the corresponding, measured thermal load of a workpiece. Since this fitting is based upon experimental data that naturally includes radiation, the latter is at least to a certain degree included in the parameters determined for the boundary zone model.

3.3 GP-FEM Linkage

The previously described models are linked via a network interface, using a shared protocol and data encoding which is independent of both systems. To the GP simulation, the FE simulation appears as a black box, while the FE simulation receives input from and provides output to the network interface. This allows for both systems to be developed and modified independently of each other. Although the GP simulation can be executed without linking to the FE simulation, predicting thermo-mechanical workpiece deformations—and by extension performing compensations of the same—requires the use of the FE simulation. A schematic overview of the interface is shown in Fig. 6. On a technical level, both simulation systems are implemented using the C++ programming language.

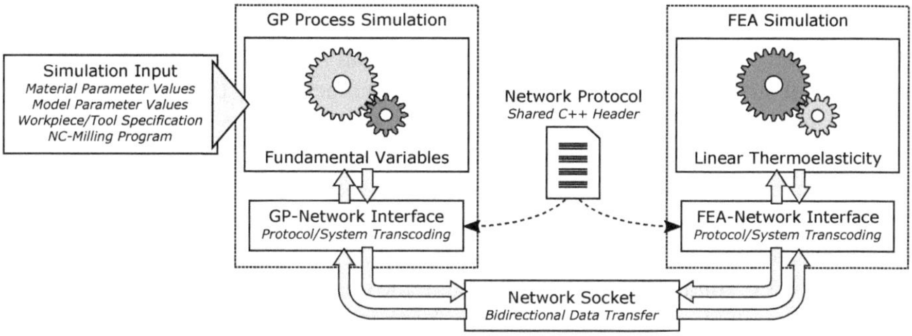

Fig. 6 The GP and FE simulation systems exchange data via a network interface. A shared protocol is used to keep the remote systems independent of each other

Both the FEM and GP simulation maintain a system-specific representation of their data model. During the initialization phase of the hybrid simulation system, the GP simulation connects to the FE simulation. Once the connection has been established, the GP simulation transmits all information pertinant to the FE computation of the thermomechanical material response. This includes the geometry specification of the workpiece and the milling tool, material properties of the workpiece, and the initial boundary conditions, such as clamping surfaces and initial material and environment temperature.

Subsequently, the GP simulation executes the simulation of the machining process. For each simulation step, the shape of the uncut chip is determined. The GP simulation steps are discretized in time corresponding to the tooth feed f_z. By sampling the uncut chip thickness, the GP simulation computes the cutting forces and the heat input, which can be used to update the optional local FD model.

As the FE simulation maintains its own data model, which includes the state of the workpiece and the definition of the milling tool. An update of the FE simulation requires only the specification of the tool engagement situation, which includes the tool pose as well as the heat input associated with that tool pose. The GP simulation transmits the tool engagement situation for every simulation step. An update of the FE simulation model consists of the following steps: First, the (inversely deformed) tool envelope is subtracted from the most recent state of the workpiece to determine the current workpiece geometry, see (17). Subsequently, the thermal state of the workpiece is updated with respect to heat input into the workpiece, heat diffusion within the workpiece, and convection. Finally, the resulting thermomechanically induced workpiece deformation is updated.

The FE simulation can be updated synchronously or asynchronously with the GP simulation. An asynchronous update of the FE simulation is achieved by queuing multiple tool engagement situations. When the update is triggered, the tool poses of all queued tool engagement situations are subtracted from the last state of the workpiece as formally defined in (17). The heat input is computed as the average heat input over all queued tool engagement situations.

The benefit of an asynchronous update is a computation speed up: Although the update of the workpiece with respect to the material removal is realized using CSG methods, at some point this implicitly defined geometry has to be discretized and meshed in order to compute the workpiece deformation due to thermal loading. Within the sequence of update steps performed by the FE simulation, the meshing is relatively expensive. While a considerable speed up is achieved by asynchronous updates of the FE simulation, this also introduces discretization errors. Further details are discussed in Sect. 3.4.

Once the FE simulation has been updated, the deformation field of the workpiece material is available to the GP simulation. With respect to the problem domain, the surface deformation of the workpiece is of high interest. Hence, the surface points of the FE mesh along with their associated temperatures and deformation vectors are transmitted back to the GP simulation by default. The deformation of the workpiece in the region of the tool engagement is considered to determine a deformation vector which is used to displace the tool for the subsequent GP simulation step (or steps, in case of asynchronous updates of the FE simulation).

For the validation experiments presented in this paper, a simple text file was exchanged that contained exactly the same information as if the online coupling was employed. This offline coupling is particularly useful to determine (discretization) parameters for the FE simulation independently of the GP simulation. This way, the deformations computed by the FE simulation are neglected when computing the heat input in the GP simulation. If the accuracy of the simulation results is insufficient using this approach, the deformations should be taken into account when computing the heat input, which was not the case for the presented application.

3.4 FEM Update Intervals

As described in Sect. 3.3, the FE simulation receives input from the GP simulation, updates its local model, and subsequently provides the GP simulation with the prediction of the deformed workpiece, i.e. a meshed representation of the workpiece, including temperature and deformation data for each mesh node. While the update cycle implies a synchronous update, where the FE simulation is updated for every single simulation step of the GP simulation, a non-synchronous update of the FE simulation is also possible. In the latter case, the FEM update is triggered after m steps of the GP simulation. When synchronously updating the FE simulation, the following steps are executed:

1. Subtract the (inversely deformed) tool envelope from the current workpiece shape.
2. Generate a mesh for the updated workpiece shape.
3. Solve the linear-thermoelastic problem to determine the heat distribution and deformations.

Experiments have shown that steps 2 and 3 together require about 80% of the total computation time during a single update step. When adopting a non-synchronous update policy (of every m steps), m tool positions are queued by the FE simulation until the update is triggered. In the first step of the update, these m tool positions are used to determine the tool envelopes, which are then subtracted in one step from the previous workpiece shape. Subsequently, steps 2 and 3 are executed as before.

This approach has a marginal effect on the computation time required for step 1. The lower update frequency of the FEM, and by extension the execution of steps 2 and 3, immediately results in a significant performance boost. At the same time, this policy introduces discretization artifacts into the simulation, because the FE simulation is updated with an average of the heat input over time. Specific peaks may be filtered out by this approach for a large m.

4 Validation of the Simulation Approach

In order to validate the presented hybrid simulation system, the machining of a workpiece was simulated and manufactured. The workpiece material was aluminum 7075, with a dimension of 70 mm × 70 mm × 20 mm. These workpieces were milled with a pocket including a reference ledge of a defined width of 0.5 mm. Due to the thermomechanical distortion occurring during the milling process, it was expected that both, the manufactured and simulated ledge, would deviate from the defined width by some measurable magnitude.

4.1 Experimental Setup

The workpiece was machined using a cylindrical cutter with two teeth and a diameter of 12 mm on a Deckel Maho DMU 50 eVolution machine tool. Figure 7 shows the manufactured workpiece, along with the respective manufacturing strategy. The workpiece was manufactured using the process parameter values listed in Table 1.

The pocket of the workpiece, shown in Fig. 7a, was machined to a total depth of 15 mm. This depth was chosen to ensure that no workpiece deformation is attributable to high clamping forces once the pocket material was removed. The reference ledge was machined at half of its respective maximum depth.

In order to validate the temperature prediction thermocouples (Type K) were used. Using a combination of form and force closure, the stock probes were clamped with a contact area of approximately 70 mm^2 on both sides. Both, the placement of the thermocouples as well as the final experimental setup, is shown in Fig. 8.

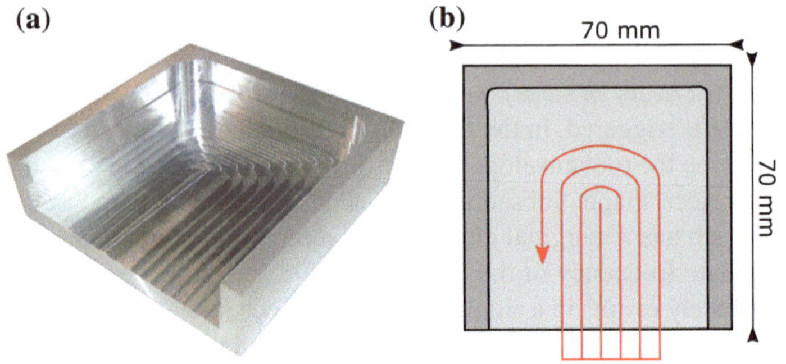

Fig. 7 **a** Workpiece with a single pocket machined using an initial notching and subsequent outward spiraling tool path. **b** Milling path used for machining the workpiece

Table 1 Process parameter values used in machining and simulation

Parameter	Value
Spindle speed n	$14000\,\text{min}^{-1}$
Tooth feed f_z	0.07 mm
Feed F	$1960\,\text{mm min}^{-1}$
Axial depth of cut a_p	1.5 mm
Radial depth of cut a_e	3.0 mm

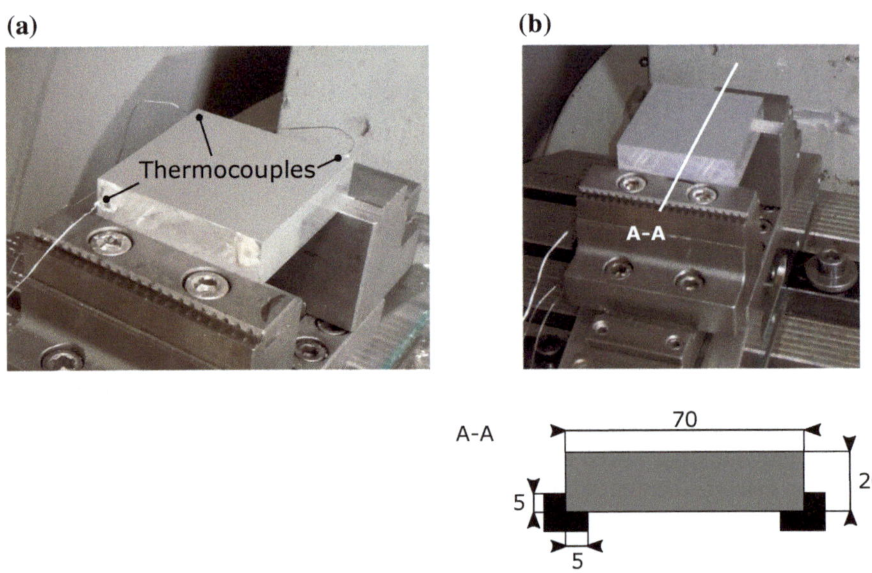

Fig. 8 **a** Type K thermocouples were placed into premachined slots at the side walls of the probes, and affixed using a two-component adhesive. **b** The probe was clamped along the two opposing sides, with the clamping providing support as well, as shown in cross-section A-A

4.2 Simulation of the Experiment

For the simulation of the experiment described in Sect. 4.1, a reference coordinate system is assumed, so that the initial workpiece is given by

$$\hat{\Omega} = \Omega_0 := (-35\,\text{mm}, 35\,\text{mm})^2 \times (-10\,\text{mm}, 10\,\text{mm}). \tag{24}$$

The workpiece itself is clamped in a bench vise so that

$$\begin{aligned}\Gamma_D := \partial\Omega_0 \cap \Big(&[-35\,\text{mm}, 35\,\text{mm}] \times [-35\,\text{mm}, -30.6\,\text{mm}] \times [-10\,\text{mm}, -5.6\,\text{mm}] \cup \\ &[-35\,\text{mm}, 35\,\text{mm}] \times [30.6\,\text{mm}, 35\,\text{mm}] \times [-10\,\text{mm}, -5.6\,\text{mm}]\Big)\end{aligned}$$

as depicted in Fig. 8. The material parameter values for the cuboidal workpiece composed of the aluminium alloy AL7075, which are necessary for (20), are given in Table 2.

For the temporal discretization of the analyzed milling process, the parameter $\psi = 3/4$ is chosen. Furthermore, the time step size is $k_n \equiv 0.0428572\,\text{s}$. This time step size results from an aggregation of the time steps used in the geometric process simulation for the prediction of cutting forces as described in Sect. 3.4. For the angle condition in the definition of the contact area of the workpiece and the milling tool in direction of the milling tool movement in (11) the parameter $\tau = 1/10$ is chosen. For the spatial discretization of the domain, an initial mesh is defined that consists of $31 \cdot 31 \cdot 9$ hexahedrons as depicted in Fig. 10a. For the octree mesh refinement towards the milling envelop surface, two refinements are pursued before the application of the marching volume polytopes algorithm. The initial temperature of the workpiece and the ambient temperature are chosen as $\theta_0 = \bar{\theta}_a = 298.15\,\text{K}$.

Table 2 Values for the various model parameters based up the aluminium alloy AL7075 as employed in (4)–(7)

Symbol	Name	Value
E	Young's modulus	$71 \times 10^9\,\text{kg}\,\text{m}^{-1}\text{s}^{-2}$
ν	Poisson number	0.33
ρ	Density	$2810\,\text{kg}\,\text{m}^{-3}$
α	Coefficient of thermal expansion	$23.1 \times 10^{-6}\,\text{K}^{-1}$
κ	Thermal conductivity	$115\,\text{kg}\,\text{ms}^{-3}\text{K}^{-1}$
c	Thermal capacity	$862\,\text{m}^2\text{s}^{-2}\text{K}^{-1}$
a_{air}	Thermal conductivity to air	$-145\,\text{kg}\,\text{s}^{-3}\text{K}^{-1}$
a_{clamp}	Thermal conductivity to clamp	$-338\,\text{kg}\,\text{s}^{-3}\text{K}^{-1}$
f_{air}	Heat transfer ratio to air	0.55
f_{clamp}	Heat transfer ratio to clamp	0.2

Table 3 Settings used in the GP simulation

Symbol	Parameter	Value
k_1	Constant energy coefficient	0.12
k_2	Linear energy coefficient	0.18
k_3	Exponential energy coefficient	−26.0

The determination of the fraction of the mechanical cutting work μW for the heat input at the (changing) contact area as utilized in (12) is based upon the GP simulation using the parameters given in Table 3. The corresponding model parameters of the heat input model were calibrated as described in Sect. 3.2.2. Furthermore, the parameters for the heat transfer to air f_{air} and to the clamp of the bench vise f_{clamp} are obtained as described in Sect. 3.2.3.

4.3 Results

In Fig. 9, the simulated and measured temperatures of the workpiece are depicted for the points

$$P_0 := (-30\,\mathrm{mm}, 34\,\mathrm{mm}, 5\,\mathrm{mm}), \quad P_1 := (-30\,\mathrm{mm}, -34\,\mathrm{mm}, 5\,\mathrm{mm}),$$
$$P_2 := (30\,\mathrm{mm}, 34\,\mathrm{mm}, 5\,\mathrm{mm})$$

with respect to the domain (24) as also illustrated in Fig. 8. The characteristics and magnitudes of the temperature curves from the simulation match those for

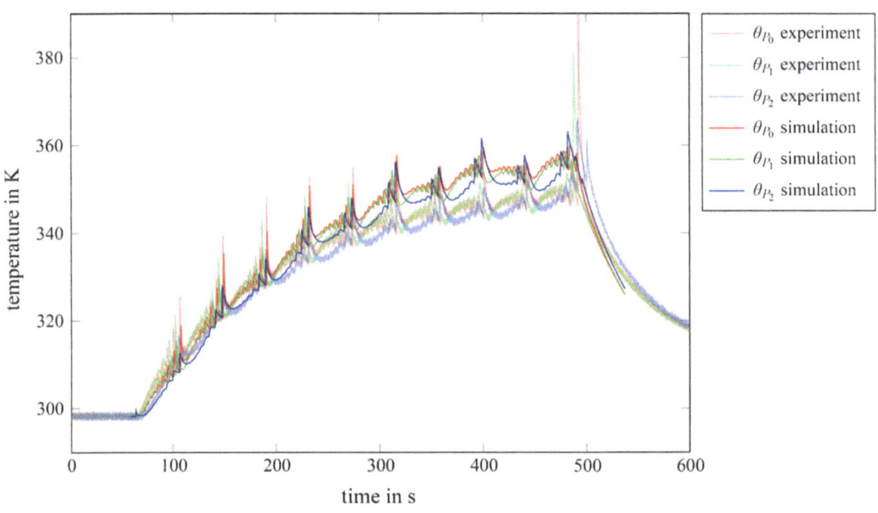

Fig. 9 Simulated and measured temperatures at the three measurement positions shown in Fig. 8

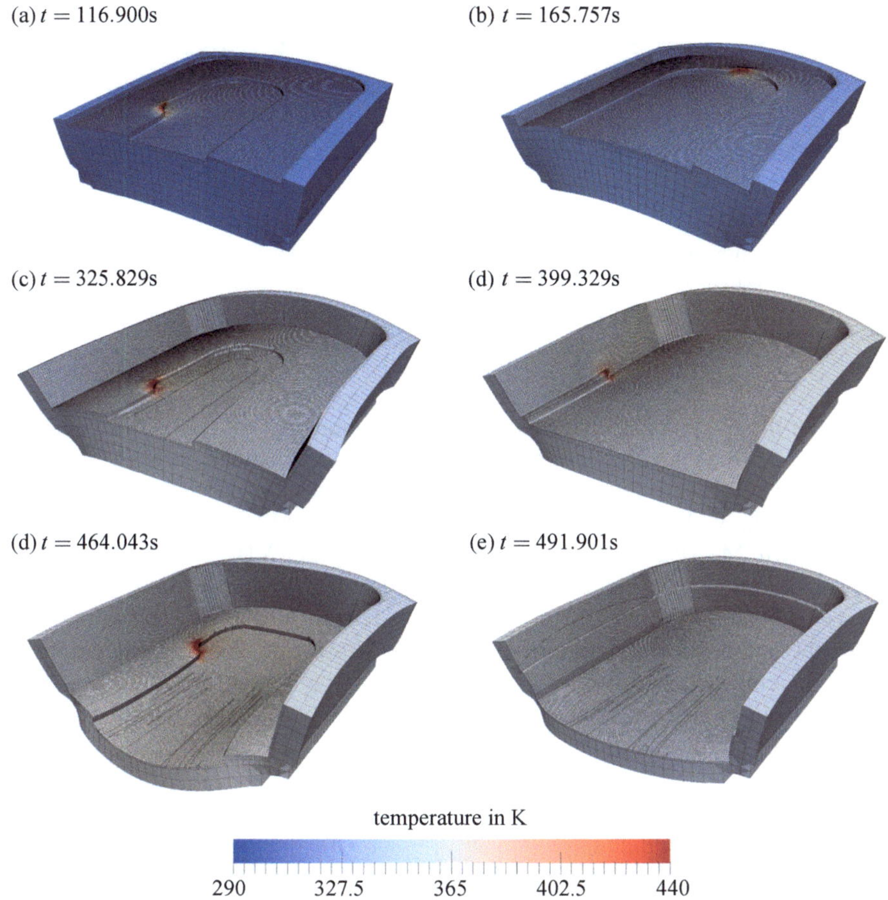

Fig. 10 Illustration of the deformed workpiece during the simulation of the milling process. For visualization purposes, the displacement is scaled by a factor of 200. The color maps display the temperature of the workpiece in K

the physical experiment, while the absolute temperatures differ slightly. Figure 10 depicts the corresponding thermomechnical deformations for various time steps to illustrate the simulation of the milling process. As expected from the bending of the workpiece during the simulation, Fig. 11 visualizes the differences for the surface of the milled workpiece from the ideal surface. Also note the same pattern for the milling error in Fig. 11b as in the surface reflections in Fig. 7a. Similar to the temperature curves, the characteristics and magnitude of the computed milling error agree with the measurements from the physical experiment, but differ in absolute values. For a more extensive discussion of the numerical results obtained for the considered milling process, see [8]. To improve the accuracy of the computed deformations and thus of the computed milling error with respect to the physical experiment, further

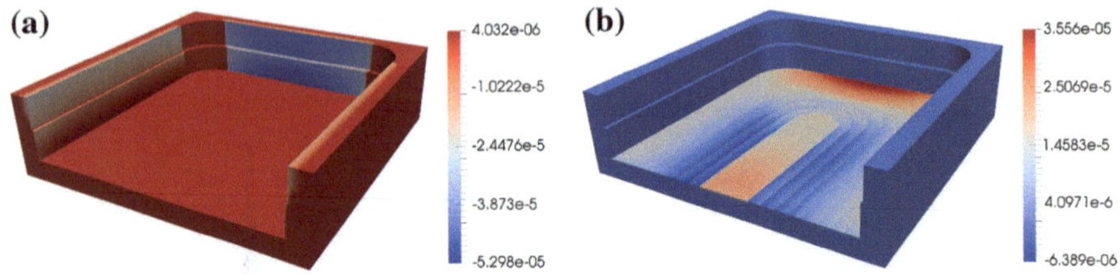

Fig. 11 Illustration of the milling error in m for the inner (plane) walls (**a**) and the bottom (**b**) of the milled pocket resulting from the deformations of the workpiece during the simulation of the milling process. *Positive and negative values* denote areas where too much and too little material was removed, respectively

parameter fitting and possibly an extension of the modeling for the boundary condition related to the clamping in the linear elastic equation may be necessary.

In order to measure the magnitude of the thermomechanical deformation of the workpiece, the width of the resulting step was measured once the workpiece had cooled down. Measurements were taken using a Keyence VHX-5000 digital microscope and a Mahr PCV contour measurement device. Multiple samples were taken at various locations. Each measurement showed a width smaller than the intended width of 0.5 mm, with deviations in the range of 30–90 μm. The magnitude of these values was confirmed with both measuring devices, as shown in Fig. 12. As described in more detail in Sect. 5.2, the size of the step and the width of the pocket were analyzed using a tactile coordinate measurement machine.

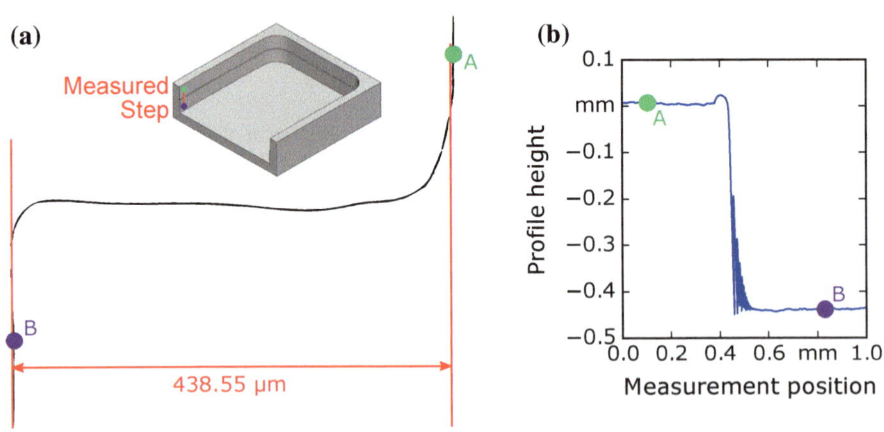

Fig. 12 **a** Using the Keyence VHX-5000 digital microscope, the distance between the vertical walls was determined to be in the range of 30–90 μm. **b** These measurements were confirmed using a Mahr PCV contour measurement device

5 Compensation of Thermomechanical Workpiece Deformations

In order to reduce the shape deviations resulting from thermomechanical deformations of the workpiece, a compensation strategy was developed. The implemented compensation approach does not strive for an optimal solution. Rather, the intention is to generate a path which satisfies the tolerances specified by the user, while taking into account the deformation at a particular time in the area of the tool-workpiece interface. Assuming that the deformation is known, either by experiment or by simulation, then an improvement may be achieved by displacing the tool opposite to the local deformation direction.

5.1 Deformation of the NC Milling Path

The time discretization of the GP process simulation corresponds to the time per tooth feed, as specified in the NC program. Therefore, it is possible to calculate the actual tool position with a very high resolution.

When using the FE simulation, the local deformation at the tool-workpiece contact area can be calculated by determining which vertices of the FE mesh are in immediate proximity to the tool shape. As the deformation and temperature for all vertices of the FE mesh are known, once these tool-workpiece interface vertices have been determined, it is possible to approximate the local deformation by calculating the average deformation vector for all those candidates (Fig. 13). For each simulation step, the tool position as well as the determined material deformation are saved to a file. That file is subsequently processed to generate an NC program compensated for thermal deformation.

In order to generate the compensated NC program, new tool positions are generated by applying the deformation vector to the original position. Subsequently, a

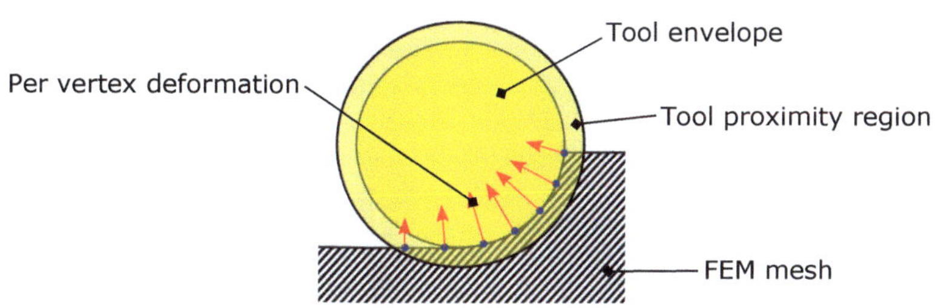

Fig. 13 The tool envelope is enlarged such as to include the proximity region. Vertices of the FE mesh, which are contained within the resulting proximity region hull, are considered to be in the tool-workpiece contact area region. By calculating the average per-vertex deformation for all this vertices, the local machining error is approximated

filter is applied to remove non-essential tool positions: For positions p_{i-1}, p_i, p_{i+1}, the position p_i is removed from the collection of tool positions if the distance between p_i and $\hat{p}_i$, its projection onto the line passing through p_{i-1} and p_{i+1}, is smaller than some user-defined threshold value. This threshold value should not be larger than the machining accuracy defined in the machine tool controller.

Once the redundant points have been removed, the remaining points are written to an NC program as linear motion commands.

5.2 Validation of the Deformed Milling Path

In order to validate the compensation method, it was applied to the NC programs used to manufacture the workpiece shown in Fig. 7. Due to the deviations of the workpiece deformations from the measured values predicted by the FE simulation, the NC programs were deformed based on measured errors instead of the simulation results in order to validate the applicability of the compensation approach.

Three features of the part were analyzed and optimized using the proposed approach. According to the CAD model, the width of the pocket is specified to be 61 mm in the upper area and 60 mm in the lower area. Consequently, the reference step should be 0.5 mm. The workpiece was machined using the original NC path and the resulting deviations were measured using a Zeiss tactile coordinate measurement machine. The width of the upper area was measured at the Z-coordinates −2, −4, and −6 mm and the width of the lower area was measured at the Z-coordinates −9, −11, and −13 mm, as shown in Fig. 14, resulting in mean widths of 60.751 and 59.91 mm, which corresponds to deviations of −248 µm and −90 µm, respectively. The measured deviations of the reference step were −86 µm at the left side, −78 µm at the right side, and −51 µm at the back. Concludingly, the width

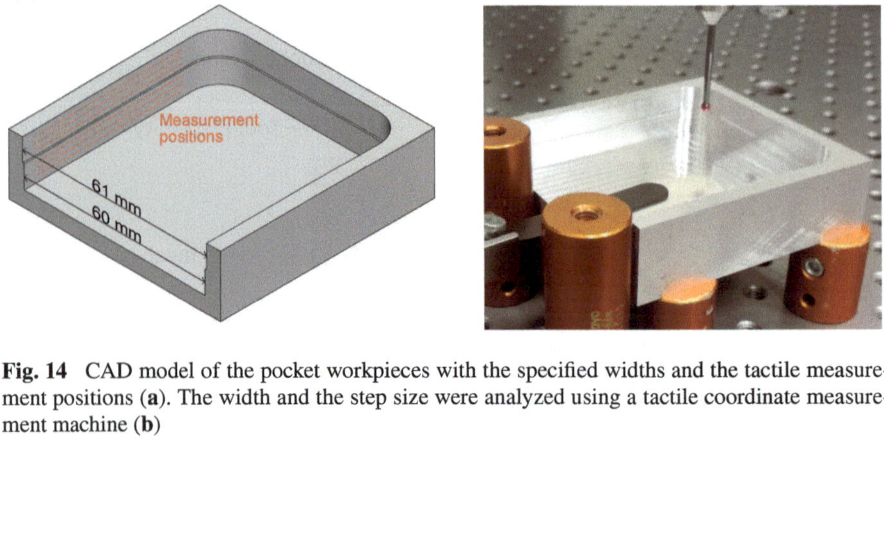

Fig. 14 CAD model of the pocket workpieces with the specified widths and the tactile measurement positions (**a**). The width and the step size were analyzed using a tactile coordinate measurement machine (**b**)

Table 4 Tactile measurement results

Feature	Expected (mm)	Original		Compensated	
		Measured mm	Deviation (μm)	Measured (mm)	Deviation (μm)
Width top	61.0	60.752	−248	61.005	5
Width bottom	60.0	59.910	−90	60.010	−10
Step left	0.5	0.414	−86	0.496	−4
Step right	0.5	0.422	−78	0.499	−1
Step back	0.5	0.449	−51	0.502	2

of the pocket is too small, which corresponds to the observation that the walls of the pockets are bend outwards due to the thermal deformation and, thus, not enough material is removed in the milling process.

In order to compensate the deviations, the NC path has been offset outwards by 0.124 mm for the upper area and 0.045 mm for the lower area. This way, more material should be removed symmetrically at all sides, resulting in an increasement of the width of the pocket of 0.248 mm and 0.09 mm, respectively. Using the corrected NC path, the workpiece was machined again and the resulting shape error were measured using the coordinate measurement machine. The measurement results are listed in Table 4. The mean measured width of the pocket was 61.005 mm in the upper area and 60.01 mm in the lower area. The measured errors of the step were decreased to −4 , −1 , and 2 μm. Concludingly, the machining errors resulting from the thermal deformation could be decreased significantly.

6 Summary and Outlook

This paper presents a hybrid simulation system for the prediction and compensation of thermomechanically induced workpiece deformations in dry NC milling processes. The simulation system consists of a geometric-physical process simulation for the prediction of, e.g., cutting forces and heat input, and a FE simulation based upon the fictitious domain method in combination with a (higher-order) finite element method for the computation of the thermomechanically induced workpiece deformation. The two components are linked via a network interface, with the GP simulation providing the boundary conditions for the FE simulation, and the latter providing the deformation information to the GP simulation.

An approach for the compensation of the form deviations resulting from the thermomechanical deformation of the workpiece was presented. The deviations are compensated by deforming the NC path of the milling tools described in the NC programs. In order to validate the presented approach, machining experiments were conducted for a simple test workpiece while measuring the workpiece temperatures.

The workpiece was cooled down to the environment temperature after machining and the form errors were measured using a tactile coordinate measurement machine at discrete points. The same machining process was simulated using the presented simulation systems. While the simulated workpiece temperatures show a good agreement with the measurement results, the predicted workpiece deformations show the correct characteristics and magnitude, but differ in absolute values from the measurements. To improve the accuracy of the computed deformations further parameter fitting is necessary. For this reason, the compensation approach was applied based on the measurements instead of the simulations in order to demonstrate the general applicability of the presented compensation strategy. After machining the workpiece with the adapted NC programs a significant reduction of the form errors were measured.

To increase the accuracy of the presented compensation approach, the compensation could be applied iteratively in order to further reduce the errors caused by the thermomechanical deformations. For the presented use case, the thermomechanical deformations of the workpiece have been neglected for the computation of the process forces and, thus, the heat input. The CSG model of the GP simulation could be extended to take these deformations into account in order to increase the accuracy of the simulation results.

Acknowledgements This paper is based on the investigations and findings of the project *Simulation of Thermomechanical Deformations in NC Milling* (SCHR 1244/2-3) and (ZA 427/3-3) of the priority program SPP 1480 (CutSim), which is kindly supported by the German Research Foundation (DFG).

References

1. Abouridouane, M., Klocke, F., Döbbeler, B.: Analytical temperature prediction for cutting steel. Annals CIRP **65**(1), 77–80 (2016)
2. Altintas, Y., Kersting, P., Biermann, D., Budak, E., Denkena, B., Lazoglu, I.: Virtual process systems for part machining operations. CIRP Annals-Manuf. Technol. **63**(2), 585–605 (2014)
3. Arrazola, P.J., Özel, T., Umbrello, D., Davies, M., Jawahir, I.S.: Recent advances in modelling of metal machining processes. Annals CIRP **62**(2), 695–718 (2013)
4. Astrakhantsev, G.: Method of fictitious domains for a second-order elliptic equation with natural boundary conditions. USSR Comput. Math. Math. Phys. **18**(1), 114–121 (1978). doi:10.1016/0041-5553(78)90012-5
5. Boley, B., Weiner, J.: Theory of Thermal Stresses. Wiley Ltd., Inc. XVI, New York and London (1960)
6. Bollig, P., Köhler, D., Zanger, F., Schulze, V.: Effects of different levels of abstraction simulating heat sources in FEM considering drilling. Proc. CIRP **46**, 115–119 (2016)
7. Byfut, A., Schröder, A.: Unsymmetric multi-level hanging nodes and anisotropic polynomial degrees in H^1-conforming higher-order finite element methods. Comput. Math. Appl. **73**(9), 2092–2150 (2017). doi:10.1016/j.camwa.2017.02.029
8. Byfut, A., Schröder, A.: A fictitious domain method for the simulation of thermoelastic deformations in NC-milling processes. Int. J. Numer. Methods Eng. (submitted) (2017)
9. Byfut, A., Hellwig, F., Schröder, A.: Marching volume polytopes algorithm. Int. J. Numer. Methods Eng. (submitted) (2017)

10. Calvo, J., Novaga, M., Orlandi, G.: Parabolic equations in time dependent domains. J. Evol. Equ. (2016). doi:10.1007/s00028-016-0336-4
11. Demkowicz, L.: Computing with *hp*-adaptive finite elements. Vol. 1: One- and two-dimensional elliptic and Maxwell problems. With CD-ROM. Chapman & Hall/CRC Applied Mathematics and Nonlinear Science Series (2007). doi:10.1201/9781420011685
12. Demkowicz, L., Kurtz, J., Pardo, D., Paszyński, M., Rachowicz, W., Zdunek, A.: Computing with *hp*-adaptive finite elements. Vol. 2: frontiers: three-dimensional elliptic and Maxwell problems with applications. Chapman & Hall/CRC Applied Mathematics and Nonlinear Science Series (2008). doi:10.1201/9781420011692
13. Denkena, B., Brüning, J., Niederwestberg, D., Grabowksi, R.: Influence of machining parameters on heat generation during milling of aluminium alloys. Proc. CIRP **46**, 39–42 (2016)
14. Düster, A., Parvizian, J., Yang, Z., Rank, E.: The finite cell method for three-dimensional problems of solid mechanics. Comput. Methods Appl. Mech. Eng. **197**(45–48), 3768–3782 (2008). doi:10.1016/j.cma.2008.02.036
15. Evans, L.: Partial Differential Equations. American Mathematical Society, Providence, RI (1998)
16. Fluegge, S.: Encyclopedia of Physics. Vol. VIa/2: Mechanics of Solids. Springer, Berlin (1972)
17. Glänzel, J., Herzog, R., Ihlefeldt, S., Meyer, A., Unger, R.: Simulation-based correction approach for thermo-elastic workpiece deformations during milling processes. Proc. CIRP **46**, 103–106 (2016)
18. Glowinski, R., Pan, T.W., Periaux, J.: A fictitious domain method for dirichlet problem and applications. Comput. Methods Appl. Mech. Eng. **111**(3–4), 283–303 (1994). doi:10.1016/0045-7825(94)90135-X
19. Green, A.: A note on linear thermoelasticity. Mathematika **19**, 69–75 (1972)
20. Haupt, P.: Continuum Mechanics and Theory of Materials. Springer, Berlin (2002)
21. Hughes, J.F.: Computer Graphics: Principles and Practice (1995)
22. Hughes, T.: The Finite Element Method—Linear Static and Dynamic Finite Element Analysis. Dover Publications Inc (2000)
23. Jawahir, I.S., Brinksmeier, E., M'Saoubi, R., Aspinwall, D.K., Outeiro, J.C., Meyer, D., Umbrello, D., Jayal, A.D.: Surface integrity in material removal processes: recent advances. Annals CIRP **60**(2), 603–626 (2011)
24. Joliet, R., Byfut, A., Kersting, P., Schröder, A., Zabel, A.: Machining of lightweight frame structures. Proc. CIRP **8**, 402–407 (2013). doi:10.1016/j.procir.2013.06.124
25. Joliet, R., Byfut, A., Kersting, P., Schröder, A., Zabel, A.: Validation of a heat input model for the prediction of thermomechanical deformations during NC milling. In: 14th CIRP Conference on Modeling of Machining Operations (CIRP CMMO) (2013). doi:10.1016/j.procir.2013.06.124
26. Karaguzel, U., Bakkal, M., Budak, E.: Modeling and measurement of cutting temperatures in milling. Proc. CIRP **46**, 173–176 (2016)
27. Karniadakis, G., Sherwin, S.: Spectral/*hp* Element Methods for CFD. Oxford University Press, USA (2005). doi:10.1093/acprof:oso/9780198528692.001.0001
28. Kersting, P., Odendahl, S.: Capabilities of a process simulation for the analysis of five-axis milling processes in the aerospace industry. In: 18th International Seminar on High Technology (2013)
29. Kienzle, O.: Die Bestimmung von Kräften und Leistungen an spanenden Werkzeugen und Werkzeugmaschinen. VDI-Z 94 (1952)
30. Parvizian, J., Düster, A., Rank, E.: Finite cell method.*h*- and *p*-extension for embedded domain problems in solid mechanics. Comput. Mech. **41**(1), 121–133 (2007). doi:10.1007/s00466-007-0173-y
31. Pavlicek, F., Beer, Y., Weikert, S., Mayr, J., Wegener, K.: Design of a measurement setup and first experiments on the influence of CO_2-cooling on the thermal displacement on a machine tool. Proc. CIRP **46**, 23–26 (2016)
32. Puls, H., Klocke, F., Döbbeler, B., Peng, B.: Multiscale modeling of thermoelastic workpiece deformation in dry cutting. Proc. CIRP **46**, 27–30 (2016)

33. Putz, M., Schmidt, G., Semmler, U., Oppermann, C., Bräunig, M., Karagüzel, U.: Modeling of heat fluxes during machining and their effects on thermal deformation of the cutting tool. Proc. CIRP **31**, 611–614 (2015)
34. Rossi, F., Baizeau, T., Moureaux, C.: Thermal finite difference modeling of machining operations in polymers. Proc. CIRP **46**, 234–237 (2016)
35. Salençon, J.: Handbook of Continuum Mechanics—General Concepts. Thermoelasticity, Springer, Berlin (2001)
36. Schulze, V. (ed.): 15th Conference on Modelling of Machining Operations—Sessions on Thermal Effects in Machining Processes, vol. 31, Procedia CIRP, Elsevier (2015)
37. Schulze, V., Uhlmann, E., Mahnken, R., Menzel, A., Biermann, D., Zabel, A., Bollig, P., Ivanov, I.M., Cheng, C., Holtermann, R., Bartel, T.: Evaluation of different approaches for modeling phase transformations in machining simulation. Prod. Eng.—Res. Dev. Annals Ger. Acad. Soc. Prod. Eng. **9**(4), 437–449 (2015). doi:10.1007/s11740-015-0618-7
38. Schweinoch, M., Joliet, R., Kersting, P.: Predicting thermal loading in NC milling processes. Prod. Eng. **9**(2), 179–186 (2015). doi:10.1007/s11740-014-0598-z
39. Schweinoch, M., Joliet, R., Kersting, P., Zabel, A.: Heat input modeling and calibration in dry NC-milling processes. Prod. Eng.—Res. Dev. Annals Ger. Acad. Soc. Prod. Eng. **9**(4), 495–504 (2015). doi:10.1007/s11740-015-0621-z
40. Schweinoch, M., Joliet, R., Kersting, P., Zabel, A.: Heat input modeling and calibration in dry NC-milling processes. Prod. Eng. Res. Dev. **9**(4), 495–504 (2015c). doi:10.1007/s11740-015-0621-z
41. Surmann, T., Ungemach, E., Zabel, A., Joliet, R., Schröder, A.: Simulation of the temperature distribution in NC-milled workpieces. Adv. Mater. Res. **223**, 222–230 (2011). doi:10.4028/www.scientific.net/AMR.223.222
42. Vos, P., van Loon, R., Sherwin, S.: A comparison of fictitious domain methods appropriate for spectral/*hp* element discretisations. Comput. Methods Appl. Mech. Eng. **197**(25–28), 2275–2289 (2008). doi:10.1016/j.cma.2007.11.023
43. Šolín, P., Segeth, K., Doležel, I.: Higher-Order Finite Element Methods. Chapman & Hall/CRC Press, London/Boca Raton (2003)
44. Wanigarathne, P.C., Kardekar, A.D., Dillon, O.W., Poulachon, G., Jawahir, I.S.: Progressive tool-wear in machining with coated grooved tools and its correlation with cutting temperature. Wear **259**(7), 1215–1224 (2005)
45. Wiederkehr, P., Siebrecht T.: Virtual machining: capabilities and challenges of process simulations in the aerospace industry. Proc. Manuf. **6**, 80–87 (2016). doi:10.1016/j.promfg.2016.11.011. 16th Machining Innovations Conference for Aerospace Industry - MIC
46. Wilmanski, K.: Continuum Thermodynamics—Part I: Foundations (Series on Advances in Mathematics for Applied Sciences). World Scientific Publishing Company (2008)
47. Young, D., Melvin, R., Bieterman, M., Johnson, F., Samant, S., Bussoletti, J.: A locally refined rectangular grid finite element method: application to computational fluid dynamics and computational physics. J. Comput. Phys. **92**(1), 1–66 (1991). doi:10.1016/0021-9991(91)90291-R
48. Zabel, A.: Prozesssimulation in der Zerspanung - Modellierung von Dreh- und Fräsprozessen. Habilitation, TU Dortmund (2010)
49. Zimmermann, M., Schindler, S., Steinmann, P., Aurich, J.C.: Compensation of thermomechanically induced workpiece and tool deformations during dry turning. Proc. CIRP **46**, 31–34 (2016)

Coupling Analytical and Numerical Models to Simulate Thermomechanical Interaction During the Milling Process of Thin-Walled Workpieces

S. Wimmer, J. Loehe and M.F. Zaeh

Abstract Lightweight design is continually gaining in importance within the engineering sector, thus leading to a wide array of low rigidity components that are very demanding regarding the milling process in case manufactured by cutting. Within this chapter a new approach to model thermomechanical effects that occur during the milling process of thin-walled structures is introduced. For this purpose, an empirical process heat model was developed, which is calibrated by deformation measurements. In conjunction with well-established analytical cutting force models, the proposed process heat model was coupled with a numerical model of the workpiece to predict process induced deformations. This allows for the design of counter measures in order to either reduce occurring workpiece deflections or to compensate the resulting geometrical error. It was shown in experiments that both, an optimization of process parameters as well as a tool path adaption, can be successfully applied to avoid the violation of tolerance specifications without affecting the productivity of the milling operation.

Nomenclature

a	Area of active surfaces associated to the heat flux in vector λ_Δ
a_e	Cutting width
a_p	Cutting depth
b_w	Remaining wall thickness
c	Model constant
d	Tool diameter
f_z	Feed per tooth
i, j	Counter of independent parameters
J_T	Cost function of thermal regression models
k	Number of regression model
k_c	Specific cutting energy

S. Wimmer (✉) · J. Loehe · M.F. Zaeh
Institute for Machine Tools and Industrial Management, Technical University of Munich (TUM), Boltzmannstraße 15, 85748 Garching, Germany
e-mail: Sepp.Wimmer@iwb.mw.tum.de

D. Biermann and F. Hollmann (eds.), *Thermal Effects in Complex Machining Processes*, Lecture Notes in Production Engineering,
DOI 10.1007/978-3-319-57120-1_14

m_k	Row size of vectors and matrices
n_k	Column size of matrices
n	Rotational speed of the tool
p	Position
P_c	Cutting power
Q_w	Material removal rate
$\dot{q}$	Heat flux
R^2	Coefficient of determination
t	Time
u_y	Total deformation of the workpiece (in y-direction)
u_T	Thermally induced deformation of the workpiece
u_{dyn}	Dynamic component of the deformation
u_{stat}	Static component of the deformation
u_{plast}	Plastic component of the deformation
v_c	Cutting speed
v_f	Feed
x	Independent parameter
z	Number of flutes
z_{MP}	Z-position of the measuring point (MP)
β	Helix angle of the tool
Δ	Difference of two regression models (k)
γ	Angular position of the leading cutting edge
$\boldsymbol{\lambda}$	Vector of independent parameters
$\boldsymbol{\varphi}$	Vector of linear model constants
$\boldsymbol{\Omega}$	Matrix of quadratic and linear interacting model constants

1 Motivation

Short production cycles and a high level of cost efficiency are among the main goals of production technologies. For metal cutting operations in particular, advancements of cutting materials, drive and control systems as well as improvements regarding the structural behavior of machine tools ensure a steady increase in reliability and reproducibility. At the same time lightweight design is gaining in importance throughout the engineering sector, which leads to highly complex low-rigidity parts with stringent requirements concerning precision and machining quality. During the machining process of thin-walled structures, deformations caused by thermal and mechanical effects can be observed that may substantially compromise the workpiece quality. Considering, for example, large structural components for aerospace applications with unit costs of several hundred thousand euros, the possibility of machining the part “right-first-time” can prove crucial.

High costs and the poor environmental compatibility of cooling lubricants promote the use of dry machining in industrial applications [1, 2]. The omission of cooling lubricants leads to high temperatures within the cutting zone and thus to an increased tool wear and larger thermally induced deformations of the workpiece. While materials with considerable heat conduction rates, e.g. aluminum alloys, allow for a quick heat dissipation from the cutting zone, materials with limited heat conduction capabilities, e.g. titanium alloys, exhibit significant temperature gradients during the cutting process, thus aggravating the problem of thermally caused deformations [3].

The complexity and interdependencies of these effects hamper a process optimization via trial-and-error approaches, which are generally associated with a cost intensive experimental effort [4]. In order to face the challenges, simulation methods that allow for a prediction of workpiece deformations are becoming more important and facilitate the design of optimization algorithms or compensation measures [5]. Within the framework of the research project "Modeling, Simulation and Compensation of Thermal Effects for Complex Machining Processes" (SPP 1480) funded by the German Research Foundation (DFG), the Institute for Machine Tools and Industrial Management at Technical University of Munich explicitly studied the helical end milling process of low-rigidity components. To accurately predict the occurring deformations during the finishing process, analytical models that describe forces and heat flux out of the process are coupled with a discrete structural model of the workpiece. With this information it is possible to select appropriate cutting parameters or to determine the expected geometrical surface error in order to apply a correction of the initial NC-path.

2 Objective and Approach

A reliable and time efficient prediction of thermo-mechanical effects during a helical end milling process is still limited as there is no universal analytical description for the complex interactions involved. Detailed numerical models of the chip formation mechanism require a high level of computational power and are only capable of dealing with short time periods. In this project a physically motivated analytical approach was developed to describe heat flux as well as cutting forces for the contemplated milling process. These analytical component models are then coupled with a finite element based model of the workpiece to enable a holistic simulation of the milling process. Based on the simulation models, measures can be taken to reduce process induced deformations; Fig. 1 illustrates the procedure.

It can be distinguished between two general approaches: Process-related and compensation strategies. It is favorable to primarily pursue process-related approaches due to their simple implementation on common machine tools. There might be situations, however, in which this is not sufficient in order to achieve a desired level of machining quality. In that case compensation techniques have to be

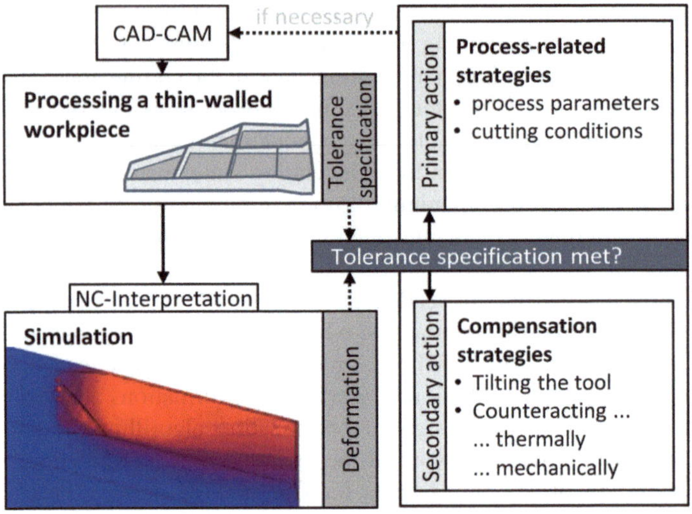

Fig. 1 Procedure to reduce process caused deformations [6]

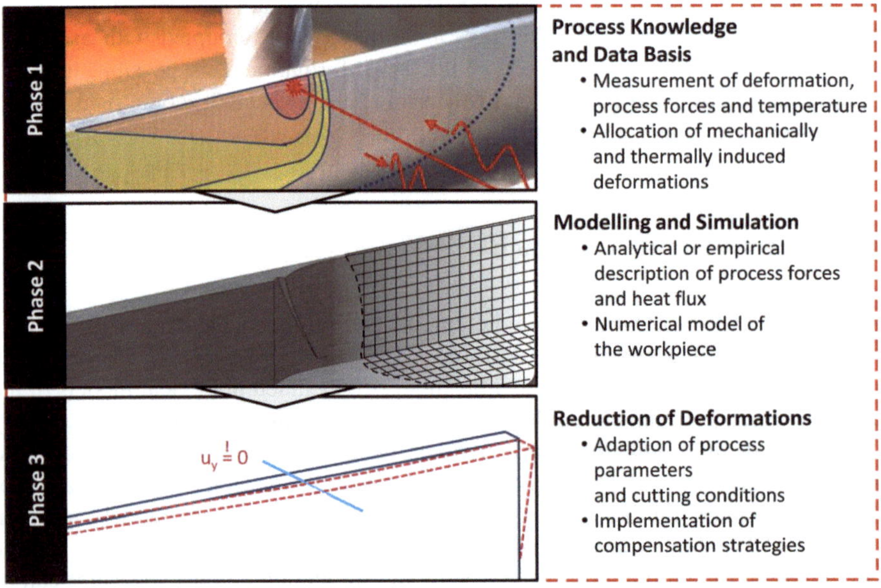

Fig. 2 General approach divided into three phases

applied. To accomplish the aforementioned goals, the project itself was divided into three progressive phases as shown in Fig. 2.

Starting with an extensive design of experiments (DoE), the reference process is studied in order to identify the critical influencing parameters accountable for the deformation behavior of the workpiece. For the experiments, a finishing process was assumed for two reasons: First, the final surface finish is produced within this step and in-process deformations negatively affect the shape accuracy. Second, during the finishing process, the workpiece is already close to its final shape, therefore the wall-thicknesses are generally the lowest, which is particularly critical in terms of workpiece deformations.

To maintain a reasonable expenditure for the experiments, straight thin-walled plates without stiffeners were used. In doing so, large workpieces and thus high material costs can be avoided. Despite the simplicity of this geometry, it is most critical in terms of heat conduction and mechanical rigidity. The lack of supporting structures keeps the workpiece stiffness at a minimum while hampering heat dissipation, thus leading to large temperature gradients. Extensive milling experiments were conducted for two different materials, a common steel alloy 1.1730 (normalized) and a high performance aluminum alloy AW 7075 (T651).

The primary objective of the milling experiments was to measure the workpiece deformation during the milling process and to split the total deformation into a mechanically and a thermally induced proportion, respectively. To accomplish this, a specific experimental setup was designed based on the application of a 1D-Scanning-Laser-Doppler-Vibrometer (LDV) which is depicted in Fig. 3. As visual access to the workpiece is mandatory for an optical measurement system,

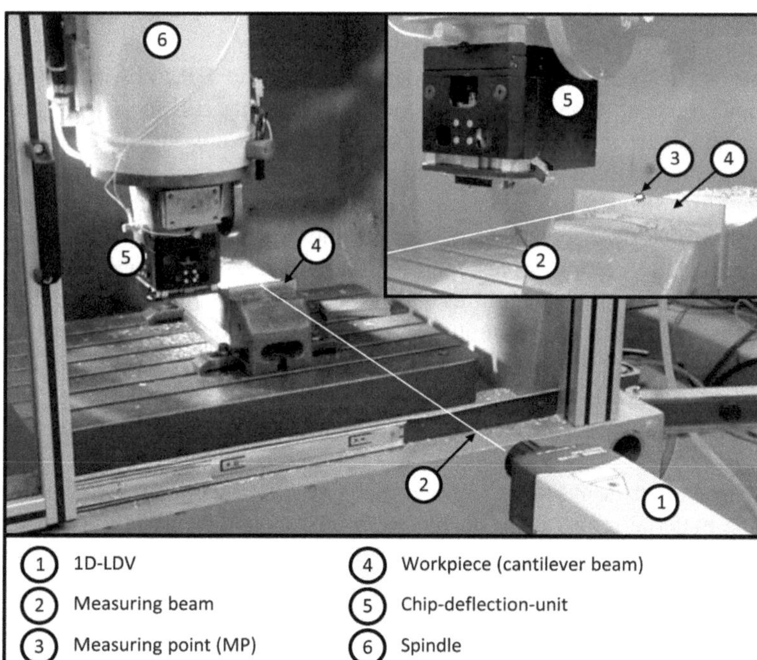

Fig. 3 Experimental set-up for the measurement of in-process deformations [7]

milling experiments were conducted on a three-axis portal milling machine with an open structure.

In order to enhance the quality of the measurement results, some disturbance sources had to be eliminated. During the in-process measurement, low frequency modes from the machine mount, which are excited by acceleration processes, may distort the signal. To make sure the workpiece deformations can be viewed in isolation, a modal analysis of the machine tool was carried out prior to the experiments. The rigid body modes of the machine in the range from 17 to 21 Hz were then filtered out from the original measurement signal. Chips can also disrupt the measuring beam which renders the signal unusable. To deflect the chip flow away from workpiece and laser beam, a chip-deflection-unit was mounted beneath the spindle. It contacts the workpiece over a stripe of Kevlar textile, which has no influence on the workpiece behavior [3, 7].

Unlike cutting forces, the heat flux cannot be recorded directly [8]. Thus, the heat flux must be determined indirectly via its resulting effects: The workpiece temperature or its deformation behavior. The measured in-process deformations are key to calibrate the process heat model by solving this so called inverse heat expansion problem (IHEP) [6]. While the temperature in the contact zone of tool and workpiece is hardly accessible for measurements, the temperature on the backside of the workpiece leads to an ill conditioning of the inverse heat transfer problem (IHTP) [9]. Using the deformation to determine the actual heat input by solving the IHEP yields better results and was therefore chosen to calibrate the empirical model. Thermocouples were also installed to gather temperature data during the milling experiments for validation purposes.

Figure 4 shows the general procedure to solve the IHEP by comparing two regression models. One regression model represents all measured parameter set-ups, the other one emulates these set-ups while applying differently dimensioned heat fluxes. Each regression model is represented as a function of the cutting power and of a heat distribution coefficient. Solving the resulting equation yields the heat flux that describes the process with the best fit [6, 7].

To predict the cutting forces for the simulation, a well-established analytical model [10] is used, which requires the determination of material specific cutting coefficients. To ensure an accurate representation of the actual process forces, a series of milling experiments was carried out, using the identical material and tool combination as for the deformation experiments. A full factorial design of experiments with repetitive measurements was applied to validate the cutting force model.

The simulation is based on a finite element (FE) model of the workpiece, in which the material removal is considered in incremental steps. For each time-step, the workpiece geometry is automatically created and transferred into an FE-mesh. Subsequently initial and boundary conditions are applied and the current deformation is calculated. Figure 5 schematically depicts the sequence for a single iteration.

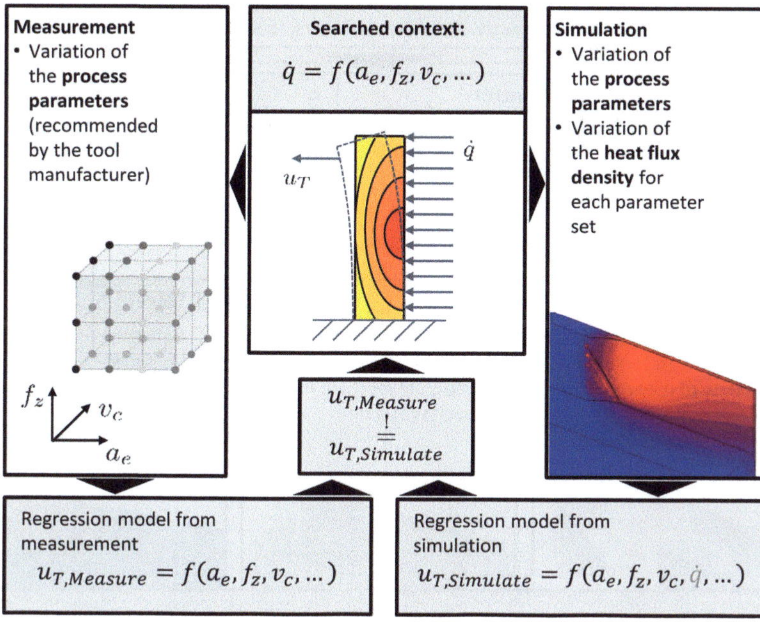

Fig. 4 General procedure to solve the IHEP based on measured deformations [7]

The simulation model requires input data regarding tool geometry, process parameters as well as tool trajectory. In order to describe the tool geometry of a helical end mill,

- diameter d,
- number of flutes z and
- helix angle β

are stored in a library for each tool. Microgeometrical features such as cutting edge radius and coating, for example, are considered within the analytical model. Based on a given

- tool trajectory,
- spindle speed n and
- feed per tooth f_z

the angular position of the leading cutting edge φ and in consequence the incremental chip volume can be calculated as a function of the elapsed time t. Via a Component Object Model (COM) interface this data is passed from the control software Matlab to the CAD software Catia, where the current volume model of the workpiece is generated by subtracting the chip volume from the previous workpiece model. Initiated by the control program, the workpiece model is imported into Hypermesh. The volume model is subdivided for an appropriate meshing: While generally large elements are applied, the subsection near the contact zone is

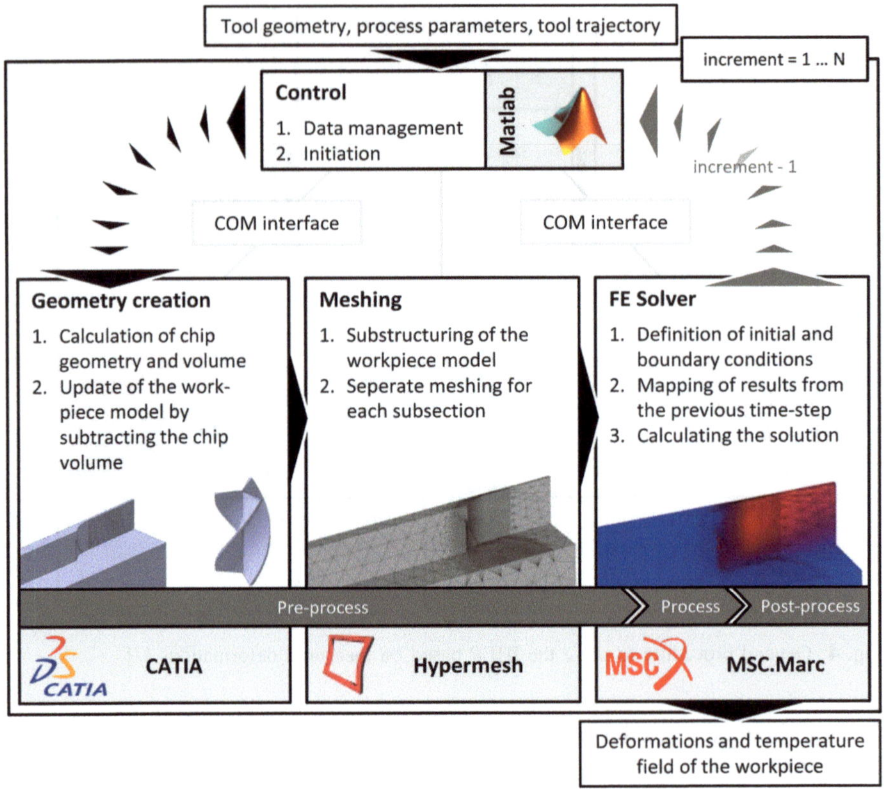

Fig. 5 Simulation sequence for one iteration (image sources: Math Works, Dassault Systémes, Altair, MSC)

provided with a considerably refined mesh to account for the steep temperature gradient. The reconstruction of the workpiece volume in each iteration facilitates an exact representation of the contact zone in the FE model. However, the mesh constantly changes between consecutive time steps. In order to apply the existing temperature field as an initial condition for the following calculation a mapping algorithm must be implemented that is capable of projecting the solution from one FE mesh onto another one. Finally, the boundary conditions, which are determined via analytical models for cutting forces and heat flux, are applied using the non-linear FE solver MSC.Marc. Again, this step is triggered by the control program via the COM interface. The output generated by the thermomechanical FE simulation are workpiece deformations and temperature fields that can be predicted for a wide range of process parameters. In doing so, a data basis is generated to facilitate the design of strategies to reduce process induced deformations without the need of extensive testing.

There are a number of measures that can be adopted to reduce thermally or mechanically induced deformations, which include [11]

- process adjustments,
- tool-trajectory adaption,
- local heating, e.g. using an integrated laser source,
- local cooling, e.g. using compressed air,
- counteracting forces, e.g. via a leading bezel.

While general process adjustments or a tool trajectory adaption can be implemented using a common machine tool, the other approaches require specific attachments within the working space. The strategy pursued in this project is therefore based on the first two measures, applied in two stages as depicted in Fig. 1. If in-process deformations result in a violation of tolerance specifications, the first step is to review the process itself and to adjust certain parameters to reduce process forces and temperature gradients without sacrificing productivity. The influence of critical process parameters can be determined using the simulation model. It was also deduced for certain materials from extensive milling experiments. However, depending on the level of maturity of the considered milling operation, the room for improvement might be limited. If process adjustments are not sufficient to meet the desired quality requirements, the processing result can be further improved by carrying out a tool-path adaption. In order to generate an optimized tool-trajectory, the geometrical error can be measured directly, e.g. using a coordinate measuring machine. This, however, is a very time consuming and costly task, especially when large and high-priced parts are considered, as an error compensation can only take place after one or more scrap parts have been machined. Another way is to predict the anticipated surface error using a process simulation. In this case an accurate and realistic representation of the milling process is mandatory, which particularly requires boundary conditions like process forces and heat flux to be valid. Coupling analytical models with an FE model provides a much-needed bridge between modeling accuracy on the one hand and computational efficiency on the other hand.

3 Project Results

One of the main advantages of the experimental setups introduced earlier (Fig. 3) is the possibility to split up mechanically and thermally induced deformations when plastic deformations are negligible [7]. An exemplary measurement sequence is depicted in Fig. 6. Choosing a high enough temporal resolution, the deformation signal can be matched with the angular position of the tool which is illustrated in Fig. 7.

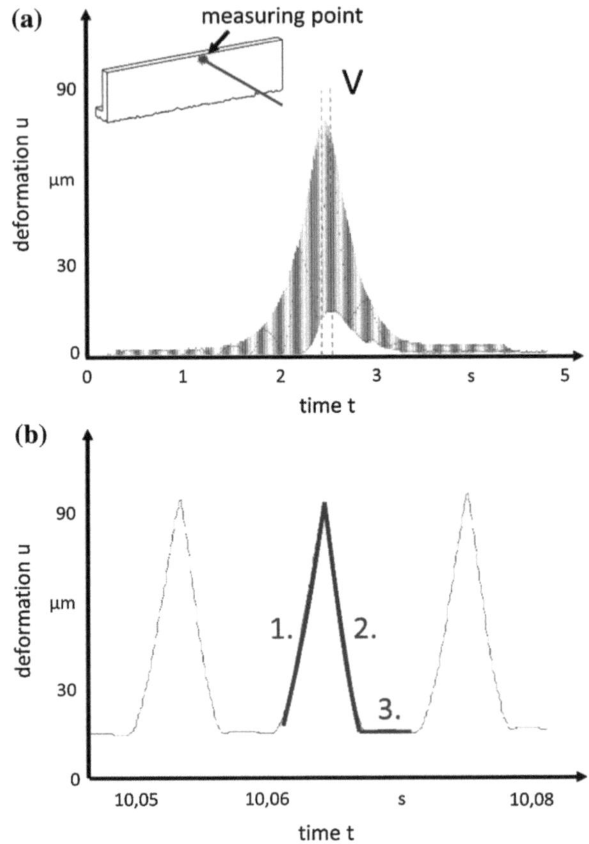

Fig. 6 **a** Deformation sequence of an AW 7075 sample with a remaining wall-thickness of 0.75 mm for process parameters $a_e = 1$ mm, $a_p = 10$ mm, $f_z = 0.1$ mm and $n = 2000$ min^{-1}; **b** Enlarged section V of one tool rotation

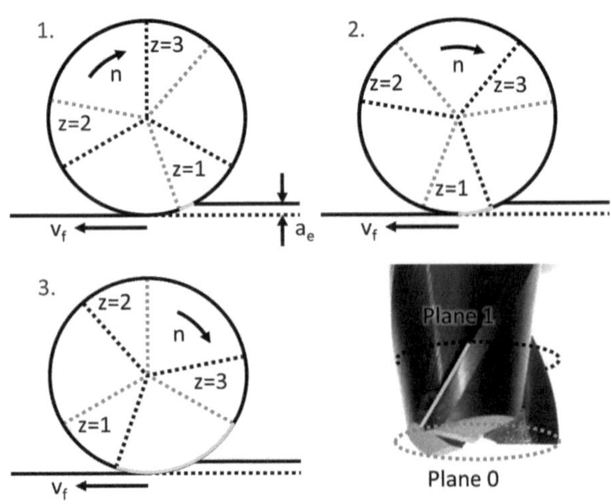

Fig. 7 Angular positions of the tool for one characteristic signal peak [11]

For one rotation of the three-fluted tool the signal shows three characteristic peaks that can be assigned to the cutting edges, respectively. For each flute three recurring states can be identified:

- Rising edge (1.): The cutting edge comes into contact with the workpiece at the level of plane 0. Subsequently, the contact length gradually increases, which also leads to an increase of the workpiece deflection.
- Falling edge (2.): As soon as the maximum chip thickness is reached, the contact length between tool and workpiece decreases and the workpiece deflection vanishes accordingly.
- Plateau (3.): During this phase, there is no contact between tool and workpiece. Despite that, deformations can be observed which can either be attributed to plastic behavior or result from an irregular temperature field.

Due to the fact that the measured deformation goes back to zero a few seconds after the tool passes the measuring point, plastic material behavior can be ruled out. Therefore, the lower envelope curve of the signal describes a purely thermally induced deformation. The slightly asymmetric shape bears a certain resemblance to thermal camera images and can be explained with the inert behavior of heat dissipation as shown in Fig. 8. When the contact zone approaches the measuring point, the workpiece temperature quickly reaches peak temperature in this spot. After the chip formation is finished, however, the cooldown is significantly slower. The lower envelope curve of the deformation signal shows a similar behavior, thus confirming the assumption of it being thermally induced [3].

For consistency reasons, identical solid carbide end milling cutters were used for both materials and parameter sets recommended for steel finishing are also used for

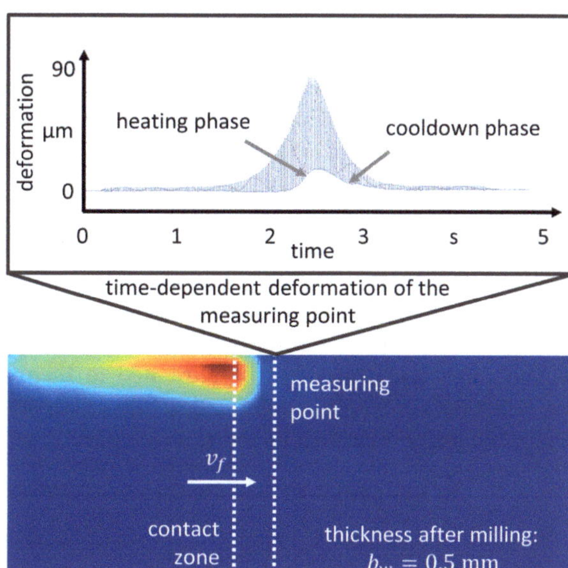

Fig. 8 Correlation of temperature field and thermally induced deformation measured on the back of the AW 7075 workpiece

Table 1 Levels of varying independent process parameters [7]

Parameters	Levels			
a_e in mm	0.10	0.30	0.50	0.70
f_z in mm	0.10	0.14	0.18	
v_c in $\frac{m}{min}$	140.00	170.00	200.00	
b_w in mm	0.75	1.50		

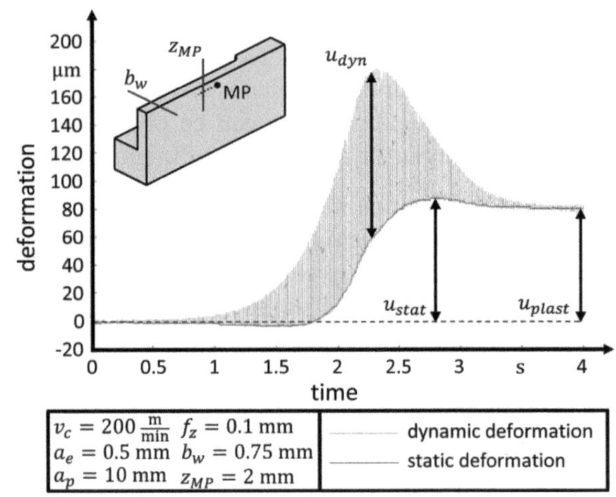

Fig. 9 Deformation sequence of a 1.1730 sample with plastic behavior separated into static and dynamic components [7]

aluminum. To find out any interdependency between process parameters, a fully crossed factorial DoE was selected, containing the whole spectrum of process parameters. Any combination was carried out a total of three times in a random order to allow for a statistical analysis of the results. The cutting depth a_p was invariably set to 10 mm, while other parameters were varied as shown in Table 1. As the measurement results could easily be influenced by wear mechanisms, tools were frequently changed. In between the experiments, the cutting edges were analyzed for any sign of deterioration, to ensure that the tools were always in perfect condition.

Due to the low yield strength of the normalized steel alloy, plastic deformations can be observed for a wide range of the parameter spectrum while the aluminum alloy shows no plastic behavior throughout all milling experiments. A typical deformation sequence of a steel sample with plastic behavior is depicted in Fig. 9. The presence of plastic deformations means, that the lower envelope curve of the signal can no longer solely be attributed to thermal effects and is called “static deformation”. “Dynamic deformation” describes the part of the total deformation that completely returns to its initial shape once the tool is no longer in contact with the workpiece.

Table 2 Constants of the static and the dynamic regression models for steel (1.1730) and aluminum (AW 7075) [7]

	Steel 1.1730_{stat}	Steel 1.1730_{dyn}	Aluminum $AW\,7075_{stat}$	Aluminum $AW\,7075_{dyn}$
c_0	−15.5	−49.2	–	−14.5
c_1 in $\frac{\text{min}}{\text{mm}}$	–	1.58	–	1.1
c_2 in $\frac{\text{min}^2}{\text{mm}^2}$	–	−0.0045	–	−0.0033
c_3 in $\frac{1}{\text{mm}}$	229.3	66.6	29.015	121.06
c_4 in $\frac{1}{\text{mm}^2}$	−226.6	−127.4	–	−167.5
c_5 in $\frac{1}{\text{mm}^2}$	743.5	641.7	–	695.5
R^2	0.9146	0.8662	0.8049	0.9420

For the evaluation of the separate signal components a generic quadratic regression model with linear interaction terms was used, for which an iterative model reduction was performed by removing insignificant factors. To assess the influence of specific factors and their combinations on the coefficient of determination, a large number of models with different factor combinations were created and compared to each other [7]. The general resulting regression model is shown in Eq. 1.

$$u_{stat/dyn} = c_0 + c_1 v_c + c_2 v_c^2 + c_3 a_e + c_4 a_e^2 + c_5 f_z a_e. \tag{1}$$

Table 2 shows the constants c_i for the steel and aluminum alloys. Some constants are insignificant for certain models, generally insignificant values are not listed.

As Fig. 10 indicates, the cutting speed has no significant influence on the dynamic deformation of both materials for the evaluated parameter spectrum. Given the respective constants for the static regression model this means that, in the present case, the heat source is independent of the cutting speed.

For most industrial users, the increase of either productivity or manufacturing accuracy is often in the center of attention. To refer to the productivity of a specific cutting operation, the material removal rate Q_w is often introduced. Taking a look on the previously discussed effect of the cutting speed during the milling experiments, the material removal rate may be increased by up to 143% compared to a minimum material removal rate, without affecting the accuracy. On the contrary, both, the cutting width and the feed per tooth, have an effect on the deformation behavior of either material, which is depicted in Fig. 11.

While the dynamic deformation is similar for steel and aluminum, the static component is substantially different. The latter observation can be explained by the presence of plastic deformations for the steel alloy, which is no issue when considering aluminum. For that reason, only the static deformations of the aluminum samples can be attributed to purely thermal effects. Due to the offsetting effects of a

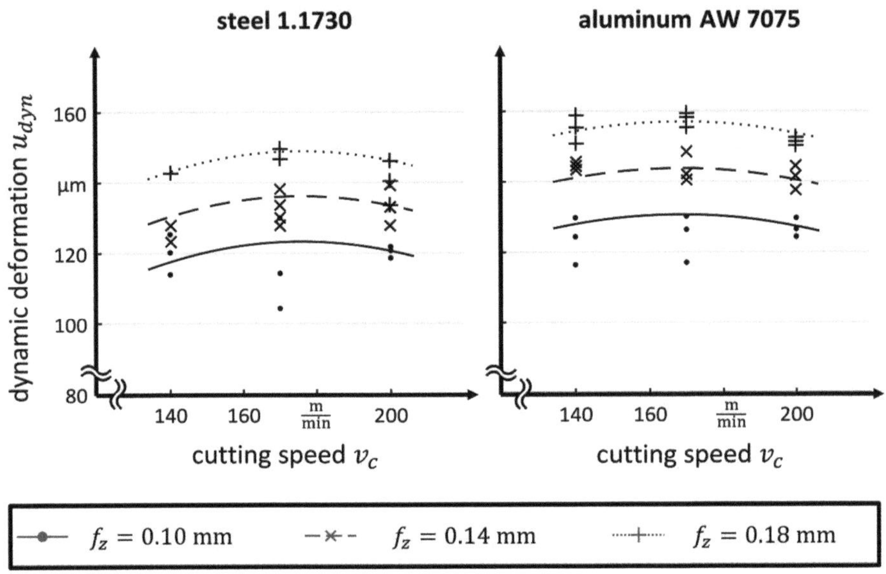

Fig. 10 Dynamic deformation component u_{dyn} in dependence of v_c and f_z for $a_e = 0.5$ mm and $b_w = 0.75$ mm [7]

lower Young's modulus on the one hand and lower specific cutting forces on the other hand, the dynamic component of aluminum approximately corresponds with the behavior of the steel alloy.

The analysis shows a clear interdependency of cutting width and feed per tooth, which makes it difficult to manually adjust those parameters in order to increase the productivity as it may have negative side effects on the workpiece deformation. Using a model based optimization, however, promises a high potential to increase the process efficiency while satisfying constraints regarding tolerance specifications. For the analyzed parameter spectrum, Q_W could be increased by up to 180% considering f_z and by up to 700% considering a_e in case the higher deformations are still bearable [7].

The discussion shows that the results from the milling experiments provide a valid basis for directly optimizing specific milling operations in terms of efficiency. Additionally, the results can also be used to calibrate the proposed process heat model [6]. Using the introduced method to identify the thermally induced deformation, it can be expressed as a function of the process parameters:

$$u_{T,Measure} = f(a_e, f_z, v_c, \ldots). \tag{2}$$

As this is not sufficient to identify the actual heat input $\dot{q}$, a numerical simulation is used: Each parameter setup from the milling experiments is built up virtually and loaded with different heat fluxes. Similar to the measurement data, a regression analysis is then applied to the simulation results:

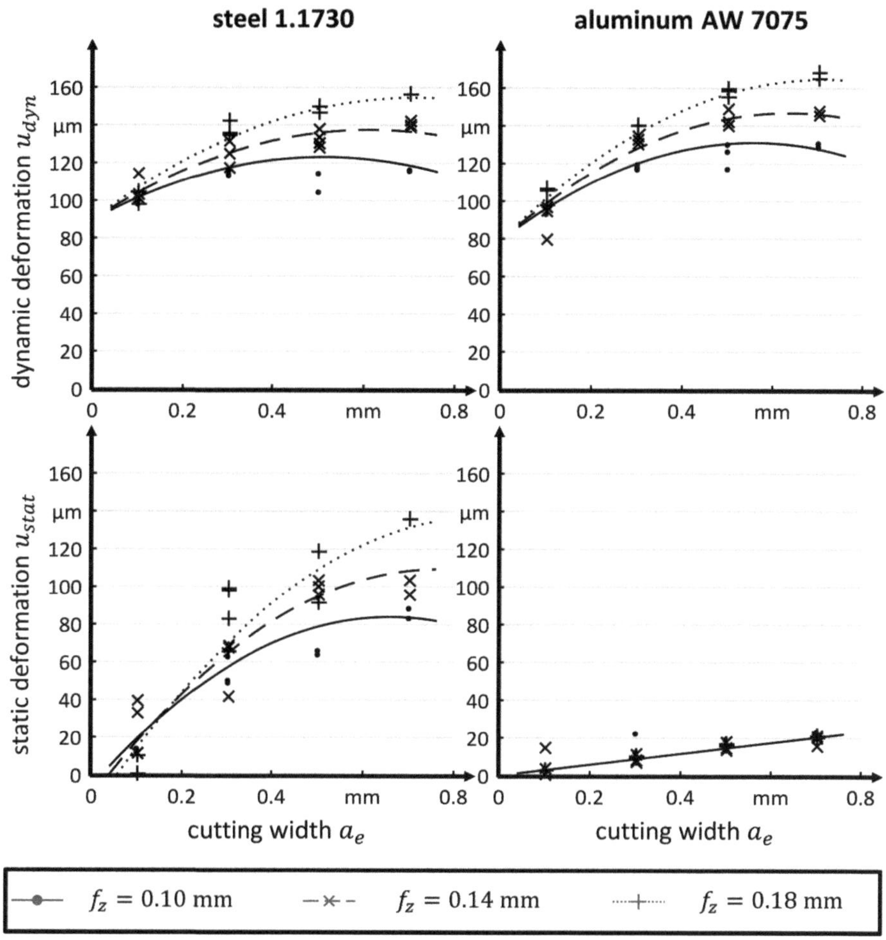

Fig. 11 Dynamic (*top*) and static (*bottom*) deformation component in dependence of a_e and f_z for $v_c = 170\ \frac{\text{m}}{\text{min}}$ and $b_w = 0.75$ mm [7]

$$u_{T,Simulate} = f(a_e, f_z, v_c, \dot{q}, \ldots). \tag{3}$$

Comparing both regression models (2) and (3) results in the following equation:

$$0 \stackrel{!}{=} u_{T,Simulate} - u_{T,Measure} = u_{T,\Delta}. \tag{4}$$

Solving Eq. (4) for the heat flux $\dot{q}$ yields the process related heat input into the workpiece. Depending on the model's complexity, however, there might be more than one solution satisfying the condition. Using regression models, the heat flux can be determined for a wide range of process parameters. This means that there is no detailed information required about different interaction zones where shearing, friction and separation processes take place, which affect the distribution of the heat flow [12–14].

Based on the nomenclature of Siebertz et al. [15] the reduced regression model (1) for one measuring point (p, e.g. MP in Fig. 9) at the time t can be written in its general form for the measured ($k=1$) and the simulated deformations ($k=2$), respectively:

$$\begin{aligned} u_{T,k}(t,p) = {} & c_{0,k}(t,p) \\ & + c_{1,k}(t,p)x_1 + \cdots + c_{i,k}(t,p)x_i \\ & + c_{1,1,k}(t,p)x_1x_1 + \cdots + c_{i,j,k}(t,p)x_ix_j \end{aligned} \tag{5}$$

with $i \leqq j$.

According to pre-tests, a quadratic model with linear interaction terms is appropriate for a milling process. For simplification, Eq. (5) is transformed into matrix form:

$$u_{T,k}(t,p) = \boldsymbol{\lambda}_k^T \boldsymbol{\Omega}_k(t,p)\boldsymbol{\lambda}_k + \boldsymbol{\lambda}_k^T \boldsymbol{\varphi}_k(t,p) + c_{0,k}(t,p), \tag{6}$$

with

$$\begin{aligned} \boldsymbol{\lambda}_k &= \begin{bmatrix} x_1 \\ x_2 \\ \vdots \\ x_i \end{bmatrix}; \ \boldsymbol{\varphi}_k(t,p) = \begin{bmatrix} c_{1,k}(t,p) \\ c_{2,k}(t,p) \\ \vdots \\ c_{i,k}(t,p) \end{bmatrix}; \\ \boldsymbol{\Omega}_k(t,p) &= \begin{bmatrix} c_{1,1,k}(t,p) & c_{1,2,k}(t,p) & \cdots & c_{1,j,k}(t,p) \\ 0 & c_{2,2,k}(t,p) & \cdots & c_{2,j,k}(t,p) \\ \vdots & \vdots & \ddots & \vdots \\ 0 & 0 & \cdots & c_{i,j,k}(t,p) \end{bmatrix}. \end{aligned} \tag{7}$$

While only the number of independent parameters determines the size of vector λ_1 $(m_1 \times 1)$, the number of varying parameters, e.g. multiple heat sources, have to be taken into account additionally for $\boldsymbol{\lambda}_2$ $(m_1 \times 1)$:

$$\boldsymbol{\lambda}_2 = \underset{m_2 \times 1}{\begin{bmatrix} \boldsymbol{\lambda}_1 \\ \boldsymbol{\lambda}_\Delta \end{bmatrix}}. \tag{8}$$

A similar extension is applied for vector $\boldsymbol{\varphi}_k$ and for the square matrix $\boldsymbol{\Omega}_k$. In order for the matrix dimensions to fit, $\boldsymbol{\varphi}_1$ and $\boldsymbol{\Omega}_1$ must also be expanded by zeros, which results in $\widetilde{\boldsymbol{\varphi}}_1$ and $\widetilde{\boldsymbol{\Omega}}_1$:

$$\widetilde{\boldsymbol{\varphi}}_1 = \underbrace{\begin{bmatrix} \boldsymbol{\varphi}_1 \\ \mathbf{0} \end{bmatrix}}_{m_2 \times 1}; \; \widetilde{\boldsymbol{\Omega}}_1 = \underbrace{\begin{bmatrix} \boldsymbol{\Omega}_1 & \mathbf{0} \\ \mathbf{0} & \mathbf{0} \end{bmatrix}}_{m_2 \times n_2}. \tag{9}$$

This allows for $\boldsymbol{\lambda}_1$ being replaced with $\boldsymbol{\lambda}_2$ without any further restrictions. For $k=1$, Eq. (6) with (8) and (9) is now given as:

$$u_{T,1}(t,p) = \boldsymbol{\lambda}_2^T \widetilde{\boldsymbol{\Omega}}_1(t,p)\boldsymbol{\lambda}_2 + \boldsymbol{\lambda}_2^T \widetilde{\boldsymbol{\varphi}}_1(t,p) + c_{0,1}(t,p). \tag{10}$$

Comparing both regression models for $k=1$ and $k=2$ according to Eq. (4) renders the following expression:

$$\begin{aligned} 0 &\overset{!}{=} \boldsymbol{\lambda}_2^T \boldsymbol{\Omega}_\Delta(t,p)\boldsymbol{\lambda}_2 + \boldsymbol{\lambda}_2^T \boldsymbol{\varphi}_\Delta(t,p) + c_{0,\Delta}(t,p) \\ &= u_{T,\Delta}(t,p,\boldsymbol{\lambda}_2). \end{aligned} \tag{11}$$

As previously stated, there are one or more $\boldsymbol{\lambda}_2$ satisfying the condition for all discrete points in time t and position p. This can also be expressed by a cost function:

$$J_T(\boldsymbol{\lambda}_2) = \sum_{t=1}^{\bar{t}} \sum_{p=1}^{\bar{p}} \left(u_{T,\Delta}(t,p)\right)^2. \tag{12}$$

Searching for $\boldsymbol{\lambda}_2$ that minimizes the cost function (12) represents a typical optimization problem with three constraints:

$$\min_{\boldsymbol{\lambda}_2} (J_T). \tag{13}$$

The first constraint requires $\boldsymbol{\lambda}_1$ to be a known quantity of $\boldsymbol{\lambda}_2$, thus only $\boldsymbol{\lambda}_\Delta$ will be varied during the optimization.

$$C\mathbf{1}\colon \boldsymbol{\lambda}_1 = [a_e \quad f_z \quad v_c \quad \ldots]^T. \tag{14}$$

The second constraint allows only heat sources to be a feasible solution and excludes heat sinks from the solution space.

$$C\mathbf{2}\colon \boldsymbol{\lambda}_\Delta \geq \mathbf{0} \text{ (elementwise)}, \tag{15}$$

with $\mathbf{0}$ representing a zero vector with matching dimensions. Finally, the power balance of the cutting process must be feasible:

$$C\mathbf{3}\colon \boldsymbol{\lambda}_\Delta^T \boldsymbol{a} \leq P_c. \tag{16}$$

Here, $\boldsymbol{a}$ represents the area of active surfaces of each heat flux from $\boldsymbol{\lambda}_\Delta$. The total energy of all those surfaces impinged with heat fluxes must never exceed the cutting

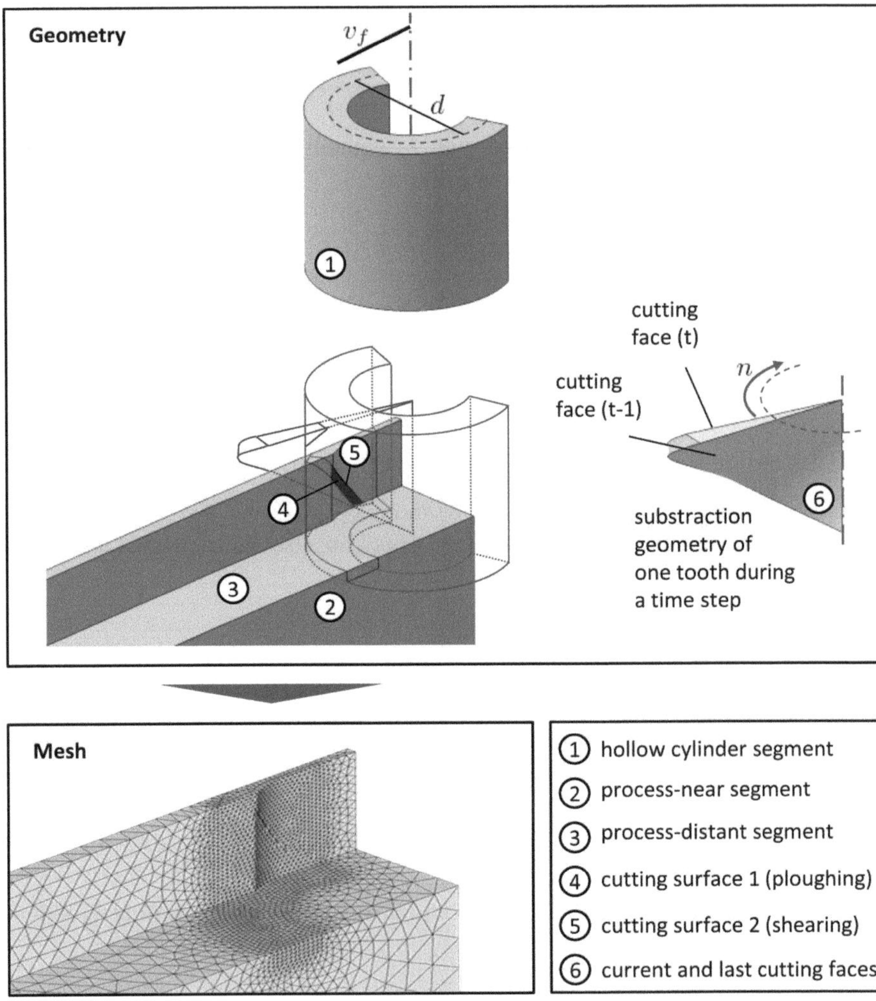

Fig. 12 Schematic representation of geometry preparation and meshing [6]

power P_c. In fact, with only the heat input into the workpiece being considered, the cutting power cannot be reached completely. The actual cutting power can be calculated by means of the specific cutting energy k_c or the respective cutting force F_c [10, 16, 17]:

$$P_c = k_c \cdot Q_w = F_c \cdot v_c. \tag{17}$$

In order to generate all necessary data to set up the optimization problem (13), a simulation model is required that is fed with multiple combinations of process parameters and heat fluxes. Figure 12 visualizes the geometry preparation and meshing of the workpiece model. To provide a high degree of flexibility regarding the workpiece geometry, an automatically generated tetrahedron mesh is used. This also enables a detailed modelling of the contact zone, where the chip formation takes place [6].

A hollow cylinder segment (Fig. 12, no. 1) separates the workpiece into process-near (Fig. 12, no. 2) and process-distant (Fig. 12, no. 3) segments. Due to high expected temperature gradients, large deformations and complex geometry features in the contact zone, a refined mesh is used for the process-near region. By extrapolation of the current and the last cutting face (Fig. 12, no. 6), the cut edge on the workpiece is modeled in detail. In doing so, two surfaces are generated: One surface (Fig. 12, no. 4) represents the area where a heat flux is applied, which covers the ploughing effect during one time step. The second surface (Fig. 12, no. 5) represents the edge of the workpiece built by the cutting plane where the respective heat flux is caused by shearing, friction and separation within the shearing zone [12]. However, a large portion of this heat is removed with the chip during the milling process. The chip itself is not modeled and thus the respective heat fraction is not applied onto the workpiece model. For each time step, the current workpiece geometry is completely regenerated and meshed. Pre-states, like for instance the temperature field, are transferred during the initialization of every single iteration step via a mapping algorithm.

With the proposed numerical model it is possible to simulate the workpiece deformations for the regression model (6) with $k = 2$ and, thus, to solve the optimization problem (13). The simulation of different process parameter combinations and heat sources can thereby be parallelized to reduce the time required as the detailed model is very demanding in terms of computational effort. Depending on the availability of computational resources, only a limited number of parallel runs can be started at a time. Therefore, the search for a correct representation of the heat source has to be narrowed down step by step. Exemplary results of global deflection and temperature field for a heat input of 5% of the total cutting power are shown in Fig. 13. The predicted peak temperatures within the cutting zone for a heat input of 5% show good agreement with the peak temperatures recorded by thermocouples positioned at the back of the workpiece, close to the tool-workpiece contact point as depicted in Fig. 14. As expected, the heat can freely flow downwards and there is a heat accumulation at the top edge. Thus, the maximal temperature is measured with thermocouple number three. Despite the low wall-thickness, a temperature gradient can be observed between the machined surface and the back of the workpiece. Therefore, the thermocouples were placed in small bores. In order to derive the regression model, the whole spectrum of process parameters has to be simulated successfully, which is very time-consuming.

After the identification of an appropriate representation of the heat source, the simulation can directly be used for a parameter optimization in order to reduce in-process deformations. In contrast to cost-intensive milling experiments, a large number of parameter combinations can be evaluated using parallel computing, given an adequate computing capacity is available. The main advantage here is that the detailed numerical model has to be used only once for the calibration of the process heat model. Subsequently, the identified heat source can serve as a

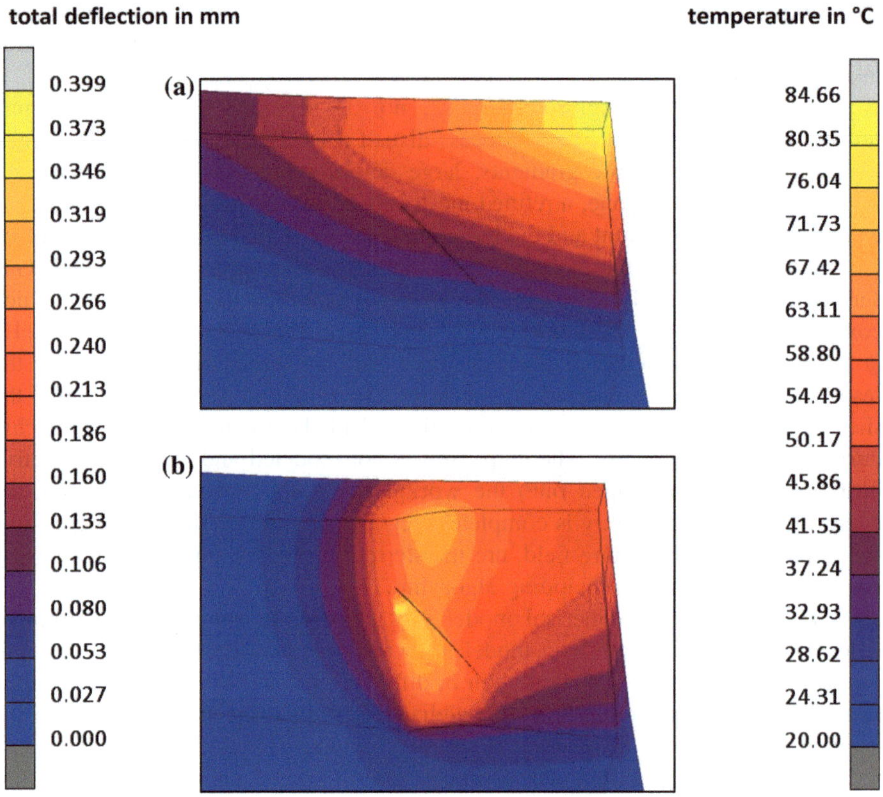

Fig. 13 Total deflection **a** and temperature field **b** of a 1.1730 sample with a remaining wall-thickness of 1.5 mm for process parameters $a_e = 0.6$ mm, $a_p = 10$ mm, $f_z = 0.14$ mm and $v_c = 170\ \frac{\text{m}}{\text{min}}$

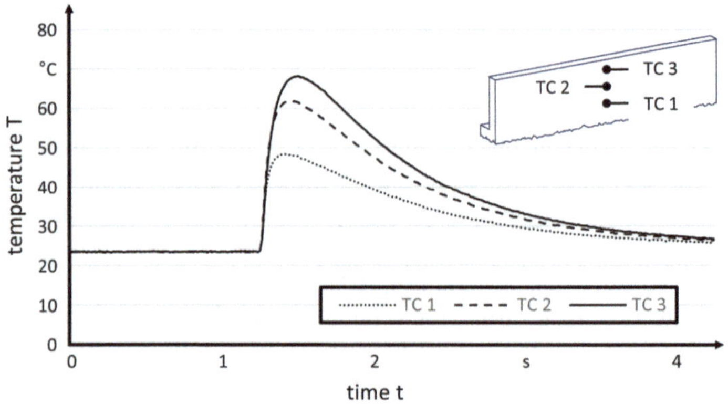

Fig. 14 Temperatures recorded with thermocouples (TC) on the back of a 1.1730 sample with a remaining wall-thickness of 1.5 mm for process parameters $a_e = 0.6$ mm, $a_p = 10$ mm, $f_z = 0.14$ mm and $v_c = 170\ \frac{\text{m}}{\text{min}}$

boundary condition for a simplified structural model to further increase the computational efficiency.

As already introduced, the achievable reduction may not be sufficient to prevent the violation of tolerance specifications in any case. For that reason, an alternative approach is considered that is not intended to reduce deformations, but to predict and reduce the expected surface error. There are similar concepts in literature [18–20] trying to compensate geometrical errors in peripheral end-milling of thin-walled structures based either on measurements or simulation models, which have proven the feasibility of this approach. However, the application is often impractical due to a high effort for measurements, a lack of accuracy in the error prediction or the complexity of the surface error to be compensated.

While the Laser-Doppler-Vibrometer (LDV) can precisely detect the in-process deformation at a specific measuring point, the surface error results from the deflection of the contact point between tool and workpiece right where the chip is removed. This contact point is nonstationary and moves from bottom to the top of the machined surface during the cutting action. This means that the deformation measured in one specific point does not directly correlate with the shape and the magnitude of the geometrical error. Using a coordinate measuring machine (CMM) or a laser scanning microscope, the actual surface error after the milling experiments (Fig. 3) can be analyzed. Additionally to the variation of process parameters (Table 1), the experiments were repeated with different tools to also evaluate different engagement conditions. Figure 15 depicts surface profiles for a

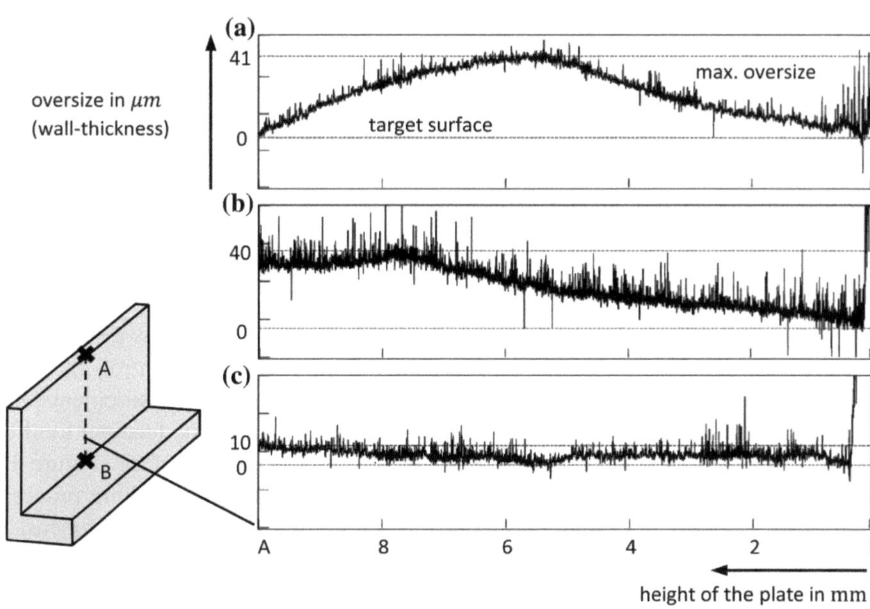

Fig. 15 Surface profiles after the finishing process of an AW 7075 sample with a remaining wall-thickness of 1.5 mm for process parameters $a_e = 0.6$ mm, $a_p = 10$ mm, $f_z = 0.1$ mm and $v_c = 140\ \frac{\text{m}}{\text{min}}$ [21]

Table 3 Tools used for the milling experiments [21]

Tool no.	Diameter (mm)	No. of flutes	Helix angle [°]
1	16	3	30
2	10	6	45

finishing process with constant process parameters, while Table 3 gives an overview of the tools used for the experiments.

Figure 15a shows the finished surface using tool number 1. Due to the helix angle, the cutting edge gradually comes into contact in point B. Thus, the workpiece is steadily being deflected, which leads to an increasing oversize on the machined surface. After a certain tool rotation, the cutting edge gradually exits the workpiece, the workpiece forms back and the oversize decreases towards the top edge. Subsequently, there is a phase without any contact between tool and workpiece (Fig. 7) where the process forces reach zero. This behavior causes a convex surface with the maximal error being located about half way up the workpiece height. Considering a common helical end mill on a five-axis machine tool, no feasible tool path can be found to facilitate an error compensation within a single cut. In fact, a complex machining strategy is required to nullify the surface error.

The finished surface of an identical process but with tool number 2 is shown in Fig. 15b. By changing the tool diameter, the number of flutes as well as the helical angle, more than one cutting edge is in mesh simultaneously. Consequently, the process forces are distributed more evenly, which effectively limits the spring back of the workpiece. In this case, the surface error can be approximated with a straight line and a good compensation within one cut is achieved by tilting the tool accordingly. The finished surface after the tool path adaption with tool number 2 is depicted in Fig. 15c. Without an introduction of a complex machining strategy, the maximal form error is reduced by about 75%. A small error remains near the top edge of the workpiece, which can be attributed to the nonlinear interdependency between cutting conditions and deformation.

According to the milling experiments, it is crucial that more than one cutting edge is in mesh at a time to prevent the spring back behavior of a thin-walled workpiece. Considering helical end mills, this is mainly influenced by the tool geometry and the cutting depth a_p. However, the cutting width a_e has a significant impact on the effective length of the cutting edge in mesh and thus controls the error characteristic to a certain extent. With regard to the deformation measurements, the influence of the cutting speed v_c is expectedly negligible, while the feed per tooth f_z directly correlates with the maximal deviation from the target surface. Figure 16 illustrates the influence of a_e and f_z on the surface error after the finishing process.

In order to avoid a complex tool path replanning and to efficiently compensate the form error, a two-pronged approach is suggested. In a first step, the process should be designed in a way that an approximately linear error characteristic can be expected. Secondly, a process-simulation is to be used for the error prediction which iteratively accounts for process-deformation interactions. While it is generally possible to derive an adapted tool path based on directly measuring the surface

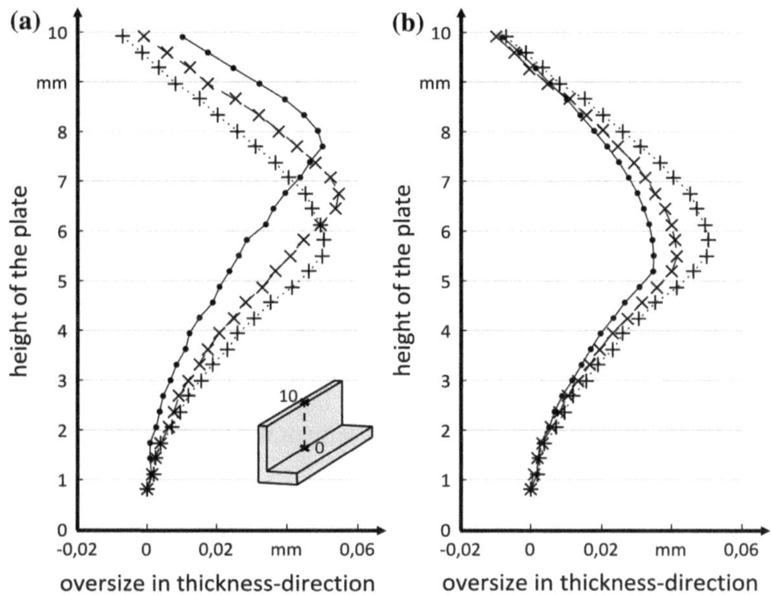

Fig. 16 Surface error of AW 7075 samples with a remaining wall-thickness of 1 mm for tool no. 1 and process parameters $a_p = 10$ mm and $v_c = 140\ \frac{\mathrm{m}}{\mathrm{min}}$; **a** with $f_z = 0.18$ mm; **b** with $a_e = 0.6$ mm

error of workpieces, it is tedious and requires the machining of numerous scrap parts. With a simulation model, which is capable of reproducing the above mentioned effects, in-process deformations can be calculated prior to the machining. By evaluating the deformation of each tool-workpiece contact point in time, the geometric error can be predicted. The positioning of the tool axis can then be determined iteratively for any position in feed direction and subsequently assembled into a corrected tool path. As a modified tool path in turn changes the cutting conditions, the whole procedure has to be reiterated several times until convergence is achieved [21]. Due to the iterative character of this approach and high demands regarding the temporal resolution, model efficiency is crucial. Therefore, a reduced model should be used for the determination of an adapted tool path. The process simulation introduced for the identification of the heat source, while generally capable of predicting the necessary workpiece deformations, is not ideal for that purpose due to the high level of detail of the structural model.

4 Conclusion

The objective of the project was to simulate the deformation behavior of thin-walled structures induced by thermomechanical effects during finishing processes with helical end mills. To achieve this, an empirical process heat model was developed and coupled with well-established process force models as well as a numerical structural model of the workpiece.

To calibrate the process heat model and to identify the heat source, two regression models, one obtained from milling experiments and the second one obtained via simulation, have to be compared. As the heat flux cannot be measured directly, there is no direct access to the actual heat source of the milling process. While the heat flux is usually determined inversely by measuring temperatures, in this project deformation measurements were chosen to establish a well-conditioned inverse heat expansion problem (IHEP). Milling experiments with two different materials were conducted for a wide range of process parameters in order to set up the experimental regression model. The simulative regression model, however, additionally requires the consideration of varying fictive heat fluxes, which leads to a large number of simulation runs. Due to a high level of detail of the structural model and strict requirements regarding the temporal resolution, the solution of the heat process model is very time consuming.

The advantage of this approach is, that the heat source can be determined very accurately and is valid for a wide range of process parameters. It can then be used on reduced numerical models to simulate workpiece deformations more efficiently, while the highly detailed and computationally expensive model is only used for solving the IHEP.

Measures to directly reduce in-process deformations or to compensate for the resulting surface error were also investigated. The simplest way to reduce process related workpiece deformations is to find an optimal set of process parameters. Individual measures were already deduced from the milling experiments, while a parameter optimization based on a simulation of workpiece deformations seems to be a very promising approach to further improve the efficiency of milling operations.

Apparently, workpiece deformations cannot be avoided entirely and there might be cases in which reduction measures are not sufficient to meet certain tolerance specifications. Therefore, a compensation approach was considered that does not require any special applications and thus can be applied on a common five-axis machine tool. In this case the objective is not to reduce any occurring deformations, but to predict them in order to adapt the tool path to compensate for expected surface errors. However, it can be problematic to apply this strategy, as some surface error characteristics require a complex tool trajectory, which increases the machining time. By appropriately designing the process, the geometrical error can be controlled, such that a simple tool path adaption is sufficient, which was proven experimentally. While it is possible to derive a tool path correction from measuring the actual surface error in principle, it is very tedious and requires the machining of numerous scrap parts. By sing a suitable process simulation with a high temporal resolution to predict deformations, an adapted tool trajectory can be calculated before machining any workpiece. The proposed process heat model allows the identification of an accurate heat source for a milling operation, which can then be used as a boundary condition for a specifically designed process simulation to determine a compensated tool path.

Acknowledgements This paper is based on the investigations and findings of the project *Coupling analytical and numerical models to simulate thermomechanical interaction during the milling process of thin-walled workpieces* (ZA 288/38-1) of the priority program SPP 1480 (CutSim), which was kindly supported by the German Research Foundation (DFG).

References

1. Weinert, K., Inasaki, I., Sutherland, J.W., Wakabayashi, T.: Dry machining and minimum quantity lubrication. CIRP Ann.—Manuf. Technol. **53**(2), 511–537 (2004)
2. Sölter, J., Gulpak, M.: Heat partitioning in dry milling of steel. CIRP Ann.—Manuf. Technol. **61**(1), 87–90 (2012)
3. Loehe, J., Zaeh, M.F., Roesch, O.: In-process deformation measurement of thin-walled workpieces. In: Procedia CIRP 1, pp. 546–551 (2012)
4. Philippe, D., Hascoët, J.Y.: Active integration of tool deflection effects in end milling. Part 2. Compensation of tool deflection. Int. J. Mach. Tools Manuf. **46**(9), 945–956 (2006)
5. Arrazola, P.J., Özel, T., Umbrello, D., Davies, M., Jawahir, I.S.: Recent advances in modelling of metal machining processes. CIRP Ann.—Manuf. Technol. **62**(2), 695–718 (2013)
6. Loehe, J., Zaeh, M.F.: A new approach to build a heat flux model of milling processes. Procedia CIRP **24**, 7–12 (2014)
7. Loehe, J., Wimmer, S., Hairer, M., Zaeh, M.F.: An experimental study on the deformation behavior of thin-walled workpieces. In: Proceedings of the 17th International Conference on Machine Design and Production, Bursa, 12–15 July 2016
8. Klocke, F., Lung, D., Puls, H.: FEM-Modelling of the thermal workpiece deformation in dry turning. Procedia CIRP **8**, 240–245 (2013)
9. Frąckowiak, A., Botkin, N.D., Ciałkowski, M., Hoffmann, K.H.: A fitting algorithm for solving inverse problems of heat conduction. Int. J. Heat Mass Transf. **53**(9–10), 2123–2127 (2010)
10. Altintas, Y.: Manufacturing Automation: Metal Cutting Mechanics, Machine Tool Vibrations, and CNC Design. Cambridge University Press, Cambridge (2000)
11. Löhe, J., Zäh, M.F.: Fräsbearbeitung dünnwandiger Werkstücke—Simulation und Kompensation thermomechanischer Wechselwirkungen. ZWF **107**(7–8), 528–532 (2012)
12. Warnecke, G.: Fertigungstechnische Berichte—Band 2—Spanbildung bei metallischen Werkstoffen. Technischer Verlag Resch KG, Gräfelfing b. München (1974)
13. Dawson, P.R., Malkin, S.: Inclined moving heat source model for calculating metal cutting temperatures. J. Eng. Ind. **106**, 179–186 (1984)
14. Lazoglu, I., Ulutan, D., Dinc, C.: 3D temperature fields in machining. In: Procedings of the 3rd International CIRP HPC Conference. Dublin (2008)
15. Siebertz, K., Van Bebber, D., Hochkirchen, T.: Statistische Versuchsplanung—Design of Experiments (DoE). Springer, Berlin (2010)
16. Denkena, B., Tönshoff, H.K.: Spanen—Grundlagen, 3rd edn. Springer, Berling Heidelberg (2011)
17. König, W., Essel, K., Witte, L.: Spezifische Schnittkraftwerte für die Zerspanung metallischer Werkstoffe. Verlag Stahleisen MBH, Düsseldorf (1982)
18. Wan, X.J., Hua, L., Wang, X.F., Peng, Q.Z., Qin, X.P.: An error control approach to tool path adjustment conforming to the deformation of thin-walled workpiece. Int. J. Mach. Tools Manuf. **51**(3), 221–229 (2008)
19. Ratchev, S., Liu, S., Huang, W., Becker, A.: An advanced FEA based force induced error compensation strategy in milling. Int. J. Mach. Tools Manuf. **46**, 542–551 (2006)

20. Chen, W., Xue, J., Tang, D., Chen, H., Qu, S.: Deformation prediction and error compensation in multilayer milling processes for thin-walled parts. Int. J. Mach. Tools Manuf. **49**, 859–864 (2009)
21. Wimmer, S., Zäh, M.F.: Fräsbearbeitung dünnwandiger Werkstücke—Prognose und Kompensation von Formabweichungen. Werkstattstechnik online 2016(9), submitted and accepted (2016)

Modeling, Simulation and Compensation of Thermal Effects in Gear Hobbing

M. Beutner, I. Kadashevich, B. Karpuschewski and T. Halle

Abstract Gear hobbing of large ring gears is industrially performed using coolant, assuming that dry hobbing would cause geometrical deviations. To achieve the economic benefits of dry hobbing, the process and its influences have to be determined to ensure high gear quality. Temperature measurements of gears were conducted to detect influences of the cutting parameters on the workpiece quality. A chip formation simulation revealed the correlation of the heat flux into the workpiece and the chip volume. Furthermore, the hobbing process is kinematically described using quaternion algebra. In this context the gear is represented by the dexel modelling technique. Coupling the model with a finite element simulation, the simulation framework is able to calculate deviations of the gears induced by the thermo-mechanical load during gear hobbing.

1 Introduction

The growing market of wind turbines, but even so cranes and tunnel boring machines, requires large ring gears for rotational motions. These gears are manufactured by hobbing which is a multi-contact process with defined cutting edges. Compared to automobile transmission gears, the quality needed for large ring gears is lower. Thus, hobbing can be conducted as a final process step.

Due to the behavior of cutting processes, heat occurs and partly dissipates into the workpiece. This causes thermal deviations which are amplified by the thin shape of the ring gears. Industrially this problem is solved using coolant to reduce the heat in the cutting zone. Therefore, the aim of the project is to establish dry hobbing to avoid the disadvantage of coolant and to fulfil modern environmental demands.

M. Beutner (✉) · I. Kadashevich · B. Karpuschewski · T. Halle
Otto-von-Guericke University of Magdeburg, Magdeburg, Germany
e-mail: martin.beutner@ovgu.de

D. Biermann and F. Hollmann (eds.), *Thermal Effects in Complex Machining Processes*, Lecture Notes in Production Engineering,
DOI 10.1007/978-3-319-57120-1_15

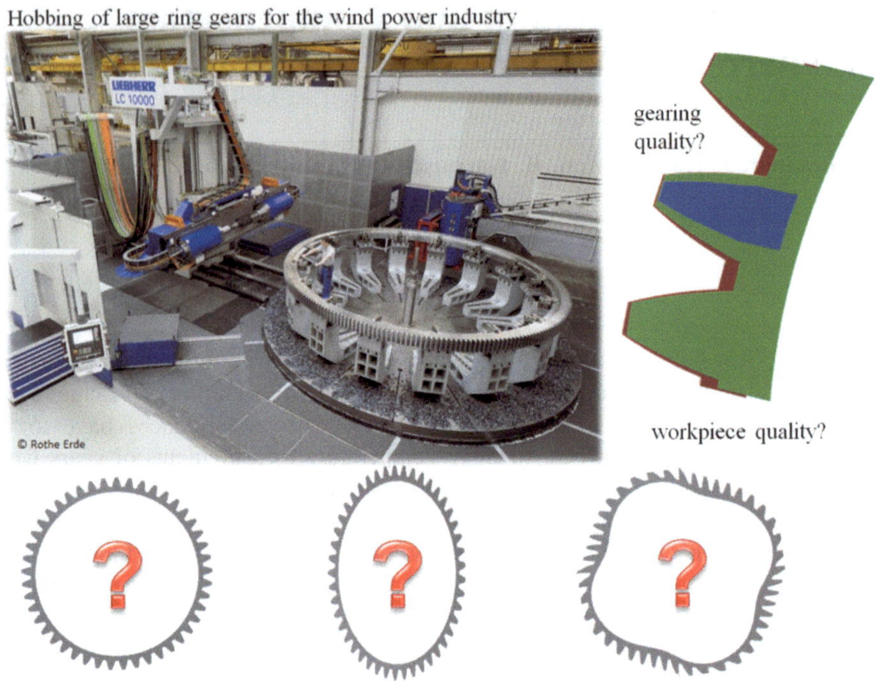

Fig. 1 Geometrical deviations caused by gear hobbing

As illustrated in Fig. 1, it is assumed that the dissipated heat causes geometrical deviations which are unknown for large ring gears. Not only the workpiece quality, but also the wear of the hob, is a reason to avoid dry hobbing for large gears in industry.

2 Hobbing of Large Ring Gears

Hobbing is the most dominant manufacturing process to produce external spur gears [1–4]. Kinematically hobbing is based on the principal of a worm drive, with a hob representing the worm and the gear as a worm gear. The advantage of this tool is the possibility to manufacture workpieces with different number of teeth, helix angles and profile displacements [5]. For different modulus (m_n) diverse hobs and hob concepts are available. As illustrated in the upper left corner of Fig. 2 for small gears with a modulus up to 4 mm full material hobs are used consisting of powder metallurgical high speed steel or cemented tungsten carbide. The multiple teeth of a hob are generated by geometrical superposition of the enveloping helix (worm) and the number of gashes (axial flutes). This tool concept changes for large modulus gears, like ring gears. In this context, the tool base body is made of a

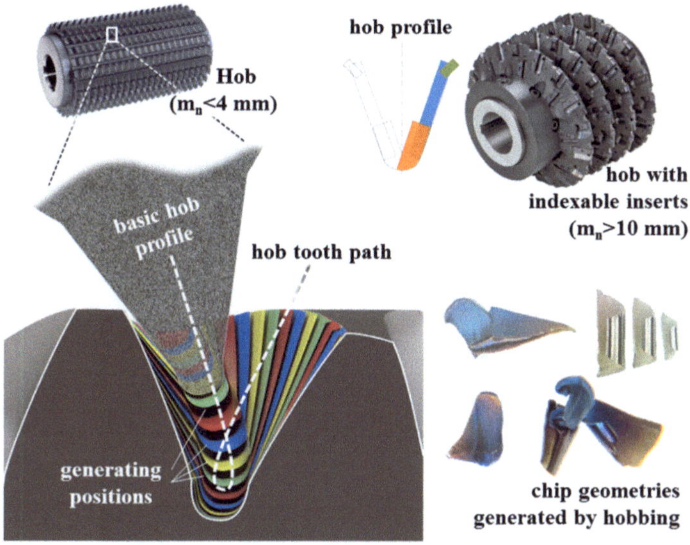

Fig. 2 Specific conditions of gear hobbing

quenched and tempered steel and the cutting teeth are represented by different consecutively applied indexable inserts. Thus, the hob profile is achieved by conjunction of the cutting edge of the inserts (Fig. 2 orange, green and blue).

The straight flanks of the hob profile are not able to generate an involute gear gap without the kinematic motion, shown in Fig. 2 bottom left. In detail, the coupling of hob and workpiece is fixed by the ratio of number of starts (hob) and teeth of the workpiece. Due to this motion the chip geometries vary along the path of a tooth. In hobbing the different relative positions of a tooth are called generating positions (GP), see bottom of Fig. 3. Differing in chip thickness, length and volume, the chip formation causes a variable load on the tool. Hence, the wear of a hob tooth is hard to predict because it is affected by gearing parameters, substrate, coating and cutting parameters. Furthermore, the heat production and heat distribution is influenced, too. Compared to other cutting technologies, the quantification of the process values (like forces and temperatures), is hard to realize due to the continuous rotation of hob and workpiece. In hobbing several teeth are cutting the workpiece at the same time. Using a well-established analogy test, it is possible to separate the multiple teeth engagements down to a single hob tooth [4, 6, 7]. This so called fly cutting test is able to manufacture the entire gear geometry like a full hob, see Fig. 3.

As illustrated in Fig. 3, a single fly cutting tooth is generated by wire erosion. That way, the geometrical parameters, the coating and the manufacturing process are comparable. A single tooth is clamped in a holder and setup on a hobbing machine. Using a different kinematic, the single tooth takes place in the same generating positions as all teeth of the entire hob. Therefore, the load and wear is concentrated on

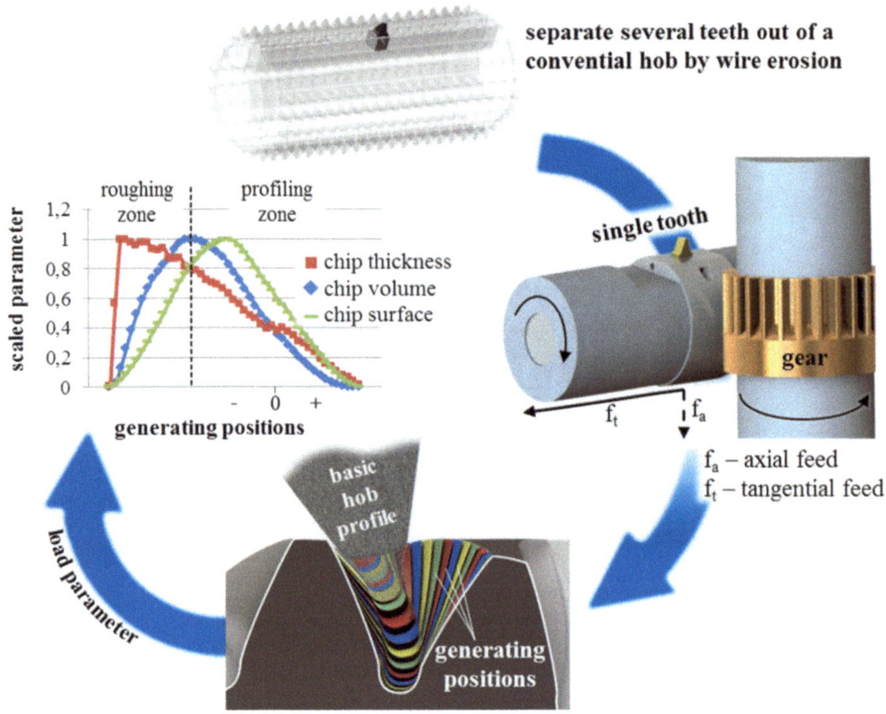

Fig. 3 Analogy test for gear hobbing

this single tooth. Beneficially, it is possible to save test time and material. Hereby the multiple teeth contacts are separated successively, which enables a detailed analysis of the single tooth engagements. A detailed analysis of the load (chip thickness, cutting length, etc.) is mostly done by an interpenetration calculation. Like illustrated in the diagram of Fig. 3, these load parameters pass through the GP. Depending on the case of gearing, the amount of the GP, the areas of roughing and finishing zone alters. In hobbing the classification of positive and negative GP is done by the shortest distance to the workpiece axis (GP 0), where the tooth is symmetrically positioned in the gap (middle of the profiling zone).

3 Temperature Measurement

For workpiece temperature measurement a new device was developed to fit the main challenges, reusability and suitability for rotational motions, see Fig. 4.

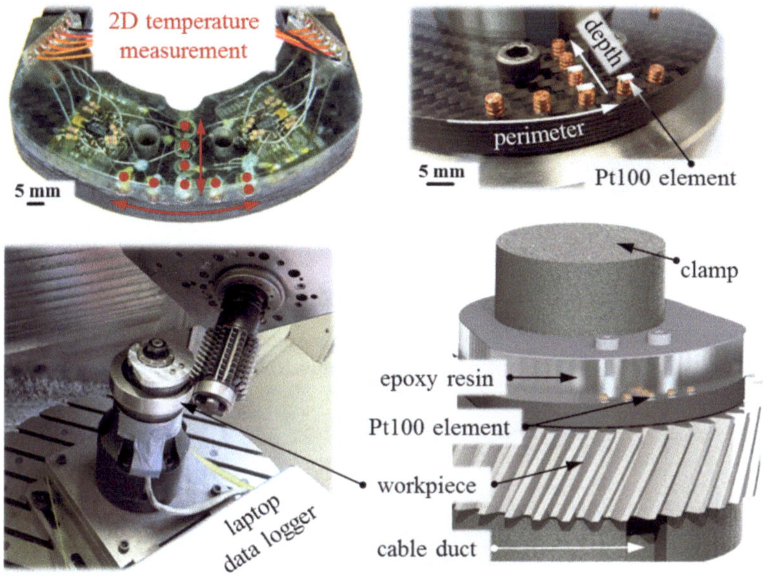

Fig. 4 Setup for temperature measurement on workpiece surface

Reusability was fulfilled using a carbon base layer with pressed in copper bolts. Carbon was chosen to thermally separate the bolts from each other to get a two dimensional temperature matrix of the workpiece surface, one in radial and one in perimeter direction, top right Fig. 4, the first one to measure the heat propagation into the depth of the gear, the other one to detect differences between the generation positions. The measurement itself was done by resistance thermometers (type PT 100 SMD 1206) which were applied on the top of the copper bolts. Epoxy resin was used to fix wiring and electronics. The data logger and laptop, both using battery power, were stored in an aluminum case on the workpiece table.

The entire setup was calibrated on a heatable copper plate. The heating system which was placed underneath the plate and the calibrated temperature sensor (Norma PT100 FT2) were linked together in a control loop to generate precise temperature values. Getting the real temperature of the copper plate, it is possible to create a calibration equation for each sensor.

Like mentioned before, one dimension of the measurement is in the depth direction of the gear. For all examined experiments the nearest sensor measured the maximal temperature, top part of Fig. 5. With increasing distance from the cutting zone (gap) the measured temperature decreases. By finishing the hobbing process, the gear got temperature balanced as every sensor curve asymptotically reaches the same value. Within the same experiment the results for the perimeter direction are on the same level. For cutting processes it is known that the temperature gradient is very high which results in a very fast decreasing temperature from the cutting zone [8, 9]. The closest sensor of the measurement device is located 2 mm away from the

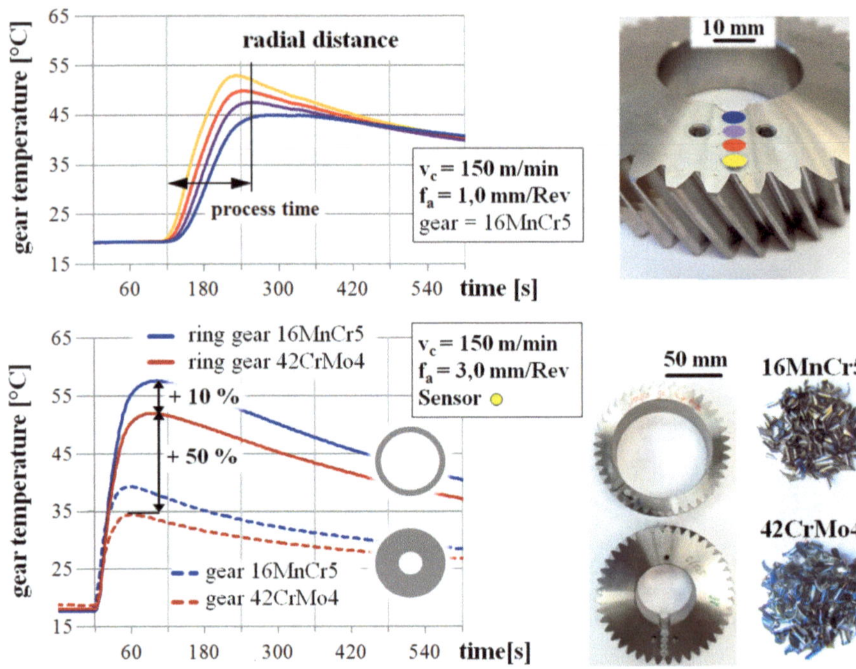

Fig. 5 Comparing gear temperatures for different geometries

cutting zone. In addition, the gear and the copper bolt act as a low pass filter which increases the inertia as well. This can be accepted because the focus of the present work is on the gear temperature itself, not on the temperature in the cutting zone.

The object of the research project is the distortion behavior of ring gears like those used in wind turbines. To examine the influence of the gear geometry an analogy ring gear and typical gear geometry are compared to each other, see Fig. 5. For both gear types the gearing stays constant. Considering the material removal, the ring gear loses 21% and the normal gear 12% of its weight. The change of the geometry results in a higher temperature on the workpiece surface. For both the case hardening steel (16MnCr5) and the quenched and tempered steel (42CrMo4) the temperature increases by nearly 50%. Keeping the boundary conditions constant, the experiment shows that a doubled material removal yields in an increase of the temperature of 50%, too.

Comparing the two steel types against each other, it is obvious that the temperature level of 42CrMo4 is lower. On the other hand the material strength is increased from 500 N/mm^2 (16MnCr5) to 750 N/mm^2 (42CrMo4). This leads to an increase of the torque which was recorded by the servotrace of the machine. Having a higher torque it is not comprehensible that the gear temperature decreases. So it is assumed that additional to the increase of the process force also the ratio of the heat input into the workpiece changed. Comparing the tempering colors of the chips this assumption was supported.

Further experiments for cutting parameters were conducted, see Fig. 6. In gear hobbing the main technological parameters are the cutting speed and axial feed. Being directly linked to the axial feed also the chip thickness alters: from 0.13 mm to 0.25 mm and to 0.30 mm. Therefore, the chip geometry, the chip volume and the amount of chips changes, but the removed material of a gap stays constant. By increasing the axial feed, the surface temperature of the gear decreases from $f_a = 1$ mm/rev to $f_a = 3$ mm/rev by 23% and from $f_a = 1$ mm/rev to $f_a = 5$ mm/rev by 32%. Similarly, the amount of chips which is an indication for the contacts of tool and workpiece decreases as well. Having 1809 chips for $f_a = 1$ mm/rev it is reduced to 656 for $f_a = 3$ mm/rev and 425 for $f_a = 5$ mm/rev. The ratio of chips is different compared to the temperature but the relation of the ratio is quite the same (81–87%) (Fig. 6).

A similar behavior of the temperature is detected by varying the cutting speed. Here, the temperature increases the most at low cutting speeds (from 50 to 100 m/min) like the gradient from 1 mm/rev to 3 mm/rev for the axial feed. Increasing both, axial feed and cutting speed, the gear temperature decreases as well as the gradient. That means, duplicating the cutting speed from 50 to 100 m/min yields in a temperature decrease of 25% but from 100 to 200 m/min by just 14.7%.

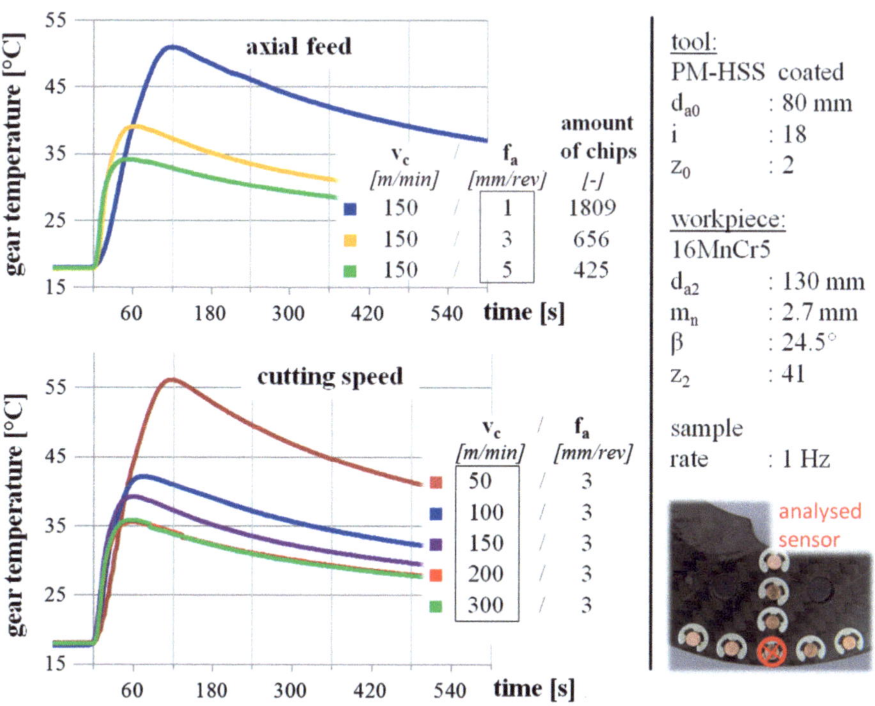

Fig. 6 Experimental results for cutting parameter variation

Reaching high cutting speeds (200 and 300 m/min), nearly the same temperatures were detected. In contrast to the axial feed the cutting speed does not influence the chip geometry and tool contacts, just the process time is shorter. Therefore, the chip formation and the frequency of tooth engagements are influenced. For 50 m/min every 16.7 ms a tooth cuts material from the workpiece, but for 300 m/min this time decreases to 2.7 ms. Hence, more heat is cut off because the time to disappear into the workpiece decreases (Fig. 6).

In addition to the temperature measurement, also the torque of the process was recorded. The machine torque of the drive spindle (hob) was extracted from the servotrace of the machine. In difference to the gear temperature, the torque increases with increasing axial feed from 1 mm/rev to 3 mm/rev by 137% and from 3 mm/rev to 5 mm/rev by 63%. The relation of the ratios is nearly the same with 117% (137–63%) for the torque and 114% for the gear temperature. But nevertheless, an increasing torque and a decreasing gear temperature are of reverse behavior. According to this reason, it is assumed that the energy of the process splits up in a different way by changing cutting parameters. Using the plotted values it is possible to calculate the necessary mechanical work of the process via torque and angle of rotation. That way the mechanical work for 3 mm/rev to 5 mm/rev is quite the same with 99 kJ. For 1 mm/rev it increases by 22% to 121 kJ. The identical work and the different workpiece temperatures for 3 and 5 mm/rev indicate the assumption as well (Fig. 7).

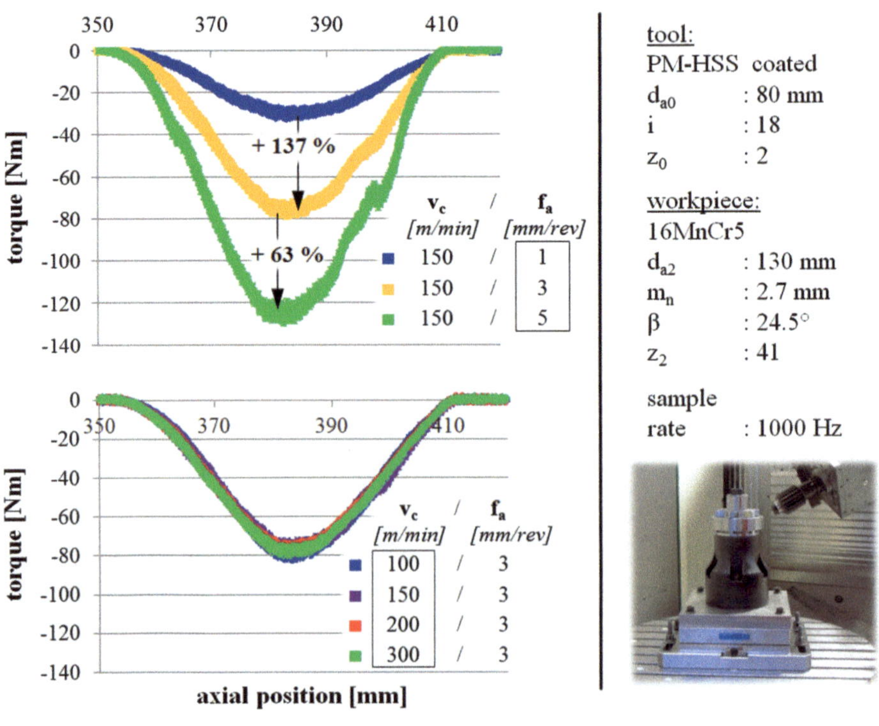

Fig. 7 Torque measurement for axial feed and cutting speed

4 Temperature Measurement in Industry

To validate the results of the explained experiments, the opportunity was given to measure the temperature of an industry large ring gear. The gearing parameters are listed in Table 1. Typical applications for these components (diameter larger than 3 m) are pitch and yaw drives for wind turbines.

To measure the temperature of the ring gear, ten sensors (again type PT 100 SMD 1206) were applied in the workpiece. The sensors were mounted on a carbon carrier surrounded by thermal grease and a brass capsule. According to this preparation the sensor was electrically encapsulated. The positioning was chosen to get information in three dimensions, along the width of the gear, along the axial pitch and in the direction of the cutting depth, see Fig. 8 bottom right.

To simplify the adjustment the sensors were spread on different teeth of the gear. Again, the whole measurement system was power supplied by battery to be independent from the circular motion of the workpiece.

The experiments were conducted using a hob with indexable inserts. To cut the gap accurately between the sensors the relative position of the hob and the workpiece was crucial. It was realized by an angular adjustment of the workpiece. Cutting gears with a large modulus, the load of the tool would be too high for one cut that is why the gear is cut in two steps with different cutting depths, see Table 1.

The experiments were conducted using a hob with indexable inserts. To cut the gap accurately between the sensors the relative position of the hob and the workpiece was crucial. It was realized by an angular adjustment of the workpiece. Cutting gears with a large modulus, the load of the tool would be too high for one cut that is why the gear is cut in two steps with different cutting depths, see Table 1.

The main aim of this experiment was the comparison of dry and wet hobbing using water based emulsion as coolant. It is obvious that the measured temperature

Table 1 Industry ring gear parameters

Parameter		Unit
Modulus	22	mm
Tip diameter	3119	mm
Number of teeth	139	–
Helix angle	0	°
Width of the gear	151.8	mm
Material	42 CrMo4 (*AISI 4140*)	–
Cutting speed	130	m/min
Axial feed	3.0	mm/Rev
Max. chip thickness	0.24	mm
Cutting depth	49.5 (*33/16.5*)	mm
Cutting strategy	Up-cut hobbing	–

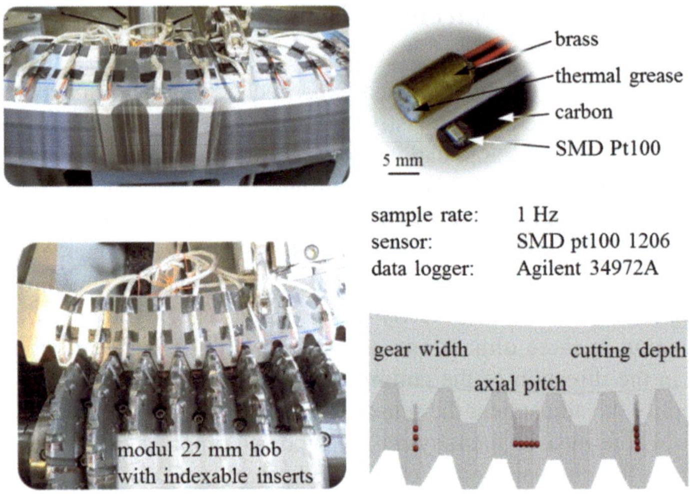

Fig. 8 Temperature measurement of large ring gear in industry

is significant higher for dry cutting, Fig. 9. For both experiments all sensors are close together which means that the gears were heated up homogenously. Looking at the peak temperature of both experiments, the propagation of the temperature in dry hobbing is larger. The actively chilled coolant inhibits the heat dissipation but for dry cutting the effects of different chip geometries (GP) becomes visible. The coolant is also responsible for the temperature drop at the beginning of the wet hobbing process. Furthermore the hob gets closer to the sensors for the second cut, that way the temperature increases compared to the first cut. But in wet cutting this effect is inhibited by the coolant, as it is the same level as for the first cut. The magnification of the single graph shows the sample rate of 1 Hz and the rotation of the workpiece.

Analyzing the different directions of the sensors, more details become evident. The rotations of the workpiece can be identified as well, see Fig. 10. Comparing the temperature in the direction of the cutting depth, the sensor at the tip of the tooth (blue) detected the highest thermal load. Increasing the cutting depth the plotted temperature decreases because more heat can dissipate into the workpiece. More interesting are the graphs in the direction of the axial pitch because the chips (GP) vary in that direction as illustrated in Fig. 2. As a characteristic the leading flank of the workpiece is cut from tip to root and the trailing flank is cut vice versa. Here, most of the workpiece material (large chip thicknesses) is cut on the part of the leading flank (rouging zone), see Fig. 3. After cutting most of the material out of the gap the involute geometry is approximated by the GP of the profiling zone. They are characterized by decreasing chip thicknesses and chip volumes as illustrated in the diagram of Fig. 3. This behavior corresponds to the plotted graphs of the sensors. Using Matlab to visualize the temperature field this effect becomes more obvious.

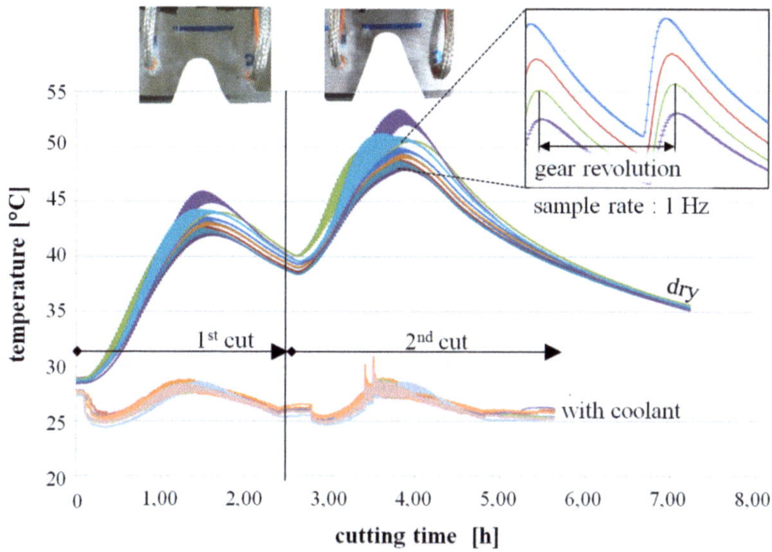

Fig. 9 Temperature plot of all sensors over total process time

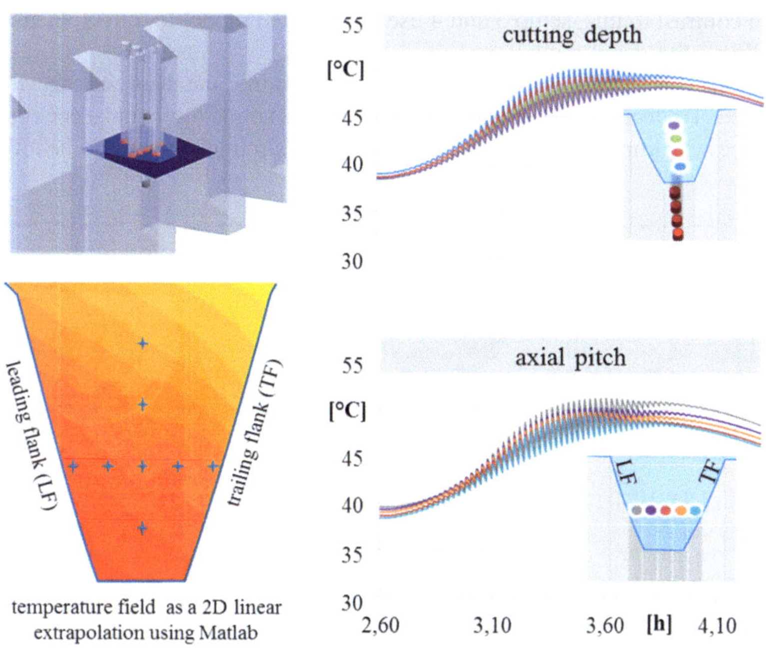

Fig. 10 Temperature evolution in two directions of the tooth

5 Simulation Setup

To give an overview about the holistic simulation approach, the different steps are briefly introduced, see Fig. 11. The first step (setup 1) in the simulation approach was the construction of the hobbing and fly cutting process in CAD environment as a fully parametrical model. As a result chip volumes were calculated. Moreover, this model served as input for the FE chip formation simulation which was done with DEFORM 3D. For different generation positions, the chip formation, chip temperature and cutting force were analyzed. Using MSC Marc Mentat for heat propagation the results of setup 1 were taken in account [10].

Setup 2 extends the focus from the chip scale to a whole gap of the gear [11, 12]. The full gear and the entire hobbing process were described with a dexel model to generate the geometry. As a substitution the removed material of the dexel is replaced by heat sources to apply thermal load to the workpiece. After several chips the model is exchanged to the FE solver (Abaqus) to calculate the deformation of the gear. This loop exchange is conducted until the gear is fully cut. Setups 3 and 4 are representing a fully coupled FE-Dexel model with heat transfer and thermo-mechanical deformations. The initial setups 1 and 2 are build backgrounds for the setups 3 and 4. Unfortunately the simulation of single chips is very time consuming and does not allow to simulate the full hobbing process in a reasonable amount of time. In contrast to this, setup 3 and 4 use a simplified model for effective numerical simulation—the dexel model.

The FE calculation was performed to determine the heat impact into the workpiece. To decrease the calculation time, only a single tooth and a section of the workpiece was implemented. Comparing forces, chip thicknesses and heat flows, a

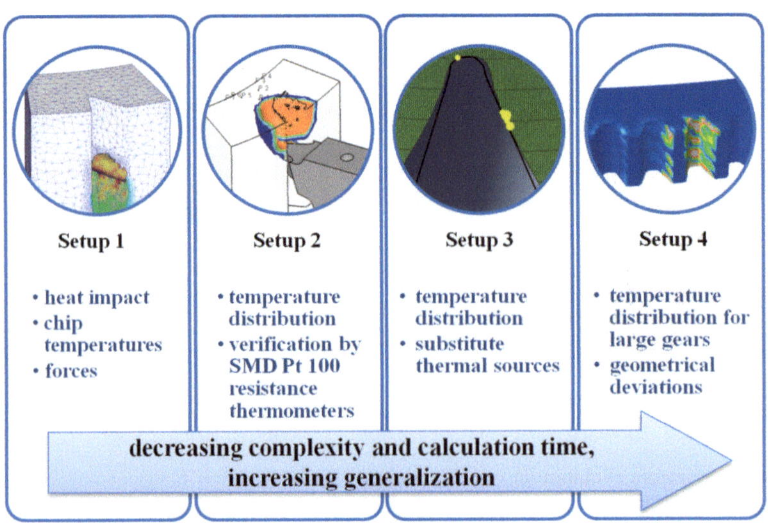

Fig. 11 Simulation approach

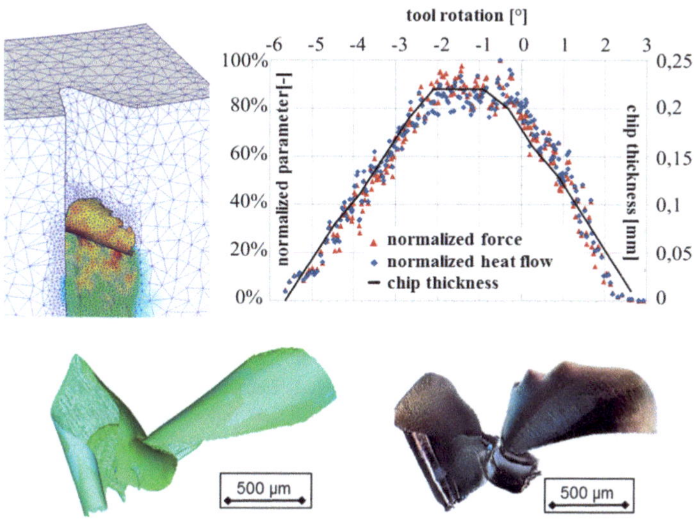

Fig. 12 Correlation of chip thickness and heat flow

correlation of chip thickness and heat flow into the workpiece was found, see Fig. 12. Correspondingly, this relation was used to create heat substitution sources within the dexel model, which means that the thermal load put on the workpiece surface increases with increasing chip thickness. Comparing the shape of the simulated and a real chip, a high correlation is visible. This is beneficial because the chip geometry is quite complex compared to milling or turning.

6 Simulation Backgrounds—Dexel

A new model for geometrical description of the workpiece is introduced. In addition to the classic dexel representation [13, 14], which only describes the body surface, the additional temperature data as well as deformations were added to the model description. This enables a transfer of data between geometrical dexel structure and the FE results.

The mathematical description of the modified dexels is simplified by the advantages of the polar coordinate systems (axial symmetry) (Fig. 13).

7 Simulation Backgrounds—Kinematic

To describe the hobbing process, quaternions were used instead of rotational matrixes. Consequently a faster numerical calculation is possible.

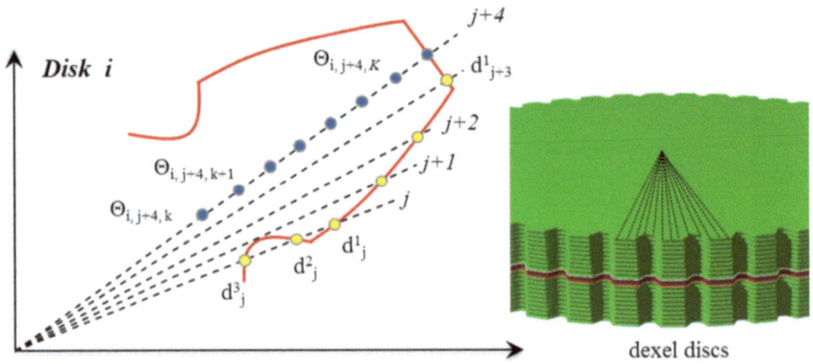

Fig. 13 Discretization model for one disk i (i = 1…I) Each point d_{mj} (m = 1, 2, 3) describes the workpiece shape and is constructed by j rays (j = 1…J) outcoming from the disk center; the discretization of the temperature field goes along the each j ray (k = 1…K) constructing ϕ_{ijk}

The description of tool/gear and geometry/kinematic is started from the partially linear discretization of the cutting tooth geometry and propagation of this geometry along the rotational axis of the hob by means of quaternion rotation and shift. As mentioned the workpiece is modelled before the process as a set of dexel disks. For each disk the contact of the cutting tool (one or more teeth) with the dexels is calculated separately. This allows parallelization within the dexel model by a number of dexel disks. To simplify the kinematic description, the intermediate steps of transformation are omitted. Equation (1) gives the position of the hob for each time increment (rotation angle) in the coordinate system of the unmoved part. For the simulation the tool moves around the workpiece to simplify the calculation. To represent simulated results, the workpiece can be rotated with known speed.

$$\xi = T \bullet [d + f_{ax}] \bullet T^* + R \bullet f(x) \bullet R^* \quad (1)$$

$$T = q(\omega_2);\ R = q(\omega_2) \bullet q(\zeta) \bullet q(\omega)$$
$$R^* = q^*(\omega) \bullet q(\zeta)^* \bullet q(\omega 2)^*$$

Equation (1) has following parameters: ω—tool rotation angle; ω_2—the workpiece rotation; quaternions T, R give the direction and amount of rotation for the tool and the workpiece; ζ—the tilt angle, d—the distance between the hob and gear center, f_{ax}—axial feed, f(x) tooth profile; $q(\omega_2)$, q (ζ), q(ω)—rotation quaternions. All variables and operations in Eq. 1 are quaternions. The * symbol notes the conjugate quaternions and the • represents a multiplications of quaternions. Angle ω_2 is linked to the tool rotation ω depending on the number of starts and the number of teeth. Equation (1) is numerically effective and can be easily implemented in any programming code. In Fig. 14 the gear geometry, tooth cutting profile, as well as calculated traces on the dexel plane are illustrated. A typical dexel discretization of the workpiece (every disk) with a diameter of 130 mm is of about 2000 dexel (elements) in perimeter direction. The tooth profile is given as a continuous curve

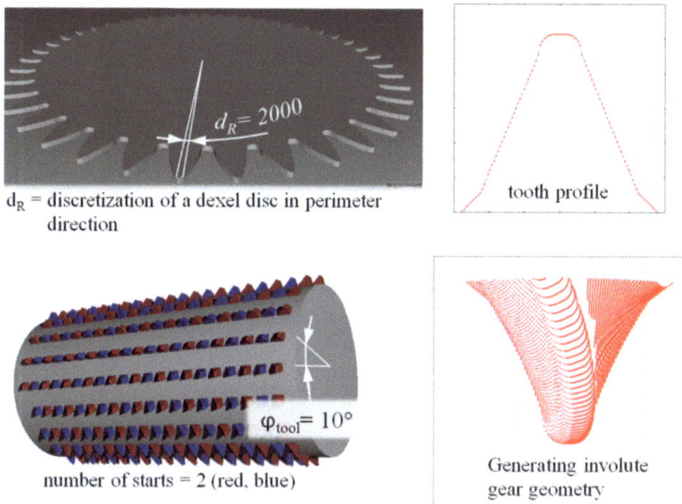

Fig. 14 Dexel-kinematics and the model geometry

and discretized with proper resolution as shown in the Fig. 14. The full kinematic is propagated by Eq. (1) for all teeth of the gear.

8 Simulation Backgrounds—FE-Dexel Coupling

The coupled FEM-dexel algorithm is a set of subroutines and controlling scripts developed in Institut für Werkstoff- und Fügetechnik at University Magdeburg which is described below:

For the **dexel model** following utilities (C++ code) are used:

- load, save and generating single hobbing step (controlled by tool rotation angle) for the dexel array,
- prepare and save calculation step for FE system based on the dexel model and the previous temperature distribution (can be loaded as a node value or imported from the Abaqus database from pervious step),
- import from Abaqus report to the deformed dexel model.

For Abaqus:

- Python script to start problem, process result and export temperature and deformations for every node,
- some scripts for visualization.

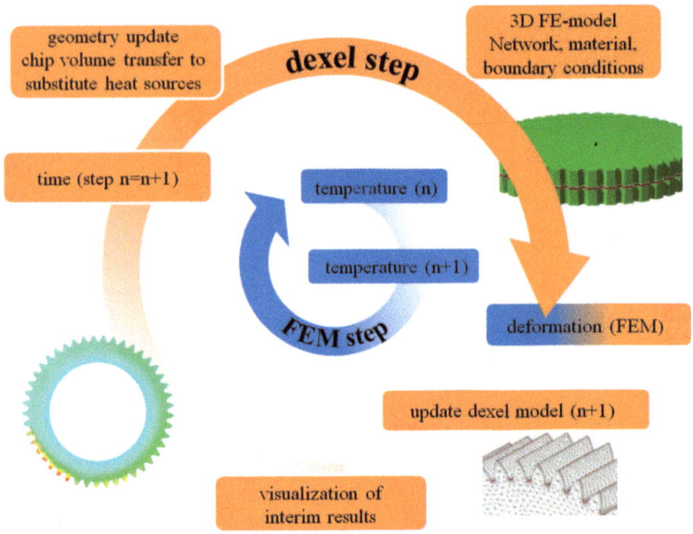

Fig. 15 Coupled dexel/FE model routine

The model of coupling is graphically represented in Fig. 15, where it is shown how the FE step is covered by many dexel procedures. The simulation can be started from an unmachined state or any known initial dexel geometry with $\omega = \omega_0$. This is necessary for hobbing with different cutting depths. The end of the simulation process can be automatically controlled and completed when one of the following conditions is realized: zero chip volume, max achieved ω, stabilized temperature of the workpiece.

9 FE-Dexel Coupling: Remarks/Benefits

The main benefit of the dexel approach is the quaternion description utilizing only one equation. The classical kinematic uses seven axis each is having a coordinate system [3]. During numerical simulations additional recalculations of matrixes/quaternions are not necessary. The equidistant rotation of the hob enables the usage of the pre-calculated quaternions. This results in a very fast implementation. The dexel model simulation is fully multiprocessing and parallelized. For example the numerical hobbing simulation of a the small gear (d_a = 130 mm) takes about 30 min on Intel i7-3632QM-2.20 GHz processor, with a dexel resolution of 1440 and 30 dexel slices and 1° tool motion resolution, this follows to 1.080.000 cuts/chips removal.

Using Abaqus, the main disadvantage is the time consuming loading of the input data which is only available in text format. In future, this part should be optimized.

10 Simulation Results

As it was shown in [12, 15], the amount of the removed material is proportional to the heat flux affecting the gear. Due to this assumption, the removed material is substituted by an external heat sources (yellow and blue dots), shown in Fig. 16.

Referring to the experimental results for large ring gear (Fig. 9), the simulation shows an acceptable agreement, see Fig. 17. The simulation results are averaged comparing to the measurement, while simulation was simplified to reduce calculation time.

Various simulations were performed for analogy gears, showing a good correlation in the temperature fields compared to experiments. Beside the heat transfer and the temperature, the deformation of the gear during cutting and after cooling is most interesting. Figure 18 shows typical thermal deformations in the dexel model for the coupled simulation.

The difference to the "ideal" profile is shown in Fig. 18. The tooth is bending to the left which causes an over cut due to the material extensions.

Compared to the simulation, the measurement of the ring gears showed a diverging behavior. The measurement was conducted matching the circularity of the inner bore of the gear before and after hobbing. Parallel to the process the surface temperature was recorded. As mentioned above, the largest gap occurred by varying the axial feed. To determine the repeatability of the process, the central point was done threefold. Within the diagram of Fig. 19 the gear temperature and the relative deviation was compared to each other.

Applying the same cutting parameters and therefore the same gear temperature, the measurement of the circularity reveals large deviations. Furthermore, the comparison of the axial feed variation shows inconsistent results. Only the

(a) (b)

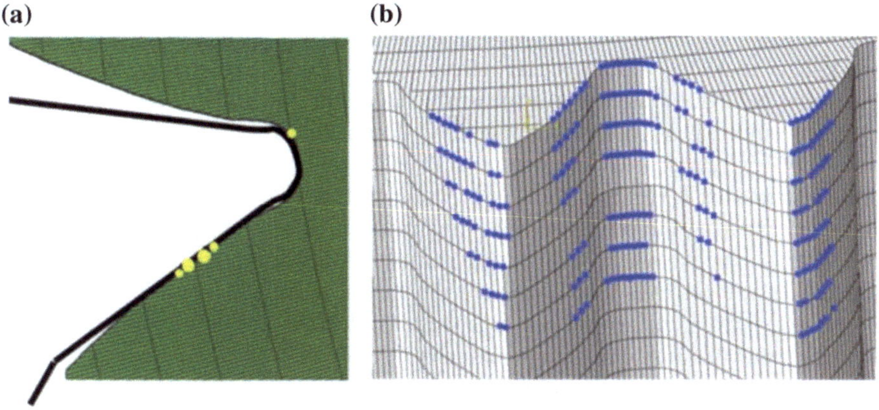

Fig. 16 External heat sources/FE Networks. **a** method of placing the external heat sources in positions, where chips have been removed; **b** the heat sources of one FE calculation step

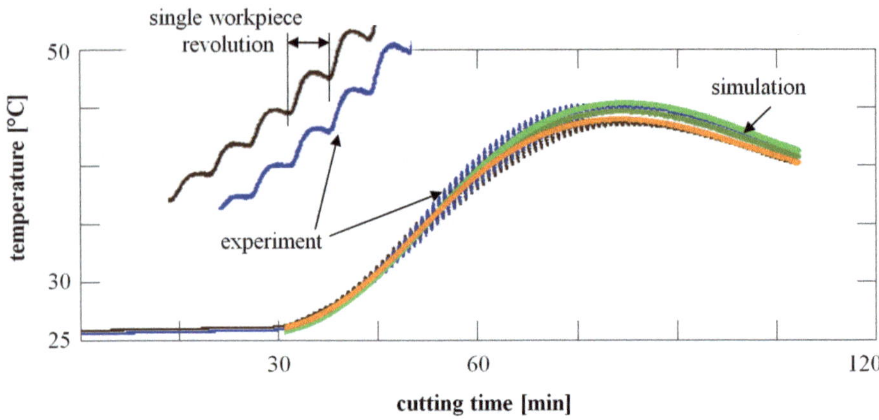

Fig. 17 Simulation for large ring gears (1651320 iterations, a 30 Gb,)

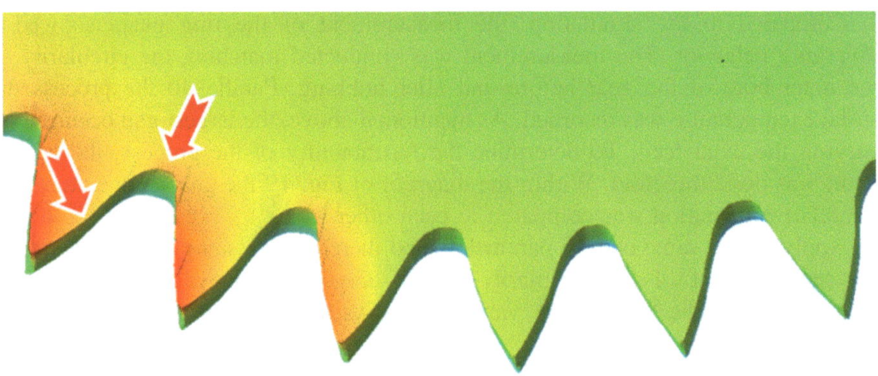

Fig. 18 Thermal deformations 1000x zoomed, the arrows show direction of deformations

workpiece material with a higher strength (42CrMo4) showed reliable results. A similar behavior was detected measuring the large ring gear in industry.

The resulting gear quality of wet and dry hobbing of the industry test is shown in Table 2. This gear type is designed as an IT 12 gear. The quality boundaries for that are illustrated in Table 2 as well. Only the circularity of the dry gear exceeds the limits. Comparing the results of both ring gears this is the only evident difference. All other geometry values and gearing parameters are acceptable. Furthermore, no tendencies are obvious and moreover the higher gear temperature (Δ = 40 °C) seems not to affect the geometry which was also detected for the analogy ring gears. Knowing that this was just a single industry test, it comes apparent that dry hobbing is possible for the demanded application.

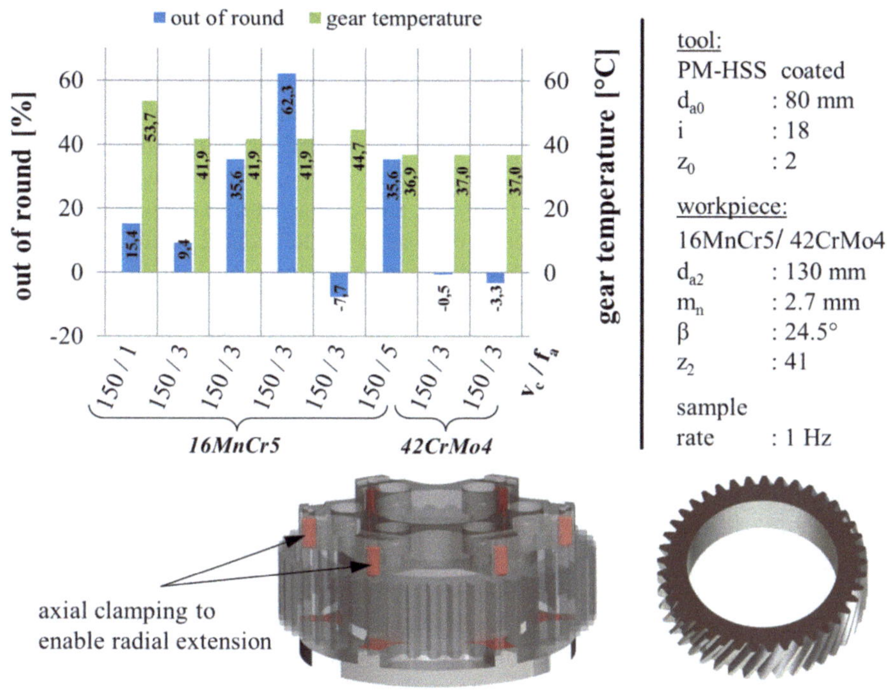

Fig. 19 Out of roundness compared to gear temperature

Table 2 Gear quality of a large ring gear

Parameter	Quality tolerance [μm]	With coolant [μm]	Dry [μm]
Flatness	<100	79.6	65.4
Circularity top	<100	93.9	196
Circularity bottom	<100	105.1	178.2
Total profile deviation	<320	169.5	220.6
Total helix deviation	<200	104.3	94.6
Total pitch deviation	<800	164	181
Pitch line runout	<450	128.3	122.7

11 Conclusion

The complex process of gear hobbing is widely used to produce gears in different sizes and geometries. Its characteristic of multiple simultaneous tooth engagements and variable chip geometries is complex and the rotation of tool and workpiece complicate research access as well. Normally hard finishing is the last step of the process chain for a gear but for large ring gears the same quality is not needed and hobbing is able to generate the required quality. Therefore, hobbing is the quality defining process. Assuming that dry hobbing will lead to geometrical deviations, the industry only manufactures with coolant.

As a first step of the project, an analogy test was used to determine chip geometries as a function of the GP to separate and to simplify the hobbing process. Concerning this, the analogy test was build up in a FE program. Thereby, a correlation of chip thickness and heat flux into the workpiece was found.

Leaving the scale of single chip formation and change over to the whole hobbing process, a different modelling technique was used to reduce calculation times. Using dexel model and quaternion algebra to generate the gear geometry, the result of the FE was implemented via substituting heat sources, which means that the chip thickness is replaced by a thermal load on the gear surface. This load causes deviations which were calculated with Abaqus. Furthermore, this coupled model exchanges the gear constantly, as described before. Being able to investigate a large ring gear in industry, it was possible to compare the data of experiment and simulation, which fit quite well. More temperature measurements were conducted with analogy ring gears to evaluate the gear temperature for different cutting parameters. In the end, increasing cutting speed or axial feed lead to decreasing gear temperatures.

Analyzing the geometrical results, mainly gear quality and circularity deviations were detected for both analogy ring gears and large ring gears in industry. Despite significant temperature differences the results showed that the temperature is not the dominating influence on the resulting geometry. Therefore, the assumption that dry hobbing will lead to geometrical deviations, which were made at the beginning of the project and which is widespread in industry, has to be reconsidered. Hence, an application of dry hobbing could be possible but other influences like tool wear and the gearing itself have to be considered. But statistically more investigations in industry should be conducted because the experience also revealed differences between ring gears in industry and research.

Acknowledgements The authors would like to thank the Deutsche Forschungsgemeinschaft (DFG) for funding this work within the Priority Program SPP1480.

References

1. Klocke, F., et al.: Fertigungsverfahren 1 - Drehen, Fräsen, Bohren. 8. Auflage. Springer, Berlin (2007)
2. Bouzakis, K.D., et al.: Manufacturing of cylindrical gears by generating cutting processes: A critical synthesis of analysis methods. CIRP Ann. Manuf. Technol. **2**(2), 676–696 (2008)
3. Sulzer, G.: Leistungssteigerung bei der Zylinderradherstellung durch genaue Erfassung der Zerspankinematik. Dissertation, TH Aachen (1973)
4. Hipke, M.: Wälzfräsen mit pulvermetallurgisch hergestelltem Schnellarbeitsstahl. Dissertation, University Magdeburg (2011)
5. Radzevich, S.P.: Dudley's Handbook of Practical Gear Design and Manufacture, 2nd edn. CRC Press (2012)

6. Bouzakis, K.D., Friderikos, O., Tsiafis, I.: FEM-supported simulation of chip formation and flow in gear hobbing of spur and helical gears. CIRP J. Manuf. Sci. Technol. **1**, 18–26 (2008)
7. Stuckenberg, A.: Vermeidung von Oberflächendefekten beim Wälzfräsen. Dissertation, RWTH Aachen (2014)
8. Müller, B.: Thermische Analyse des Zerspanens metallischer Werkstoffe bei hohen Schnittgeschindigkeiten. Dissertation, RWTH Aachen (2004)
9. Damaritürk, H.S.: Temperaturen und Wirkmechanismen beim Hochgeschwindigkeitsfräsen von Stahl, Technische Hochschule Darmstadt. Dissertation, Carl Hanser Verlag München Wien, Darmstadt (1990)
10. Beutner, M., Stark, S., Halle, T., Karpuschewski, B.: Messtechnische Erfassung und Simulation von thermischen Bearbeitungseinflüssen beim Wälzfräsen In: GETPRO 4. Kongress zur Getriebeproduktion, 05./06. März 2013, Würzburg, pp. 378–390 (2013)
11. Stark, S., et al.: Experimental and numerical determination of cutting forces and temperatures in gear hobbing. Key Eng. Mater. **504–506**, 1275–1280 (2012)
12. Stark, S., et al.: Heat flux and temperature distribution in gear hobbing operations. Procedia CIRP **8**, 456–461 (2013)
13. Eisenstöck, R.: Visualization of algebraic surfaces using the Dexel model. In: 9th European Workshop on Computational Geometry CG'93, pp. 48–51 (1993)
14. Hanson, A.J., et al.: Visualizing Quaternions—Series in Interactive 3D Technology. Morgan-Kaufmann/Elsevier (2005)
15. Kadashevich, I., Beutner, M., Karpuschewski, B., Halle, T.: A novel simulation approach to determine thermally induced geometric deviations in dry gear hobbing. Procedia CIRP **31**, 483–488 (2015)

Modelling and Simulation of Internal Traverse Grinding—From Micro-thermo-mechanical Mechanisms to Process Models

R. Holtermann, S. Schumann, A. Menzel and D. Biermann

Abstract This contribution deals with the modelling and simulation of Internal Traverse Grinding (ITG) using electroplated cubic Boron Nitride (cBN) wheels. This abrasive process fulfils the industrial demands for an extensive rate of material removal along with a good surface quality while minimising the number of manufacturing processes. To overcome one drawback of ITG in terms of a highly concentrated thermal load on the workpiece surface, a multi-scale simulation framework that combines different modelling methods in a hybrid framework is presented. In this context, a geometric-kinematic simulation is combined with a finite element analysis which focuses on the thermo-mechanical response of a single cBN grain being in contact with a hardened workpiece. Via a special scale-bridging scheme, the results of both the former simulations are used to compute a thermo-mechanical load compound acting as a boundary condition in a process-scale finite element model. The latter is then used to capture thermally induced geometrical errors during ITG and to develop compensation strategies accordingly.

R. Holtermann (✉) · A. Menzel
Institute of Mechanics (IM), TU Dortmund, Leonhard-Euler-Straße 5,
44227 Dortmund, Germany
e-mail: raphael.holtermann@udo.edu

A. Menzel
e-mail: andreas.menzel@udo.edu

S. Schumann · D. Biermann
Institute of Machining Technology (ISF), TU Dortmund, Baroper Straße 303,
44227 Dortmund, Germany
e-mail: schumann@isf.de

D. Biermann
e-mail: biermann@isf.de

A. Menzel
Division of Solid Mechanics, Lund University, P.O. Box 118, 22100 Lund, Sweden
e-mail: andreas.menzel@solid.lth.se

D. Biermann and F. Hollmann (eds.), *Thermal Effects in Complex Machining Processes*, Lecture Notes in Production Engineering,
DOI 10.1007/978-3-319-57120-1_16

1 Introduction

Internal Traverse Grinding (ITG) with electroplated cubic Boron Nitride (cBN) wheels has proven to be capable of the demanding material removal requirements in the context of internal hard-finishing. To be specific, a surface roughness of $R_z = 2\,\mu m$ at a material removal rate of $Q_w = 100\,mm^3/s$ is achievable, cf. [1, 2]. This is based on the usage of electroplated cBN grinding wheels that have a special geometry which consists of two separate zones, cf. Fig. 1. The roughing zone of the tool has a conical shape and therefore enables a high material removal rate. The finishing zone is cylindrical and can optionally be touch-dressed to improve the finished surface smoothness. During grinding, both workpiece and tool rotate in the same direction and the tool passes the workpiece in axial direction with a constant feed velocity v_{fa} after the radial stock removal $a_{e,tot}$ has been applied to the tool spindle. Hence, a combined manufacturing process is designed that allows for both high material removal rates and good finishing quality in just one axial stroke. One drawback of ITG is the high thermal load on the workpiece, causing geometrical errors of the finished part on the one hand, and phase transformations resulting in e.g. white layers on the workpiece surface, cf. [3] on the other. The thermal influence on the complex meso-scale chip formation resulting from ITG can examplarily be observed in Fig. 2a, where an etched cross-section of a segmented chip resulting from the present abrasive process is depicted.

1.1 Simulation Framework

To capture the complex thermo-mechanical mechanisms of the abrasive process, a multi-scale simulation framework has been proposed and developed, aiming at a realistic, but still computationally efficient modelling of ITG. Furthermore, the frame-

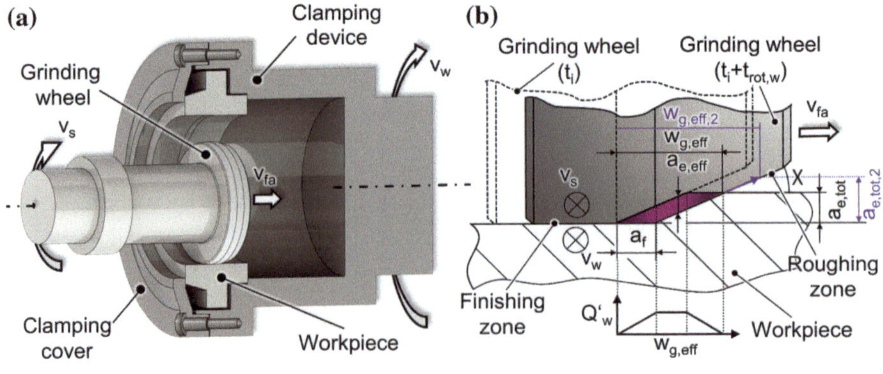

Fig. 1 Fundamentals of ITG: **a** Process kinematics; **b** Cross-section of the undeformed chip and resulting specific removal rate Q'_w along the effective width of the grinding wheel $w_{g,eff}$

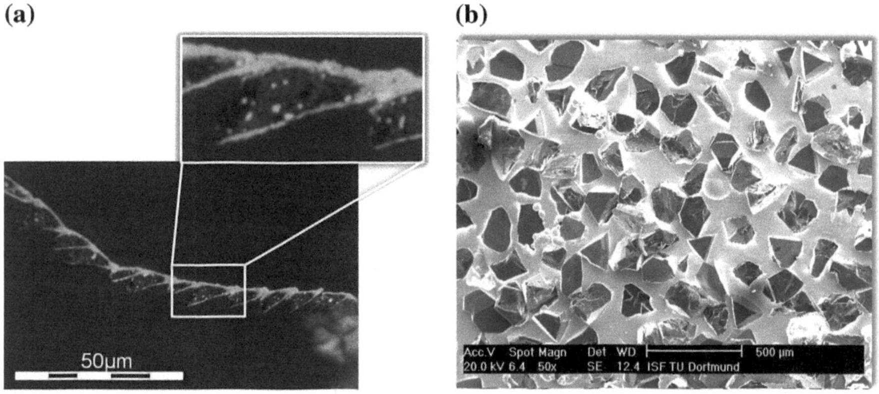

Fig. 2 Microscopic analyses: **a** Ground and etched chip obtained from experiments. The *white areas* imply regions of austenite-martensite phase transformation, commonly referred to as white layers (taken from [4] with kind permission); **b** SEM micrographs of a grinding wheel surface with single layer electroplated abrasive grit of grain size B181

work was designed to target the reduction of thermally induced geometric errors of the finished workpiece.

The simulation framework consists of three major components: First, a kinematic simulation which models the grinding wheel surface, see Fig. 2b, based on experimentally determined measurements as well as on subsequent numerical analysis, is used to calculate the transient penetration history of every grain intersecting with the workpiece, cf. Sect. 3. Secondly, a plane-strain, h-adaptive finite element model incorporating elasto-plastic hardening, thermal softening as well as ductile damage is used to model the proximity of one single cBN grain during grinding and to capture the complex thermo-mechanical material feedback on the meso-scale; compare Sect. 4.1. For the third component of the framework, the results of the preceding two simulations are—by application of a special scale-bridging scheme, see Sect. 4.2—combined into a macro-scale thermo-mechanical load compound. This load is then applied as a boundary condition in the process model, presented in Sect. 5. This process model realistically simulates the thermo-mechanical workpiece response on a macro-scale and therefore enables the development of compensation strategies to improve the manufacturing accuracy of the investigated abrasive process by minimising the geometrical error of the finished part.

2 Experiments and Thermo-Mechanical Loading During ITG

In order to calibrate and validate the components of the simulation framework, experimental investigations were carried out. Figure 3 shows the experimental setup within an internal grinding machine IC400 by Danobat Overbeck. The electroplated

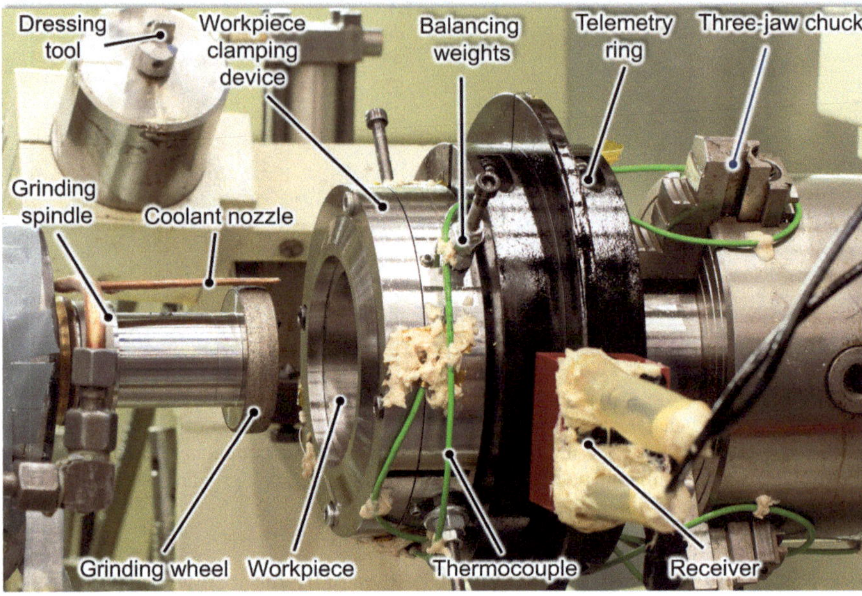

Fig. 3 Illustration of the experimental setup

cBN grinding wheel with its aggressive topography and large chip space ensures high-speed grinding conditions with high achievable material removal rates during ITG. Therefore, the grinding grains have high protrusion heights and are strongly bonded in a nickel layer. The grinding wheels manufactured by August Rueggeberg GmbH & Co. KG are mounted in a grinding spindle, which is attached to a three-component force dynamometer provided by Kistler Instrumente AG to measure the mechanical load onto the workpiece during the process. The tools are dynamically and finely balanced at the nominal speed achieving a high balancing quality $q_{\mathrm{bal}} \leq$ G2.5, which is essential when introducing high-speed grinding conditions. To improve the surface quality of the workpiece, the finishing zone was touch-dressed by a single diamond representing laboratory conditions. For cooling and for reduction of friction, grinding oil was used, the temperature of the latter was measured prior to and after the process.

The workpiece is solely axially clamped in a device depicted in Fig. 1a. The device is attached to a three-jaw chuck of the workpiece spindle of the internal grinding machine. The workpiece has external holes in which thermocouples of Type K with a diameter of $d = 0.25$ mm are positioned. To ensure a good heat conduction between the workpiece and the thermocouples, heat conducting paste was applied. A telemetry made by Manner Sensortelemetrie GmbH was used to transfer the measuring data from the rotational workpiece system to the measuring computer. Therefore, the thermal loading onto the workpiece can be measured during ITG.

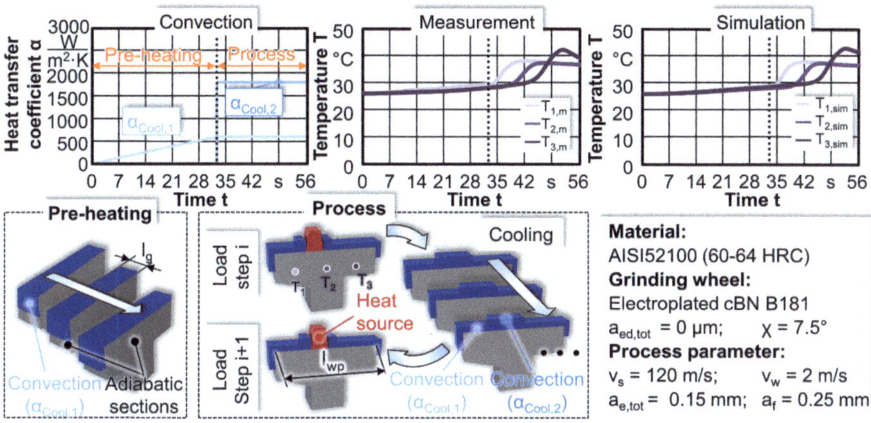

Fig. 4 Finite element model to simulate the thermal load during ITG—basic structure, pre-heating and comparison between simulation and measurement at different thermocouple positions T_1, T_2 and T_3

In the bottom section of Fig. 4, a finite element model is shown, which is used to simulate the thermal loading during ITG. It is fundamentally motivated by the classical Carslaw and Jaeger heat source theories [5, 6]. Nevertheless, it is adapted to the time consuming calculation of internal grinding and the constraints of the internal grinding machine. For detailed descriptions of the finite element model used, the reader is referred to [7–9]. The workpiece is represented by a volume segment, which is bounded in axial direction by the length of the workpiece and in circumferential direction by the contact length l_g. This value can be calculated according to [10] as

$$l_g = \sqrt{a_{\mathrm{e,eff,max}}\, d_{\mathrm{s,eff}}}, \tag{1}$$

where $a_{\mathrm{e,eff,max}}$ is the maximum effective depth of cut and $d_{\mathrm{s,eff}}$ the equivalent diameter. The latter can be determined by

$$d_{\mathrm{s,eff}} = \frac{d_s\, d_w}{d_s - d_w}, \tag{2}$$

where d_s is the diameter of the grinding wheel and d_w is the diameter of the workpiece. The coolant is represented by convective boundary conditions. Related to the process kinematics, the FE model is loaded by a heat source with the heat flux density $\mathbf{q}$, when the grinding wheel makes contact with the workpiece. Afterwards, only convective boundary conditions are applied. After one revolution of the workpiece, the volume segment is loaded again by moving the heat source in axial direction with the axial feed.

Prior to the process phase, the rotating grinding wheel is swirling the coolant within the grinding machine. In addition to this, the tool approaches the turning workpiece and a high-pressure pump is switched on. This results in a pre-heating of the workpiece regarding the higher temperature of the coolant in comparison to the workpiece. In order to model the described pre-heating of the workpiece, the heat transfer coefficient $\alpha_{\text{Cool},1}$ is increased stepwise, matching measured and simulated temperature curves prior to the process phase, cf. Fig. 4 (top: convection, measurement and simulation) and [7, 9].

The magnitude of the heat flux density which is transported into the workpiece can be estimated according to the theory of Lowin [11] as

$$||\mathbf{q}|| = \frac{F_\text{t}\, v_\text{s}\, K_\text{w}\, K_\text{v}}{A_\text{con}} = \frac{F_\text{t}\, v_\text{s}\, K_\text{w}\, K_\text{v}}{w_\text{g,eff}\; l_\text{g}}, \tag{3}$$

where F_t is the tangential force component magnitude and v_s is the grinding wheel velocity magnitude. K_v is the heat transformation factor which defines the amount of mechanical energy which is transformed into heat and is set to 1 for the presented work. A_con is the area of the contact zone and can be achieved with a multiplication of the effective width of the grinding wheel $w_\text{g,eff}$ and the contact length l_g. The heat dividing factor K_w defines the heat which enters the workpiece and is used to calibrate the heat input. For this purpose, in the presented work this factor was deduced with a linear optimisation matching the measured and the simulated temperatures. In the top section of Fig. 4, it can be clearly observed that there is a good agreement between the measurements and the simulated temperature curves after the calibration.

In order to simulate ITG more precisely, a modified process power theory of Lowin was developed with two-dimensional [7, 8] and three-dimensional [9] characteristics. In Fig. 5a, the specific tangential force and the specific material removal rate along the width of the grinding wheel are shown. Under the simplification that higher specific material removal rates give rise to the forces during ITG, a local formulation of the tangential forces can be used, which are derived according to the

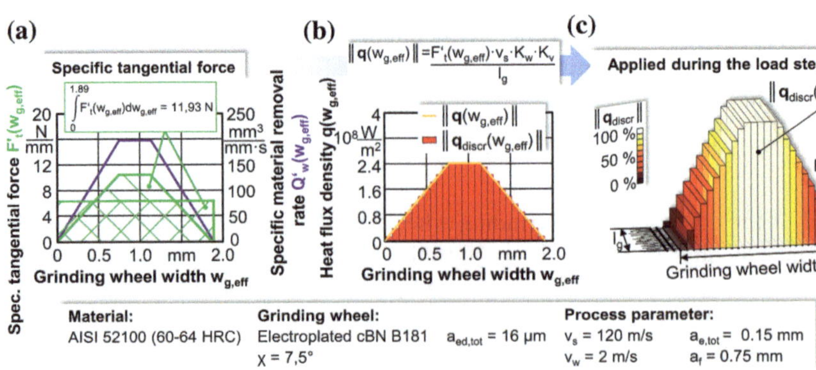

Fig. 5 Derivation of the thermal load for the FE model during ITG: **a** Distribution of the tangential force; **b** Derivation of the modified process power theory; **c** Estimation of the local thermal load $||\mathbf{q}_\text{discr}(w_\text{g,eff}, l_\text{g})||$ regarding the discretisation of a FE model

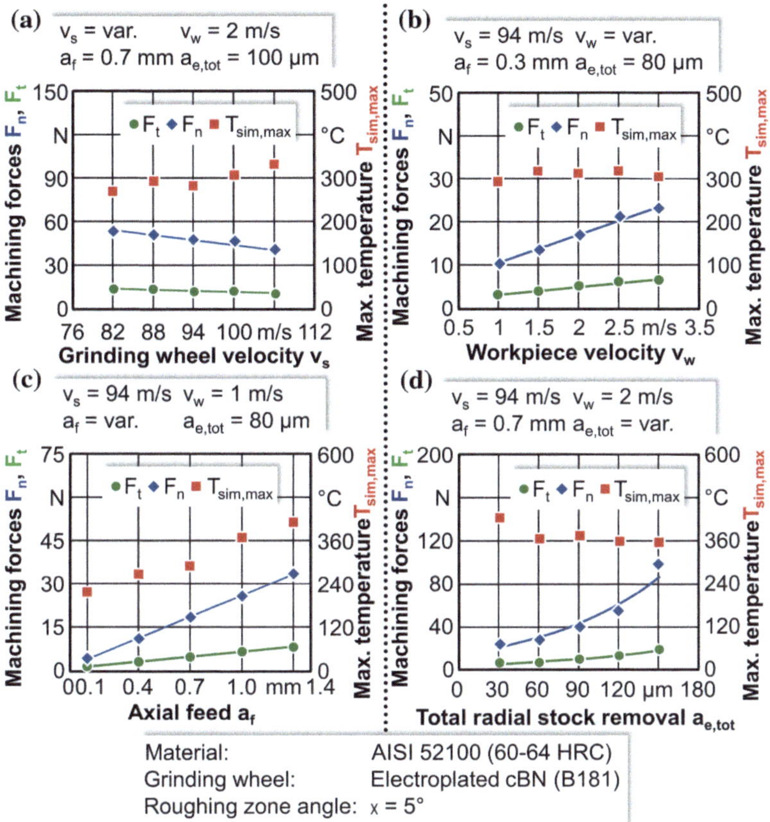

Fig. 6 Influences of the process parameters on the thermo-mechanical load of ITG. **a** Grinding wheel velocity; **b** Workpiece velocity; **c** Axial feed; **d** Total radial stock removal

cross-section of the undeformed chip during ITG. Integrating this formulation in a modified variant of the process power theory of Lowin leads to a local thermal load distribution $||\mathbf{q}(w_{\mathrm{g,eff}})||$, see Fig. 5b. The discrete formulation of the thermal load distribution $||\mathbf{q}_{\mathrm{discr}}(w_{\mathrm{g,eff}})||$, respectively using an uniform distribution along the contact length l_{g} $||\mathbf{q}_{\mathrm{discr}}(w_{\mathrm{g,eff}}, l_{\mathrm{g}})||$, can be applied for thermal FE simulations [7, 8], cf. Fig. 5c.

Using experimental investigations and the presented simulation framework, the ITG can be analysed in detail. In Fig. 6 the influences of the process parameters of ITG on the thermo-mechanical load are shown. The workpiece velocity v_{w}, the axial feed a_{f} as well as the total radial stock removal $a_{\mathrm{e,tot}}$ directly raise the material removal rate Q_{w}, which is defined as

$$Q_{\mathrm{w}} = v_{\mathrm{w}}\, a_{\mathrm{e,tot}}\, a_{\mathrm{f}}. \tag{4}$$

According to this relationship, the machining forces rise due to an increase of these three process parameters. The material which comes in the area of operation is removed by numerous engagements of individual grinding grains, Fig. 6b, c. As soon as the grinding wheel velocity increases, the material removal is performed by a higher number of grain engagements. Thus, the single grain load and the single grain chip thickness are smaller. This results in decreasing machining forces at higher grinding wheel velocities (cf. Fig. 6a).

When rising the grinding wheel velocity, only a small increase of the temperature within the contact zone can be observed. According to the higher velocity, the amount of kinematic cutting edges rises, resulting in a decrease of the single grain chip thicknesses. Every chip formation is carried out with elastic and plastic deformation of the workpiece material, as well as, friction which raises the input of thermal energy into the workpiece. According to the higher amount of grain engagements, the material removal induces more energy input, resulting in the small increase of the temperature within the contact zone observable in Fig. 6a.

The thermal load onto the workpiece nearly remains constant when varying the workpiece velocity (Fig. 6b). The total arising process power magnitude P can be calculated according to

$$P = F_t \, v_s. \tag{5}$$

The process power rises with higher workpiece velocities in contrast to the contact time between workpiece and grinding wheel t_c [12], which can be calculated as

$$t_c = \frac{l_g}{v_w}, \tag{6}$$

where l_g is the contact length. Therefore, these opposing effects extinguish each other, resulting in the stationary simulated temperatures varying the workpiece velocity in the investigated process parameter space.

Increasing the axial feed results in higher thermal loads during ITG, compare Fig. 6c. On the one hand, the process power rises, thereby increasing the thermal load. On the other hand, the axial feed has a strong influence on the macro-scale engagement situation and the cross-section of the undeformed chip, resulting in higher effective depth of cut $a_{e,eff}$ as well as in a rise of the contact time t_c. The latter is caused by an extension of the contact length l_g. A detailed analysis of this process parameter is documented in [13]. The mentioned reasons explain the higher temperatures within the contact zone when increasing the axial feed.

The temperature within the contact zone is nearly constant when raising the total radial stock removal (Fig. 6d). An exception is the lowest value, where the temperature slightly increases. When the total radial stock removal increases, the area of the contact zone, especially the effective width of the grinding wheel $w_{g,eff}$, raises, cf. Fig. 1b ($a_{e,tot,2} > a_{e,tot}$). The acting section of the grinding wheel enlarges so that the additional material removal is conducted by the kinematic cutting edges of the supplementary engaging section of the grinding wheel, see Fig. 1b (purple draw-

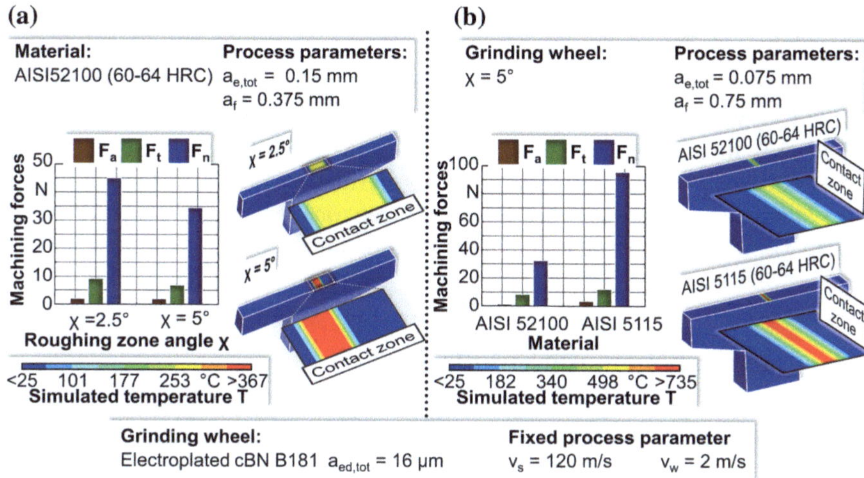

Fig. 7 Influences of the roughing zone angle (**a**) and the machined material (**b**) on the thermo-mechanical load during ITG

ing and arrow). This results in the quasi stationary level of the simulated temperature in the considered process parameter space. Using a very small total radial stock removal, the raising of the temperature is founded in the decrease of the single grain chip thicknesses regarding the cross-section of the undeformed chip. This causes a less effective machining, which results in a higher thermal load due to more friction occurring during the chip formation.

The thermo-mechanical load on the workpiece when varying the roughing zone angle and the workpiece material is shown in Fig. 7a. Using larger roughing zone angles, the global mechanical load on the workpiece decreases. Conversely, the thermal load onto the workpiece is lowered by smaller roughing zone angles. Considering the chip cross-section of the undeformed chip during ITG, a smaller roughing zone angle results in twice the effective width of the grinding wheel, or the effective depth of cut $a_{e,tot}$, respectively. The effective width of the grinding wheel can be calculated as

$$w_{g,eff} = a_f + \frac{a_{e,tot}}{\tan(\chi)}. \tag{7}$$

When decreasing the roughing zone angle, more grains are active during grinding to remove the material, so that the single grain chip thickness is lowered significantly resulting in a smaller thermal load. In contrast to the global mechanical load, the specific forces or the local mechanical load of the workpiece, respectively, increase with larger roughing zone angles, according to the longer effective width of the grinding wheel ($\chi = 2.5° : F'_t = 2.41\,\text{N/mm}$, $F'_n = 11.89\,\text{N/mm}$; $\chi = 5° : F'_t = 3.30\,\text{N/mm}$, $F'_n = 16.61\,\text{N/mm}$). A detailed description of the influence of the roughing zone angle on the thermo-mechanical load during ITG is documented in [8].

The comparison of the ground materials AISI 52100 and AISI 5115, which were hardened to 60–64 HRC, is shown in Fig. 7b. The mechanical load acting on the workpiece is three times higher when grinding AISI 5115 than AISI 52100. The mechanical energy which is transformed to thermal energy leads to significantly higher simulated temperatures within the contact zone when machining AISI 5115 in comparison to the other material under consideration.

3 Geometric-Kinematic Simulation

In order to estimate a chronological sequence of the meso-scale engagement situation of the grinding grains, a geometric-kinematic simulation of ITG was developed. Dexel elements, which are defined by a starting point, a direction and a length, are used to represent the workpiece. Arranging these elements in a radial grid, the internal surface of the workpiece can be modelled.

The grinding wheel is built up with Constructive Solid Geometry (CSG) elements consisting of geometrical primitives, like cylinders or cones, which can be manipulated by Boolean operations [14]. By following this approach, the topography of the grinding wheel can be implemented in the simulation, representing the cBN grains with an intersection of an octahedron, a hexahedron and a tetrahedron according to the morphology of this grain type, see Fig. 8 and [15, 16]. The statistical distribu-

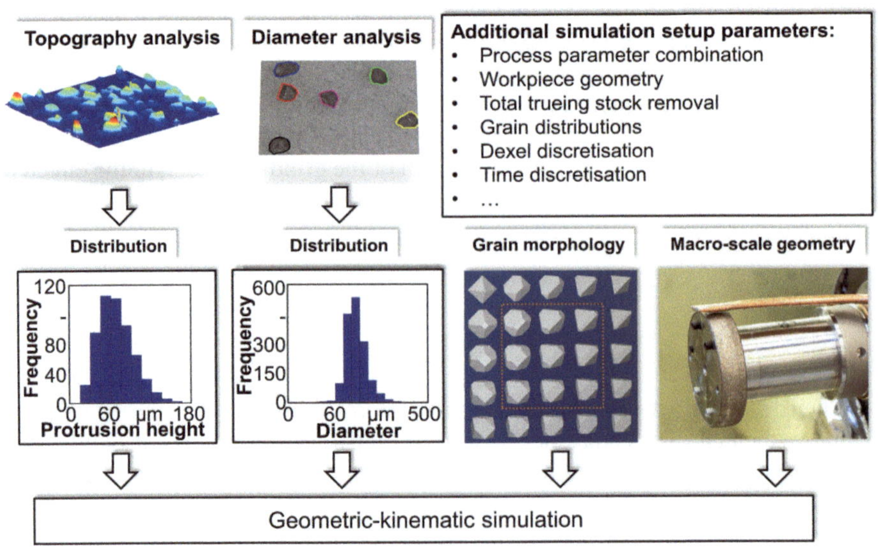

Fig. 8 Input values for the geometric-kinematic simulation

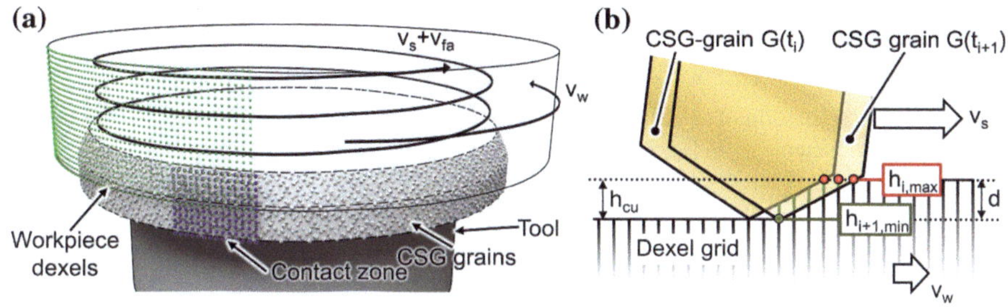

Fig. 9 Geometric-kinematic simulation procedure. **a** Kinematics during the simulation; **b** Calculation of the undeformed single grain chip thickness

tions of the grain protrusion height and of the grain diameter are input functions for the simulation. Both distributions are based on extensive measurements and numerical analysis, cf. [9, 17, 18]. Additional setup parameters are the time and dexel grid discretisation, the process parameters as well as geometrical constrains of the workpiece and the grinding wheel, cf. Fig. 8. The touch-dressing of the finishing zone of ITG can be carried out by an intersection of the CSG grains with a cylinder according to the experimentally applied dressing radius. A detailed description of the geometric-kinematic simulation is documented in [9, 18].

During a simulation, the workpiece and the tool model are moved relative to each other according to the process kinematics of ITG, see Fig. 9a. When a CSG grain makes contact with a dexel element, the latter is trimmed to the intersection point, thus the numerical material removal is carried out. To guarantee an efficient simulation procedure, only a section of the workpiece of 16×8 mm is modelled. Using such type of simulation, different analysis methods can be used. The undeformed single grain chip thickness can be calculated, cf. Fig. 9b, subtracting the minimum height after the cut from the maximum height before the cut of the involving dexels.

Furthermore, the rake angle of the single grain engagement can be calculated regarding to the grain shape and the engagement condition, cf. [19]. In a more detailed analysis the material removal in the contact zone can be estimated, which can be used, e.g., for the determination of the thermal load distribution of finite element simulations [9]. Another important result of this simulation is the frequency distribution of engagement conditions over time, see Fig. 10, which will be invoked in the scope of the scale bridging scheme presented in Sect. 4.2. Another exemplary application is the grain engagement related to a small section of the workpiece, which, in turn, can be investigated to estimate the chronological sequence of the occurring undeformed single grain chip thicknesses. By linking this information with a meso-scale FE model, the differences in the chip formation can be modelled with respect to the heated and deformed material behaviour influenced by previous grain engagements, see [20].

In summary the geometric-kinematic simulation can be invoked to efficiently calculate the transient engagement situations of the cBN grains during ITG. The drawback of such a simulation approach is that only a pure penetration calculation can be

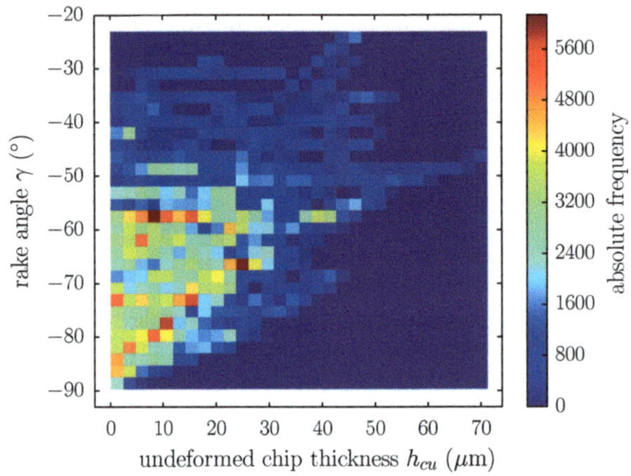

Fig. 10 Two-dimensional histogram or absolute frequency distribution $f(h_{cu}, \gamma)$ with a resolution of 30 bins per parameter

conducted, neglecting elastic and plastic deformation during the single grain engagement and the chip formation. To model these effects, a meso-scale single grain FE model, which is described in the following section, is used and combined with the geometric-kinematic simulation.

4 Finite Element Modelling of ITG

This section deals with the finite element modelling approach developed in the scope of SPP 1480. As outlined above, a single grain finite element model is invoked to compute the thermo-mechanical response caused by one single cBN grain, cf. Sect. 4.1. This measure is then transferred to the macro scale by the bridging approach presented in Sect. 4.2.

4.1 Single Grain Finite Element Model

To predict the amount of thermal energy per cBN grain transported into the workpiece material during grinding, a parametric, two-dimensional plane strain FE model reflecting the proximity of one single grain on the meso-scale is considered. The parametric model design enables an efficient automation of the simulation framework, which is important when aiming at an automatic bridging of the meso-scale thermo-mechanical results to be used as thermo-mechanical loads on the macro-scale. The basic setup including the relevant parameters and boundary conditions of the model is depicted in Fig. 11. The grain is modelled as a rigid conductor that reflects cBN material parameters, whereas the workpiece consists of hardened AISI 52100, modelled as described in the following Sect. 4.1.1.

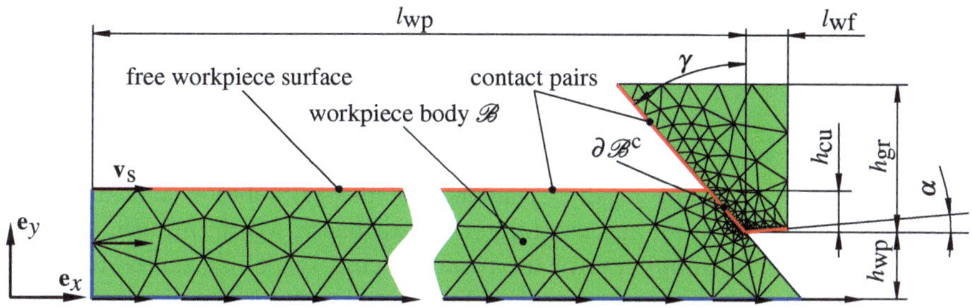

Fig. 11 Boundary value problem for the single grain model. The *red* marked edges represent contact pairs, the *blue* marked edges denote the regions that are loaded with Dirichlet boundary conditions, such as the grinding wheel circumferential velocity $\mathbf{v}_s$

The workpiece is moved with a constant velocity of v_s in $\mathbf{e}_x$ direction, applied as a Dirichlet boundary condition at the blue edges depicted in Fig. 11. The grain itself is fixed by Dirichlet boundary conditions and therefore only allows for changes in temperature. The grain geometry is simplified to two straight lines—reflecting the rake face and the wear flat area which is comparable to the flank face in orthogonal cutting simulations—connected by a small radius. In general, more sophisticated grain shapes exported from the topography analysis can also be used, compare [17], but have proven to be unnecessary for the parametric design of the model. The red edges in Fig. 11 depict corresponding contact pairs. In addition to the contact between grain and workpiece, self contact of the latter surface is also enabled since the chip generated usually comes into contact with the free workpiece surface during the chip formation process. A Lagrangian formulation is used to describe the boundary value problem which, in turn, is solved by an explicit time integration scheme using the commercial software Abaqus/Explicit.

Following the parametric modelling approach, we parametrise the rake face geometry for a plane strain assumption solely by the rake face angle γ. The next two major parameters to describe the single grain cutting conditions are the undeformed chip thickness h_{cu} and the cutting velocity v_s. This parameter tuple $\mathscr{P}_{meso} = \{h_{cu}, \gamma, v_s\}$ will subsequently be used to evolve the bridging scheme presented later.

In analogy to an orthogonal cutting model, we furthermore introduce the flank, or relief angle α along with the flank face length, or wear flat area length l_{wf}, respectively. In the present context however, we set $\alpha = 0°$ and $l_{wf} = h_{cu}$ to remain constant throughout the rest of this contribution. All remaining geometric measures as well as all simulation parameters—such as overall simulation time, time incrementation and remeshing parameters—which are necessary to entirely define the underlying boundary value problem are subsequently defined as a function of the above-mentioned parameters set $\mathscr{P}_{meso}$.

The widely known main issues in the field of modelling of high speed abrasive or cutting processes are finite strains and high strain rates along with a likely occurrence of local workpiece bulk separation. To overcome related algorithmic problems, commonly employed methods are, e.g., a prescribed damageable or deleteable element layer, respectively, or a remeshing algorithm to compute the structural response accordingly. Since the first method inherently induces a pre-defined deformation pattern regarding chip deformation, cf. [21], we employ an h-adaptive remeshing scheme developed in [22].

This eases to overcome mesh dependency of the FE solution and allows for the calculation of the onset and evolution of shear bands which, in turn, is essential for the predictive simulation of the resulting chip shape. Currently, besides a Zienkiewicz and Zhu error estimate, cf. [23], two additional, more physically motivated indicators are used to trigger local mesh refinement. These are the equivalent plastic strain $\varepsilon_{\mathrm{p}}(t)$ on the one hand, and its rate, i.e. $\dot{\varepsilon}_{\mathrm{p}} = \partial \varepsilon_{\mathrm{p}} / \partial t$ on the other hand. After a mesh refinement step, the nodal as well as the integration point variables are mapped from one mesh to another using the superconvergent patch recovery technique proposed by Zienkiewicz and Zhu, cf. [24].

In detail, the adaptivity method can be used in a very flexible manner, so the field variables on which the error estimation and mesh refinement is based, can be chosen freely from a sensible combination of variables. Let $\vartheta(\mathbf{x})$ denote an arbitrary field variable of any dimension resulting from a finite element approximation of polynomial order p and a characteristic element size h in the neighbourhood $\mathcal{N}_x$ of the spatial point $\mathbf{x}$. The Zienkiewicz and Zhu (Z^2) error estimate $\mathbf{e}_\vartheta$ referring to $\vartheta(\mathbf{x})$ is defined as

$$\mathbf{e}_\vartheta = \widehat{\vartheta} - \vartheta, \tag{8}$$

where $\widehat{\vartheta}$ denotes a better solution of the finite element problem in terms of a Z^2 superconvergent node patch recovery, see [23]. The latter is computed by the determination of a polynomial function $P(\mathbf{x}, \mathbf{a}) = \mathbf{p} \cdot \mathbf{a}(\mathbf{x})$ of at least order $p + 1$, that approximates the field function $\vartheta(\mathbf{x}) \in \mathcal{N}_x$. In the present case of two-dimensional linear finite elements, $\mathbf{p} = [1, x_1, x_2, x_1^2, x_1 x_2, x_2^2]$ represents the minimum approximation. The parameter vector $\mathbf{a}$ to entirely define P is, in turn, determined by a least square optimisation which uses the values of ϑ at the n integration points $\mathbf{x}_k^{\mathrm{GP}}$ of the element patch $\mathcal{N}_x$ surrounding the nodal position $\mathbf{x}$. This yields the following non-linear minimisation problem

$$\mathbf{a} = \arg\min \left(\sum_{k=1}^{n} \left[\vartheta(\mathbf{x}_k^{\mathrm{GP}}) - P(\mathbf{x}_k^{\mathrm{GP}}, \mathbf{a}) \right]^2 \right) \tag{9}$$

to be solved for every component ϑ of $\vartheta = \vartheta_i \, \mathbf{e}_i$. The recovered solution is finally computed via $\widehat{\vartheta}(\mathbf{x}) = P_i(\mathbf{x}, \mathbf{a}) \mathbf{e}_i$. Knowing this error estimate $\mathbf{e}_\vartheta$ now at every node position of the workpiece body $\mathbf{x} \in \mathcal{B}$, the new local element size $\overline{h}$ is finally estimated by

$$\overline{h} = h \left[\frac{||\mathbf{e}_\vartheta||}{||\mathbf{e}_\vartheta||_{\max}} \right]^{-\frac{1}{p}}, \tag{10}$$

based on the assumption $||\mathbf{e}_\vartheta|| \sim h^p$, cf. [22]. In this context, $||\mathbf{e}_\vartheta||_{max}$ denotes the defined maximum error L^2 norm with respect to the variable ϑ. As already mentioned above, the definition of a local maximum error magnitude

$$||\mathbf{e}_\vartheta||_{\mathscr{N}_x} < ||\mathbf{e}_\vartheta||_{max} \tag{11}$$

to define the local mesh refinement strategy can be extended in a flexible way to be suitable for a wide range of applications.

4.1.1 Material Modelling

To model the behaviour of hardened AISI 52100, an isotropic thermo-elastic and thermo-viscoplastic material model that is fully thermo-mechanically coupled is invoked. We assume large plastic strains reflecting the typical behaviour of ferrous metal based materials in the regime of manufacturing processes. We make use of a multiplicative split of the deformation gradient

$$\mathbf{F} = \mathbf{F}_\mathrm{e} \cdot \mathbf{F}_\mathrm{p}, \tag{12}$$

so that related logarithmic strains additively decompose as

$$\varepsilon = \varepsilon_\mathrm{e} + \varepsilon_\mathrm{p} \quad \text{with} \quad \varepsilon_\mathrm{e} = \frac{1}{2} \ln \left(\mathbf{F}_\mathrm{e} \cdot \mathbf{F}_\mathrm{e}^{\,\mathrm{t}} \right). \tag{13}$$

defining, e.g., the elastic strains. Assuming isotropic material behaviour, the Cauchy stress tensor takes the form

$$\sigma = \left[\lambda_0 \,\mathrm{tr}\left(\varepsilon_\mathrm{e} \right) - \left[3\,\lambda_0 + 2\,\mu_0 \right] \alpha_0\, \theta \right] \mathbf{I} + 2\,\mu_0\, \varepsilon_\mathrm{e} \quad \text{and} \quad \mathbf{q} = -\lambda\, \nabla_\mathbf{x}\, \theta, \tag{14}$$

is the heat flux vector according to the isotropic version of Fourier's law. Here, λ_0 and μ_0 denote the Lamé constants, α_0 is the thermal expansion coefficient, λ characterises the isotropic thermal conductivity, and θ represents the absolute temperature.

The constitutive model is restricted to an isotropic hardening behaviour, neglecting the Bauschinger-Effect of kinematic hardening. To account for the plastic material response, the well-established Johnson-Cook approach is used, cf. [25], in combination with an isotropic von Mises type associated flow rule for the evolution of the yield stress

$$\sigma_y^{\mathrm{JC}} = \Big[[A + B\left[1 - \exp(-\varepsilon_{\mathrm{p}} N)\right]\Big] \left[1 + C \ln\left(1 + \frac{\dot{\varepsilon}_{\mathrm{p}}}{\dot{\varepsilon}_{\mathrm{p},0}}\right)\right] \left[1 - \left\langle \frac{\theta - \theta_0}{\theta_m - \theta_0} \right\rangle^M\right] \tag{15}$$

with $\langle \bullet \rangle$ denoting the McCauley operator defined as $\langle \bullet \rangle \equiv \frac{1}{2}\left[\bullet + |\bullet|\right]$. Consequently, the material model incorporates strain and strain rate dependent work hardening along with thermal softening. The material parameters $A, B, C, M, N, \dot{\varepsilon}_{\mathrm{p},0}$ and θ_0 are taken from [26]. The full set of material parameters used in this section is depicted in Table 1, cf. Appendix.

As widely used in high speed machining simulations, we introduce a local ductile damage variable d representing the local amount of micro-cracks and micro-voids present in the workpiece material. In the present framework, we restrict the damage model to an isotropic and a local approach evaluated solely at the material point level of the body under consideration. Although this technique inhibits the drawback of mesh-dependency and is most elegantly solved by a non-local gradient-enhanced finite-element framework, cf. [22, 27, 28], we incorporate adaptive remeshing to overcome this issue as outlined above. The internal damage variable d evolves governed by the saturation type equation (saturation value $\overline{d}$)

$$\dot{d} = \vartheta_d \left\langle \varepsilon_{\mathrm{p}} - \varepsilon_{\mathrm{p}}^d \right\rangle \left[\overline{d} - d\right] \dot{\varepsilon}_{\mathrm{p}} \quad \text{with} \quad \varepsilon_{\mathrm{p}}(t) = \varepsilon_{\mathrm{p}}|_{t=0} + \int_0^t \sqrt{\frac{2}{3}\dot{\varepsilon}_{\mathrm{p}} : \dot{\varepsilon}_{\mathrm{p}}}\,\mathrm{d}t. \tag{16}$$

In this context, ϑ_d is the saturation rate, $\overline{d}$ denotes the saturation value and $\varepsilon_{\mathrm{p}}^d$ represents the critical accumulated plastic deformation activating the damage evolution. The accumulated plastic strain is denoted by ε_{p}. Following this approach, the yield stress results in

$$\sigma_y = [\,1 - d\,]\,\sigma_y^{\mathrm{JC}}. \tag{17}$$

An example yield surface as a function of temperature and plastic strain is depicted in Fig. 12.

Lastly, the generation of heat in the single grain model is based on plastic work on the one hand and frictional contact between workpiece and grain on the other. It is assumed that only a constant fraction η of the plastic dissipation contributes to the local mass specific energy balance [29, 30], i.e.

$$\Delta r = \eta\,\sigma : \dot{\varepsilon}_{\mathrm{p}} \tag{18}$$

is added to the thermal energy balance. Furthermore, the influence of friction or surface traction, respectively, contributes to the local energy balance in terms of a rate of work

$$\int_{\partial \mathscr{B}^{\mathrm{c}}} \dot{\varphi} \cdot \mathbf{t}\,\mathrm{d}a = \int_{\partial \mathscr{B}^{\mathrm{c}}} \dot{\varphi} \cdot \sigma \cdot \mathbf{n}\,\mathrm{d}a = \int_{\partial \mathscr{B}^{\mathrm{c}}} \tau\,\dot{\xi}\,\mathrm{d}a \tag{19}$$

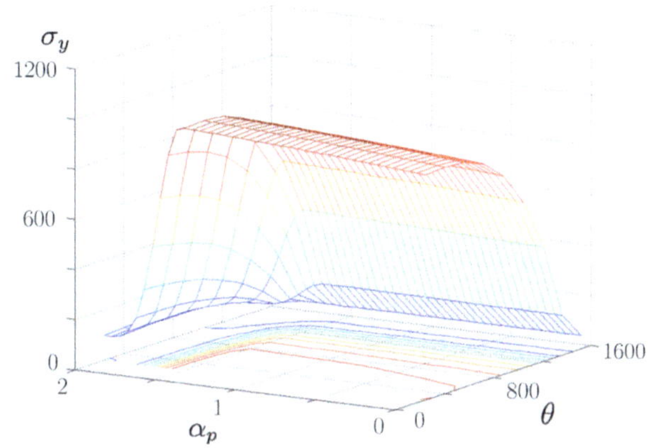

Fig. 12 Exemplary yield surface representation as a function of temperature θ (K) and plastic strain $\alpha_p = \epsilon_p$ (no unit), illustrating the strain and temperature depending hardening and softening behaviour at a strain rate of $\dot{\epsilon}_p = 10000\,\mathrm{s}^{-1}$

with $\partial \mathscr{B}^c$ representing the current contact surface $\partial \mathscr{B} \cap \partial \mathscr{B}_{\text{grain}}$, so the intersection of both workpiece boundary $\partial \mathscr{B}$ and grain boundary $\partial \mathscr{B}_{\text{grain}}$. The spatial velocity field vector is denoted by **v** and **t** refers to the surface traction vector, respectively. The incremental slip between the two contact surfaces is represented by $\dot{\xi}$ and the absolute shear stress value reads τ.

4.1.2 Results

Figures 13 and 14 depict some exemplary results of the single grain finite element model in terms of von Mises equivalent stress and temperature contour plots. Here, the h-adaptive mesh refinement is well observable when comparing the depicted mesh in the deformed configuration with the reference configuration mesh displayed in Fig. 11. Since mesh refinement is triggered mainly by the plastic strain, the work-

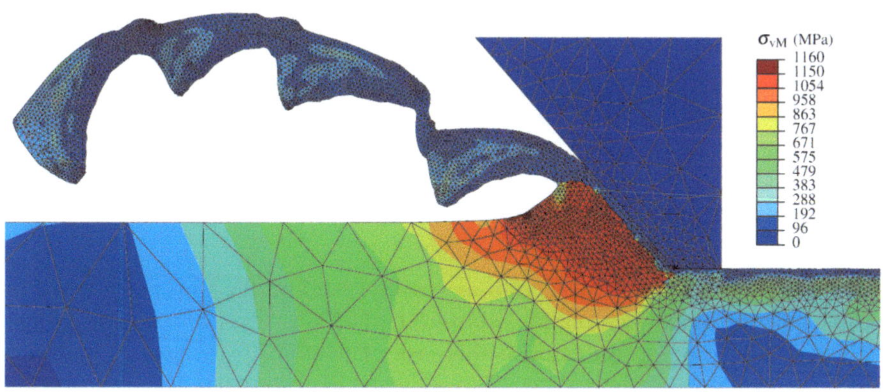

Fig. 13 Von Mises equivalent stress for $\gamma = -40°$, $h_{cu} = 30\,\mu\text{m}$, $v_s = 120\,\text{m/s}$

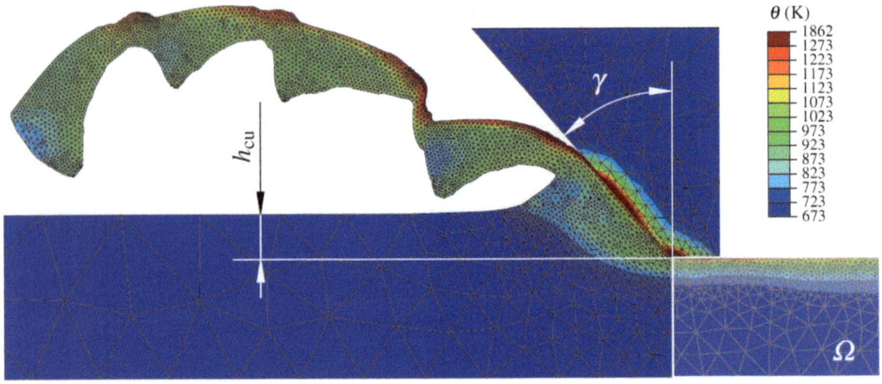

Fig. 14 Temperature contour plot for $\gamma = -40°$, $h_{\mathrm{cu}} = 30\,\mu\mathrm{m}$, $v_{\mathrm{s}} = 120\,\mathrm{m/s}$, extended with the major parameters h_{cu} and γ. The white shaded area represents the reference volume Ω for the heat source determination

piece regions exceeding an equivalent stress of $\sigma_{\mathrm{vM}} \geq 680\,\mathrm{MPa}$ evince a tendency to a finer mesh. Since the entire red area bounded by the primary and secondary shear zone exhibits the minimum mesh size, the entire chip as well as the resulting workpiece surface also retain this high mesh density. The minimum element size is set to $1/30\,h_{\mathrm{cu}}$ and the maximum element size corresponds to approximately h_{cu}.

The effect of the ductile damage implemented is also conveniently observable in terms of a softened resulting workpiece surface along with a totally softened chip evincing a serrated structure. This chip structure is also observable in experiments, see Fig. 2, and proves the correctness of the modelling approach invoked here. For a more detailed discussion on chip formation, see [17]. Moreover, an investigation on multiple grain engagements is presented in [20].

An exemplary result of the resulting single grain specific cutting force is depicted in Fig. 15, where the chip segmentation is well perceptible. Until a shear band in the primary shear zone is entirely build up, the force components increase, whereas a sudden decrease is observable when the chip segment is sliding on the fully established shear band. For the scale bridging approach presented in the following section, a time averaging of F'_{t} and F'_{n} is carried out, depicted as the horizontal lines in Fig. 15.

4.2 Bridging Approach

To calculate the macroscopic thermo-mechanical load on the workpiece based on the two types of simulations outlined above, a homogenisation—or scale-bridging—scheme is necessary to accomplish this task. Therefore, two physical measures need to be defined to entirely describe the targeted load compound. Concerning the mechanical load, the mean values $\langle F'_{\mathrm{t}} \rangle$ and $\langle F'_{\mathrm{n}} \rangle$ of the meso-scale specific forces $F'_{\mathrm{t}}(t)$ and $F'_{\mathrm{n}}(t)$ constitute the obvious physical quota to be bridged to the macro-

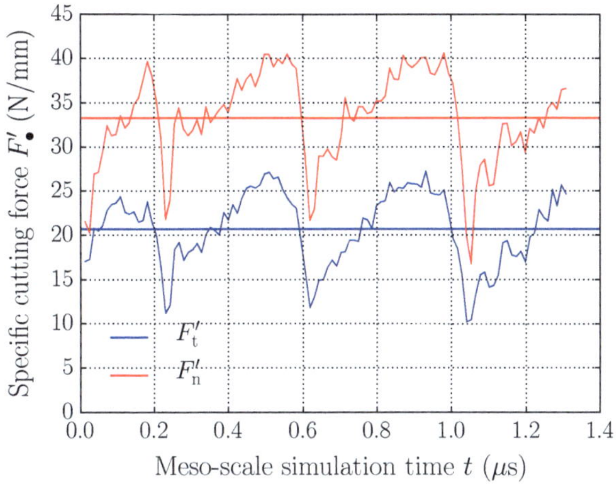

Fig. 15 Specific force component evolutions $F'_t(t)$ and $F'_n(t)$ per unit plane strain thickness for $\gamma = -55°$, $h_{cu} = 5\,\mu\text{m}$, $v_s = 120\,\text{m/s}$. Referring to Fig. 11, the total reaction force is composed as $\mathbf{F}'_c = F'_t\,\mathbf{e}_x + F'_n\,\mathbf{e}_y$. The periodic increase and decrease of the force components is due to the chip segmentation occurring during the simulation. The horizontal lines depict the time-averaged mean values of the force components $\langle F'_t \rangle$ and $\langle F'_n \rangle$, respectively

scale, compare Fig. 15. Regarding the heat induction, one first needs to define a thermal measure that is independent of the meso-scale model size and that represents the influence of all process parameters, accordingly. Furthermore, an invariance against the reference temperature θ_0 is desirable as well, since we are solely interested in the thermal energy increment as a function of the major parameters.

For this reason, a heat source term r is defined as a function of the three major parameters in terms of

$$r(h_{cu}, \gamma, v_s) =: \frac{\Delta Q}{\Delta V_{chip}}, \tag{20}$$

representing the amount of thermal energy ΔQ induced into the workpiece with respect to the ground unit volume ΔV_{chip} of machined workpiece material. To compute the heat energy Q induced in the machined workpiece bulk, we evaluate the volume integral

$$Q(h_{cu}, \gamma, v_s, t) = \int_{\Omega} \varrho \int_{\theta_0}^{\theta} c(\theta)\, \mathrm{d}\theta\, \mathrm{d}v \tag{21}$$

in terms of a finite element discretisation

$$Q^{FE}(h_{cu}, \gamma, v_s, t) = \sum_{i=1}^{n_e} \varrho_i\, v_i\, \frac{1}{2} \sum_{k=1}^{n_\theta} \left[\theta_{k+1} - \theta_k\right] \left[c(\theta_{k+1}) - c(\theta_k)\right] \tag{22}$$

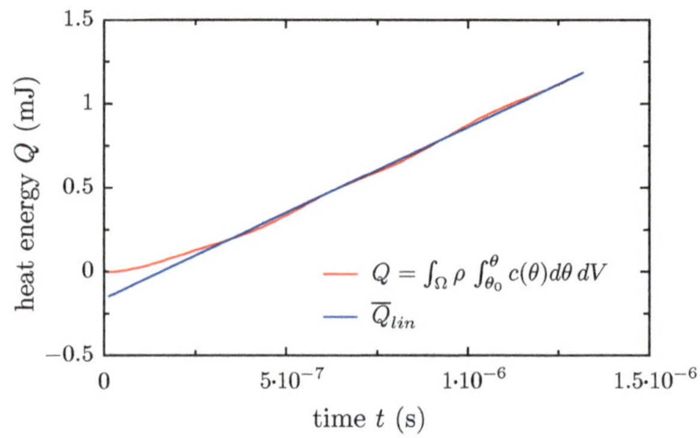

Fig. 16 Heat energy evolution $Q(t)$ for $\gamma = -55°$, $h_{\text{cu}} = 5\,\mu\text{m}$, $v_{\text{s}} = 120\,\text{m/s}$. The slight periodic variation of the energy term results from the periodic chip segmentation occurring during the simulation

for each time step over the remaining workpiece volume Ω according to Fig. 14. In this context, $c(\theta)$ denotes the current heat capacity depending on the absolute material temperature $\theta(t)$ and ρ is the spatial mass density. The number of elements of volume v_i within the region Ω is represented by n_e, the number of sample points defining the discrete temperature dependency of the heat capacity $c(\theta_k)$ is denoted by n_θ. Because the undeformed chip thickness h_{cu}, the plane strain thickness Δz_{ps} of the meso-scale model as well as the cutting speed v_{s} remain constant over the meso-scale simulation time, the evolution of $Q(t)$ is assumed to be approximately linear in time, which is clearly observable in Fig. 16, where an exemplary evolution of $Q(t)$ is displayed. In this figure, the red line represents the evolution of the heat energy Q according to (21) induced by the abrasive process, and the blue line depicts a linear regression function thereof that was calculated by least-square-minimisation, resulting in a function $\overline{Q}_{lin}$. As clearly observable, the error of the regression scheme in comparison to the original data is small and thus, the linear regression constitutes a highly efficient averaging scheme to approximate the induced thermal power P during the grinding simulation. In order to establish a functional dependency between P and r, we make use of the linear dependency between the above-mentioned meso-scale parameters and the chip volume

$$\Delta V_{chip} = h_{\text{cu}}\, \Delta z_{\text{ps}}\, \Delta x = h_{\text{cu}}\, \Delta z_{\text{ps}}\, v_{\text{s}}\, \Delta t \tag{23}$$

or material removal rate

$$Q_{\text{w}} = \frac{\Delta V_{\text{chip}}}{\Delta t}, \tag{24}$$

respectively. This is then inserted into the approximated thermal process power induced into the workpiece

$$P = \dot{Q}_{lin} = \frac{\text{d}Q}{\text{d}t} \approx \frac{\Delta \overline{Q}_{lin}}{\Delta t}, \quad \text{yielding} \quad r(h_{\text{cu}}, \gamma, v_{\text{s}}) = \frac{\Delta \overline{Q}_{lin}}{\Delta V_{chip}} = \frac{P}{h_{\text{cu}}\, \Delta z_{\text{ps}}\, v_{\text{s}}}, \tag{25}$$

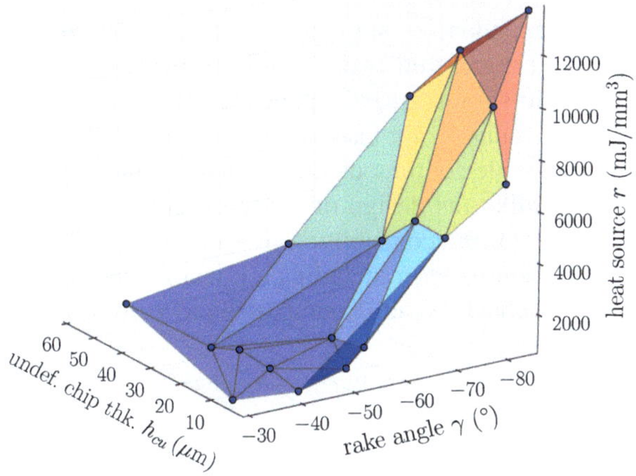

Fig. 17 Delaunay triangulation of the heat source term r, see (25), as a function of rake angle γ and undeformed chip thickness h_{cu}, resulting from the meso-scale simulations. Every vertex point (*blue dots*) of the triangulation constitutes one finite element simulation, resulting in the depicted surface. The colour of the triangular surface elements corresponds to the slope of the respective triangle. Taken from [18]

which constitutes the desired transferable heat measure. In the above equations, $\Delta x = v_{\mathrm{s}}\,\Delta t$ represents the displacement in direction of v_{s} during a time increment Δt whereas Δz_{ps} denotes the plane strain thickness of the two-dimensional finite element model.

At this point, we can now examine the resulting form of the heat source term r as well as the specific force components F'_{t} and F'_{n} as a function of the meso-scale parameters h_{cu} and γ. The referring results are displayed in Figs. 17 and 18, where each blue vertex point represents one meso-scale simulation that has been carried out.

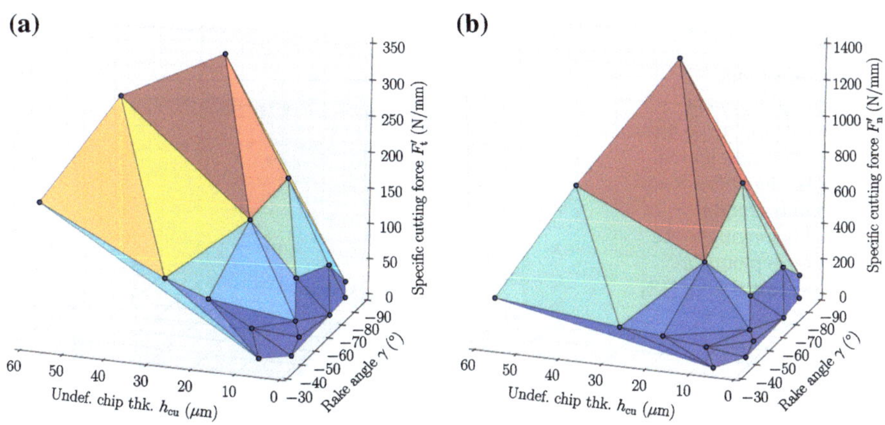

Fig. 18 Delaunay triangulations of the specific averaged single grain cutting force components $\langle F'_{\bullet}\rangle$ per unit plane strain thickness: **a** specific tangential component F'_{t}; **b** specific normal component F'_{n}. The colour of the triangular surface elements corresponds to the slope of the respective triangle

With this discrete functional dependencies $r(h_{\text{cu}}, \gamma, v_{\text{s}})$ and $F'_{\bullet}(h_{\text{cu}}, \gamma, v_{\text{s}})$ at hand, the scale-bridging is performed in the following way: At first, the latter discrete functional dependencies are converted to continuous functions $\overline{r}$ and $\overline{F}'_{\bullet}$ by applying a weighted regression over the ranges of h_{cu} and γ, according to the results of the kinematic simulation, compare Fig. 10. These approximated functions are subsequently evaluated for each time increment of the kinematic simulation, resulting in a macro-scale continuous time signal for the referring measure.

Regarding the design of the regression function, we will invoke an exponential approach for the contribution of γ in combination with a quadratic term for h_{cu} for the heat source term including a linear coupling term since the dependency of r on h_{cu} is obviously weak in comparison to γ; to be specific

$$\overline{r}(h_{\text{cu}}, \gamma) \approx a_0 + a_1\, h_{\text{cu}} + a_2\, h_{\text{cu}}^2 + a_3\, \gamma + a_4\, \exp(a_5\, [-\gamma - a_6]) + a_7\, h_{\text{cu}}\, \gamma. \tag{26}$$

The specific force regression is designed by analogy, i.e.

$$\overline{F}'_{\bullet}(h_{\text{cu}}, \gamma) \approx a_0 + a_1\, h_{\text{cu}} + a_2\, \exp(a_3\, [-\gamma - a_4]) + a_5\, \gamma + a_6\, h_{\text{cu}}\, \gamma. \tag{27}$$

In addition, we will take into account the distribution information $f(h_{\text{cu}}, \gamma)$ from the kinematic simulation, cf. Fig. 10. To do so, we overlay this distribution with the sample points and calculate a Voronoi tessellation [31] based on the latter in order to define influencing areas associated with each of the sample points. This Voronoi tessellation is then altered to only yield finite cells and to exactly coincide with the histogram boundaries, cf. Fig. 19.

With the finite Voronoi cells at hand, we can now calculate a normalised distribution of weighting factors for the set of sample points situated within the histogram area. This is accomplished by the evaluation of

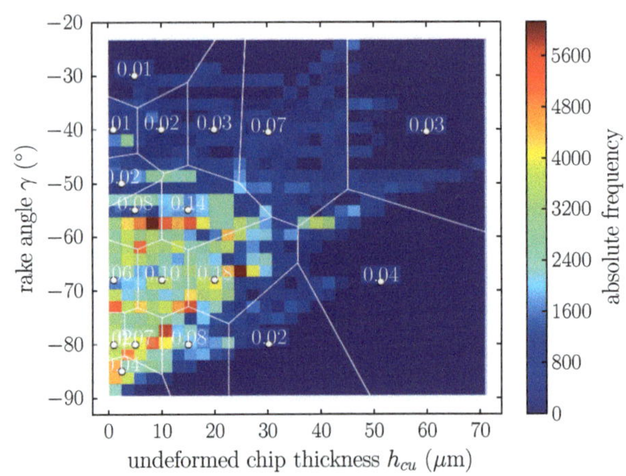

Fig. 19 Absolute frequencies $f(h_{\text{cu}}, \gamma)$ of engagement conditions with finite Voronoi tessellation overlay. The *numbers* denote the according weights w_i of every cell belonging to the sample points plotted as *white dots*. Taken from [18]

$$w_i = \frac{\displaystyle\int_{\Omega_i} f_k \, \mathrm{d}A}{\displaystyle\int_{\Omega_{\mathrm{hist}}} f_k \, \mathrm{d}A}, \qquad \text{such that} \quad \sum_{i=1}^{m} w_i = 1 \tag{28}$$

for each of the m Voronoi cell regions Ω_i within the total histogram area Ω_{hist}. Furthermore, f_k denotes the absolute frequency of a certain parameter combination $f_k(h_{\mathrm{cu}}, \gamma) \in \Omega_i$. In this context, d$A$ denotes an area element of a Voronoi cell. The resulting weighting factors are depicted in Fig. 19 in terms of the white numbers within the cells. With these weights $w_i \in \{w_1, w_2, \ldots, w_n\}$ at hand, we can now continue moving towards the calculation of the regression functions.

To calculate the coefficients a_j of $[a_1, a_2, \ldots, a_n] = \mathbf{a}$ in Eqs. (26) and (27), we invoke a weighted non-linear least-square optimisation to fit the above-mentioned functions to the set of sample points from the meso-scale finite element simulations via

$$\mathbf{a} = \arg\min_{\mathbf{a}} \sum_{i=1}^{m} w_i \left[r_i(h_{\mathrm{cu}}, \gamma) - \overline{r}_i^{\bullet}(h_{\mathrm{cu}}, \gamma; \mathbf{a}) \right]^2, \tag{29}$$

where r_i denotes the heat source term resulting from the meso-scale simulation carried out at sample point i, and $\overline{r}_i$ represents the above-mentioned regression function (26). The minimisation problem for F'_{t} and F'_{n} is carried out in analogy. The scalar m denotes the number of weights w_i, equal to the number of sample points in the present case. As a result from this procedure we retrieve the coefficients depicted in Table 2 (cf. Appendix) and the according function plots shown in Figs. 20 and 21, respectively.

To complete the above-mentioned coupling approach in terms of calculating a heat source term on the macro-scale, we evaluate the regression function $\overline{r}$ in the framework of the kinematic simulation. In detail, by superposing and averaging these results over a certain amount of tool revolutions, we finally retrieve a macroscopic thermo-mechanical load profile that can then be applied to a macro-scale process model as a boundary condition in terms of a moving load.

As stated above, at every time step t_i of the kinematic simulation, the regression functions concerning the grain-wise thermal

$$P_{\mathrm{mac}}(t_i) = \sum_{k=1}^{n_k} \left[\overline{r}(h_{\mathrm{cu}}(t_k), \gamma(t_k)) \, Q_{\mathrm{w}}(t_k) \right]_{t_k = t_i} \tag{30}$$

and mechanical contribution

$$\tilde{F}_{\bullet}(t_i) = \sum_{k=1}^{n_k} \left[\overline{F}'_{\bullet}(h_{\mathrm{cu}}(t_k), \gamma(t_k)) \, \Delta\bar{z}(t_k) \right]_{t_k = t_i} \tag{31}$$

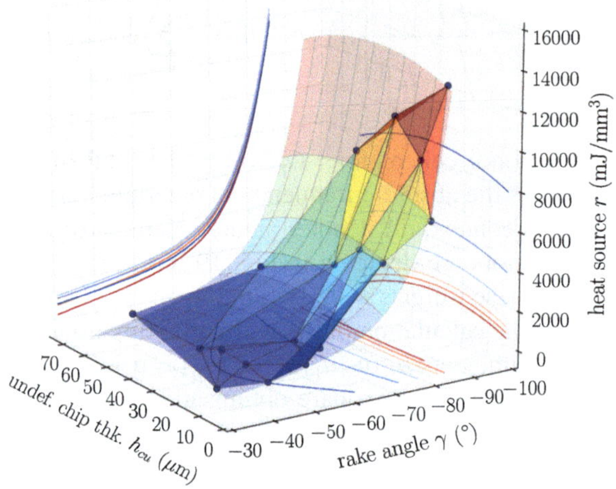

Fig. 20 Exponential regression function $\bar{r}$, including a contour line projection onto each base plane. In addition, the triangulated original data, see Fig. 17 is displayed. The squared error sum of the unweighed regression is $R^2 = 0.953$. The original heat source sample points are plotted as blue dots along with the Delaunay triangulation of sample points to enable a comprehensive comparison between original data and approximation. Taken from [18]

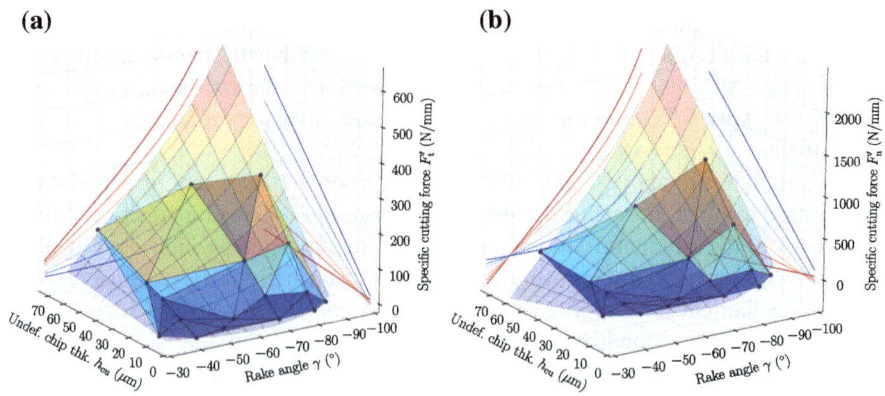

Fig. 21 Results of the regression scheme when invoking the exponential approximation (27) with respect to the specific tangential force component (**a**) and the specific normal force component (**b**)

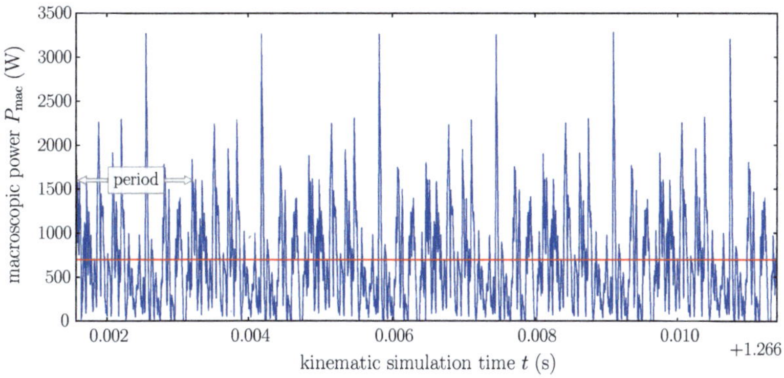

Fig. 22 Overall estimated macroscopic thermal power $P_{mac}(t)$ over kinematic simulation time t according to [18]. The red line constitutes the average thermal power $\langle P_{mac} \rangle = 699.741$ W induced over the depicted time span $\Delta T_{kin} = 0.00981$ s representing six grinding wheel revolutions. The period, i.e. the duration of one exemplary grinding wheel revolution, is indicated by the white annotation box. Note, that the blue graph depicts a steady-state time interval, neglecting the initial 1.266 s of the process run-up time and also neglecting the time span until the grinding wheel has entirely passed through the workpiece. The grain engagements and the corresponding thermal power repeat periodically

are evaluated and superposed for all the n_k grains that are in contact simultaneously with the workpiece at time t_k, resulting in a transient expression for the thermal power—depicted in Fig. 22—and the mechanical load that is induced in the workpiece. In this context, the current meso-scale material removal rate is denoted by Q_w, and $\Delta\bar{z}(t_k)$ denotes the approximated engagement width of the grain currently in contact at the kinematic simulation time t_k.

By averaging these time-dependent measures over a certain time ΔT, or a number of time steps n_n, respectively, we retrieve

$$\langle P_{mac} \rangle = \frac{1}{\Delta T} \sum_{i=1}^{n_n} \Delta t_i \, P_{mac}(t_i) \quad \text{and} \quad \langle \tilde{F}_\bullet \rangle = \frac{1}{\Delta T} \sum_{i=1}^{n_n} \Delta t_i \, \tilde{F}_\bullet(t_i) \quad \forall \quad t_i \in \Delta T \tag{32}$$

as the average thermal power induction and the process force components that can subsequently be used in the macro-scale finite-element simulation described in the following section.

5 Process Model and Compensation Strategies

To capture the deformation behaviour of the workpiece during grinding, an axisymmetric finite element model that includes the entire workpiece body and the clamping cover is used, see Fig. 23. Since it could be shown that the clamping mechanism and especially the clamping cover has a not negligible influence on the deformation of the workpiece, the clamping mechanism is included in the simulation. As shown in Fig. 1, the workpiece is in contact only at two ring-shaped surfaces at the tool entrance and exit faces when clamping is applied. To model this, the clamping device is assumed to be rigid, whereas the clamping cover is modelled as a deformable solid body, cf. Fig. 23. During grinding, the thermo-mechanical load compound $\mathscr{L}$ that substitutes the grinding wheel and which is based on the results of the preceding Sect. 4.2, is moved along the inner workpiece surface with the axial feed velocity v_{fa}.

The macro-scale simulation of ITG is then generally divided into three phases:

1. Application of clamping in terms of the traction loading in terms of $\overline{\sigma}_z^{\text{cov}}$ on the bold load surface, cf. Fig. 23.
2. Application of the moving thermo-mechanical load compound $\mathscr{L}$ in terms of transient Neumann boundary conditions: Simultaneously, the grinding coolant is implemented as a second Neumann load that acts on the entire inner workpiece surface with heat transfer coefficient $\alpha_{\text{c}} = 3000\,\text{W}/(\text{mK})$ and with a fluid temperature of $\theta_{\text{c}} = 309.85\,\text{K}$. The latter was measured in experiments. At a first stage,

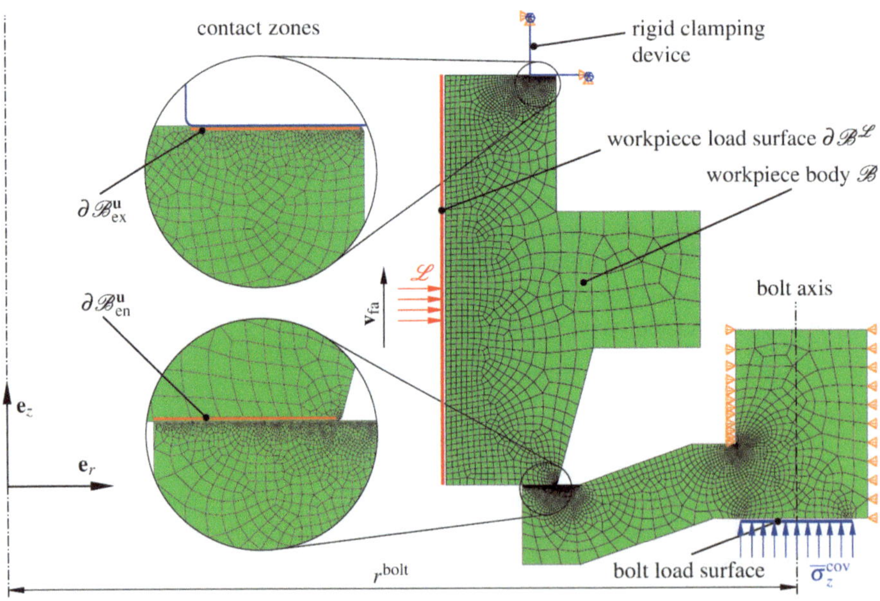

Fig. 23 Workpiece model including the clamping cover and contact at both clamping faces. The magnified regions depict the referring contact zones. The clamping device is assumed to be rigid and displayed as *blue lines*. The mechanical Dirichlet boundary conditions are represented by the orange symbols. During clamping, the blue marked surface is loaded with a stress $\overline{\sigma}_z^{\text{cov}}$ reflecting the maximum tension that can be applied to the clamping cover by the 4 M6 bolts of grade 8.8 that tighten the cover in the experimental setup. The thermo-mechanical load compound $\mathscr{L}$ moves with the axial feed velocity v_{fa} along the workpiece load surface marked in *red*. The *orange* marked surfaces represent the workpiece contact boundaries

this is carried out without any material removal to determine the general thermo-mechanical response of the workpiece-clamping cover system, see Fig. 25. The modelling of material removal as an extension to this first stage is based on a continuous remeshing scheme which is established as follows:
The axially moving tool region $\mathscr{G}(t_n)$ is iteratively intersected with the deformed configuration of the workpiece body $\mathscr{B}(t_n)$ at given time points $t_n \in [t_1 \dots t_{\max}^{\text{grd}}]$, so that the axial tool protrusion per time increment Δt equals the axial feed a_f, cf. Fig. 1. After this alteration of the mesh, all variables from the deformed configuration of the previous load step t_n are interpolated to the new intersected mesh, which then acts as reference configuration for the next thermo-mechanical load step t_{n+1}. The load compound $\mathscr{L}(t_{n+1})$ is applied as a superposition of heat flux $\mathbf{q}$ and traction σ_L that reflects the normal grinding force component F_n. Both are assumed to only act on the effective grinding wheel engagement width $w_{\text{g,eff}} = a_\text{f} + a_{\text{p,eff}}$, cf. Figs. 1 and 26.
3. Removal of the clamping boundary conditions and cooling the workpiece to reference temperature via a convective film cooling that acts on the entire boundary of the workpiece. In fact, only at one single node, a Dirichlet boundary condition restricting only the axial motion, is applied to ensure numerical stability during cooling.

Figure 24 shows the mechanical response during the application of the clamping stress $\overline{\sigma}_z^{\text{cov}}$ and proves to be a good agreement of the radial deformation of the inner workpiece surface between simulation and experiment. The subsequent Fig. 25 displays the deflection of the workpiece surface as well as the contact zones during the application of the above-mentioned load compound without material removal. The axial and radial displacements $u_\bullet^{\text{cl}}$ observed here are applied as transient Dirichlet boundary conditions in the context of the remeshing-based material removal simulation, which substitutes the clamping cover part by the latter boundary conditions. This model simplification is justified by the results shown in Fig. 27b, where a good agreement of the workpiece inner surface profile between experiments and material removal simulation is displayed (compare the blue curves). The workpiece contour was recorded by a coordinate measuring device Prismo Vast by Carl Zeiss AG.

The only effect that cannot be captured by the simulation at this stage is the strong local minimum observable in the measurement data for $z \geq 25$ mm. A similar effect is visible at the entry phase of the tool for $z \leq 10$ mm, where predominantly the curvature of the measured curve deviates from the simulated one. However, in the steady state range $10\,\text{mm} \leq z \leq 25$ mm the general tendency of the experiments can be captured well by the material removal simulation and therefore qualifies the latter for the development of compensation strategies presented in the next section.

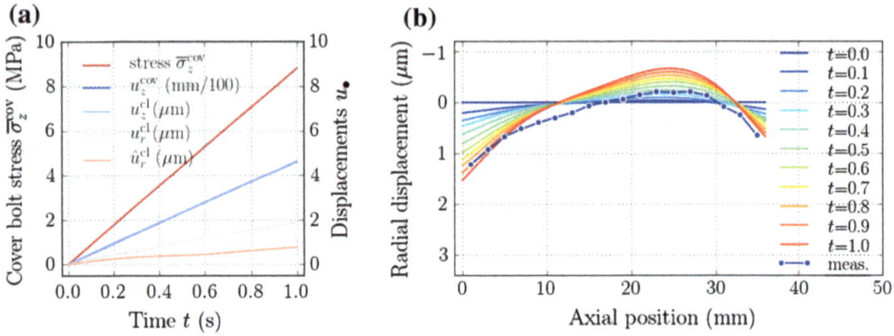

Fig. 24 Results of the clamping stress test including the cover and contact at both workpiece clamping faces when enforcing zero displacement Dirichlet boundary conditions on both inner and outer radial clamping cover surfaces. The radial displacement of the workpiece clamping nodes is free and only restricted by a penalty tangential contact constraint enforcement, implementing a friction coefficient of $\mu_{\text{con}} = 0.1$. The radial displacement of the workpiece clamping region at the exit face $\partial\mathscr{B}^{\mathbf{u}}_{\text{ex}}$ is denoted by $\hat{u}_r^{\text{cl}}$. The line labelled with "meas." represents the measured displacement of the inner workpiece surface after applying the clamping

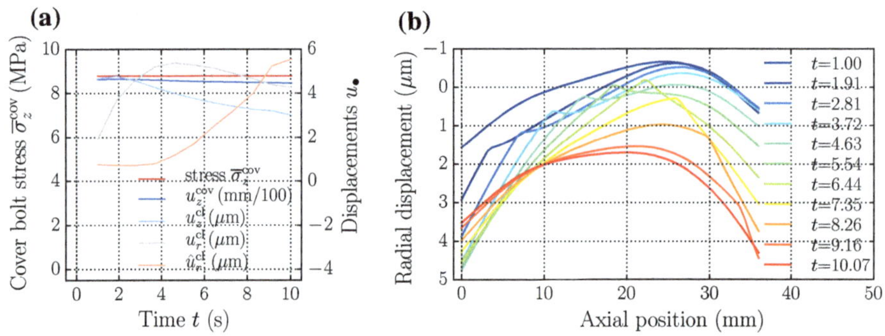

Fig. 25 Workpiece inner surface and clamping face displacements during the thermo-mechanical loading phase. This displays the follow-up simulation step referring to Fig. 24. The radial displacement of the workpiece clamping nodes is free and only restricted by a penalty tangential contact constraint enforcement, implementing a friction coefficient of $\mu_{\text{con}} = 0.1$ according to a steel-steel contact situation

5.1 Compensation Strategies

To minimise the radial error observable in Fig. 27b, the so far straight—or simply parallel to the workpiece axis $\mathbf{e}_z$—tool trajectory $\mathscr{T}(t) = \left[r^{\text{i}}_{\text{nom}} - r^{\mathscr{G}}_{\text{eff}}\right]\mathbf{e}_r + v_{\text{fa}}\, t\, \mathbf{e}_z$ will be corrected by $\Delta\mathscr{T}(t)$, which has to be calculated accordingly. In this context, $r^{\text{i}}_{\text{nom}}$ represents the nominal inner workpiece radius, $r^{\mathscr{G}}_{\text{eff}}$ denotes the effective tool radius which includes the abrasive layer of the tool and which is therefore defined by the most protruding grain situated on the tool body. At a time increment t_n, the current tool region is denoted by $\mathscr{G}_n$. A superscript of zero denotes a reference

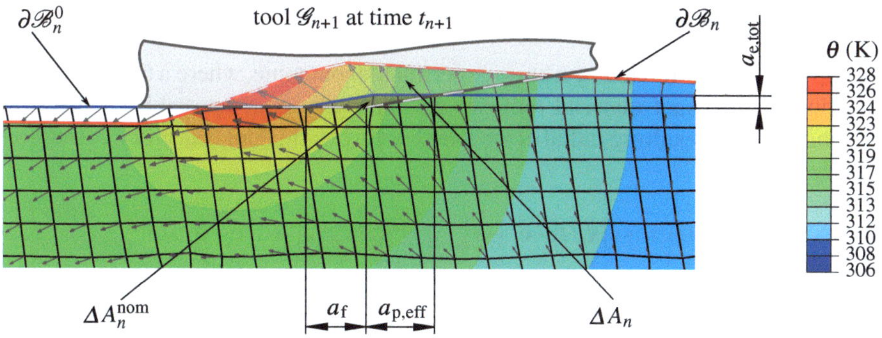

Fig. 26 Effective and nominal amount of removed material in the process zone at time t_n. The displayed mesh represents the reference configuration $\mathbf{X}_n$, whereas the *arrows* represent the displacement vectors $\mathbf{u}_k$ pointing at the spatial configuration $\mathbf{x}_n$. The *blue* marked surface represents the reference model boundary $\partial\mathscr{B}_n^0$, whereas the *red line* depicts the spatial boundary $\partial\mathscr{B}_n$. The effective amount of material removed ΔA_n is emphasised by the light shaded area bounded by a *grey dashed line*, the nominal amount of ground material $\Delta A_n^{\mathrm{nom}}$ is represented by the *dark shaded area*. An isotropic displacement scaling factor of 3000 is applied

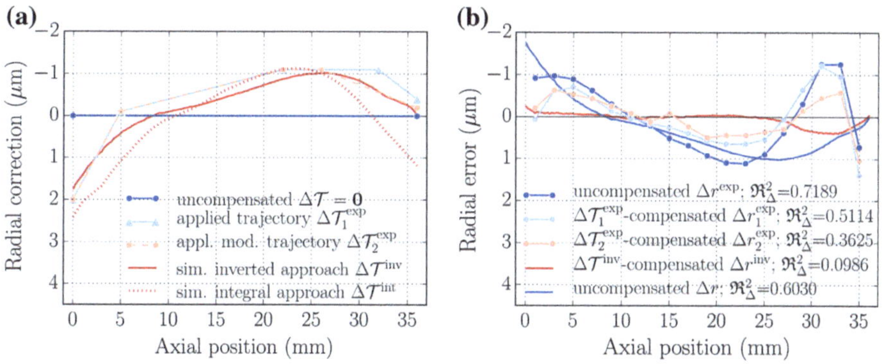

Fig. 27 Tool compensation trajectories $\Delta\mathscr{T}^{\bullet}$ (**a**) versus radial workpiece surface errors $\Delta r^{\bullet}$ (**b**). Lines with markers denote experimental outcome, lines without markers represent simulation results. The experimental results are normalised with respect to their arithmetic average. The light blue trajectory $\Delta\mathscr{T}_1^{\mathrm{exp}}$ refers to a process simulation neglecting the transient clamping effect observable in Fig. 25

configuration, e.g. referring to the undeformed configuration of the workpiece body $\mathscr{B}_n^0$ at t_n in terms of the remeshing material removal scheme.

In order to calculate the tool path correction $\Delta\mathscr{T}(t)$, the following approaches are taken into account.

5.1.1 Inversion of the Resulting Surface Profile After Cooling ($\Delta\mathscr{T}^{\text{inv}}$)

This approach is motivated by a basic experimental outcome, where a transient determination of the exact material removal discrepancies is not possible. It therefore uses the inverted resulting radial error after cooling as the tool trajectory correction, namely

$$\Delta\mathscr{T}^{\text{inv}}(t) = -\left[\mathbf{x}_k(t)\cdot\mathbf{e}_r - r^{\text{i}}_{\text{nom}}\right] \;\forall\; \mathbf{x}_k = r_k\mathbf{e}_r + z_k\mathbf{e}_z \in \partial\mathscr{B}^{\mathscr{L}} \tag{33}$$

$$= -\Delta r(t) \tag{34}$$

after cooling the workpiece to ambient temperature, cf. Fig. 27b. With $\dot{z}(t) = v_{\text{fa}} =$ constant, we can parametrise the above trajectory equation in space instead of time and retrieve the incremental formulation

$$\Delta\mathscr{T}^{\text{inv}}_n = -\left[r_n(z) - r^{\text{i}}_{\text{nom}}\right] = -\Delta r_n. \tag{35}$$

This compensation approach is straightforward, easily applicable to real processes and is therefore applied in the present work. The resulting trajectory is displayed by the red curve $\Delta\mathscr{T}^{\text{inv}}$ in Fig. 27a.

5.1.2 Compensation Based on Integral Material Removal ($\Delta\mathscr{T}^{\text{int}}$)

This method is based on the target nominal amount of material ΔV^{nom}_n removed at time step t_n, which in general can directly be determined from the process parameters in combination with the tool geometry $\mathscr{G}$, see violet area in Fig. 1. In the present axisymmetric model, the nominal removed material volume ΔV_n is represented by a trapezoidal area $\Delta A^{\text{nom}}_n = a_{\text{f}}\, a_{\text{e,tot}} = 0.05\,\text{mm}^2$, cf. the dark shaded area in Fig. 26. This nominal value represents the ideal amount of material removed during the process and can be compared with the effective amount of material removed that can be determined from the simulation, see the light shaded area bounded by the light grey dashed line in Fig. 26. This effective amount of material removed at time t_n

$$\Delta A_n = \mathscr{G}_{n+1} \cap \partial\mathscr{B}_n \tag{36}$$

can be expressed as the integral over all load surface nodes $\mathbf{x}^{\partial}_n$ in spatial configuration at time t_n

$$\Delta A_n =: \int_{\Delta z} \left[\mathbf{X}^{\partial}_{n+1} - \mathbf{x}^{\partial}_n\right]\cdot\mathbf{e}_r\,\text{d}z. \tag{37}$$

Here, Δz represents the axial integration range, and the inner surface nodes $\mathbf{X}^{\partial}_{n+1} \in \partial\mathscr{B}^{\mathscr{L}}_{n+1}$ in the updated and modified reference configuration result from the intersection

$$\partial \mathscr{B}_{n+1}^{\mathscr{L}} = \partial \left[\mathscr{B}_n \setminus \mathscr{G}_{n+1} \right] \cap \partial \mathscr{B}_n^{\mathscr{L}}, \tag{38}$$

which can be computed straightforwardly at t_n.

Taking the following constraint into account,

$$\Delta A_n \overset{!}{=} \Delta A_n^{\text{nom}} \tag{39}$$

$$\Rightarrow a_\text{f}\, \Delta_r \mathscr{T}_n = \left[\Delta A_n^{\text{nom}} - \Delta A_n \right] = \left[a_\text{f}\, a_\text{e,tot} - \Delta A_n \right], \tag{40}$$

the incremental compensated tool trajectory follows as

$$\Delta \mathscr{T}_n^{\text{int}} = \frac{1}{a_\text{f}} \left[\Delta A_n^{\text{nom}} - \Delta A_n \right] \tag{41}$$

$$= \frac{1}{a_\text{f}} \left[\Delta A_n^{\text{nom}} - \int_{\Delta z} \left[\mathbf{X}_{n+\Delta n}^{\partial} - \mathbf{x}_n^{\partial} \right] \cdot \mathbf{e}_r \, \text{d}z \right] \forall\, t_n \in \left[t_1,\, t_{\max}^{\text{grd}} \right]. \tag{42}$$

This basically represents a discrete forward differential of the amount of material removed with respect to time while invoking a step size of Δn. Following this approach, $\Delta z = a_\text{f}\, \Delta n$ holds and a choice of $\Delta n > 1$ can be made to smoothen the resulting trajectory. For $\Delta n = 8$, the red dotted curve $\Delta \mathscr{T}^{\text{int}}$ in Fig. 27a is obtained.

5.2 Results

The numerical and experimental results of the process model for both uncompensated and compensated tool trajectories are summarised in Fig. 27. To quantify the quality of the resulting radial error, a residual norm

$$\mathfrak{R}_\Delta^2 = \frac{\displaystyle\int_{z_{\min}}^{z_{\max}} \sqrt{\left[r^\text{i}(z) - r^\text{i}_{\text{nom}} \right]^2} \, \text{d}z}{\displaystyle\int_{z_{\min}}^{z_{\max}} \text{d}z} = \frac{\displaystyle\int_{z_{\min}}^{z_{\max}} \sqrt{\left[\Delta r^\text{i}(z) \right]^2} \, \text{d}z}{\displaystyle\int_{z_{\min}}^{z_{\max}} \text{d}z} \tag{43}$$

has been defined. Based on the simulated trajectory $\Delta \mathscr{T}^{\text{inv}}$, both trajectories $\Delta \mathscr{T}_1^{\text{exp}}$ $\Delta \mathscr{T}_2^{\text{exp}}$ have been deducted, basically only differing for $z \geq 26$ mm. The referring radial error results are displayed in the colour corresponding to the tool trajectory applied. Both experimentally applied compensation curves $\Delta \mathscr{T}_1^{\text{exp}}$ and $\Delta \mathscr{T}_2^{\text{exp}}$ have been motivated by the simulated trajectory $\Delta \mathscr{T}^{\text{inv}}$, whereas the compensation trajectory $\Delta \mathscr{T}_1^{\text{exp}}$ (light blue curve) is based on a slightly different boundary conditions during the simulated material removal, published in more detail in [32]. Therefore, the compensation trajectory $\Delta \mathscr{T}_1^{\text{exp}}$ (orange dashed curve) yields more accurate results.

In comparison to the uncompensated radial error Δr^{exp}, both compensated curves Δr_1^{exp} and Δr_2^{exp} show an improvement in shape and residual norm, respectively. Especially in the axial range $z \geq 12\,\text{mm}$, Δr_2^{exp} exhibits a decrease in radial error of approximately $0.6\,\mu\text{m}$ in average, which proves the correctness of the simulation-based compensation strategy. However, the enormous radial error reduction predicted by the simulation, cf. the red curve Δr^{inv}, cannot be reached in the experiments. The reason for this is due to the modelling assumptions that, at this stage, do not take the grinding machine spindle compliances into account. Moreover, the substitution of the clamping device and the clamping cover by Dirichlet boundary conditions during the material removal simulation may constitute another error source.

6 Conclusions and Outlook

In this contribution, a multi-scale framework targeting the thermo-mechanical simulation of Internal Traverse Grinding was presented. Multiple modelling approaches have been combined and validated against experiments, yielding an efficient, innovative simulation framework. Based on measurements of the grinding wheel topography, a geometric-kinematic simulation has been established to predict the engagement conditions of every cBN grain of the grinding wheel in detail. The output of this simulation is combined with the results of a single grain thermo-mechanically coupled finite element model via a sophisticated scale-bridging scheme. The results of this combined simulation scheme are subsequently used in a macro-scale process model that is capable to predict the geometrical error of the functional workpiece surface after grinding. Moreover, compensation strategies to minimise the latter manufacturing errors have been presented and validated, successfully finalising the complex simulation task.

Bearing future research in mind, a three-dimensional extension of some simulation framework components constitutes a valuable increase in the overall modelling accuracy. A three-dimensional single grain model provides a more detailed insight into the meso-scale thermo-mechanical response of the workpiece during grinding. Especially material removal effects such as chip splitting or the non-constant friction around the grain circumference cannot be captured by the current plane strain model. In this context, however, the parametrisation scheme for the bridging quantities such as the heat source term r and the grinding force components $F_\bullet$ will have to be extended accordingly, since applying the same grain geometry in an altered orientation of the grain will result in a completely different workpiece response.

Regarding the process model, the effect of the spindle stiffness shall be included by an extension to a three-dimensional model of the workpiece in combination with elastic boundary conditions applied, thereby capturing this obviously not negligible effect.

Acknowledgements Financial support by the Deutsche Forschungsgemeinschaft (DFG) within the context of Priority Program 1480 (project IDs: ME 1745/7-1-3; BI 498/23-1-3) is gratefully acknowledged.

Appendix

In the following, parameters used within the simulation framework presented are summarised.

Table 1 Material parameters for the materials used in the single grain finite element model. The workpiece consists of 100Cr6/AISI 52100 bearing steel hardened to 60–64 HRC, the grain consists of monocrystalline cubic boron nitride (cBN). The Johnson-Cook parameters are taken from [33] and abbreviated with JC. Further parameters are adopted from [34, 35]

part	parameter	symbol	value	ref. temp.
workpiece	Young's modulus	E^{wp}	208 GPa	293.15 K
	Poisson's ratio	ν^{wp}	0.28	293.15 K
	inelastic heat fraction	η	0.9	293.15 K
	mass density	ρ_0^{wp}	7810 kg/m^3	293.15 K
	thermal conductivity	λ^{wp}	52.5 W/(m K)	293.15 K
	thermal expansion coefficient	α_0^{wp}	$1.19 \cdot 10^{-5}$	293.15 K
	initial JC yield stress	A	688.17 MPa	293.15 K
	initial JC hardening modulus	B	150.82 MPa	293.15 K
	JC strain rate hardening parameter	C	0.04279	293.15 K
	JC thermal softening parameter	M	2.7786	293.15 K
	JC exponential hardening parameter	N	0.3362	293.15 K
	JC reference temperature	θ_0	300.0 K	
	JC melting temperature	θ_m	1697.15 K	
	JC reference strain rate	$\dot{\varepsilon}_{\mathrm{p},0}$	1.0 s^{-1}	293.15 K
	damage saturation rate	ϑ_d	5.0	293.15 K
	damage threshold plastic strain	$\underline{\varepsilon}_{\mathrm{p}}^{d}$	2.0	293.15 K
	saturated damage value	$\underline{d}$	0.9	293.15 K
	specific heat	$c^{\mathrm{wp}}(\theta)$	473.75 J/(kg K)	293.15 K
			487.836 J/(kg K)	373.15 K
			517.285 J/(kg K)	473.15 K
			530.089 J/(kg K)	523.15 K
			400.768 J/(kg K)	573.15 K
			572.343 J/(kg K)	623.15 K
			588.988 J/(kg K)	673.15 K
			651.728 J/(kg K)	773.15 K
			710.627 J/(kg K)	873.15 K
			773.367 J/(kg K)	973.15 K
			1588.988 J/(kg K)	1023.15 K
			626.120 J/(kg K)	1073.15 K
			550.576 J/(kg K)	1173.15 K
cBN grain	mass density	ρ_0^{cBN}	4084 kg/m^3	293.15 K
	specific heat	c^{cBN}	558 J/(kg K)	293.15 K
	thermal conductivity	λ^{cBN}	40 W/(m K)	293.15 K

Table 2 Coefficients resulting from the regression analysis. More detailed heat regression data is published in [18]

coefficient	heat source term $\overline{r}$	spec. tan. force component $\overline{F}'_{\mathrm{t}}$	spec. nor. force component $\overline{F}'_{\mathrm{n}}$
a_0	-486.1174	52.5528	264.7994
a_1	62.6139	-4.1119	-30.0815
a_2	-1.2414	7.4918	16.9073
a_3	22.8165	0.0744	0.0843
a_4	0.1280	49.7697	42.7850
a_5	36.6064	1.4093	7.3304
a_6	-30.7391	-0.1689	-0.7920
a_7	-0.1809	–	–

References

1. Marschalkowski, K., Biermann, D., Weinert, K.: On the characteristics of high-performance internal traverse grinding using electroplated CBN wheels. In: Aoyama, T., Takeuchi, Y. (eds.) Proceedings of the 4th CIRP International Conference on High Performance Cutting (CIRP HPC 2010), vol. 1, pp. 393–398 (2010)
2. Marschalkowski, K.: Beitrag zur Prozessentwicklung für das Hochleistungsinnenrund-Schälschleifen mit galvanisch gebundenen CBN-Schleifscheiben. Ph.D. thesis, Institut für Spanende Fertigung, TU Dortmund (2010)
3. Klocke, F., Brinksmeier, E., Weinert, K.: Capability profile of hard cutting and grinding processes. CIRP Ann. Manuf. Technol. **54**(2), 22–45 (2005)
4. Schulze, V., Uhlmann, E., Mahnken, R., Menzel, A., Biermann, D., Zabel, A., Bollig, P., Ivanov, M., Cheng, C., Holtermann, R., Bartel, T.: Evaluation of different approaches for modeling phase transformations in machining simulation. Prod. Eng. **9**(4), 437–449 (2015)
5. Jaeger, J.C., Carslaw, H.S.: Conduction of Heat in Solids. Oxford University Press, London (England) (1959)
6. Jaeger, J.C.: Moving sources of heat and the temperature at sliding contacts. J. Proc. R. Soc. NSW **76**(3), 203–224 (1942)
7. Schumann, S., Holtermann, R., Biermann, D., Menzel, A.: Hochleistungs-Innenrundschälschleifen—Thermomechanische Betrachtung in Abhängigkeit vom radialen Gesamtaufmaß. Diam. Bus. **11**(2), 36–43 (2013)
8. Schumann, S., Holtermann, R., Biermann, D., Menzel, A.: Lokale Betrachtung des Innenrundschälschleifens - Modellhafte Ermittlung der thermomechanischen Belastung in Abhängigkeit des Schruppzonenwinkels. wt - Werkstattstechnik online **103**(1), 493–498 (2013)
9. Schumann, S., Siebrecht, T., Kersting, P., Biermann, D., Holtermann, R., Menzel, A.: Determination of the thermal load distribution in internal traverse grinding using a geometric-kinematic simulation. In: Procedia CIRP, vol. 31, pp. 322–327 (2015). 15th CIRP Conference on Modelling of Machining Operations (15th CMMO)
10. Klocke, F.: Manufacturing Processes 2—Grinding, Honing, Lapping. Springer (2009)
11. Lowin, R.: Schleiftemperaturen und ihre Auswirkungen im Werkstück. Ph.D. thesis, RWTH Aachen (1980)
12. Holtermann, R., Schumann, S., Menzel, A., Biermann, D.: Ansätze zur Modellierung und Simulation des Innenrundschälschleifens. Diam. Bus. **40**, 30–41 (2012)
13. Schumann, S., Würz, E., Biermann, D., Holtermann, R., Menzel, A.: Wärmeeintrag beim Hochleistungs-Innenrundschälschleifen beherrschen - Ermittlung der thermischen Werkstückbelastungen mittels FEM. VDI-Z Integrierte Produktion **154**(I-2012), 31–33 (2012)

14. Foley, J.D., Feiner, S.K., Hughes, J.F., Phillips, R.L.: Introduction to Computer Graphics. Addison-Wesley Longman Publishing Co., Boston (United States of America) (1994)
15. Bailey, M., Hedges, L.: Die Kristallmorphologie von Diamant und ABN. Industrie-Diamanten-Rundschau **3**(29), 126–129 (1995)
16. Jackson, M.J., Davim, J.P.: Machining with Abrasives. Springer, Berlin (2011)
17. Holtermann, R., Schumann, S., Menzel, A., Biermann, D.: Modelling, simulation and experimental investigation of chip formation in internal traverse grinding. Prod. Eng. **7**(2–3), 251–263 (2013)
18. Holtermann, R., Menzel, A., Schumann, S., Biermann, D., Siebrecht, T., Kersting, P.: Modelling and simulation of internal traverse grinding: bridging meso- and macro-scale simulations. Prod. Eng. **9**(4), 451–463 (2015)
19. Holtermann, R., Schumann, D., Menzel, A., Biermann, D.: A hybrid approach to the modelling and simulation of grinding processes. In: Proceedings of the 11th World Congress Computational Mechanics (WCCM XI), pp. 1932–1937 (2014)
20. Holtermann, R., Schumann, S., Zabel, A., Biermann, D., Menzel, A.: Numerical determination of process values influencing the surface integrity in grinding. Procedia CIRP **45**, 39–42 (2016)
21. Hortig, C., Svendsen, B.: Simulation of chip formation during high-speed cutting. J. Mater. Process. Technol. **186**(1–3), 66–76 (2007)
22. Hortig, C.: Local and non-local thermomechanical modeling and finite-element simulation of high-speed cutting. Ph.D. thesis, TU Dortmund, Institute of Mechanics (2011)
23. Zienkiewicz, O.C., Zhu, J.Z.: The superconvergent patch recovery and a posteriori error estimates. Part 2: error estimates and adaptivity. Int. J. Numer. Methods Eng. **33**(7), 1365–1382 (1992)
24. Zienkiewicz, O.C., Zhu, J.Z.: The superconvergent patch recovery and a posteriori error estimates. Part 1: the recovery technique. Int. J. Numer. Methods Eng. **33**(7), 1331–1364 (1992)
25. Johnson, G.R., Cook, W.H.: A constitutive model and data for metals subjected to large strains, high strain rates and high temperatures. In: Proceedings of 7th International Symposium on Ballistics, vol. 1, pp. 541–547 (1983)
26. Huang, Y., Liang, S.Y.: Force modelling in shallow cuts with large negative rake angle and large nose radius tools application to hard turning. Int. J. Adv. Manuf. Technol. **22**(9–10), 626–632 (2003)
27. Waffenschmidt, T., Polindara, C., Menzel, A., Blanco, S.: A gradient-enhanced large-deformation continuum damage model for fibre-reinforced materials. Comput. Methods Appl. Mech. Eng. **268**, 801–842 (2014)
28. Waffenschmidt, T., Polindara, C., Menzel, A.: A gradient-enhanced continuum damage model for residually stressed fibre-reinforced materials at finite strains. In: Lenarz, T., Wriggers, P. (eds.) Biomedical Technology, pp. 19–40. Springer International Publishing (2015)
29. Simulia: Abaqus documentation. Dassault Systèmes, version 6.14 edition (2014)
30. Zaera, R., Rodríguez-Martínez, J.A., Rittel, D.: On the Taylor-Quinney coefficient in dynamically phase transforming materials. Application to 304 stainless steel. Int. J. Plast. **40**, 185–201 (2013)
31. Aurenhammer, F.: Voronoi diagrams—a survey of a fundamental geometric data structure. ACM Comput. Surv. **23**(3), 345–405 (1991)
32. Biermann, D., Holtermann, R., Menzel, A., Schumann, S.: Modelling and simulation of thermal effects in internal traverse grinding of hardened bearing steel. CIRP Ann. Manuf. Technol. **65**(1), 321–324 (2016)
33. Ramesh, A., Melkote, S.N.: Modeling of white layer formation under thermally dominant conditions in orthogonal machining of hardened AISI 52100 steel. Int. J. Mach. Tools Manuf. **48**(3–4), 402–414 (2008)
34. Wang, Z., Al-Zkeri, I.: Determination of flow stress data for AISI 52100 using machining tests. Technical Report Report No. ERC-NSM-07-R-08, Engineering Research Center for Net Shape Manufacturing, The Ohio State University 1971 Neil Avenue Columbus, Ohio 43201 (2007)
35. Irretier, A.: Schlussbericht SFB570—Teilprojekt C1 "Stoffwertebestimmung". Technical report, Universität Bremen (2009)

MIX
Papier aus verantwortungsvollen Quellen
Paper from responsible sources
FSC® C105338

If you have any concerns about our products,
you can contact us on
ProductSafety@springernature.com

In case Publisher is established outside the EU,
the EU authorized representative is:
Springer Nature Customer Service Center GmbH
Europaplatz 3, 69115 Heidelberg, Germany

Printed by Libri Plureos GmbH
in Hamburg, Germany